Value Addition and Product Diversification in Sugarcane

Giriyapura Shivalingamurthy Suresha
Gopalareddy Krishnappa
Murali Palanichamy
Huskur Kumaraswamy Mahadeva Swamy
Hari Kuppusamy
Hemaprabha Govindakurup
Editors

Value Addition and Product Diversification in Sugarcane

Springer

Editors
Giriyapura Shivalingamurthy Suresha
Biochemistry (Plant Science)
ICAR-Sugarcane Breeding Institute
Coimbatore, Tamil Nadu, India

Murali Palanichamy
Agricultural Economics
ICAR-Sugarcane Breeding Institute
Coimbatore, Tamil Nadu, India

Hari Kuppusamy
Agricultural Microbiology
ICAR-Sugarcane Breeding Institute
Coimbatore, Tamil Nadu, India

Gopalareddy Krishnappa
Genetics and Plant Breeding
ICAR-Sugarcane Breeding Institute
Coimbatore, Tamil Nadu, India

Huskur Kumaraswamy Mahadeva Swamy
Genetics and Plant Breeding
ICAR-Sugarcane Breeding Institute
Coimbatore, Tamil Nadu, India

Hemaprabha Govindakurup
Genetics and Plant Breeding
ICAR-Sugarcane Breeding Institute
Coimbatore, Tamil Nadu, India

ISBN 978-981-97-7230-8 ISBN 978-981-97-7228-5 (eBook)
https://doi.org/10.1007/978-981-97-7228-5

This Springer imprint is published by the registered company Springer Nature Singapore Pte Ltd.
The registered company address is: 152 Beach Road, #21-01/04 Gateway East, Singapore 189721, Singapore

Preface

Sugarcane is a highly efficient C4 biomass crop that converts 6–7% of sunlight into biomass and is grown in more than 80 countries primarily for table sugar production. Due to its high biomass and sugar-yielding ability, sugarcane is considered the prime feedstock for second-generation biofuels, such as lignocellulose-derived bioethanol and advanced biofuels like biodiesel. Sugarcane is one of the most important agro-based industrial crops and the primary source of sugar and jaggery. Although sugarcane production has been increasing over several decades, it has only recently gained importance for sugarcane product diversification and bioenergy applications.

In the era of renewable energy and expanding product portfolios, there is a need for an up-to-date textbook that covers recent technological interventions in value addition and diversification for bioenergy applications. This book covers comprehensive information on sugarcane by-products, preservation, functional foods, bioethanol, liquid fuels, and energycane. It also covers important topics like biopharming/molecular pharming and genomics for product diversification. The book also includes technological advancements in the area of lignocellulosic pathway engineering for biofuel production, synthesis of high-value biochemicals, and production of biodegradable and recyclable materials. Bioenergy policy, economic feasibility, sugarcane agribusiness and value chain management, and market intelligence will also be discussed in this book.

As energy security gains global attention, this book will give information and prospects of green fuels and their applications. Since many countries are producing surplus sugar, it is the need of the hour to diversify sugarcane for the production of value-added food products, fuels, and feed to sustain the crop, farmers, and industries. In the present socio-economic context, it is imperative to get maximum benefit for all the stakeholders involved in sugarcane industry. This could be possible through better utilization of sugarcane and its by-products for the production of high-value products, including bioenergy. In this context, the proposed book offers a great opportunity to bring together recent technological advancements in the areas of value addition and product diversification and to provide more insight into the role of sugarcane in green fuels and global energy security in a sustainable manner. This book features 18 chapters contributed by global experts from all over the world in the field of sugarcane research.

This book will cater to the needs of various stakeholders including students, researchers, policymakers, and academicians, who are working in the areas of functional foods, nutraceutical research, bioenergy, and synthetic biology. It is also designed for personnel in the sugar and food industry, green fuels, chemical industries, and textile and recycling industries. It will also serve as an excellent guide for young entrepreneurs, rural youth, women, and progressive sugarcane farmers to take up sugarcane as a profitable business.

We are assertive that this book will be a landmark, as this kind of compilation on value addition and product diversification in sugarcane is the first of its kind and the chapters contain updated information as well as the views conveyed by the researchers based on their vast experience and expertise in value addition in sugarcane. We thank all the eminent authors and oblige their valuable contributions towards enhancing the quality of this book. Undoubtedly, *Value Addition and Product Diversification in Sugarcane* is a timely and much warranted book considering the significance of value addition to achieving future demand.

Coimbatore, Tamil Nadu, India — Giriyapura Shivalingamurthy Suresha
Coimbatore, Tamil Nadu, India — Gopalareddy Krishnappa
Coimbatore, Tamil Nadu, India — Murali Palanichamy
Coimbatore, Tamil Nadu, India — Huskur Kumaraswamy Mahadeva Swamy
Coimbatore, Tamil Nadu, India — Hari Kuppusamy
Coimbatore, Tamil Nadu, India — Hemaprabha Govindakurup

Contents

Editors and Contributors

About the Editors

Giriyapura Shivalingamurthy Suresha, Ph.D. graduated from the University of Agricultural Sciences, Bengaluru, India. He did his Ph.D. at ICAR-Indian Agricultural Research Institute, New Delhi, in the discipline of Biochemistry. Currently, he has been working as a Senior Scientist at the ICAR-Sugarcane Breeding Institute, Coimbatore, for the past 14 years. He has published more than 70 articles which include research papers, book chapters, conference proceedings, popular articles, and technical bulletins, among others. He has developed and disclosed eight novel, innovative value-added technologies, including the production of freeze-dried and spray-dried sugarcane juice powder, cane jam, liquid jaggery, cane dietary fibre food products, frozen sugarcane juice-based value-added products, and cane wine. He has filed two patents on cane jam technology and Bt 62 biopesticide formulation. Dr. Suresha was awarded the Indo-Australian Career Boosting Gold Fellowship (IACBGF) in 2013 by the Department of Biotechnology, Government of India, Early Career Research Award in 2017 by the Department of Science and Technology (DST)-Science and Engineering Research Board (SERB), Government of India, and ICAR-SBI Foundation Day Award 2023 under the scientific category for his outstanding contribution at ICAR-Sugarcane Breeding Institute, Coimbatore. He is a life member of the Society for Sugarcane Research and Development and the Society for Plant Biochemistry and Biotechnology.

Gopalareddy Krishnappa, Ph.D. Scientist, ICAR-SBI, Coimbatore, completed his Bachelor's degree in 2006 and Master's degree in 2008 from UAS, Bangalore, and a Doctorate degree in genetics from ICAR-IARI, New Delhi, in 2016. Dr. Reddy developed 17 wheat varieties and 29 wheat genetic stocks to benefit a multitude of stakeholders including farmers, consumers, researchers, and industry. His research is largely focused on the development of high-yielding wheat varieties with improved quality attributes. Dr. Reddy utilized modern breeding tools, such as genomic selection, MARS, MAS, and GWAS, to complement conventional breeding techniques, thereby dissecting complex traits in wheat. He has published more than 50 research and review articles in high-impact national and international journals, 5 books, 40 book chapters, and 50 technical or popular articles. Dr. Reddy was conferred with the Nanaji Deshmukh Award, AIASA Young Scientist Award 2020, Best Worker 2020, NEEDEF Young Scientist Award 2021, and Dr. VS Mathur Memorial Award 2022.

Murali Palanichamy, Ph.D. graduated from TNAU, Coimbatore, and completed his post-graduation and Ph.D. from NDRI, Karnal. He has more than 15 years of experience in sugarcane economics and is currently an Officer-in-Charge of the Sugarcane Agribusiness Centre. He is working on impact assessment and cost and returns, techno-economic feasibility studies on value-added technologies at the institute, and studies on the impact of sugarcane varieties developed by the institute. He has published more than 30 publications and filed 4 patents and 2 trademarks. He is a national facilitator for the National Institute of Agricultural Marketing, a member of the Core Research Group-Impact Assessment of ICAR Technologies, Officer-in-Charge of the Agricultural Knowledge Management Unit, Nodal Officer for the KRISHI Portal and AEBAS, and Consulting Editor for Sugar Tech (Springer, India). He is a life member of Agricultural Economics Research Review, New Delhi, Indian Society of Coastal Agricultural Research, Kolkata, Society for Sugarcane Research and Promotion, and Society for Sugarcane Research and Development.

Huskur Kumaraswamy Mahadeva Swamy, Ph.D. graduated from the University of Agricultural Sciences, Bengaluru, with a Master's degree from GBPUA&T, Pantnagar, and Doctoral degree from PJTSAU, Telangana. He has more than 30 peer-reviewed publications and has delivered oral and poster presentations in international meetings. He contributed to the development of low soil P tolerance and BLB-resistant rice in the background of Improved Samba Mahsuri. He is currently working on molecular markers, pre-breeding, and the development of improved varieties of sugarcane. He was awarded the Young Scientist Award by Bharatiya Sugar in 2021.

Hari Kuppusamy, Ph.D. currently serving as Principal Scientist and Officer-in-Charge of Intellectual Property and Technology Management Unit at ICAR-SBI and completed his Master's degree from TNAU, Coimbatore, and his Doctorate from ICAR-IARI, New Delhi. He is involved in the development of many technologies including a soil moisture indicator, a spray, and freeze-dried sugarcane juice. His major areas of interest are biofertilizers, biocontrol agents, water conservation, biofuel, and value-added products from sugarcane. He has published over 80 research articles in reputable journals, over 150 popular book chapters, and other publications. He has filed ten Indian patents and four designs. He has received the Sir Venkataraman Award 2011, won the first prize in the Best Research/Innovation Adoption of New Technology for Water Conservation category at the National Water Awards 2019, presented by Ministry of Jal Shakti, Government of India, and Winner of TERI-IWA-UNDP Water Sustainability Award 2021 for excellence in water use efficiency in agricultural sector in 2022, among others.

Hemaprabha Govindakurup, Ph.D. has over 35 years of experience in sugarcane and is currently serving as Director ICAR-Sugarcane Breeding Institute, Coimbatore, India. She has completed her Ph.D. at Tamil Nadu Agricultural University, Coimbatore, India. Being specialized in genetics and plant breeding, she has developed 26 sugarcane varieties. She has published more than 70 research articles, 17 books, and presented over 160 articles in seminars/symposia and 56 manuals/book chapters/pamphlets. She is a fellow of the National

Academy of Agricultural Sciences (NAAS), the Society for Sugarcane Research and Development, and the Society for Sugar Research and Promotion. She is the recipient of the Sir T.S. Venkataraman Award for outstanding research in sugarcane agriculture.

Contributors

G. J. Abhishek ICAR-National Bureau of Plant Genetic Resources, New Delhi, India

V. P. Adarsh Microbial Processes and Technology Division, CSIR-National Institute for Interdisciplinary Science and Technology, Thiruvananthapuram, Kerala, India

Academy of Scientific and Innovative Research (AcSIR), Ghaziabad, India

Adewale George Adeniyi Department of Chemical Engineering, Landmark University, Omu-Aran, Nigeria

Department of Chemical Engineering, University of Ilorin, Ilorin, Nigeria

M. Alagupalamuthirsolai Division of Crop Production, ICAR-Sugarcane Breeding Institute, Coimbatore, Tamil Nadu, India

Amaresh ICAR-Sugarcane Breeding Institute, Coimbatore, Tamil Nadu, India

J. Anitha Department of Biochemistry (PG and Research), Kongunadu Arts and Science College (Autonomous), Coimbatore, India

S. Anusha Division of Crop Production, ICAR-Sugarcane Breeding Institute, Coimbatore, Tamil Nadu, India

C. Appunu ICAR-Sugarcane Breeding Institute, Coimbatore, Tamil Nadu, India

Devi Priya Arumugam Biocatalysts Laboratory, Department of Agricultural Microbiology, Tamil Nadu Agricultural University, Coimbatore, India

R. Arunkumar Division of Crop Production, ICAR-Sugarcane Breeding Institute, Coimbatore, Tamil Nadu, India

Athulya Microbial Processes and Technology Division, CSIR-National Institute for Interdisciplinary Science and Technology, Thiruvananthapuram, Kerala, India

Academy of Scientific and Innovative Research (AcSIR), Ghaziabad, India

G. Avinash ICAR-Indian Agricultural Statistics Research Institute, New Delhi, India

K. Chandran ICAR-Sugarcane Breeding Institute Research Centre, Kannur, Kerala, India

V. G. Dhanya ICAR-Sugarcane Breeding Institute, Regional Centre, Kannur, India

C. Dilsha ICAR-Sugarcane Breeding Institute Research Centre, Kannur, Kerala, India

Kandhalu Sagadevan Dinesh Babu Department of Botany and Microbiology, Hemvati Nandan Bahuguna Garhwal University, Chauras Campus, Tehri Garhwal, Uttarakhand, India

K. Elayaraja ICAR-Sugarcane Breeding Institute, Coimbatore, India

Ebuka Chizitere Emenike Department of Pure and Industrial Chemistry, Nnamdi Azikiwe University, Awka, Nigeria

Giriyapura Shivalingamurthy Suresha Biochemistry (Plant Science), ICAR-Sugarcane Breeding Institute, Coimbatore, Tamil Nadu, India

Valan Rebinro Gnanaraj Microbial Processes and Technology Division, CSIR-National Institute for Interdisciplinary Science and Technology, Thiruvananthapuram, Kerala, India

Academy of Scientific and Innovative Research (AcSIR), Ghaziabad, India

R. Gomathi Division of Crop Production, ICAR-Sugarcane Breeding Institute, Coimbatore, Tamil Nadu, India

R. Gopi ICAR-Sugarcane Breeding Institute Research Centre, Kannur, Kerala, India

Hemaprabha Govindakurup Genetics and Plant Breeding, ICAR-Sugarcane Breeding Institute, Coimbatore, Tamil Nadu, India

Oviya Govindaraj Biocatalysts Laboratory, Department of Agricultural Microbiology, Tamil Nadu Agricultural University, Coimbatore, India

P. Govindaraj ICAR-Sugarcane Breeding Institute, Coimbatore, India

Huskur Kumaraswamy Mahadeva Swamy Genetics and Plant Breeding, ICAR-Sugarcane Breeding Institute, Coimbatore, Tamil Nadu, India

Kingsley O. Iwuozor Nigeria Sugar Institute, Ilorin, Kwara State, Nigeria

Department of Pure and Industrial Chemistry, Nnamdi Azikiwe University, Awka, Nigeria

Kalaiselvi ICAR-Sugarcane Breeding Institute, Coimbatore, India

Keerthana Kamaraj Division of Crop Improvement, ICAR-Sugarcane Breeding Institute, Coimbatore, India

Baskaran Kannan Agronomy Department, Institute of Food and Agricultural Sciences, University of Florida, Gainesville, FL, USA

Elumalai Karpagam Division of Crop Improvement, ICAR-Sugarcane Breeding Institute, Coimbatore, India

Lakshmi Kasirajan Division of Crop Improvement, ICAR-Sugarcane Breeding Institute, Coimbatore, Tamil Nadu, India

C. Keerthi ICAR-Sugarcane Breeding Institute Research Centre, Kannur, Kerala, India

Gopalareddy Krishnappa Genetics and Plant Breeding, ICAR-Sugarcane Breeding Institute, Coimbatore, Tamil Nadu, India

V. Krishnapriya Division of Crop Production, ICAR-Sugarcane Breeding Institute, Coimbatore, Tamil Nadu, India

Ravinder Kumar ICAR-Sugarcane Breeding Institute, Regional Centre, Karnal, India

Shwetha Kumari ICAR-Sugarcane Breeding Institute, Coimbatore, India

Hari Kuppusamy Agricultural Microbiology, ICAR-Sugarcane Breeding Institute, Coimbatore, Tamil Nadu, India

R. Lavanya ICAR-Sugarcane Breeding Institute, Coimbatore, Tamil Nadu, India

B. Mahendran ICAR-Sugarcane Breeding Institute Research Centre, Kannur, Kerala, India

Shraddha Maitra Department of Agricultural and Biological Engineering, University of Illinois Urbana-Champaign, Urbana, IL, USA

R. T. Maruthi ICAR-Sugarcane Breeding Institute, Coimbatore, Tamil Nadu, India

Rangappa Thimmaiah Maruthi ICAR-Sugarcane Breeding Institute, Coimbatore, Tamil Nadu, India

Reshma M. Mathew Microbial Processes and Technology Division, CSIR-National Institute for Interdisciplinary Science and Technology, Thiruvananthapuram, Kerala, India

Academy of Scientific and Innovative Research (AcSIR), Ghaziabad, India

M. R. Meena ICAR-Sugarcane Breeding Institute, Regional Centre, Karnal, India

Kehinde Temitope Micheal Department of Mechanical Engineering, Ladoke Akintola University of Technology, Ogbomosho, Nigeria

Taiwo Temitayo Micheal Department of Agricultural Engineering, Ladoke Akintola University of Technology, Ogbomosho, Nigeria

Varucha Misra ICAR-Indian Institute of Sugarcane Research, Lucknow, Uttar Pradesh, India

K. Mohanraj Division of Crop Improvement, ICAR-Sugarcane Breeding Institute, Coimbatore, Tamil Nadu, India

P. M. Namratha ICAR-Sugarcane Breeding Institute, Coimbatore, Tamil Nadu, India

P. S. Neetha Microbial Processes and Technology Division, CSIR-National Institute for Interdisciplinary Science and Technology, Thiruvananthapuram, Kerala, India

Academy of Scientific and Innovative Research (AcSIR), Ghaziabad, India

Gauri Nerkar ICAR-Sugarcane Breeding Institute, Coimbatore, Tamil Nadu, India

M. Nisha ICAR-Sugarcane Breeding Institute Research Centre, Kannur, Kerala, India

Aswini Nunavath ICAR-National Institute of Biotic Stress Management, Raipur, India

Cletus Chinedu Onwuegbusi Department of Pure and Industrial Chemistry, Nnamdi Azikiwe University, Awka, Nigeria

Murali Palanichamy Agricultural Economics, ICAR-Sugarcane Breeding Institute, Coimbatore, Tamil Nadu, India

Neha Patwa ICAR-Sugarcane Breeding Institute, Regional Centre, Karnal, India

Adhini S. Pazhany Division of Crop Improvement, ICAR-Sugarcane Breeding Institute, Coimbatore, Tamil Nadu, India

P. Prakash ICAR-Central Tuber Crops Research Institute, Thiruvananthapuram, Kerala, India

P. T. Prathima Division of Crop Improvement, ICAR-Sugarcane Breeding Institute, Coimbatore, Tamil Nadu, India

Thakku R. Ramkumar Department of Plant and Soil Sciences, Institute of Genomics for Crop Abiotic Stress Tolerance, Texas Tech University, Lubbock, TX, USA

R. Ranjani Department of Biochemistry (PG and Research), Kongunadu Arts and Science College (Autonomous), Coimbatore, India

Biji Raphy Microbial Processes and Technology Division, CSIR-National Institute for Interdisciplinary Science and Technology, Thiruvananthapuram, Kerala, India

Academy of Scientific and Innovative Research (AcSIR), Ghaziabad, India

Meena Sankar Microbial Processes and Technology Division, CSIR-National Institute for Interdisciplinary Science and Technology, Thiruvananthapuram, Kerala, India

Academy of Scientific and Innovative Research (AcSIR), Ghaziabad, India

Sheelamary Sebastiar Division of Crop Improvement, ICAR-Sugarcane Breeding Institute, Coimbatore, India

R. Selvakumar Department of Science and Humanities, Sri Krishna Polytechnic College, Kovaipudur, Tamil Nadu, India

Kalai Selvi ICAR-Sugarcane Breeding Institute, Coimbatore, Tamil Nadu, India

P. J. Sherin Department of Biochemistry (PG and Research), Kongunadu Arts and Science College (Autonomous), Coimbatore, India

Vijay Singh Department of Agricultural and Biological Engineering, University of Illinois Urbana-Champaign, Urbana, IL, USA

S. Solomon Chandra Sekhar Azad University of Agriculture & Technology, Kanpur, Uttar Pradesh, India

AthiraRaj Sreeja-Raju Microbial Processes and Technology Division, CSIR-National Institute for Interdisciplinary Science and Technology, Thiruvananthapuram, Kerala, India

Academy of Scientific and Innovative Research (AcSIR), Ghaziabad, India

V. Sreenivasa ICAR-Sugarcane Breeding Institute, Coimbatore, India

Santhoshkumar Subramaniam Biocatalysts Laboratory, Department of Agricultural Microbiology, Tamil Nadu Agricultural University, Coimbatore, India

T. C. K. Sugitha Karunya Institute of Technology and Sciences, Coimbatore, India

Rajeev K. Sukumaran Microbial Processes and Technology Division, CSIR-National Institute for Interdisciplinary Science and Technology, Thiruvananthapuram, Kerala, India

Academy of Scientific and Innovative Research (AcSIR), Ghaziabad, India

A. Swaliha ICAR-Sugarcane Breeding Institute, Coimbatore, Tamil Nadu, India

Prathima P. Thirugnanasambandam ICAR-Sugarcane Breeding Institute, Coimbatore, Tamil Nadu, India

Prathima Perumal Thirugnanasambandam Division of Crop Improvement, ICAR-Sugarcane Breeding Institute, Coimbatore, India

Linus Anietie Umoren Department of Pure and Industrial Chemistry, Nnamdi Azikiwe University, Awka, Nigeria

Sivakumar Uthandi Biocatalysts Laboratory, Department of Agricultural Microbiology, Tamil Nadu Agricultural University, Coimbatore, India

Vinayaka ICAR-Sugarcane Breeding Institute, Coimbatore, India

V. Vinu Division of Crop Improvement, ICAR-Sugarcane Breeding Institute, Coimbatore, Tamil Nadu, India

ICAR-Indian Institute of Spices Research, Kozhikode, Kerala, India

Rajesh Yarra Department of Plant and Agroecosytem Sciences, University of Wisconsin-Madison, Madison, WI, USA

Sugarcane By-Product-Based Industries in Asian Countries

1

S. Solomon and Varucha Misra

Abstract

Sugarcane stands out as an exceptional converter of solar energy into valuable biomass and sugar. A diverse array of by-products are generated during sugarcane processing, including bagasse, molasses, press mud, and lesser-known residues like trash, wax, fly ash, biochar, etc. These by-products, often rich in fibres, lignin, pentosans, and pith, present a unique opportunity for conversion into high-value products through chemical, biochemical, and microbial technologies. The Indian sugar industry, on identifying the economic potential of these by-products, has established sugar-agro-industrial complexes where bioethanol, bioelectricity, compressed biogas (CBG), and various value-added products are produced. Amid global moves in sugar markets, an increased demand for bio-based fuels, and consumer preferences favouring sustainable products, the sugar industry is experiencing a significant transformation. The utilization of sugarcane by-products not only augments economic viability but also aligns with broader goals of sustainable resource utilization and waste reduction. This review underscores the diversity of sugarcane by-products originating from the sugar industry, showcasing their role as raw materials for diverse sectors such as paper board, textile, ethanol, CBG, and bioplastic manufacturing.

Keywords

By-products · Molasses · Bagasse · Press mud · Straw · Ethanol · Bioproducts

---

S. Solomon (✉)
Chandra Sekhar Azad University of Agriculture & Technology, Kanpur, Uttar Pradesh, India

V. Misra
ICAR-Indian Institute of Sugarcane Research, Lucknow, Uttar Pradesh, India

G. S. Suresha et al. (eds.), *Value Addition and Product Diversification in Sugarcane*, https://doi.org/10.1007/978-981-97-7228-5_1

1.1 Introduction

Global trends, such as fluctuating sugar markets, the demand for alternative bio-based fuels, and consumer preferences for bio-based products, are driving the sugar industry towards diversification for sustainability (Solomon and Swapna 2022; Moreno et al. 2020). Asian countries, including the South-East Asian region, are important sugar producers, with Asia expected to contribute roughly 36% of the world's sugar production. South-East Asia has a share of approximately 10% of the world's sugar production. Sugar crops are valuable biomass feedstocks for renewable fuel and green energy, offering various main products and by-products (Zabed et al. 2014). Sugarcane is among these sugar crops, which not only serves as a key source of sugar but also generates significant by-products, including bagasse, molasses, and press mud (Wani et al. 2023). The presence of fermentable sugars in sugarcane juice has an added advantage over grain crops (Solomon 2011). The diversification of sugarcane by-products represents a significant frontier in sustainable and economically viable practices within the sugar processing industry (Solomon and Swapna 2022). Global recognition prevails among countries engaged in cane sugar production, acknowledging the profitability of diversification. Rather than relying solely on sugar production, sugarcane-producing countries have realized the potential of extracting value from the various by-products of the sugar industry (Singh and Solomon 1995; Solomon and Singh 2005; Yadav and Solomon 2006; Manohar Rao 2008). Recognizing the immense potential inherent in sugarcane-derived by-products, there is a growing emphasis on exploring innovative avenues for their diversification. This transformative approach seeks to harness the latent value in sugarcane by-products, thus propelling the industry towards a circular and environmentally conscious paradigm. Recently, there has been a notable shift towards the use of alternative resources for fibrous raw materials due to factors such as the depletion of trees in many countries, a shortage of wood resources, and a heightened awareness of sustainability issues. Agricultural residues have emerged as a significant and crucial alternative resource in response to these challenges (Gharehkhani et al. 2017; Bhardwaj et al. 2019). Globally, the rate of paper consumption is rapidly increasing, with the current world production of pulp and paper standing at approximately 400 million tonnes. Projections indicate that in 2025, there will be a shortfall of 12 million tonnes (mt) of wood to meet the paper requirement (Kulkarni 2013). The bagasse obtained after sugarcane processing, which serves as a raw material for paper production, could be one of the solutions to this shortfall. Unlocking novel applications and markets for sugarcane by-products to other industries enhances the sugar industry's economic resilience and long-term sustainability. Additionally, these diversification endeavours contribute substantially to the broader goals of sustainable resource utilization and waste reduction (Solomon 2011).

1.2 Beyond the Sweetness: Unveiling the Versatile By-Products of Sugarcane from the Sugar Industry

The sugarcane industry is an inexhaustible source of biomass and by-products (Raza et al. 2021; Solomon 2011). Transforming these by-products into value-added products is crucial for fostering a sustainable circular bioeconomy. Bagasse (25–30% of cane), press mud (3–5% of cane), and molasses (3.5–5% of cane) are the three primary by-products obtained after sugarcane crushing, post-clarification, and centrifugation, respectively, in the sugar processing industry (Solomon 2011). These co-products from the sugar industry find diverse applications due to the various components present in them (Table 1.1). In the South African sugar industry, the processing of 8.46 tonnes of sugarcane results in the production of 0.56 tonnes of filter press, 0.38 tonnes of molasses, and 2.4 tonnes of bagasse, requiring 17,000 m^3 of water (Mashoko et al. 2010). The production of 1 kg of sugar generates approximately 0.3 kg of molasses and 1.25 kg of bagasse (Botkin et al. 2012). In the Thai sugar industry, 1 tonne of sugarcane produces 103.6 kg of sugar and 45.2 kg of molasses, whereas in the Mauritius sugar industry, 1 tonne of sugar production generates 0.27 tonnes of molasses as a by-product (Prasertsri 2007).

1.2.1 Bagasse and Its Diverse Uses

Bagasse, the residue left after sugarcane juice extraction, holds significant value as a by-product. It is cost-effective and has minimal production and transportation expenses. Consequently, it is highly prized, primarily for its role as a substitute for fossil fuel and wood in steam and electrical energy generation. Through anaerobic digestion facilitated by microbes, bagasse can yield biogases, thus contributing to the biofuel sector (Osman et al. 2006). Additionally, bagasse is utilized in the production of cellulosic ethanol and furfural. Furfural serves as a solvent for refining lubricating oils, wood resins, and vegetable oils. Furfuryl alcohol, derived from furfural, acts as a raw material for furan polymers, modified formaldehyde products, urea-based polymers, anti-corrosive coatings, fragrances, and solvents for resins

Table 1.1 Composition of the primary by-products of sugarcane

Primary sugarcane by-products	Composition	References
Bagasse	Dissolved sugars (2–5%); water (45–50%); fibres (40–45%); cellulose (36%); pentosans (26.0%); lignin (20.0%); ash (2.2%)	Sahu (2018)
Molasses	Sucrose (35%); water (20%); fructose (9%); glucose (7%); reducing sugars (3%); carbohydrates (4%); nitrogenous compounds (4.5%); non-nitrogenous acids (5%), ash (12%); others (5%)	Teclu et al. (2009)
Press mud	Moisture (75–80%); sugar (2–5%); fibres (5–10%)	Sahu (2018)

and dyes. The National Sugar Institute, Kanpur, India, is also synthesizing chloromethyl furfural and lactic acid from sugarcane bagasse hemicellulose (Sharkara 2022). Bagasse's versatility and eco-friendly nature make it a valuable resource in various industrial applications (Santos et al. 2020). In the field of biotechnology, bagasse can be utilized for the synthesis of sugar, oligosaccharides (Xue et al. 2016), acids (like succinic, lactic, citric, etc.), and enzymes (such as lipase, cellulase, xylanase, etc.) (Kumar et al. 2003; Borges and Pereira 2011; Sindhu et al. 2016). It also serves as a good growth medium for mushrooms due to the high concentration of nitrogen (Sardar et al. 2020; Akter et al. 2022). Moreover, bagasse is also utilized as microalgae biomass since it works as a rich carbon source. Oleaginous microalgae are well-known for producing oils that can be converted to biofuels (Nguyen et al. 2018). Bagasse can even be converted to biochar, which has diverse applications (Saleh and Hedia 2018) like against nitrate leaching in calcaric dark red soil (Kameyama et al. 2012), and can be used as a soil fertilizer to enhance soil productivity, soil porosity, and water retention (Inyanga et al. 2010). Furthermore, bagasse is used as a cost-effective adsorbent material for eliminating pesticides from the surroundings (Toledo-Jaldin et al. 2020). The application of bagasse is also being seen in the production of carbon-based nanomaterials (Oliveira Alves et al. 2012), economical cellulosic nanofibres (Shahi et al. 2020), nanostructured films (Pereira et al. 2007), and various bioadsorbents for effectively removing heavy metals (such as copper, cadmium, lead, nickel, selenium, zinc, etc.) as well as phenols and dyes from wastewater (Al-Arni 2018; Moubarik and Grimi 2015). Silicon nanoparticles are also produced using sugarcane bagasse (Seroka et al. 2022). Moreover, bagasse is employed as a material for manufacturing components used in acoustic protection (Dos Santos et al. 2014). Recently, bagasse has also been utilized in the production of bio-compressed natural gas (bio-CNG) (Munagala et al. 2022).

Agglomerated products are also produced from bagasse and are used in the construction material industry (Pozzer et al. 2020; Cangussu et al. 2023; Abdulkadir et al. 2014). These are also used for making green roofs (Araújo and Colombo 2021; Araújo et al. 2022) and ceramic tiles (Rainho Teixeira et al. 2015). Bagasse is additionally utilized to produce recyclable and biodegradable composite materials. These materials have a role in the automotive industry, for example, polypropylene reinforced with natural fibres derived from sugarcane bagasse, novel alternatives to talc-filled polypropylene, and polypropylene reinforced with glass fibres (Luz et al. 2010; Oliver-Ortega et al. 2019). Fibreboards (medium to dense) are also being prepared from straw, cereals, and bagasse (Yadav and Solomon 2006), owing to their fibrous texture and low density. Bagasse not only adds lightness but also enhances the mechanical characteristics of fibreboards (Rainey and Clark 2004; Delgado and Casanova 2001). Bagasse also finds application in paper pulp, toilet paper, and hardboard production (Hamzeh et al. 2013; Kiatkittipong et al. 2009). The fibres found in sugarcane by-products are characterized by their dense walls, variable lengths, and predominantly blunt ends (Pfafi and Sisko 1995). These fibres offer versatile applications, proving useful in the production of a range of paper products. When mixed with different proportions of wood pulp, these sugarcane fibres can be utilized to manufacture printing and writing paper, Bristol board, tissue paper, glassine and greaseproof paper, duplex and triplex paper, corrugating medium, linerboard, wrapping and bag papers, and newsprint (Bian et al. 2019). It is also utilized in making cutlery, disposable cups, bowls, plates, and glasses. High-quality wrapping and magazine paper can be efficiently produced using depithed bagasse as the primary raw material. Bagasse pulp serves as the foundation for creating various types of papers, each tailored to specific requirements set by the paper industry. To manufacture paper, approximately 3.4 tonnes of bagasse are needed for every tonne of paper produced (Poopak and Reza 2012).

Bagasse, with its rich nutrient content, serves as fodder for cattle (Okano et al. 2007; Luz et al. 2010). It is rich in lignin content, which may cause difficulties in the ruminal digestion of cellulose and hemicellulose, resulting in a lower nutritional value. Conversely, when bagasse undergoes pretreatment, whether it is through mechanical means like shredding, chemical processes involving urea and ammonia (Balgees et al. 2016), alkaline hydrogen peroxide treatment, hydrolysis (Lee et al. 2019), biological pretreatment through fermentation with fungi (Olagunju et al. 2014), or co-digestion with poultry manure (Shahowna et al. 2013), its nutritional value is significantly enhanced (Khan et al. 1992). Importantly, pretreated bagasse does not impede the activity of cellulolytic bacteria in the cattle rumen, ensuring that it does not compromise the digestibility of the fodder (Rabelo et al. 2015).

Bagasse ash has the potential to be utilized as an agricultural soil amendment, although it lacks sufficient mineral nutrients, making it a low-value fertilizer (Balakrishnan and Batra 2011). To enhance its nutrient content, it can be combined with sludge from cane juice filtration or vinasse from ethanol distillation. Bagasse ash serves as a substitute for sand and an additive in brick, cement, mortar, and concrete production (Kazmi et al. 2016; Arif et al. 2016; Abdulkadir et al. 2014; de Siqueira and Cordeiro 2022; Faria et al. 2012; Anjos et al. 2020; Kulkarni et al. 2022). Incorporating bagasse ash into cement-based materials enhances mechanical performance and reduces greenhouse gas (GHG) emissions from cement factories. This practice not only lowers construction costs but also mitigates waste disposal challenges, thus preventing soil and air pollution (Tripathy and Acharya 2022). The carbonation of sugarcane bagasse and the incorporation of clay and sugarcane molasses are used for the production of eco-friendly briquettes for household use, producing minimal smoke and odour during combustion (Onchieku et al. 2012). Moreover, Rovani et al. (2018) reported the production of silica nanoparticles using sugarcane bagasse fly ash and the utilization of the product as an adsorbent. Kumari et al. (2023) developed a novel green synthetic method wherein sugarcane bagasse ash is employed for the production of silica nanoparticles.

Bagasse fibres and ash find application as additives in the manufacturing of clay, glass, and ceramic products (Faria et al. 2012; Janbuala et al. 2018). These can also be utilized in the texstile industry (Samantra et al. 2015). Bagasse ash, raw bagasse, and bagasse fibres are used as alternative reinforcements in natural rubber, yielding eco-friendly composite materials with improved mechanical properties (Barrera Torres et al. 2021). Moreover, bagasse finds application in the production of ceramic membrane filters, which are used in various separation operations (Batra and Tewari

2006). Although these filters offer superior stability under extreme pH and temperature conditions, they are more expensive compared to polymer membrane filters.

1.2.2 Value-Added Products from Molasses

Molasses, a viscous substance, is another by-product generated during the process of sugar production (Zhang et al. 2021). It is produced at a rate of 40 kg per tonne of cane. It is characterized by a substantial sugar content ranging between 45% and 55%. Ethanol, the most common product of molasses, is produced globally on an industrial scale (Kabeyi 2022). It yields between 280 and 320 L/tonnes. Ethanol serves various purposes, including potable consumption and as a green, clean energy source. The growing awareness of the drawbacks associated with fossil-based fuels has positioned ethanol as a significant biofuel, particularly in the automotive industry. A worldwide trend is emerging towards the adoption of ethanol-blended automobile fuels, with Asian countries also joining this shift. The escalating use of ethanol, coupled with its efficient blending with traditional automobile fuels, is driven by the increasing demand to meet the targets outlined in the national biofuel policies of nations globally. Projections indicate a substantial rise in the number of vehicles in ASEAN (Association of Southeast Asian Nations) countries, China, and India by 2035, consequently increasing the demand for automobile fuels by 15–350%. Meeting these targets solely with conventional fossil fuels seems challenging. Moreover, the emphasis on cleaner and environmentally friendly biofuels has propelled the exploration of bioethanol and biodiesel blends for automobiles. By 2030, it is anticipated that Asian countries may fulfil 3–14% of their fuel demands through biofuels. Notably, Malaysia, India, and Indonesia are expected to contribute significantly, with Malaysia meeting 3%, India meeting 12%, and Indonesia meeting 11% of their respective auto fuel demands through biofuels. Sugarcane molasses has emerged as a prime feedstock for bioethanol production, particularly in Asian and South-East Asian nations where sugarcane is a vital crop. Sugarcane stands out as the second-most productive feedstock, yielding approximately 4500 L/ha, with cellulosic ethanol from bagasse offering an exciting prospect. The abundance of bagasse produced by sugar mills, coupled with advancing technologies for cellulose and lignin degradation, positions cellulosic ethanol as a promising avenue for ethanol production. Ethanol produced from molasses plays a crucial role as a raw material for various chemicals and diverse products. The primary intermediaries, ethylene and acetaldehyde, serve as key building blocks in the production of a multitude of compounds.

Extensive applications of molasses have been reported in different industries. It is utilized in fungi and bacteria cultivation for fermentation processes. The National Sugar Institute, Kanpur, is also making efforts to extract essential oil from sugarcane molasses (NSI Kanpur Newsletter 2023). Molasses is also used in the production of laccase by *Pycnoporus sanguineus* for dye decolourization (Marim et al. 2016). It is also utilized in the manufacturing of various chemicals on commercial and non-commercial scales. The chemical production occurs through ethanol by oxidation, dehydration, or modification in certain reactions. A few notable chemicals derived from molasses include dextran, sorbitol, lactic acid, oxalic acid, tartaric acid, maleic and fumaric acid, citric acid, succinic acid, glycerol, monosodium glutamate, L-lysine, β-carotene, butanol, acetone, aconitic acid, and itaconic acid (Table 1.2). In East Asian countries, the production of chemicals from molasses has a significant role. Furthermore, bacterial fermentation of diluted molasses, containing ample nutrients, has been demonstrated to yield lactic acid (John et al. 2006). Lactic acid is neutralized with lime to facilitate the production of calcium lactate, a compound employed in various applications. Calcium lactate plays a role in the production of baking powder and is a component of poultry feeds. Additionally, calcium salt and iron salt play roles in the pharmaceutical industry and the medical sector by serving as blood coagulants and anaemia controllers, respectively. Beyond these applications, lactic acid opens avenues for the production of essential derivatives, including methyl lactate, and acrylonitrile, utilized in industries such as leather and textiles (Delavier 1967; La Fromageot 1965). Moreover, the utilization of molasses, in conjunction with chemicals like glycerol, kojic acid, and β-carotene, contributes to the development of diverse products.

Interestingly, ephedrine hydrochloride, a component found in cough syrups, is commercially produced from molasses in certain countries, with India being one of them. Molasses functions as the raw material for the extraction of phytohormones such as cytokinin. In Cuba, an innovative application involves the production of explosives using cane molasses and ammonium nitrate, showcasing the diverse range of bioproducts that can be derived from sugarcane (Solomon 2011).

Molasses also serves as a fundamental raw material for the production of broad-spectrum biocides, potent inhibitors targeting bacteria such as *Leuconostoc* like enzyme inhibitor (IFOPOL). This application has proven effective in post-harvest management, notably utilized by sugar factories in Cuba. Furthermore, the cultivation of *Trichoderma* on sugarcane molasses has demonstrated adverse effects on the growth of soil-borne plant pathogens (Rodríguez-León et al. 1999). Additionally, *Xanthomonas campestris* utilizes molasses as a carbon source in the production of xanthan gum, a versatile substance widely employed in industries such as clothing, paint, ceramics, and petroleum (El-Salam et al. 1994).

Molasses is also utilized as a valuable feed for animals. It is a highly palatable energy feed, providing numerous applications in livestock nutrition. It serves as a source of dietary sugars, is rich in sucrose content, and offers a valuable blend of energy and minerals. It enhances the palatability of the ration, making it more appealing to animals. Molasses is employed to mitigate the dustiness of a ration, improving its overall feed quality. It acts as a binder for pelleting, contributing to the formation and stability of pelleted feeds (Mordenti et al. 2021). Cane molasses plays a role in stimulating rumen microbial activity and promoting efficient digestion and microbial protein synthesis. Beyond its known nutritional components, molasses may provide unidentified factors that contribute to the overall animal health. Molasses is being converted into feed blocks in many countries where grazing takes precedence over confined breeding. The ease of incorporating molasses into feed blocks has proven effective in consistently lowering the incidence of

Table 1.2 Utilization of sugarcane molasses in the production of some chemicals

Chemical produced	Substrate	Uses	Industry	References
Oxalic and tartaric acid	Molasses + nitric acid	In leather tanning and garment production	Leather and cloth	Sethy (2013)
Maleic acid	Molasses + *Rhizopus nigricans*	In fumaric acid and glyoxylic acid production	Chemical	Roa Engel et al. (2008)
Fumaric acid	Maleic acid (using molasses) + HCl solution (10–20%)	In pharmaceutical products and to confer acidic flavour in food items	Pharmaceutical and food	Roa Engel et al. (2008)
Acetic acid	Molasses + *Acetobacter*	In pharmaceuticals, fibres, plastics, adhesives, explosives; in the production of acetic anhydride, cellulose acetate, vinyl acetate, etc., on a large scale	Pharmaceutical, chemical, paint	Chandrasekaran (2012)
Butanol	Molasses + *Clostridium acetobutlicum*	In petroleum products; acetate and phthalate salts of butanol as solvents; as a hydraulic fluid, as a plasticizer and as an intermediate for production of many chemicals	Petroleum	Dwivedi (1995)
Acetone	Molasses + *Clostridium acetobutlicum*	As a nail polish remover; a solvent for many plastic and synthetic fibres; a cleaner in tools, in varnishes and paints; a polyester resin thinner	Cosmetic	
Citric acid	Molasses + *Aspergillus niger*	As a metal cleaner, a penetrating agent in leather tanning, and in plasticizer and resin production	Pharmaceutical, textile, confectionary, printing, chemical, leather food, and beverage	Show et al. (2015), Gould (1947), Hatt (1961)
L-lysine	Molasses + *Corynebacterium glutamicum*	As an ingredient in fish meal, soybean cake, and rapeseed cake	Food	Bhopal and Gudre (2001), Anaya-Reza and Lopez-Arenas (2017)
Ethylene	Molasses	In chemical synthesis	Chemical	Patarau (1982)

Chemical produced	Substrate	Uses	Industry	References
Polyethylene	Molasses	In film sheets, packaging materials, injection moulding	Plastic	Patarau (1982)
Poly(3-hydroxyalkanoate)	Molasses + *Pseudomonas mendocina* CH50	As bioplastics	Plastic	Basnett et al. (2020)
Polylactic acid	Molasses	In plastic degradation	Plastic and packaging	Jayasekara et al. (2022)
Polystyrene	Molasses	As disposable plastic; packaging material; license plate frames; smoke detectors; plastics	Packaging	Patarau (1982), Ji et al. (2023)
Ethyl chloride	Molasses	As a pain reliever	Medical and sports	Patarau (1982)
Diethyl ether	Molasses	As a solvent in laboratories; as a general anaesthetic	Medical and chemical	Patarau (1982)
2,3-Butanediol	Molasses + *Bacillus subtilis*	For printing inks; in perfumes, moistening and softening agents; in pharmaceuticals and explosives	Pharmaceutical, cosmetic, packaging	Deshmukh et al. (2015)
Dextran	Molasses + *Leuconostoc mesenteroides*	In sugar syrup; in confectionery and as a stabilizer in ice creams etc.	Sugar and food	Solomon (2011)
Inositol (vitamin B)	Molasses + *Aspergillus terreus*	As an animal and microbial growth factor; in microbiological and pharmaceutical preparations	Pharmaceutical	Evans and Hibbert (1947)
Monosodium glutamate (Ajinomoto)	Molasses + *Micrococcus glutamicus*	As a flavour intensifier	Food	Tracy (2019)

management costs (Tiwari et al. 1990; Garg and Gupta 1992). This practice provides a practical and cost-effective solution for enhancing feeding strategies in environments where grazing is a predominant aspect of livestock management. Furthermore, molasses is processed into powder form, which is utilized as an organoleptic, energetic, and flavouring supplement in animal feed. Its easy formulation, handling, transport, and storage help cattle farmers and feed manufacturers (Santos et al. 2020).

Additionally, molasses powder finds application in the steel-making industry, where it is used to coat forms for casting steel, thus demonstrating its versatility and value across various sectors (Santos et al. 2020). Notably, explosive compounds like nitromiel are derived from sugarcane molasses, showcasing the multifaceted applications of this agricultural by-product (Bhopal and Gudre 2001). Furthermore, molasses has also found its place in the food industry where it is used for sweetening and flavouring foods and as an ingredient in baking, cooking, and producing several yeasts like baker's yeast, torula yeast, feed yeast, etc. Synthetic rubber is also produced from molasses. Furthermore, molasses can be repurposed as a carbon feedstock for the production of polyhydroxyalkanoates (PHAs) through microbial fermentation. Molasses has the potential to undergo fermentation into glycerol or glycerine, which finds applications in the manufacturing of resins, cellophane, cosmetics, food and beverages, explosives, tobacco, and the nitroglycerin industry (Solomon 2011).

1.2.3 Press Mud and Its Uses

Press mud, a by-product formed during the cane juice decantation process in sugar and/or alcohol production, is the residue that is removed during the treatment stage (Poria et al. 2022). Press mud is also known as filter cake or press cake. Each year, the sugar industry generates approximately 500,000 tonnes of press mud. It is rich in lignocellulosic components and serves as a potent organic fertilizer, providing essential nutrients to plants and helping them defend against soil-borne diseases (Rolz et al. 2010). It is recognized as an excellent alternative to chemical fertilizers due to the shift towards organic farming and eco-friendly practices in the farming sector. Bio-compost fertilizers, derived from sugar press mud, offer a balanced blend of organic and inorganic nutrients, promoting the growth of high-quality, healthy crops (Zhu et al. 2007). Furthermore, these fertilizers enhance soil fertility by improving its microbial quality, a factor crucial for plant health (Bulluck et al. 2002). Press mud is particularly rich in nitrogen and phosphorous, stimulating nitrogen-fixing bacterial activities and creating a favourable environment for earthworm growth due to its high moisture content (50–70%) (Domınguez and Edwards 1997). The use of press mud resulted in improved yields of various crops such as bananas, eucalyptus, guava, vegetable seedlings (Rabelo et al. 2015), and sugarcane (Basanta et al. 2007; Barzaga et al. 2004). Press mud with cow manure and vermicompost is reported to be a good nutrient enhancer of soil (Prakash and Karmegam 2010). Furthermore, it serves as a structural soil amendment, countering soil degradation issues like erosion, compaction, crusting, and cracking, adjusting pH levels, enhancing drainage, and fostering the natural growth of beneficial bacteria and microorganisms (Guerra and Vazquez 2003). Furthermore, filter cake can be transformed into biochar, serving as an effective soil amendment. In addition to its soil-enriching qualities, press mud serves as a valuable reservoir of sugarcane wax. Cuba has pioneered a technology to extract this wax from press mud. The application of press mud has been reported in various sectors, including food, pharmaceuticals, chemicals, cosmetics, cleaning, and polishing. Sugarcane filter cake serves as a valuable resource in the sugar industry when utilized with bagasse as a fuel-producing raw material. In doing so, not only does the sugar industry benefit but also environment-related concerns are reduced (Ochoa et al. 2010). Furthermore, the Central Building Research Institute, Roorkee, has developed a method of calcination through which press mud can be utilized for making cementitious material (Tran 2012; Yadav and Solomon 2006). This converted material can be used as building lime, expanding the range of its utility in construction projects. Additionally, incinerated sugarcane filter cake was also used as a component in the production of lightweight foamed concrete, showcasing its diversity in various industrial contexts (Makul and Sua-iam 2016). Furthermore, press mud is also being utilized as an agglutinant in the production of combustible briquettes prepared from sugarcane straw and bagasse (Alvarez and Romero 2007; El Haggar et al. 2005). Bhargava and Srivatsa (2002) also successfully utilized filter cake in a sludge furnace. Moreover, blending filter cake with bagasse as a fuel in industrial boilers within sugar factories proves to be both economically and ecologically beneficial. Besides, various other chemicals, such as *n*-triacontanol, policosanol, and citric acid, have been successfully produced from press mud, showcasing its diverse potential applications. The National Chemical Laboratory, Pune, has developed a method for preparing steroids and superior-quality wax from press mud (Mohan et al. 2021). Furthermore, sun-dried sugarcane press mud has also shown potential as a valuable feed ingredient for swine production (Sahu et al. 2021).

Sugar mills in the sugarcane-growing states of India are exploring the use of press mud, a by-product of sugarcane, in producing bio-CNG, a renewable source of transport fuel. Biogas and bio-CNG are excellent sources of bioenergy from the sugar industry, which play significant roles in bringing about self-reliance in the energy sector for the country. The anaerobic digestion of filter cake results in the production of biogas/compressed biogas. Approximately 1 kg of CNG can be produced from 25 to 30 kg of filter cake, which makes the process lucrative. India has the potential to produce around 10 MMT (million metric tonnes) of press mud/filter cake per year, which could be diverted for producing bio-CNG. The government of India launched the Sustainable Alternative Towards Affordable Transportation (SATAT) scheme in October 2018, which aims to benefit vehicle users as well as farmers and entrepreneurs. The government plans to roll out 5000 CBG plants across the country in a phased manner. This initiative is expected to generate direct employment for 75,000 people and produce 50 million tonnes (mt) of bio-manure for crops. The MoPNG Ministry, SATAT, had aimed to achieve production of 15

MMT of CBG by 2023. An investment of more than 2 lakh crore is expected to be pumped into setting up these plants

A standard 3500 TCD (tonnes crushed per day) sugar mill in India can generate about 100–125 tonnes of press mud per day. One tonne of press mud, on an average, can generate 85–100 cubic metres of raw biogas. A 100-TPD (tonnes per day) capacity biogas plant can generate 850–950 kWh of power or close to 4000 kg of purified methane gas (CH_4 >95%). A biogas plant can be operated for 365 days in a year, whereas press mud generated during the crushing season can be stored and used during the off-season. To encourage the bio-CNG sector, the central government is offering a subsidy of up to Rs 40 million.

1.2.4 Sugarcane Trash and Its Uses

Straw, material removed from sugarcane before crushing, comprises dried and fresh leaves, stalks, and plant tips (Saad et al. 2008; Moriya et al. 2007). For every tonne of cultivated sugarcane, approximately 140 kg of sugar cane straw is produced (CENBIO 2003). With the increase in the adoption of non-burning harvest practices, the availability of sugarcane straw as feedstock for lignocellulosic biomass is growing (Alonso-Pippo et al. 2020; Leal et al. 2013). Straw return/trash mulching, a common agricultural practice, is used to augment soil fertility and organic carbon content (Castioni et al. 2018; Lisboa et al. 2018), resulting in improved physical, chemical, and biological properties of the soil, controlling erosion and weeds, conserving soil moisture, and enhancing the soil microbiota (Santos et al. 2020). Retaining a portion of the straw in the soil also benefits nutrient cycling and sustaining crop yields (Cherubin et al. 2019). This multifaceted utilization of sugarcane extraneous matter not only addresses energy requirements but also contributes to sustainable agricultural practices and environmental conservation.

Furthermore, sugarcane straw can be utilized as a valuable feedstock in both thermochemical and biochemical conversion processes. This has opened up new ways for its integration into future sugarcane biorefineries, like low-value biofuels and high-value chemical productions. The production of xylitol (having significant industrial and clinical applications) from sugarcane straw is one such example (Hernández-Pérez et al. 2020). Sugarcane straw also serves as a feedstock for bioelectricity and cellulosic ethanol production in Brazil (Cherubin et al. 2019). Studies conducted in Brazil have demonstrated that combining sugarcane extraneous matter with bagasse can nearly double electricity production compared to using bagasse alone. This co-generation approach, widely practiced in countries such as Brazil, is gaining traction in Asian and South-East Asian nations. Given that approximately 600 million people in Asia lack reliable access to electricity, with a significant portion residing in South-East Asia, sugarcane-growing countries like Thailand, Indonesia, Malaysia, Vietnam, China, and India can leverage their abundant extraneous matter reserves along with bagasse to meet both the mill's power needs and contribute excess electricity to the grid. Phenolic compounds are even present in sugarcane straw, which, on purification, produces amberlite XAD-2 with improved antioxidant capacity, thus paving the way for its diverse uses in the food and cosmetic industries (Carvalho et al. 2023). Lignocellulose nanocrystals derived from sugarcane straw have opened up innovative avenues for their utilization, further emphasizing the sustainable and novel aspects of cellulose nanocrystals in materials science (Bilatto et al. 2020). Furthermore, higher levels of zeolitic materials are being produced from sugarcane straw ash (comprising aluminosilicate glass, mullite, and quartz). These zeolitic materials are efficient adsorbents and are being used for wastewater treatment (Funagaro and Reis 2014). Additionally, sugarcane straw ash is rich in silicon dioxide, which is a highly pozzolanic material (Martirena et al. 1998; Paya et al. 2002). This can be effectively utilized as a substitute for cement in concrete production. On pulverization, sugarcane straw ash can be incorporated into concrete mixtures and can replace up to 10% of the cement weight in concrete without losing its strength capacity (Ahmed et al. 2019). In the future, the innovative use of sugarcane straw as a raw material for textile fibre production is a possibility, further highlighting its potential applications in various industries.

1.2.5 Sugarcane Peel Wax and Its Uses

Sugarcane peel contains a significant amount of wax, approximately 0.95% of its dry weight. The composition of sugarcane peel wax includes a variety of compounds such as alkane, ester, fatty acids, and alcohols, among others. Notably, it contains *sec*-butyl isothiocyanates, a compound with medicinal properties. Moreover, components like heneicosane, an insect semiochemical, and edible fatty acids are present in sugarcane wax, further enhancing its value. Essential compounds like D-003 acids, policosanol, and octacosanol can also be derived from sugarcane wax, which has multifarious applications. These essential compounds have lipid-lowering effects on both humans and animals (Gnanaraj 2012). The extracted wax from sugarcane peel is a versatile raw material for diverse uses (Georges et al. 2006; Taylor 2000). It serves as a cost-effective substitute for expensive waxes like candelilla and chitosan, particularly in edible coatings (Saucedo-Pompa et al. 2009; Amarante et al. 2001; Chien et al. 2007). Sugarcane wax can even be utilized in the care products sector, such as car care, shoe, and floor items. Additionally, it is also used in the leather, additives, paint, printing ink, and plastic industries. The potential uses of sugarcane wax extend to pharmaceutical preparations, confectionery items, processed foods, carbonless papers, electrical insulation of cables and wires, and crayons and coloured pencils. M/s Godavari Biorefineries, based in Mumbai, is actively engaged in the commercial production of sugarcane wax under the trade name "Nautowax." This company is keen on exploring its applications in the cosmetic and food industries. Additionally, the Shriram Institute for Industrial Research is involved in research related to the preparation of lipstick and cosmetic pencil compositions using sugarcane wax (Singh et al. 2015). Sugarcane wax serves various industrial purposes, with its oil component being utilized for corrosion protection and crude sugarcane wax directly employed in the production of carbon paper. Furthermore, all constituents of sugarcane wax, including hard wax, oil, and resin, serve as plasticizers in the tyre industry. Additionally, it plays a crucial role in enhancing the melting and release properties of toners used in multicolour photographic copiers and printers. Refined sugarcane wax finds applications as a waterproofing emulsion for particle boards and textile treatment, as well as in hot melting glues, in the removal of castings from fibreglass moulds, and in precision casting processes (Mohan et al. 2021).

1.2.6 Biochar and Its Diverse Uses

Thermochemical processes such as the pyrolysis of bagasse produce biochar as a predominant product, existing in various forms of black carbon. Biochar has a diverse range of applications. It is incorporated into soils to enhance their structure, as soils need a balanced mix of aggregates, solids, and organic matter (humus). The inherent crystalline structure of biochar ensures the stability of soil organic matter content, making it resilient and enduring in various soil types (Sohi et al. 2010). In the case of compacted soil, its integration can enhance aeration due to its diverse porosity levels (Heitkötter and Marschner 2015). Likewise, biochar is also used as a crucial soil amendment or conditioner (Ding et al. 2016; Igalavithana et al. 2016). Biochar with compost has been found to be equally beneficial as using biochar alone in enhancing plant yields. The long-lasting presence of biochar in the environment and soils, extending for several years in fields, makes it a cost-effective alternative as a soil enrichment agent (Joseph et al. 2010). Biochar can also be transformed into activated carbon, a remarkably porous material ideal for adsorption purposes. Additionally, biochar serves as a fuel source and a means of carbon sequestration. Consequently, it emerges as a powerful tool for mitigating climate change. Furthermore, the addition of biochar to soils also enhances the production of ethylene (a plant growth and ripening hormone), leading to high crop yields (Spokas et al. 2010). In animal husbandry, biochar serves multiple purposes, including its use as an ingredient in probiotic animal feed (referred to as "carbon feed"), as a silage additive, as bedding material, or as a conditioner for manure and slurry, often in conjunction with lactobacilli (Kammann and Schmidt 2014). This results in its positive impact on animal health (Schmidt et al. 2019; Toth and Dou 2016).

Biochar has also emerged as a valuable adsorbent in wastewater treatment facilities. Decima et al. (2021) emphasized the important features of biochar, like its high specific surface area, aromatic carbon components, and micropore volumes for increasing the efficiency of wastewater treatment, especially in removing pharmaceuticals like carbamazepine. Additionally, Wang et al. (2018) underscored biochar's remarkable adsorption abilities for heavy metals such as Hg(II) and Pb(II) in solution. Likewise, biochar also functions as a significant carrier for microorganisms, forming co-sorbents like biochar with bacteria, as co-sorbents not only improve the sorption capacity for Pb(II) but also increase adsorption.

1.3 Commercially Available Sugarcane Bio-based Products

Carnegie has pioneered the creation of Biobased Xorel, a PVC-free textile that stands out as the first high-performance interior textile crafted from sugarcane-based materials (https://carnegiefabrics.com/carnegie-xorel). Green Line Paper in York, Pennsylvania, sells sugarcane bagasse products like platters, bowls, and sugarcane-based notebooks (https://greenlinepaper.com/bagasse-products/). Berry Global Inc. introduced the Bhoomi bottle, a bio-based bottle, made purely from sugarcane in New York, New Jersey, Texas, Louisiana, and California (Evansville 2020; Nagal 2021). AT&T, a telecommunications company in the United States, pioneered the adoption of plant-based packaging, incorporating materials where 30% is derived from ethanol sourced from sugarcane in the packaging of the company's wireless accessories like device cases (Pierce 2014). Another company, Servous, in Miami, Florida, manufactures and sells bagasse plates, bowls, chicken roaster containers, meat trays, produce trays, take-out food containers, and paper shopping bags (https://atyourservous.com/materials/bagasse/). Raydel in Australia manufactures Policosanol 10, which contains a Cuban sugarcane ingredient. This product is helpful for reducing overall cholesterol by increasing good cholesterol and even has an anti-ageing effect (https://www.raydel.com.au/product/policosanol-10). Love Dear, Taiwan, designed a shampoo bottle based on 96% sugarcane waste having 60% biocarbon (https://www.love-dear.com/?locale=en). M&H introduced its sugarcane plastic in collaboration with Bulldog, asserting itself as the pioneer among men's skincare brands to incorporate sugarcane as a primary raw material (https://uk.bulldogskincare.com/pages/sugarcane). Ecover, Belgium (30–100% plant-based components), is also available at the commercial level with the use of a renewable plastic named Plant-astic, obtained from sugarcane (https://www.ecover.com/faq/what-is-plantplastic/). PlantBottle™, introduced in 2009 by the Coca-Cola Company, utilizes monoethylene glycol (30%) produced from sugarcane (https://www.coca-colacompany.com/media-center/100-percent-plant-based-plastic-bottle).

Several cosmetic products using sugarcane like sugarcane massage creams (for men and women), nutrifying moisturizers, sugarcane scalp conditioning shampoos, sugarcane facial kits, etc. are being produced and sold under the brand name Oir. Procter & Gamble, the parent company of Pantene, is also using bioplastics (a material derived from sugarcane) for their shampoo and conditioner bottles with their first inauguration in the European market (Gyekye 2011). TUV Austria and Green Straws are also producing and selling sugarcane straws, an alternative to paper and plastic straws, on a commercial scale (https://thesugarcanestraw.com/). A Brazilian company, Braskem, is producing bio-polyethylene or green polyethylene using sugarcane under the brand name I'm Green™ (https://bioplasticsnews.com/im-green/). This brand is also being utilized by Johnson & Johnson, Brazil, for the packaging of cosmetics (containing 60% green plastics and 40% recycled materials), such as sunscreens (Mohan 2012; Yeomans 2011). H.J. Heinz Co. is making its ketchup bottles from plant materials with a notable 30% ethylene glycol usage derived from ethanol sourced from Brazilian sugarcane with innovation in collaboration with the

Coca-Cola Co. (Biopolymers 2011). Nestlé, Brazil, with Tetra Pak and Braskem, is incorporating polyethylene derived from sugarcane into the production of caps for box-type packaging used for two widely consumed Nestlé milk brands, Ninho and Molico (Berger 2011). RPC M&H Plastics, Brazil, is presenting a novel type of plastic with sugarcane as the fundamental ingredient, resulting in a "green" polyethylene variant. PHB Industrial S/A is utilizing 100% Brazilian technology using sugarcane bagasse to produce polyhydroxybutyrate, under the brand name Biocycle®. It finds application in various industries, including auto parts, agricultural components, credit cards, cutlery, cosmetic packaging, and more, thus showcasing its versatility and sustainable characteristics (Janeiro 2009). Dow Chemical Co Ltd. in partnership with Mitsui & Co Ltd. in Brazil produces bioplastics from ethanol obtained from sugarcane (Smock 2011; Esposito 2011). Corbion (Bonsurco, SEDEX) employs sugarcane from Thailand as a primary source for producing polylactic acid (PLA) bioplastics using cane sugar-derived lactic acid (https://bonsucro.com/corbion-achieves-first-chain-custody-certification-bioplastics/).

In India, Gunjan International, Delhi, is a leading manufacturer, supplier, and exporter of sugarcane bagasse products like paper pulp plates, disposable pulp plates, bagasse pulp bowls and cups, and sugarcane pulp plates in the global market (https://www.pulptablewareproducts.com/company-profile.html). Another company, Saattvic Ecocare Products LLP, Mumbai, is also a prominent manufacturer of bagasse disposable plates and biodegradable sugarcane bagasse plates (https://www.saattvicecocare.in/biodegradable-bagasse-plates.html). Besides, Kochi-based Saintjo Innovations in Kerala makes products under the brand name Chuk, which include sugarcane-based plates, containers, etc. of various sizes (Nandalal 2018). Furthermore, EcoHealthcare LLP in Rajasthan, India, produces sugarcane bagasse-based products under the name of Canefibre®. It includes tableware products such as bowls and takeaway containers, plates, etc., from bagasse and alternatives for plastics and paper in retail packaging (https://ecohealthware.com/canefiber-sugarcane-bagasse/). Additionally, GoEco in Rajasthan also makes bagasse plates, containers, clamshells, bowls, and meal trays (https://goeco.co.in/). Moreover, the Tamil Nadu Newsprint and Papers Limited (TNPL), India, manufactures newsprint and writing paper by utilizing bagasse, a residue derived from sugarcane (https://www.tnpl.com/). Mawana Sugars in Uttar Pradesh, India, manufactures and supplies rectified spirits, denatured spirits, fuel ethanol, organic manure, and fusel oil, which are required in various other industries like fusel oil, used as a solvent and diluent of printing inks, hydraulic fluids and gums for printing industries, spirits for chemical industries, and pharmaceutical manufacturers (https://www.mawanasugars.com/distillary.php). Dhampur Bio Organics Ltd. in India produces extra neutral alcohol (ENA) (a food-grade alcohol obtained by brewing sugarcane molasses and yeast) for the manufacturing of domestic spirits or country liquor (https://www.dhampur.com/our-products/ena). Godavari Sugar Refineries is making commercial sugarcane wax-based products under the name of Naturowax®. It is used as an additive in lubricants, as waterproof surface coatings in leather goods, as an anti-scuff agent in graphical printing inks, and for production of lipsticks, mascara, moisturizing creams, and sunblock (https://godavaribiorefineries.com/

naturowax-%C2%AE). Furthermore, Vipin Sarin, a sugarcane value-added entrepreneur, in Haryana, is manufacturing and selling several value-added products made from sugarcane, like, ganna jalebi, ganna chuski, frozen gola, cane imli, etc. (Unnithan 2021). Moreover, the Indian Institute of Science (IIT), Guwahati, has developed a novel method for a sugar substitute named xylitol from the sugarcane waste bagasse using ultrasonic fermentation (Somasekhar 2022). Godavari Sugar Refineries manufactures various chemicals, namely, 3-methoxy butanol, 1,3-butylene glycol, bio-based butanol, ethyl acetate, ethyl crotonate, ethyl lactate, etc., from sugarcane-derived ethanol. These chemicals are used as raw materials by other industries (Table 1.3).

Table 1.3 Sugarcane-based chemicals produced by Godavari Sugar Refineries

Chemicals	Uses
Acetates derivatives	Automotive Original Equipment Manufacturer (OEM) and refinish, wood furniture coatings, leather coatings, plastic coatings, aircraft coatings, cable lacquers, and various fabric coatings
Acetaldehyde diethyl acetal	Alcoholic beverages, citrus fruits, dairy products, and tropical fruits
Acetaldehyde	Manufacturing of amber note fragrance intermediates; a valuable ingredient in a wide range of food products
Bio-based acetic acid	Formulation in hair conditioners, shampoos, hair rinses, wave sets, and can be found in skincare products
Bio-based butanol	As a solvent in paints and coatings; manufacturing of butyl acetate
Cellulose acetate	Ink reservoirs in fibre tip pens
Cellulose derivatives	Acrylic paints as rheology modifiers
Ethyl acetate	An aroma enhancer in cosmetics
Ethyl lactate	Raw material with a special rum, fruit, and cream flavour; formulating printing inks, coatings, resin cleaners, paint strippers, graffiti removers, ink cleaners, etc.; as an ideal wipe solvent
Sodium sulphate	Dyeing textiles
1,3-Butylene glycol	Unsaturated polyester resins
2-Ethyl-1,3 hexanediol	Detergent compositions for removing iodine stains from textiles
3-Methyl-3-penten-2-one	Imparting fullness and subtle strength to fragrances
3-Methoxy butanol	As a solvent or co-solvent in coatings, inks, and adhesives

1.4 Diversifying Sugarcane By-Products for a Sustainable Bio-based Economy

The sugar industry's reliance on fossil fuels poses environmental challenges, and a shift to biomass-based renewable energy is seen as a more sustainable option (Owusu and Asumadu-Sarkodie 2016; Kabeyi and Olanrewaju 2022). The sugarcane industry, with its abundant supply in Asian countries, is positioned to lead a bio-based economy, producing renewable and sustainable bioproducts. Developing sustainable supply chains is crucial for the success of a sugarcane-based bioeconomy. Sugarcane and other sugar crops are not only excellent sources of renewable feedstocks for bioenergy and green fuel but also yield valuable by-products such as bagasse and molasses-based chemicals, pharmaceuticals, and bioplastics (Solomon 2011). These crops can be cultivated in large areas, including low-fertility and marginal lands, with high conversion efficiency. The well-established production, logistics, processing, and marketing systems associated with sugar crops enhance their attractiveness as raw materials for an efficient bio-based industry. Sugarcane, when combined with efficient co-product and waste utilization, offers a favourable net energy and greenhouse gas (GHG) balance, making it a preferred crop for bioethanol production. Currently, about 45% of global ethanol production is derived from sugar crops. Globally, biofuels may constitute approximately 8% of the transport mix by 2030, with an estimated 15% share in Asian countries (OECD-FAO Agricultural Outlook 2021–2030). This projection assumes rapid expansion of high-yielding sugar crops on underutilized lands and large-scale commercialization of cellulosic ethanol from crop residues. The shift towards a bio-based economy is expected to increase the share of biofuels in the Asian region. The growing emphasis on second-generation bioethanol production has stimulated the exploration of methods to convert these residues into functional bioproducts, thereby enhancing the overall sustainability of the sugarcane processing cycle (Boro et al. 2022). Efficient utilization of sugarcane by-products aligns with the principles of a circular economy, where waste is minimized and resources are maximally utilized through recycling and repurposing. In this context, converting sugarcane straw and bagasse into value-added products represents a key strategy. This approach not only mitigates the environmental impact of by-product disposal but also contributes to the economic viability of the sugarcane industry. The rise of second-generation bioethanol production involves utilizing lignocellulosic biomass, such as sugarcane residues, for ethanol production. This shift not only addresses the challenge of competition between food and fuel production but also promotes the efficient use of agricultural waste. Additionally, the development of technologies to convert these residues into functional bioproducts, beyond bioethanol, adds versatility to the sugarcane industry. The exploration of innovative approaches to repurpose sugarcane by-products can lead to the production of bio-based chemicals, materials, and other valuable commodities. This diversification not only reduces the environmental footprint associated with by-product disposal but also contributes to the creation of a more sustainable and resilient sugarcane industry (Antunes et al. 2022).

1.5 Charting the Green Horizon: Exploring the Bio-based Economies of Asian Nations

In the current global landscape, a transition towards efficiency and sustainable development is underway. The surge in demand for resources, particularly energy, is pronounced, with Asia, led by economic powerhouses like China and India, playing a pivotal role in this paradigm shift. As these nations experience rapid growth and increasing prosperity, the demand for food, fuel, feed, and other resources intensifies. However, the unchecked and unsustainable use of these resources poses challenges such as climate change, CO_2 emissions, and greenhouse gas effects, especially in the Asian region. Recognizing the need for a cleaner, sustainable standard of living, there is a compelling case for Asian countries to champion the shift to a bio-based economy. Given the substantial contribution of the sugar industry to the economies of many Asian nations, leveraging sugarcane-based bioproducts on a commercial scale has become a strategic imperative (Solomon and Swapna 2022). While sugars and sweeteners are established products, the production and utilization of other sugarcane-based bioproducts are yet to reach commercial scale in most Asian countries, barring advancements in bioethanol and bagasse-based co-generation. Bagasse-based co-generation, in particular, has gained traction, with India showcasing notable success in utilizing the excess power generated from bagasse to augment the national grid. The changing perspective on biomass, from being considered free with little commercial value to being regarded as precious, is a response to the fossil fuel crisis and climate change. In India, several sugar mills utilize bagasse-generated power, with the potential for grid-interactive power generation estimated to exceed 10,000 MW. Initiatives by entities like Godavari Refineries, engaged in diverse product lines from sugar to biofuels and chemicals, underscore the potential of sugarcane as a versatile feedstock. The refinery's foray into chemical production, utilizing ethanol as a feedstock, exemplifies the versatility of sugarcane-derived resources in various industries. Encouraging and maximizing such initiatives by different sugar mills and sugarcane processing groups is crucial for efficiently steering the country towards a robust bioeconomy. As Asia stands at the forefront of global economic growth, the responsible utilization of bio-based resources has become not only an economic necessity but also a sustainability imperative (Solomon and Swapna 2022).

Thailand has emerged as a notable player in the realm of bio-based products, showcasing significant achievements in leveraging sugarcane and agro-wastes (Bioplastics 2020). Among the key contributors is Mitr Phol Sugar Corporation, a leading sugar processing group in the country, renowned for its exemplary diversification within the sugar sector (Mitr Phol Group 2017). Mitr Phol Sugar Corporation's impressive portfolio spans various sugar categories, including raw, refined, and value-added, along with sugar syrup production. The company goes beyond traditional sugar processing, engaging in co-generation from bagasse, and manufacturing particle boards and melamine-faced panels. Furthermore, it plays a role in the production of ethanol, gasoline, and gasohol, exemplifying a comprehensive approach to utilizing sugarcane resources. While Thailand has made strides in bio-based industries, there remains untapped potential, especially in biomass-based co-generation from sugarcane biomass and other sources (https://www.mitrphol.com/offshore.php). The imperative for South-East Asian countries, including Thailand, is to intensify efforts in harnessing this potential to meet growing power demands through co-generation. Ethanol, another vital bioproduct derived from the sugar industry, boasts well-developed production, processing, and marketing systems across Asian sugar-producing nations. Apart from its use in potable applications and blending with fuels, ethanol serves as a crucial source for various chemical derivatives and pharmaceuticals. The national biofuel policies set by each country guide their ethanol production efforts to meet blending targets. China, for instance, aims at a 10% ethanol blending (E10) by 2020, whereas India has targeted a 20% blending by 2025 under its national biofuel policy. The Philippines had aimed to target a blend rate at 20% by 2020 and beyond (Ramsey et al. 2023). Thailand, with an ambitious plan to implement E20, signalling a shift from E10 and introducing E85 by 2027, necessitates an additional 9.1 billion litres per year. The country is gearing up for this challenge through its 10-year bioeconomy hub program, backed by a substantial investment of 11.3 billion. In addition to these initiatives, Thailand is already paving the way for large-scale bioplastic production, focusing on sugarcane and other crop biomass. These endeavours collectively position Thailand as a dynamic force in the bio-based economy, exemplifying a commitment to sustainable and innovative solutions (Sukyai et al. 2016).

Across Asia, countries are making significant strides in the development of bio-based products, with China standing out as a pioneer in establishing its own bioeconomy initiative. China's commitment to bio-based products is exemplified through strategic joint ventures with industry giants such as DuPont and Jilin Province New Tianlong Industry Co., resulting in the establishment of China's largest cellulosic ethanol facility. This facility, dedicated to converting agricultural biomass into ethanol, underscores China's dedication to fostering a robust bio-based economy. China has not only focused on domestic initiatives but has also made international headlines with Dragonair's pioneering commercial flight in 2015, operated with a 50:50 blend of fossil fuel and bio-based fuel from Shanghai to Hong Kong. The nation continues to plan large-scale second-generation biofuel projects, signalling significant potential for bioproducts. In response, the Chinese sugar industry must align itself with these developments, scaling up sugarcane and sugar production to ensure an abundant supply of sugarcane-based feedstocks.

Meanwhile, Malaysia has emerged as a bioeconomy trailblazer in the ASEAN region, being the first to establish a national bioeconomy initiative and the second in Asia after China. With an impressive annual growth rate of approximately 15%, investments in Malaysia's bioeconomy sector are projected to reach RM 181.2 billion by 2030, up from RM 107 billion in 2010, contributing 13.4% to the nation's gross domestic product (GDP). Leveraging its extensive oil palm diversity, Malaysia is a key player in palm-oil-based biodiesel production. The trend extends across the ASEAN countries, with nations like Indonesia, the Philippines, and Vietnam also making strides in promoting various blends of biofuels. As these countries actively pursue bio-based initiatives, the landscape of the Asian bioeconomy continues to evolve, showcasing the region's commitment to sustainable and innovative solutions. Despite the well-established sugar sector in many Asian countries, these nations have yet to emerge as major players in the bioeconomy landscape.

The Asian continent, encompassing the South-East Asian region, presents significant transformative opportunities even in harnessing the potential of bagasse for co-generation. Given that approximately 600 million people in Asia lack reliable access to electricity, with a significant portion residing in South-East Asia, sugarcane-growing countries like Thailand, Indonesia, Malaysia, Vietnam, China, and India can leverage their abundant extraneous matter reserves along with bagasse to meet both the mill's power needs and contribute excess electricity to the grid. With Asian countries poised for substantial economic development, the demand for power is expected to surge rapidly. Existing resources alone cannot meet this escalating demand. Utilizing biomass, including the abundant agricultural waste produced by these nations, could offer a clean and efficient method for electricity generation. Estimates indicate that power generation based on sugarcane bagasse could fulfil up to 25% of the power requirements in major sugarcane-producing countries. However, the full potential of this opportunity has yet to be realized in Asian and South-East Asian countries. A conservative estimate suggests that sugar, rice, and palm oil mills in the entire South-East Asian region generate more than 230 million tonnes of biomass annually, translating to a co-generation potential of 16–19 GW. Despite this vast potential, South-East Asian countries have not fully tapped into the co-generation possibilities offered by sugarcane biomass. Gradual upscaling could lead to commercial-level co-generation, where electricity is generated on a large scale and supplied to the national grid, thus addressing the power needs of the sugar-agro complexes. Brazil and India, the largest sugarcane-producing countries, have well-established systems of bagasse-based co-generation that generate energy at a commercial scale. Globally, the potential for bagasse-based co-generation has been estimated at 135,029 GWh per year. In India, there are approximately 206 co-generation units with a cumulative installed exportable capacity of 3123 MW during peak seasons. Furthermore, with the modernization of new and existing sugar mills, India has the potential to generate surplus power across all sugar factories, reaching up to 5000 MW in the foreseeable future.

In the face of volatile sugar prices and related challenges, the imperative for the sugar industry is clear: bio-based diversification is the key to ensuring productivity, profitability, and sustainability. A substantial upscaling of sugarcane and sugar production is essential, paving the way for an increased output of ancillary products that can drive bioeconomy initiatives. To achieve this transformation, overcoming challenges such as low productivity, high production costs, inadequate crop varieties, declining soil health, limited mechanization, suboptimal performance of sugar mills, and insufficient modernization is paramount. Tackling these obstacles head-on is crucial for positioning the sugar industry in Asia and the South-East Asian region as trailblazers in the global march towards a bio-based economy. In this era of change, where adaptability is key, the sugar industry must evolve into a multifaceted powerhouse, contributing not only to traditional sweeteners but also to a diverse array of bio-based products, ensuring resilience and sustainability in the face of evolving economic landscapes. The abundant supply of sugarcane in Asian and South-East Asian countries makes this region stand out as the prime candidate to lead the way in establishing a sugarcane-based bioeconomy. A wide array of renewable and sustainable bioproducts can be derived from sugarcane, making it a promising cornerstone for such an initiative. Beyond the actual production and conversion technologies, the development of sustainable supply chains is equally crucial for the success of a sugarcane-based bioeconomy.

1.6 Conclusions

The multifaceted nature of sugarcane by-products, encompassing bagasse, molasses, press mud, and various residues, underscores the sugar industry's potential to contribute significantly to the bioeconomy. By leveraging advanced technologies, sugarcane by-products are being converted into high-value products that have significant roles in other industries. This results in addressing contemporary challenges along with economic benefits. The use of bio-based products, especially from the agricultural left, has seen a rise, with respect to the management of environmental issues. The sugar industry is also not left untouched as it has also adapted to the changing dynamics. The successful utilization of sugarcane by-products emerges as a promising pathway towards a more sustainable and economically viable future. The pursuit of novel applications and markets for these by-products aligns with broader objectives of sustainable resource management and waste reduction. Commercially, there are many products available even for direct usage by humans like bagasse-made plates, bowls, notebooks, cosmetic items, etc. The textile industry is also using sugarcane-derived components for producing final products. Depending on the specific end product, the textile industry comprises 60–85% sugarcane-derived components. Although research and development endeavours are underway, especially in countries like Thailand, the standardized adoption of bagasse-based products in textile manufacturing in various Asian countries is still pending, despite the potential for creating eco-friendly garments from sugarcane-based materials.

References

Abdulkadir T, Oyejobi D, Lawal A (2014) Evaluation of sugarcane bagasse ash as a replacement for cement in concrete works. Acta Tech Corviniensis-Bull Eng 3:71–76

Ahmed A, Uche OAU, Abdulwahab MT, Amali NU (2019) Mechanical properties of sugarcane straw waste ash used as binder in concrete production. FUW Trends Sci Technol J 4(1):247–249

Akter M, Halawani RF, Aloufi FA, Taleb MA, Akter S, Mahmood S (2022) Utilization of agro-industrial wastes for the production of quality oyster mushrooms. Sustainability 14:994

Alonso-Pippo W, Luengo CA, Alonsomador MAL, Garzone P, Cornacchia G (2020) Energy recovery from sugarcane trash in the light of 2nd generation biofuels. Part 1: current situation and environmental aspects. Waste Biomass Valori 2:1–16

Alvarez R, Romero C (2007) Evaluación de mezclas y aglutinantes en la elaboración de briquetas de biomasa. In: Proceedings 8vo Congreso Iberoamericano de Ingeniería Mecánica. Cuzco, Peru

Al-Arni S (2018) Extraction and isolation methods for lignin separation from sugarcane bagasse: A review. Industrial Crops and Products 115:330–339

Amarante C, Banks NH, Ganesh S (2001) Effects of coating concentration, ripening stage, water status and fruit temperature on pear susceptibility to friction discolouration. Postharvest Biol Technol 21(3):283–290

Anaya-Reza O, Lopez-Arenas T (2017) Comprehensive assessment of the L-lysine production process from fermentation of sugarcane molasses. Bioprocess Biosyst Eng 40:1033–1048

Anjos MAS, Araujo TR, Ferreira RLS, Farias EC, Martinelli AE (2020) Properties of self-leveling mortars incorporating a high-volume of sugar cane bagasse ash as partial Portland cement replacement. J Build Eng 32:101694

Antunes F, Mota IF, Burgal JDS, Pintado M, Costa PS (2022) A review on the valorization of lignin from sugarcane by-products; from extraction to application. Biomass Bioenergy 166:106603

Araújo de Almeida M, Colombo R (2021) Construction of green roofs via using the substrates made from humus and green coconut fiber or sugarcane bagasse. Sustain Chem Pharm 22:100477

Araújo VDOD, Martins VA, Cavalheri PS, Honorio L, Lima PDM, Filho FJCM (2022) Incorporation of tannery waste and sugarcane bagasse ash in soil-cement bricks. Revista AIDIS de ingeniería y ciencias ambientales: Investigación, desarrollo y práctica. 15, 1 (abr. 2022), 154–171. https://doi.org/10.22201/iingen.0718378xe.2022.15.1.77053.

Arif E, Clark MW, Lake N (2016) Sugar cane bagasse ash from a high efficiency co-generation boiler: applications in cement and mortar production. Constr Build Mater 128:287–297

Balakrishnan M, Batra VS (2011) Valorization of solid waste in sugar factories with possible applications in India: a review. J Environ Manag 92:2886–2891

Balgees A, Elmnan A, Hemeedan AA, Ahmed RI (2016) Influence of different treatments on nutritive values of sugarcane bagasse. Glob J Anim Sci Res 2:295

Barrera Torres G, Dognani G, da Silva Agostini DL, dos Santos RJ, Carmago Cabrera F, Gutierres Aguilar CM, de Paiva FFG, Rainho Texteira S, Job AE (2021) Potential eco-friendly application of sugarcane bagasse ash in the rubber industry. Waste Biomass Valori 12:4599–4613

Barzaga JU, Perez JA, Santana JA (2004) Produccion de compost y resultados preliminares de su aplicación en áreas de la Empresa azucarera Dos Ríos. Jornada Científica por el 40 Aniversario del Instituto Nacional de Investigaciones de la Caña de Azúcar. http://www.santiago.cu/hosting/etica/Sede40/tec/t13.htm. Accessed 23 Mar 2010

Basanta R, Garcia Delgado M, Cervantes J (2007) Sostenibilidad del reciclaje de residuos de la agroindustria azucarera. Una revisión. Ciencia y Tecnología Alimentaria 5:293–305

Basnett P, Marcello E, Lukasiewicz B, Nigmatullin R, Paxinou A, Ahmad MH, Gurumayum B, Roy I (2020) Antimicrobial materials with lime oil and a poly(3-hydroxyalkanoate) produced via valorisation of sugarcane molasses. J Funct Biomater 11(2):24

Batra VS, Tewari PK (2006) A process for the preparation of inorganic membrane filters for solid-liquid and solid-gas separations. Indian Patent 1326/Del/2006, 6 Jul 2006

Berger H (2011) Brazil Braskem makes renewable PE for tetra Pak's Nestle packages. Independent Commodity Intelligence Services. https://www.icis.com/explore/resources/news/2011/07/29/9481251/brazil-braskem-makes-renewable-pe-for-tetra-pak-s-nestle-packages/. Accessed 13 Nov 2023

Bhardwaj NK, Kaur D, Chaudhry S, Sharma M, Arya S (2019) Approaches for converting sugarcane trash, a promising agro-residue, into pulp and paper using soda pulping and elemental chlorine-free bleaching. J Clean Prod 217:225–233

Bhargava R, Srivatsa S (2002) Briqueting of filter cake. In: Proceedings of the 64th Annual convention of the Sugar Technologists Association of India, Kerala

Bhopal P, Gudre RH (2001) carotene production in sugarcane molasses by *Rhodotorula glutinis* mutant. J Ind Microbiol Biotechnol 26(2):327–332

Bian H, Gao Y, Luo J, Jiao L, Wu W, Fang G, Dai H (2019) Lignocellulosic nanofibrils produced using wheat straw and their pulping solid residue: from agricultural waste to cellulose nanomaterials. Waste Manag 91:1–8

Bilatto S, Marconcin JM, Mattoso LHC, Farinas CS (2020) Lignocellulose nanocrystals from sugarcane straw. Ind Crop Prod 157:112938

Bioplastics (2020) Brochure. Thailand Board of Investment. https://www.boi.go.th/upload/content/BioplasticsBrochure.pdf

Biopolymers (2011) Heinz ketchup converts to biobased bottle. https://www.ptonline.com/articles/heinz-ketchup-converts-to-biobased-bottle Accessed 13 Nov 2023

Borges ER, Pereira N (2011) Succinic acid production from sugarcane bagasse hemicellulose hydrolysate by Actinobacillus succinogenes. J Ind Microbiol Biotechnol 38:1001–1011

Boro M, Verma AK, Chettri D, Yata VK, Verma AK (2022) Strategies involved in biofuel production from agro-based lignocellulose biomass. Environ Technol Innov 28:102679

Botkin DB, Keller EA, Rosenthal DB (2012) Environmental science. Wiley, New York

Bulluck LR, Brosius M, Evanylo GK, Ristaino JB (2002) Organic and synthetic fertility amendments influence soil microbial, physical and chemical properties on organic and conventional farms. Appl Soil Ecol 19(2):147–160

Cangussu N, Chaves P, da Rocha W, Maia L (2023) Particleboard panels made with sugarcane bagasse waste—an exploratory study. Environ Sci Pollut Res 30(10):25265–25273

Carvalho MJ, Costa JR, Pedrosa SS, Pintado M, Oliveira ALS, Madureira AR (2023) Sugarcane straw extracts: production and purification by amberlite XAD-2. Food Bioprod Process 140:189–199

Castioni GA, Cherubin MR, Menandro LMS, Sanches GM, Bordonal RDO, Barbosa LC, Franco HCJ, Carvalho JLN (2018) Soil physical quality response to sugarcane straw removal in Brazil: a multi-approach assessment. Soil Tillage Res 184:301–309

CENBIO (Brazilian Reference Center on Biomass) (2003). http://www.cenbio.br. Accessed 14 Apr 2003

Chandrasekaran M (ed) (2012) Valorization of food processing by-products: fermented foods and beverages series. CRC Press, Boca Raton, p 836

Cherubin MR, Lisboa IP, Silva AGB, Varanda LL, Bordonal RO, Carvalho JLN, Otto R, Pavinato PS, Soltangheisi A, Cerri CEP (2019) Sugarcane straw removal: implications to soil fertility and fertilizer demand in Brazil. Bioenergy Res 12:888–900. https://doi.org/10.1007/s12155-019-10021-w

Chien PJ, Sheu F, Yang FH (2007) Effects of edible chitosan coating on quality and shelf life of sliced mango fruit. J Food Eng 78(1):225–229

de Siqueira AA, Cordeiro GC (2022) Sustainable cements containing sugarcane bagasse ash and limestone: effects on compressive strength and acid attack of mortar. Sustainability 14:5683

Decima MA, Marzeddu S, Barchiesi M, Marcantonio CD, Chaivola A, Boni MR (2021) A review on the removal of carbamazepine from aqueous solution by using activates carbon and biochar. Sustainability 13(21):11760. https://doi.org/10.3390/su132111760

Delavier HJ (1967) Sugar for feeding purposes. F.O. Licht's Yearbook and Directory, Ratzeburg, pp 242–245

Delgado AV, Casanova CA (2001) Sugar processing and by-products of the sugar industry. Food and Agriculture Organization, Agricultural Services Bulletins

Deshmukh AN, Mistry SR, Ywale TB, Mahajan DM, Jain R (2015) Production of 2,3-butanediol from sugarcane molasses using *Bacillus subtilis*. Int J Adv Biotechnol Res 6(1):66–79

Ding Y, Liu Y, Liu S, Li Z, Tan X, Huang X, Zeng G, Zhou L, Zheng B (2016) Biochar to improve soil fertility, a review. Agron Sustain Dev 36:36

Domınguez J, Edwards CA (1997) Effects of stocking rate and moisture content on the growth and maturation of Eisenia Andrei (Oligochaeta) in pig manure. Soil Biol Biochem 29:743–746

Dos Santos Pegoretti T, Mathieux F, Evrard D, Brissaud D, de Franca Arruda JR (2014) Use of recycled natural fibres in industrial products: a comparative LCA case study on acoustic components in the Brazilian automotive sector. Resour Conserv Recycl 84:1–14

Dwivedi RS (1995) Development of molasses and alcohol based chemical industries. Sugarcane: agro-industrial alternatives. Oxford & IBH Publishing Co. Pvt. Ltd, New Delhi, pp 174–198
El Haggar S, El Gowini MM, Nemerow NL, Veziroglu TN (2005) Environmentally balanced industrial complex for the cane sugar industry in Egypt. In: Proceedings of international hydrogen energy congress and exhibition IHEC, Istanbul, Turkey
El-Salam MHA, Fadel MA, Muradm HA (1994) Bioconversion of sugarcane molasses into xanthum gum. J Biotechnol 33(1):103–106
Esposito F (2011) Dow, Mitsui to make bioplastics in Brazil. Plastics News. https://www.plasticsnews.com/article/20110725/NEWS/307259956/dow-mitsui-to-make-bioplastics-in-brazil. Accessed 13 Nov 2023
Evans TH, Hibbert H (1947) Bacterial polysaccharides, scientific reports series no. 6. Sugar Research Foundation, New York
Evansville I (2020) Berry global teams up with Bhoomi to launch 100%. https://www.berryglobal.com/en/news/articles/berry-global-teams-up-with-bhoomi-to-launch-100-sugarcane-based-hdpe-bottle. Accessed 13 Nov 2023
Faria KCP, Gurgel RF, Holanda JNF (2012) Recycling of sugarcane bagasse ash waste in the production of clay bricks. J Environ Manag 101:7–12
Funagaro DA, Reis TVDS (2014) Use of sugarcane straw ash for zeolite synthesis. Int J Energy Environ 5(5):559–566
Garg MR, Gupta BN (1992) Effect of supplementing urea molasses mineral block lick to straw based diet on DM intake and nutrient utilization. Asian Australas J Anim Sci 5(1):39–44
Georges P, Sylvestre M, Ruegger H, Bourgeois P (2006) Ketosteroids and hydroxyketosteroids, minor metabolites of sugarcane wax. Steroids 71(8):647–652
Gharehkhani S, Yarmand H, Goodarzi MS, Shirazi SFS, Zubir MNM, Amiri A, Solangi K, Ibrahim R, Kazi SN, Wongwises S (2017) Experimental investigation on rheological, momentum and heat transfer characteristics of flowing fiber crop suspensions. Int Commun Heat Mass Transfer 80:60–69
Gnanaraj RA (2012) Applications of sugarcane wax and its products: a review. Int J Chem Tech Res 4(2):705–712
Gould BS (1947) Chemical compounds formed from sugar by molds. Scientific reports series No. 7. Sugar Research Foundation, New York
Guerra E, Vazquez M (2003) Dynamics of the co-composting of barley waste with liquid poultry manure. J Sci Food Agric 83:166–172
Gyekye L (2011) Pantene unveils shampoo bottle made from sugarcane. https://www.packagingnews.co.uk/news/pantene-unveils-shampoo-bottle-made-from-sugar-cane-based-pet-27-04-2011. Accessed 13 Nov 2023
Hamzeh Y, Ashori A, Khorasani Z, Abdulkhani A, Abyaz A (2013) Pre-extraction of hemicelluloses from bagasse fibers: effects of dry-strength additives on paper properties. Ind Crop Prod 43:365–371
Hatt HH (1961) Sucrose and the chemical industry. In: Proceedings of the Queensland Society Sugar Cane Technologists, pp 11–19
Heitkötter J, Marschner B (2015) Interactive effects of biochar ageing in soils related to feedstock, pyrolysis temperature, and historic charcoal production. Geoderma 245–246:56–64
Hernández-Pérez AF, Chaves-Villamil AC, Arruda PVD, dos Santos JC, Felipe MDGDA (2020) Sugarcane syrup improves xylitol bioproduction from sugarcane bagasse and straw hemicellulosic hydrolysate. Watse Biomass Valori 11:4215–4224
https://atyourservous.com/materials/bagasse/. Accessed 13 Nov 2023
https://bioplasticsnews.com/im-green/. Accessed 13 Nov 2023
https://bonsucro.com/corbion-achieves-first-chain-custody-certification-bioplastics/. Accessed 13 Nov 2023
https://carnegiefabrics.com/carnegie-xorel. Accessed 13 Nov 2023
https://doi.org/10.2136/sssaspecpub63.2014.0040. Accessed 13 Nov 2023
https://doi.org/10.2136/sssaspecpub63.2014.0043.5. Accessed 13 Nov 2023
https://ecohealthware.com/canefiber-sugarcane-bagasse/. Accessed 13 Nov 2023

https://godavaribiorefineries.com/naturowax-%C2%AE. Accessed 13 Nov 2023
https://goeco.co.in/. Accessed 13 Nov 2023
https://greenlinepaper.com/bagasse-products/. Accessed 13 Nov 2023
https://thesugarcanestraw.com/. Accessed 13 Nov 2023
https://uk.bulldogskincare.com/pages/sugarcane. Accessed 13 Nov 2023
https://www.coca-colacompany.com/media-center/100-percent-plant-based-plastic-bottle. Updated 24 Feb 2023
https://www.dhampur.com/our-products/ena. Accessed 13 Nov 2023
https://www.ecover.com/faq/what-is-plantplastic/. Accessed 13 Nov 2023
https://www.gp-award.com/en/produkte/sugarcane-shampoo. Accessed 13 Nov 2023
https://www.love-dear.com/?locale=en. Accessed 13 Nov 2023
https://www.mawanasugars.com/distillary.php. Accessed 13 Nov 2023
https://www.pulptablewareproducts.com/company-profile.html. Accessed 13 Nov 2023
https://www.raydel.com.au/product/policosanol-10. Accessed 13 Nov 2023
https://www.saattvicecocare.in/biodegradable-bagasse-plates.html. Accessed 13 Nov 2023
https://www.tnpl.com/. Accessed 13 Nov 2023
Igalavithana AD, Ok YS, Usman ARA, Al-Wabel MI, Oleszczuk P, Lee SS (2016) The effects of biochar amendment on soil fertility. In: Guo M, He Z, Uchimiya SE (eds) Agricultural and environmental applications of biochar: advances and barriers 63. Wiley, New York
Inyanga M, Gao B, Pullammanappallil P, Ding W, Zimmerman AR (2010) Biochar from anaerobically digested sugarcane bagasse. Bioresour Technol 101:8868–8872
Janbuala S, Eambua M, Satayavibul A, Nethan W (2018) Effect of bagasse and bagasse ash levels on properties of pottery products. Heliyon 4:e00814
Janeiro RD (2009) Biocycle-sustainable polymer from sugar cane. PHB Industrial S/A. Bioenergy II, p 50. https://dc.engconfintl.org/cgi/viewcontent.cgi?article=1015&context=bioenergy_ii. Accessed 13 Nov 23
Jayasekara T, Surendra YW, Rathnayake M (2022) Polylactic acid pellets production from corn and sugarcane molasses: process simulation for scaled up processing and comparative life cycle analysis. J Polym Environ 30:4590–4604
Ji X, Shen Z, Xu W, Yao S, Xiong L, Zhang H, Zhou H, Chen X, Chen X (2023) Synthesis of hyper-cross-linked nonpolar polystyrene divinylbenzene resins by pendent vinyl groups and application to the decolourization of cane molasses. Appl Biochem Biotechnol 195:3406–3424
John RP, Nampoorthiri KM, Pandey A (2006) Solid-state fermentation for L-lactic acid production from agro wastes using Lactobacillus delbrucckii. Process Biochem 41(4):759–763
Joseph S, Arbestain MC, Lin Y, Munroe P, Chia CH, Hook J, Van Zwieten L, Kimber S, Cowie A, Singh BP, Lehmann J, Foidl N, Smernik RJ, Amonette JE (2010) An investigation into the reactions of biochar in soil. Soil Res 48:501–515
Kabeyi MJB (2022) Sugarcane molasses to energy conversion for sustainable production and energy transition. In: Proceedings of the international conference on industrial engineering and operations management, Istanbul, Turkey, March 7-10, IEOM Society International, pp 2299–2311
Kabeyi MJ, Olanrewaju OA (2022) Sustainable energy transition for renewable and low carbon grid electricity generation and supply. Front Energy Res 9:1–45. https://doi.org/10.3389/fenrg.2021.743114
Kameyama K, Miyamoto T, Shiono T, Shinogi Y (2012) Influence of sugarcane bagasse-derived biochar application on nitrate leaching in calcaric dark red soil. J Environ Qual 41(4):1131–1137
Kammann C, Schmidt HP (2014) Biochar in Europe. Biochar J, Arbaz, Switzerland. ISSN 2297-1114. www.biochar-journal.org/en/ct/34. Accessed 3 Dec 2014
Kazmi SMS, Abbas S, Saleem MA, Munir MJ, Khitab A (2016) Manufacturing of sustainable clay bricks: utilization of waste sugarcane bagasse and rice husk ashes. Constr Build Mater 120:29–41
Khan MF, Ali A, Muller ZO (1992) Nutritional evaluation of sugarcane bagasse-based rations treated with urea and cattle manure. Anim Feed Sci Technol 38:135–141

Kiatkittipong W, Wongsuchoto P, Pavasant P (2009) Life cycle assessment of bagasse waste management options. Waste Manag 2:1628–1633

Kulkarni HD (2013) Pulp and paper industry raw material scenario—ITC plantation a case study. IPPTA 25(1):78–89

Kulkarni P, Ravekar V, Rao PR, Waigokar S, Hingankar S (2022) Recycling of waste HDPE and PP plastic in preparation of plastic brick and its mechanical properties. Clean Mater 4:100113

Kumar D, Jain VK, Shanker G, Srivastava A (2003) Citric acid production by solid state fermentation using sugarcane bagasse. Process Biochem 38:1731–1738

Kumari S, Srivastava VP, Mohan N (2023) Silica nanoparticles from sugarcane bagasse fly ash: converting waste to wealth. Int J Sci Res 12(4). https://doi.org/10.21275/SR23425113104

La Fromageot D (1965) Saccahrose en Nutrition Animal (le Poulet). Documentation Scientifique du Cedus, Paris

Leal MRLV, Galdos MV, Scarpare FV, Seabra JEA, Walter A, Oliveira COF (2013) Sugarcane straw availability, quality, recovery and energy use: a literature review. Biomass Bioenergy 53:11–19

Lee S, Scott SD, Pekas EJ, Lee J, Park S (2019) Improvement of lipids and reduction of oxidative stress with octacosanol after taekwondo training. Int J Sports Physiol Perform 14:1297–1303

Lisboa IP, Cherubin MR, Lima RP, Cerri CC, Satiro LS, Weinhold BJ, Schmer MR, Jin VL, Cerri CEP (2018) Sugarcane straw removal effects on plant growth and stalk yield. Ind Crop Prod 111:794–806

Luz SM, Caldeira-Pires A, Ferrão PMC (2010) Environmental benefits of substituting talc by sugarcane bagasse fibers as reinforcement in polypropylene composites: ecodesign and LCA as strategy for automotive components. Resour Conserv Recycl 54:1135–1144

Makul N, Sua-iam G (2016) Characteristics and utilization of sugarcane filter cake waste in the production of light weight foamed concrete. J Clean Prod 126:118–133

Manohar Rao PJ (2008) Working of jiangmen sugar and chemical complex in China. Paper presented at the Sugar Asia 2008: an international exclusive exhibition on sugar & down stream industries, July 25–26, New Delhi, India

Marim RA, Oliveira ACC, Marquezoni RS, Servantes JPR, Cardoso BK, Linde GA, Colauto NB, Valle JS (2016) Use of sugarcane molasses by *Pycnoporus sanguineus* for the production of laccase for dye decolourization. Genet Mol Res 15(4):gmr15048972

Martirena JF, Middendorf B, Budelman H (1998) Use of wastes of the sugar industry as pozzolan in limepozzolan binders: study of the reaction. Cem Concr Res 28:1525–1536

Mashoko L, Mbohwa C, Thomas VM (2010) LCA of the South African sugar industry. J Environ Plan Manag 53(6):793–807

Mitr Phol Group (2017) Sustainability report, p 138

Mohan AN (2012) Brazilian sunscreen launched in sugarcane-based plastic bottle. Sustainability https://www.packworld.com/news/sustainability/article/13359210/brazilian-sunscreen-launched-in-sugarcanebased-plastic-bottle. Accessed 13 Nov 2023

Mohan S, Chithra L, Nageswari R, Manimozhi Selvi V, Mathialagan M (2021) Sugarcane wax—a par excellent by-product of sugar industry—a review. Agric Rev 42(3):315–321

Mordenti AL, Giaretta E, Campidonico L, Parazza P, Formigoni A (2021) A review regarding the use of molasses in animal nutrition. Animals 11:115. https://doi.org/10.3390/ani11010115

Moreno AD, Susnozas A, Oliva JM, Negro MJ (2020) Overview of bio-based industries. In: Galanakis CM (ed) Biobased products and industries. Elsevier, Amsterdam, pp 1–40

Moriya RY, Goncalves AR, Duarte MCT (2007) Ethanol/water pulps from sugar cane straw and their biobleaching with xylanase from *Bacillus pumilus*. Appl Biochem Biotechnol 137:501–513

Moubarik A, Grimi N (2015) Valorization of olive stone and sugar cane bagasse by-products as biosorbents for the removal of cadmium from aqueous solution. Food Res Int 73:169–175

Munagala M, Shastri Y, Nagarajan S, Ranade V (2022) Production of bio-CNG from sugarcane bagasse: commercialization potential assessment in Indian context. Ind Crop Prod 188(Part A):115590

Nagal V (2021) The berry global teams up with the Bhoomi to introduce the 100% sugarcane-based HDPE bottle. https://packagingguruji.com/the-berry-global-teams-up-with-the-bhoomi-to-introduce-the-100-sugarcane-based-hdpe-bottle/. Accessed 13 Nov 2023

Nandalal (2018) A green alternative: try out these containers from sugarcane waste. The New Indian Express. https://www.newindianexpress.com/cities/thiruvananthapuram/2018/may/25/a-green-alternative-try-out-these-containers-from-sugarcane-waste-1819514.html

Nguyen HC, Su CH, Yu YK, Huong DTM (2018) Sugarcane bagasse as a novel carbon source for heterotrophic cultivation of oleaginous microalga *Schizochytrium* sp. Ind Crop Prod 121:99–105

NSI Kanpur (2023) Newsletter, July Sept 2023

Ochoa PAG, Eras JJC, Gutierrez AS, Hens L, Vandecasteele C (2010) Residue from sugarcane juice filtration (filter cake): energy use at the sugar factory. Water Biomass Valori 1:407–413

OECD-FAO Agricultural Outlook 2021–2030. https://www.oecd-ilibrary.org/sites/89d2ac54-en/index.html?itemId=/content/component/89d2ac54-en. Accessed 13 Nov 2023

Okano K, Fukui S, Kitao R, Usagawa T (2007) Effects of culture length of *Pleurotuseryngii* grown on sugarcane bagasse on in vitro digestibility and chemical composition. Anim Feed Sci Technol 136:240–247

Olagunju A, Muhammad A, Aimola IA, Abdullahi SA, Danhassan MS (2014) Effect of *Lachnocladium* spp. fermentation on nutritive value of pretreated sugarcane bagasse. Int J Mod Biol Med 5:24–32

Oliveira Alves J, Soares Tenório JA, Zhuo C, Levendis YA (2012) Characterization of nanomaterials produced from sugarcane bagasse. J Mater Res Technol 1:31–34

Oliver-Ortega H, Julian F, Espinach FX, Tarres Q, Ardanuy M, Mutje P (2019) Research on the use of lignocellulosic fibers reinforced bio-polyamide 11 with composites for automotive parts: car door handle case study. J Clean Prod 226:64–73

Onchieku J, Chikamai B, Rao M (2012) Optimum parameters for the formulation of charcoal briquettes using bagasse and clay as binder. Eur J Sustain Dev 1:477

Osman ME, Khattab OH, Hammad IA, El-Hussieny NI (2006) Optimization of bio-fuel production by Saccharomyces cerevisiae isolated from sugarcane bagasse. J Am Sci 7(5):486–492

Owusu PA, Asumadu-Sarkodie S (2016) A review of renewable energy sources, sustainability issues, and climate change mitigation. Cogent Eng 3(1):1167990

Patarau JM (1982) By-products of the cane sugar industry: an introduction to their industrial usage, 2nd edn. Elsevier Scientific Publishing Company, Amsterdam, p 365

Paya J, Monzo J, Borrachero MV, Diaz P, Ordonez LM (2002) Sugarcane bagasse ash (SCBA): studies on its properties for reusing in concrete production. J Chem Technol Biotechnol 77:321–325

Pereira AA, Martins GF, Antunes PA, Conrrado R, Pasquini D, Job AE, Curvelo AA, Ferreira M, Riul A Jr, Constantino CJ (2007) Lignin from sugar cane bagasse: extraction, fabrication of nanostructured films, and application. Langmuir 23:6652–6659

Pfafi I, Sisko M (1995) Fiber Atlas: identification of papermaking fibers. Springer-Verlag, New York

Pierce LM (2014) AT&T chooses plant-based plastic packaging for telecom accessories. Packaging Digest. https://www.packagingdigest.com/smart-packaging/att-chooses-plant-based-plastic-packaging-telecom-accessories. Accessed 13 Nov 2023

Poopak S, Reza AR (2012) Environmental benefit of using bagasse in paper production—a case study of LCA in Iran. In: Global warming—impacts and future perspectives. InTech Publishing, London, pp 205–222

Poria V, Jhilta P, Rana A, Khkhar J, Singh S (2022) Pressmud: a sustainable source of value-added products. Environ Technol Rev 11(1):187–201

Pozzer T, Gauss C, Ament Barbirato GH, Fiorelli J (2020) Trapezoidal core sandwich panel produced with sugarcane bagasse. Constr Build Mater 264:120718

Prakash M, Karmegam N (2010) Vermistabilization of pressmud using Perionyx ceylanensis Mich. Bioresour Technol 101:8464–8468

Prasertsri P (2007) Thailand sugar Annuak, USDA foreign agriculture service-gain report-global agriculture information network, gain report number: TH7048

Rabelo SC, Costa ACD, Rossel CEV (2015) Industrial waste recovery. In: Santos F, Borem A, Caldos C (eds) Sugarcane agricultural production, bioenergy and ethanol. Elsevier, Amsterdam, pp 365–381

Rainey TJ, Clark NB (2004) An overview of bagasse as a resource for the Australian paper industry. Int Sugar J 1271:608–611

Rainho Teixeira S, Arenales A, de Souza AE, da Silva Magalhaes R, Vilche Pena AF, Aquino D, Freire R (2015) Sugarcane bagasse: applications for energy production and ceramic materials. J Solid Waste Technol Manag 41:229–238

Ramsey S, Williams B, Jarell P, Hubbs T (2023) Global demand for fuel ethanol through 2030. Economic Research Service US Department for Agriculture. https://www.ers.usda.gov/webdocs/outlooks/105762/bio-05.pdf?v=5239.1

Raza QUA, Bashir MA, Rehim A, Sial MU, Raza HMA, Atif HM, Brito AF, Geng Y (2021) Sugarcane industrial byproducts as challenges to environmental safety and their remedies: a review. Water 13(24):3495

Roa Engel CA, De Lima MA, Maziero P, de Azevedo ER, Garcia W, Polikarpov I (2008) Fumaric acid production by fermentation. Appl Microbiol Biotechnol 78(3):379–389

Rodríguez-León JA, Domenech F, León M, Mendez T, Rodríguez DE, Pandey A (1999) Production of spores of *Trichoderma harzianum* on sugarcane molasses and bagasse pith in solid state fermentation for bioethanol. Braz Arch Biol Technol 42(1). https://doi.org/10.1590/S1516-89131999000100010

Rolz C, de Leon R, Cifuentes R, Porres C (2010) Windrow composting of sugarcane and coffee byproducts. Sugar Technol 12(1):15–20

Rovani S, Santos JJ, Corio P, Fungaro DA (2018) Highly pure silica nanoparticles with highly adsorption capacity obtained from sugarcane waste ash. ACS Omega 3(3):2618–2627. https://doi.org/10.1021/acsomega.8b00092

Saad MBW, Oilveira LRM, Candido RG, Quintana G, Rocha GJM, Goncalves AR (2008) Preliminary studies on fungal treatment of sugarcane straw for organosolv pulping. Enzym Microb Technol 43:220–225

Sahu O (2018) Assessment of sugarcane industry: suitability for production, consumption, and utilization. Ann Agrar Sci 16:389–395

Sahu S, Patel BHM, Sarangi A, Kumar D, Mahesh MS, Upadhyay D, Dutt T (2021) Sun dried sugarcane press mud could be a prospective feedstuff for swine production. Sugar Tech 24:788–797

Saleh ME, Hedia R (2018) Mg-modified sugarcane bagasse biochar for dual removal of ammonium and phosphate ions from aqueous solutions. Alex Sci Exch J 39:74–91

Samantra KK, Basak S, Chattopadhyay SK (2015) Recycled fibrous and nonfibrous biomass for value-added textile and nontextile applications. In: Muthu S (ed) Environmental implications of recycling and recycled products. Environmental footprints and eco-design of products and processes. Springer, Singapore, pp 167–212

Santos F, Eichler P, Machado G, Mattia JD, Souza GD (2020) By products of the sugarcane industry. In: Santos F, Rabelo SC, Matos MD, Eichler P (eds) Sugarcane biorefinery, technology and perspectives. Elsevier, Amsterdam, pp 21–48. https://doi.org/10.1016/B978-0-12-814236-3.00002

Sardar H, Anjum MA, Nawaz A, Naz S, Ejaz S, Ali S, Haider ST (2020) Effect of different agrowastes, casing materials and supplements on the growth, yield and nutrition of milky mushroom (*Calocybe indica*). Folia Hortic 32:115–124

Saucedo-Pompa S, Rojas-Molina R, Aguilera-Carbo AF, Saenz-Galindo A, Garza HDL, Jasso-Cantu D, Aguilar CN (2009) Edible film based on candelilla wax to improve the shelf life and quality of avocado. Food Res Int 42(4):511–515

Schmidt HP, Hagemann N, Draper K, Kammann C (2019) The use of biochar in animal feeding. Peer J 7:e7373

Seroka NS, Taziwa RT, Khotseng L (2022) Extraction and synthesis of silicon nanoparticles (SiNPs) from sugarcane bagasse ash: a mini-review. Appl Sci 12(5):2310

Sethy A (2013) Preparation of oxalic acid from molasses. B. Tech Project Report. National Institute of Technology, Rourkela
Shahi N, Min B, Sapkota B, Rangari VK (2020) Eco-friendly cellulose nanofiber extraction from sugarcane bagasse and film fabrication. Sustainability 12:6015
Shahowna EM, Mahala AG, Mokhtar AM, Amasaib EO, Attaelmnan BA (2013) Evaluation of nutritive value of sugar cane bagasse fermented with poultry litter as animal feed. J Food Sci Technol 4:106–109
Sharkara (2022) Abstracts. Natl Sugar Inst J 53(1):32–35
Show PL, Oladele KO, Siew QY, Zakry FAA, Lan JCW, Ling TC (2015) Overview of citric acid production from Aspergillus niger. Front Life Sci 8(3):271–283
Sindhu R, Gnansounou E, Binod P, Pandey A (2016) Bioconversion of sugarcane crop residue for value added products—an overview. Renew Energy 98:203–215
Singh GB, Solomon S (1995) Sugarcane: agro-industrial alternatives. Oxford & IBH Publishing Co. Pvt. Ltd, New Delhi
Singh AK, Gogoi R, Talwar M, Tyagi AK (2015) Evaluation of the suitability of sugarcane wax for application in cosmetics. Project report, pp 1–27
Smock D (2011) Dow & Mitsui Plan Brazil Bioplastics plant. Design news. https://www.design-news.com/industry/dow-mitsui-plan-brazil-bioplastics-plant. Accessed 13 Nov 23
Sohi SP, Krull E, Lopez-Capel E, Bol R (2010) A review of biochar and its use and function in soil. Adv Agron 105:47–82
Solomon S (2011) Sugarcane by-products based industries in India. Sugar Tech 13(4):408–416
Solomon S, Singh GB (2005) Sugarcane diversification: recent developments and future prospects. In: Singh GB, Solomon S (eds) Sugarcane: agro-industrial alternatives. Oxford IBH Publishing Company, New Delhi, pp 523–541
Solomon S, Swapna M (2022) Indian sugar industry: towards self-reliance for sustainability. Sugar Tech 24:630–650
Somasekhar M (2022) Sugar substitute xylitol produced from bagasse by IIT Guwahati researchers. https://eng.ruralvoice.in/technology/sugar-substitute-xylitol-produced-from-bagasse-by-iit-guwahati-researchers.html
Spokas KA, Baker JM, Reicosky DC (2010) Ethylene: potential key for biochar amendment impacts. Plant Soil 333:443–452
Sukyai P, Yimgkamhaeng N, Lam NT, Tangsatianpan V, Watcharinrat C, Vanitjinda G, Vanichsriratana W, Sriroth K (2016) Research and development prospects for sugarcane and sugar industry in Thailand. Sugar Tech 18:583–587
Taylor AK (2000) From raw sugar to raw materials. Chem Innov 30(11):45–48
Teclu D, Tivchev G, Laing M, Wallis M (2009) Determination of the elemental composition of molasses and its suitability as carbon source for growth of sulphate reducing bacteria. J Hazard Mater 161(2–3):1157–1165
Tiwari SP, Singh UB, Mehra UR (1990) Urea molasses mineral blocks as a feed supplement: effect on growth and nutrient utilization in buffalo calves. Anim Feed Sci Technol 29:333–341
Toledo-Jaldin HP, Sanchez-Mendieta V, Blanco-Flores A, Lopez-Tellez G, Vilchis-Nestor AR, Martin-Hernandez O (2020) Low-cost sugarcane bagasse and peanut shell magnetic composites applied in the removal of carbofuran and iprodione pesticides. Environ Sci Pollut Res 27:7872–7885
Toth JD, Dou Z (2016) Use and impact of biochar and charcoal in animal production systems. In: Guo M, He Z, Uchimiya SM (eds) Agricultural and environmental applications of biochar: advances and barriers, 63. Wiley, New York
Tracy SE (2019) Tasty waste: industrial fermentation and the creative destruction of MSG. Food Cult Soc 22(5):548–565
Tran G (2012) Sugarcane pess mud. Feedipedia, a programme by INRA, CIRAD, AFZ and FAO. https://www.feedipedia.org/node/563. Last update May 27, 2015, 18: 02
Tripathy A, Acharya PK (2022) Characterization of bagasse ash and its sustainable use in concrete as a supplementary binder—a review. Constr Build Mater 322:126391

Unnithan L (2021) This startup has found innovative ways to diversify sugarcane byproducts. https://krishijagran.com/agriculture-world/this-startup-has-found-innovative-ways-to-diversify-sugarcane-byproducts/. Accessed 12 Nov 2023

Wang S, Xu Y, Norbu N, Wang Z (2018) Remediation of biochar on heavy metal polluted soils. IOP Conf Ser Earth Environ Sci 108(4):042113

Wani AK, Rahayu F, Fauziah L, Suhara C (2023) Advances in safe processing of sugarcane and bagasse for the generation of biofuels and bioactive compounds. J Agric Food Res 12:100549

Xue JL, Zhao S, Liang RM, Yin X, Jiang SX, Su LH, Yana O, Duan CL, Liu JL, Feng JX (2016) A biotechnological process efficiently coproduces two high value added products, glucose and xylo oligosaccharides from sugarcane bagasse. Bioresour Technol 204:130–138

Yadav RL, Solomon S (2006) Potential of developing sugarcane by products based industries in India. Sugar Tech 8(2–3):104–111

Yeomans M (2011) J & J go green with PE containers for its Brazilian sun care range. Cosmetics Design USA. https://www.cosmeticsdesign.com/Article/2011/12/20/J-J-go-green-with-PE-containers-for-its-Brazilian-sun-care-range. Accessed 13 Nov 2023

Zabed H, Faruq G, Sahu JN, Azirun MS, Hashim R, Boyce AN (2014) Bioethanol production from fermentable sugar juice. Sci World J 2014:957102

Zhang S, Wang J, Jiang H (2021) Microbial production of value-added bioproducts and enzymes from molasses, a by-product of sugar industry. Food Chem 346:128860

Zhu P, Ren J, Wang L, Zhang X, Yang X (2007) Long-term fertilization impacts on corn yields and soil organic matter on a clay-loam soil in Northeast China. J Plant Nutr Soil Sci 170:219–223

Historical Perspectives on Sugarcane Breeding for Value Addition

2

Hemaprabha Govindakurup and K. Mohanraj

Abstract

Sugarcane, as the primary source of sugar and jaggery, is one of the most important agro-based industrial crops, playing a pivotal role in humankind's sustenance, and is grown in more than 100 countries. Although sugarcane production has experienced a steady increase over the decades, it is only recently that sugarcane has gained importance as both a sugar and energy crop by yielding multiple products through deliberate product diversification efforts that have helped transform it into a wonder crop. In the era of renewable energy and product portfolio, there is a need for up-to-date compiled information regarding sugarcane's history for a better understanding of product diversification. This chapter covers sugarcane origin and diversification, conventional methods of sugarcane breeding, species and genus variations, breeding for multiple uses, and genetic improvement to achieve value addition. This chapter also includes focussed topics like targeted breeding for chewing canes/juice purposes, jaggery, and sugar content. With sugarcane being amenable to diverse uses, the latest information regarding energy canes, red-fleshed canes with antioxidant properties, and ornamental and vegetable sugarcane are also covered. Finally, recent topics like genome editing, genetic engineering, and induction of variation through in vitro callus culture have also been mentioned, so as to benefit the various stakeholders in the sugarcane value chain.

H. Govindakurup (✉)
Genetics and Plant Breeding, ICAR-Sugarcane Breeding Institute, Coimbatore, Tamil Nadu, India

K. Mohanraj
Division of Crop Improvement, ICAR-Sugarcane Breeding Institute, Coimbatore, Tamil Nadu, India

G. S. Suresha et al. (eds.), *Value Addition and Product Diversification in Sugarcane*, https://doi.org/10.1007/978-981-97-7228-5_2

Keywords

Sugarcane history · Genetic improvement · Value addition · *Saccharum* complex

2.1 Introduction

Sugarcane breeding has a glorious past, which has set trends in breeding with the use of wild species. Sugarcane (*Saccharum* spp. hybrids) is one of the most promising industrial crops in the world and is grown in more than 100 countries as the major source of crystal sugar and other sweetening agents for human consumption (Heinz 1987). An ancient crop, sugarcane is indigenous to India and has references in the ancient Vedic literature as Ikshu (sugarcane) in the *Atharva Veda*. Traditionally a labour-intensive crop, sugarcane was grown in developing countries of the tropics, where unskilled labour was available for its cultivation. Since 8000 BC, *Saccharum officinarum* has been cultivated as a garden crop in Indonesia and New Guinea, where the species is believed to have originated (Daniels and Roach 1987; Fauconnier 1993), and this was later supported through amplified fragment length polymorphism marker studies (Aitken et al. 2006). The evolution of sugarcane is still not clear. There is a major belief behind its selection from the sweet clones of *Saccharum robustum* and its domestication by humans for multiple purposes, such as building materials for the construction of walls, fences, and roofs and as bows in archery (Daniels and Roach 1987).

2.2 Sugarcane Breeding over the Years

Breeding has evolved over the years, starting with selection followed by domestication of species. Conventional breeding yielded several promising cultivars and adopted the principles of statistics and biometrics for experimentation and selection. Mutation breeding, tissue culture-based improvement, and molecular techniques were incorporated in due course. Presently, sugarcane breeding is in transition to its smart breeding phase (Shen et al. 2022; Xu et al. 2022) with the adoption of cutting-edge technologies in genomics, precision breeding approaches, and modern information technologies for expanding the applications of the crop to meet multiple needs for a sustainable future as well as to provide opportunities to cultivate designer sugarcane varieties with specific value-added uses in both industry and medicine.

However, refinement of breeding and selection methods expanded with the utilization of wild germplasm was hindered owing to the crop's genetic constitution as a complex polyploid. The initial efforts for the domestication of *S. officinarum* and other species did not bring major improvement; however, the discovery of seed set in sugarcane in 1887–1888 by Sotwedel (Harrison 1888) created new hope and opportunity for sugarcane improvement. Sugarcane breeding was initially accomplished mainly in Java and India. The breeding strategies used included interspecific hybridization, backcrossing, and intercrossing, and the principles of quantitative genetics and statistical approaches were incorporated from time to time to improve the experimental efficiency and the selection of improved varieties (Zhang et al. 2023). Traits such as high yield and high sugar from *S. officinarum* and vigour and stress tolerance from *Saccharum spontaneum* were critical for creating hybrids of agronomic and economic importance. Commercial hybrids such as Co 205 from India ushered in a sugar revolution and thus began a new chapter in sugarcane improvement through interspecific hybridization. The hybrid POJ 2878 became a huge success, accounting for 95% of the total sugarcane planting area in Java and has proven to be a superior parent responsible for the development of several commercial hybrids in most of the sugarcane breeding countries (Jeswiet 1930). Historic landmarks by Kobus and tri-specific hybrids involving *S. officinarum*, *S. spontaneum*, and *Saccharum barberi* such as 'Co 213', 'Co 281', and 'Co 290' by Venkataraman (Ram et al. 2022) resulted in giant leaps in sugarcane production and productivity around the globe. The sweet form, *S. barberi* was cultivated in India and these early varieties replaced the indigenous cultivars even from the first quarter of the twentieth century.

The spread of *S. officinarum* and *Saccharum sinense* was facilitated owing to its chewing value and its use as feed for pigs by the Papuans and Austronesians through migrations of the Austronesian peoples. Later on, the stalks were processed for rum production. Combustible materials, such as charcoal, biochar, etc., were also produced. The juice extracted from sugarcane was popular as garapa or caldo de cana in Brazil, chediraz in India, and aseer asab in Egypt. It was bottled and locally distributed in certain countries. Later, sugarcane juice was processed through boiling and cooling to make unrefined brown sugar in cake form, which was called jaggery or gur or khandsari, panela, and rapadura in India, Colombia, and Brazil, respectively (Kansal 1998). Processing of sugar to yield refined sugar and other products, including for biofuel production, started in due course. The industry grew in magnitude and strength as the producer of 80% of global sugar. Sugarcane has evolved as a multiutility crop of increased importance, and the need to incorporate more traits for targeted trait improvement is essential to develop designer crop varieties. Breeding has been the major factor that led to a breakthrough in sugar production. Improved breeding techniques, methods, and strategies were developed and refined over the years, primarily for developing high-yielding and sugar-rich varieties. More than 100 products are made from sugarcane. This chapter deals with the breeding approaches and achievements in developing major products from sugarcane.

2.3 The *Saccharum* Species and Related Genera

Knowledge of the genus *Saccharum* and its species and related genera is an essential prerequisite for genetic improvement. Sugarcane belongs to the genus *Saccharum* L., subtribe Saccharastra, group Saccharinae Benth and tribe Andropogoneae of Poaceae family, which includes grasses (Daniels and Roach 1987). Taxonomic and phylogenetic studies revealed sugarcane forming a closely interbreeding group referred to as '*Saccharum* complex'. The *Saccharum* complex comprises *Saccharum*, Erianthus section Ripidium, Miscanthus section Diandra, Narenga, and Sclerostachya (Daniels and Roach 1987). These genera closely interbreed in nature and are characterized by high levels of polyploidy, which is hindered in clear taxonomic grouping (Sreenivasan et al. 1987). The genus *Saccharum* traditionally comprises six species such as *S. officinarum, S. barberi, S. sinense, S. spontaneum, S. robustum*, and *S. edule* (D'Hont et al. 1998). The phenotypic characteristics and other characteristics of *Saccharum* and related species of major interest in sugarcane breeding programmes and their relevance to value addition are given in Table 2.1.

2.4 Breeding for Multiple Uses of Sugarcane

Sugarcane has played a significant role in ancient civilizations, and its influence can still be felt in modern society. From its use in religious ceremonies to its role in Ayurvedic medicines and the food industry, sugarcane has had a profound impact on human history. The production of sugar by boiling the juice is believed to have been discovered in India in the first millennium BC, when the practice of preparing sugar from palms was already in vogue. In recent times, sugarcane has been used to produce biofuels, and its demand is increasing as a renewable source of energy. Between 1961 and 2018, the global area under sugarcane cultivation increased from 8.91 to 26.27 mha, and the production increased from 44.7 MMT to 190.7 MMT (FAO 2020). Sugarcane yields several other products. For instance, in Cuba, about 31 products including raw and refined sugar, glucose, dextran, molasses, alcohol, rum, syrups, confectionery, crude wax, etc. are produced from sugarcane. The estimates showed that 100 t of sugarcane produced 14.3 t raw sugar, 27.2 t bagasse, 5.2 t filter cake, and 2.6 t molasses (Allen et al. 1997). In Brazil, 47% of sugarcane was used to produce 17.8 billion litres (BL) of ethanol, which catered to 40% of fuel for cars (Orellana and Neto 2006). In the 2010/2011 season, ethanol production increased to 54% of the total sugarcane production, from which 27.6 BL ethanol was produced (Conab 2011). Bagasse was a common fuel in kilns and later used in boilers. The fermentation and distillation of cane juice were practised to produce alcoholic beverages such as rum in many countries. Animal feed and poultry feed are also made from bagasse, which is the fibrous portion of sugarcane after the juice has been removed. Bagasse constitutes about 24–28 kg from 100 kg of processed sugarcane (Cheavegatti-Gianotto et al. 2011). Two types of fibres are present in bagasse: long fibres in the rind and shorter and softer fibres in the pith. Vinasse or spent wash is the residue left after fermentation and is used as fertilizer (Cheavegatti-Gianotto et al. 2011). Though Thailand increased ethanol production to 1.7 ML/day from 12 distilleries, the country reverted the policy towards sugar production since 2009–2010. India utilized molasses to produce 3.2 BL ethanol/year in its 300 distilleries (Gopinathan and Sudhakaran 2009). The number of distilleries increased since 2018, as the country aims at 20% ethanol blending in petrol by 2025 AD. More countries including the USA produce ethanol from sugarcane.

Table 2.1 The species and their relevance to value addition

Species	Features	Value
Saccharum officinarum	Thick stems, high sucrose content, high purity, juicy stem (around 70% juice), low fibre, productive and most intensively cultivated sugarcane for sucrose in the stalk internodes. The stem shows large variation in shape, size, and colour, the typical forms are considered to have $2n = 80$, and several atypical forms exist in different collections. Most of the clones are nonflowering or shy flowering and shy tillering	The cane juice, chewing, sweeteners, breeding for developing sugarcane varieties, and religious rituals. Some important indigenous clones of this species are Vellai, Poovan, Black Cheribon, Patta Patti, Tahiti, Seleri, IJ 76-501, La Purple, BM Hitam, Bois Rogue, Chittan, German Green, Namam, TofunaFal, Fiji B, Fiji 30, 21NG 2, 28NG 210, 57NG 243, IJ 76-419, IJ 76-420, IJ 76-556, and IJ 76-55
Saccharum spontaneum	Wide variability and adaptability to diverse climatic and edaphic conditions from riverbanks to deserts and from sea shores to high altitude Himalayan mountain terrains. Exhibits variation from short bushy grasses to vigorous, profusely tillering and tall forms. Chromosome numbers vary from $2n = 40–128$ and $x = 8$ and the most frequent count is $2n = 64$. They have stoleniferous habits with very little juice and almost no sucrose in the stalk. Profusely flowering with a good seed set, but susceptible to sugarcane smut disease	Hedges, fence, fuel, breeding as a donor for abiotic and several biotic stresses and ratoonability. Deep root system to prevent soil erosion
Saccharum robustum	Wild canes growing along the riverbanks of New Guinea. Five types of *S. robustum*, viz., Port Moresby type ($2n = 80$), Goroka type ($20 = 80$) with small stalks, red-fleshed type with thick and tall stalks ($2n = 60$), Teboe Salah ($2n = 60$) with large stalks, and Wau/Bulolo type with small stalks ($2n = 60$). The stem is long, thick, and vigorous, rich in fibre with low sugar content, and not suitable for agricultural production. The species is distinguished by its lack of rhizomes, large inflorescence, and small spikelets, survives under a wide range of temperatures, and is found from sea level to altitude up to 2700 m in the Himalayas (Mukherjee 1957). The tall red-fleshed forms are grouped under Sanguineum group	Thatching, in breeding as a donor for tallness, resistance to abiotic stress, red forms useful to develop biofortified sugarcane, with manifold higher antioxidants than normal in normal sugarcane (Chandran et al. 2020)

(continued)

Table 2.1 (continued)

Species	Features	Value
Saccharum barberi and *Saccharum sinense*	*Saccharum barberi* (Indian indigenous sugarcane) clones have many erect stalks of 2.0–2.5 cm in diameter with moderate sugar and high fibre content. *Saccharum sinense* (Chinese indigenous sugarcane) clones are taller and more vigorous than *S. barberi* and are more resilient and better adapted to low temperatures and dry conditions. *S. barberi* and *S. sinense* are often clubbed together in spite of the above differences. *Saccharum sinense* clones are low in sucrose and have high fibre content. Juice is low in purity with high starch content. Bremer (1923) classified these into five morphological groups: Sunnabile ($2n = 82$ and 116), Mungo ($2n = 82$), Nargori ($2n = 107$ and 124), Saretha ($2n = 90$ and 92), and Pansahi group ($2n = 118$). Pansahi group was later grouped to form *S. sinense*	Source of sucrose and abiotic stress resistance
Saccharum edule	*S. edule* is an unusually robust perennial herb that spreads by rhizomes. It is believed to have evolved from *S. robustum* or *S. officinarum* ($2n = 80$), though it has also been proposed that it originated through natural hybridization between *S. robustum* and *Miscanthus* (Grassl 1967). The group is not widely considered as a separate species due to its sterility	Cultivated from Borneo and Java through Melanesia to the New Hebrides for its aborted edible inflorescences. No breeding value

(continued)

Table 2.1 (continued)

Species	Features	Value
Other genera: *Erianthus*, *Sclerostachya*, *Narenga*, and *Miscanthus*	*Erianthus* includes *E. arundinaceus* ($2n = 20$, 30, 40, 60), *E. ravennae* ($2n = 20$), *E. elephantinus* ($2n = 20$), *E. bengalense* ($2n = 40$, 60), *E. longisetosus* ($2n = 30$), and *E. procerus* ($2n = 40$). *E. arundinaceus*, the only cane-forming species, is widely distributed in India, China, Myanmar, Thailand, Philippines, Indonesia, and New Guinea (Wu and Raven 1994). *E. arundinaceus* germplasm collections are largely from the Indian subcontinent or Indonesia–Papua New Guinea. *E. procerus* has large droopy silky panicles (Amalraj and Balasundaram 2006), distributed mostly in Burma, China, India, Indonesia, and New Guinea (Amalraj and Balasundaram 2006)	Erianthus grown for thatching, roofing, fencing, has immense breeding value for its biomass, stress tolerance, multiple ratoonability
	The genus *Sclerostachya* (Hack.) A. Camus consists of three species, *S. fusca* (Roxb.) A. Camus, *S. milroyi* Bor. and *S. ridleyi* (Hack.) A. Camus, distributed throughout northern India to an altitude up to 1300 m in the Himalayas. *Sclerostachya* is characterized by a glabrous sheath and leaves, a hollow stem, the absence of root eyes, poorly developed buds, and copper-brown panicles	*Sclerostachya is a* source of resistance to waterlogging
	Narenga fallax (Bal.) Bor. and *N. porphyrocoma* Bor. are the two species of the genus *Narenga* Bor., which is very close to *Sclerostachya*. *Narenga* has bearded nodes, pubescent sheath and leaf. It is widely distributed in Southeast Asia. The chromosome number in both genera is $2n = 30$ (Sreenivasan et al. 1987)	
	Miscanthus, Imperata, Vetivaria, and *Neyraudia* are also crossable with *Saccharum* and conserved in the germplasm collection	Miscanthus is a source of high biomass, pest resistance, and cold tolerance

Many countries adopt paper production from bagasse, for its ideal paper production characteristics with long cellulose fibres (1–1.5 mm) which are in between the hardwood fibres (0.7–1 mm) and the softwood fibres (2.5–5 mm) (Allen et al. 1997). The pith constituting 25–35% weight of dry bagasse is not useful for high-quality paper making (Dunlap and Callihan 1969), hence blended with wood pulp. Products like particle board, fibre board, wall panels, carton boxes, packaging materials, disposable table wares, seedling trays, etc., in several places and their uses have increased recently. Bagasse is used as an animal feed and its low digestibility is improved through steam treatment (Allen et al. 1997). Bagasse is a source of renewable energy through co-generation in sugar mills; it is burnt to produce steam to run the sugar mills and excess energy is directed to the electricity grid (Mackintosh 2000; Cheavegatti-Gianotto et al. 2011). Being an effective biosorbent, bagasse is also used in waste water management (de Matos et al. 2003). Bagasse has futuristic uses for producing 2G ethanol as a source of biofuel (Sainz and Dale 2009).

Molasses is the thick syrup left after removing sucrose from syrup in sugar mills. Generally, C molasses is used for alcohol fermentation, as a feed supplement, and as fertilizer. Trash is the plant material left after harvesting sugarcane crops, and on average 10 t/ha of trash is produced from cane (Karve et al. 2001). Earlier managing trash was a major problem, but today trash is used and incorporated into the field. In Australia, trash is mulched or baled or used as cattle feed. In India, shredding of trash is practised to increase the soil organic content, improve water conservation and soil porosity, and check weeds. Trash is also turned into solid briquettes for use as fuel (Karve et al. 2001). The waxy coating on the stalks yields sugarcane wax, used in cosmetics and pharmaceutical products. This wax is characterized by long-chain fatty acids/alcohols used for products to lower cholesterol in humans (Janikula 2002). Lipids are present in the cells of sugarcane (Allen et al. 1997). Burning of bagasse leaves sugarcane ash. Filter cake or press mud is the solid waste after filtering the cane juice and is often used as a fertilizer after proper curing (Cheavegatti-Gianotto et al. 2011). One tonne of crushed sugarcane produces 0.01 t of sugarcane ash and 0.05 t of press mud (Qureshi et al. 2000), which is a source of essential plant nutrients (Calcino 1994). Sugarcane ash blended in concrete for construction has high early strength (Sales and Lima 2010) and is used as an adsorbent in the dye industry (Kanawade et al. 2010). Other sugarcane-based phytochemicals include glycolic acid used for skin rejuvenation (Allen et al. 1997), high-density polyethylene, etc. Sugarcane trash and wild forms are used in roof thatching, hedges, and fences. Thus, all parts of the sugarcane plant are useful and lead to increasing product diversification and breeding efforts require a focus on the yield and quality of every component while focusing on the primary products such as sugar and ethanol.

2.5 Genetic Improvement to Achieve Value Addition in Sugarcane

Genetic studies in sugarcane hybrids revealed the allo-autopolyploid nature of their genomes. A classical hybrid has a chromosome contribution of 80% from *S. officinarum*, 10% from *S. spontaneum*, and 10% recombinant chromosomes of the two progenitor species which provides resistance to various diseases and abiotic stresses, as well as increased vigour, hardiness, tillering, and ratooning capabilities (D'Hont et al. 1996). Breeding programmes in sugarcane have evolved to achieve diversification and develop multipurpose varieties. Value addition for any trait depends on understanding the association between the component traits and their relative contribution to the target trait to facilitate trait-specific improvement in breeding populations. Earlier hybrids (until the nineteenth century) were mostly species cultivars of *S. officinarum*, *S. sinense*, and *S. barberi*. Later on, several projects on molecular markers, genomics, transcriptomics, and different sequencing projects are at unravelling the complexity of its genome. Several such initiatives included both parental species and hybrid genotypes aimed at accelerating precision breeding, providing better adaptation of future sugarcane varieties, and towards value addition with modern genomic technologies. Being a crop of ancient origin, the crop was grown for different products, and breeding efforts for major uses are discussed in this chapter.

2.6 Breeding for Chewing Canes/Juice

Originally, *Saccharum officinarum* is a large grass that originated in the tropics of Papua New Guinea. It was grown in home gardens throughout the Pacific and Southeast Asia primarily for chewing before it became a sweetening agent. The chewing clones are characterized by thick canes with soft tissues and less fibre content, sweetness, and juice content like Purple Mauritius, B 208 and Fiji B, waxy red, etc. (Thirumala Rao 1954; Rakkiyappan et al. 2003). Sugarcane is a traditional commodity for juice, which is relishing, nourishing, and healing as a rich source of carbohydrates, essential amino acids, vitamins (vitamin C, B1, B2, and B6), and minerals (calcium, phosphorus, iron, magnesium, potassium), acids (succinic, fumaric, malic, and citric acids) which are vital elements for good health. The holistic science of Ayurveda recommends sugarcane juice, as it supports normal liver functions and as a natural remedy for jaundice, by replenishing lost electrolytes and nutrients for a faster recovery. This is known to have anti-cancer properties and to detoxify the body. The rich calcium in the juice supports the skeletal system and prevents tooth decay. There are other therapeutic and cosmetic values for sugarcane apart from having stress-relieving properties. It is reported to cure febrile disorder. Phenolic compounds in sugarcane juice showed that the juice possessed a strong ability to protect against DNA damage induced by hydroxyl radicals (Abbas et al. 2014). The nitrite-scavenging properties of sugarcane tops would prove as a source of natural antioxidants (Sun et al. 2014).

Genetic improvement for developing chewing canes was attempted by evaluation of the *S. officinarum* clones and intercrossing of flowering forms of *Saccharum*. The origin of *S. officinarum* is intimately associated with the activities of humans, as *S. officinarum* is a cultivated or garden species that is not found in the wild (Sreenivasan et al. 1987).

Glyn (2004) gave a detailed account of the spread and wide adoption of *S. officinarum* across the globe, which was popularly called Otaheite, Bourbon, or Creole canes. According to him, when sugarcane was introduced in the tropical regions by the Portuguese and Spanish, canes were known as Creole and formed the basis of the sugar industry, while Otaheite was the first noble canes to be cultivated on a worldwide scale and was the standard variety of Brazil, British Guiana, Hawaii, India, Java, Mexico, and in parts of West Indies. It was grown in the West Indies as Bourbon, in Hawaii and Java as Lahaina, and in India as Vellai. Otaheite succumbed to diseases in 1890, specifically rind disease and red rot, caused by the fungi

Phaeocytostroma sacchari and *Glomerella tucumanensis*, respectively (Glyn 2004). Cheribon or transparent cane attained importance in the world in the late nineteenth and early twentieth centuries. The Cheribon series of canes were characterized by stripes of different colours and patterns and stem shapes and were the most valuable noble cane in breeding. Tanna or Caledonian cane also existed in a selection of striped, light, and dark forms. White Tanna accounted for 63% of the area under cane in Mauritius for its hardiness, resistance to gumming disease, and its better performance in the cool and high rainfall regions. Tanna was not a successful parent for its almost completely sterile flowers. The first commercial varieties to be raised from seed were purely of noble cane origin. In Java, EK and SW11 originated from Otaheite and Cheribon, respectively, in Queensland, Q813 and HQ409 from Badilla, a form of *S. officinarum*, H 109 in Hawaii originated from an Otaheite and a light Cheribon clone named Rose Bamboo, and D109, D145, and D625 in Guyana from striped Cheribon. Earlier Indian canes of *S. officinarum* origin were the Paunda canes, of which Striped Mauritius, Red Mauritius, and White Transparent were common. Hemja, Pathri, Chin, Katha, Saretha, Laldi, Dhaul, Mungo, Matna, Nargori, Rekra, Sarauti, Baraukh (*S. barberi*) and Uba, Cayania, Oshima, Zwinga (*S. sinense*) were predominant.

Even though *S. officinarum* accumulates very high sucrose in the cane, intraspecific improvement in *S. officinarum* for developing chewing canes did not attract much attention from breeders due to its high susceptibility to diseases (Cox et al. 2000) The significance of intraspecific improvement of *Saccharum* species before interspecific hybridisation was discussed by various authors (Brown et al. 1969; Tew 1987). Walker (1987) reported significant gains after three cycles of intercrossing and selection to develop a range of improved clones. Nair et al. (1998) utilized 13 typical and atypical clones of *S. officinarum* through polycrossing and selected 31 clones based on stalk number, cane yield, and sucrose % to constitute the cycle-1 population which were further improved up to cycle 3 stage. They reported progressive improvement in yield traits over the cycles. The improvement from cycle 1 to cycle 2 was higher compared to that from cycle 2 to cycle 3. The improved *S. officinarum* population registered up to 40% more cane yield over the base population. Also, the majority of the superior hybrids were from cycle 3, suggesting that there is scope for improvement even beyond cycle 3. Repeated cycles of intraspecific hybridization and selection can lead to generation-wise improvement in important agronomic characters in the species germplasm while maintaining a high level of genetic diversity. The study clearly demonstrated significant improvement in NMC/ha (37.18%), single cane weight (16.67%), and cane yield (61.51%) in 2 cycles of selection (Karuppaiyan et al. 2020). The improved lines of *S. officinarum* identified were PIO 90-196, PIO 90-99, PIO 90-224, PIO 88-136, and PIO 88-76 (Rakkiyappan et al. 2003).

In India, a meagre quantity (1%) of sugarcane is used for chewing and as juice for beverage purposes. The taste and quality of the juice are mainly dependent on cane maturity, and good quality juice has a high proportion of solids to water. Tenderness, juiciness, and flavour compounds are typical indices for a good quality chewing cane. Shanmuganathan et al. (2024) evaluated Sirukamboor Badila canes for the identification of flavour compounds in fresh juice using GC–MS and ICP-OES. The presence of 50 different types of flavour compounds is reported in the fresh juice. Relatively abundant quantity of hexadecanal, eicosenoic acid, hexadecanoic acid, octadecadienoic acid, methyl esters, 2,3-dihydroxypropyl ester, glycidyl oleate, fumaric acid, glycidyl palmitate, silane, octadec-9-enolide, and hexasiloxane has been reported, contributing to the typical grassy, nutty, and waxy flavour of the chewing cane.

A yellow cane variety of *Saccharum officinarum* is the most popular sugarcane cultivar grown for juice production in Malaysia and locally it is known as 'Tebu kuning'. This variety is the most popular and it produces abundant juice and less fibre and has good flavour (Siti and Baharuddin 1994). The changes in physico-chemical properties of sugarcane juice at different maturation stages were studied and the results indicated that single cane weight, cane diameter, Brix, and sucrose content increased significantly in all portions (top, middle, and bottom) up to the end of maturity. Conversely, titratable acidity, pH, juice yield, glucose and fructose contents decreased significantly during maturation (Qudsieh et al. 2001).

Chewing cane possesses a rich source of magnesium, iron, potassium, and zinc and constitutes different types of flavour compounds. Owing to the varied probiotic microbial genomes, chewing cane juice can be included in the daily intake of food in the forms of juice or raw cane and also can be commercialized as an alternative to commercially available beverages with the richness of healthy fatty acid derivatives and minor elements. Chewing canes are suited for subsistence farming and the economic genetic value is very low with less demand. Breeding only for chewing/juice purposes is limited because of the lack of commercial value in profit-making agriculture.

2.7 Improvement in Sugar Content

Sugar being a part of food/diet is consumed in different quantities in different countries and the per capita consumption is associated with the standard of living of a nation. About 120 countries grow sugarcane and 28 countries are actively involved in sugarcane improvement, primarily for sucrose, yield, and stress resistance. With the devastation of sugarcane cultivation in Indonesia, when Cheribon, the leading variety in Java, fell to Sereh disease, new sources with resistance to this disease were identified from India, and the breeders particularly Kobus in 1890 and Jeswiet in 1921 made a breakthrough with the creation of POJ2878, which became one of the most successful and historically important sugarcane varieties in the world (Brandes and Sartoris 1936). In India, Barber and Venkataraman crossed *S. officinarum* clone Vellai and a wild *S. spontaneum* clone (Coimbatore) to develop Co 205 which was hardy, high yielding, and erect, suitable for subtropical regions and released in the Punjab province in 1918 to replace the *S. barberi* clones in cultivation. Soon the variety greatly increased cane yield in northern India (Nair 2008). Co 205 served as a male parent leading to the development of 539 commercial Co varieties (Adhini et al. 2017). These landmark varieties were responsible for improved cane yield, ratooning potential, and disease resistance in several countries. Dutch workers in Java referred to this process as 'nobilization', a term which was widely adopted (Roach 1989; Edme et al. 2005). Nobilization became the norm in sugarcane breeding after the historic success of the early crosses between *S. officinarum* and *S. spontaneum*. There arose a situation of the bottleneck of germplasm utilization owing to problems of flowering and seed set caused by sterility and incompatibility barriers, and only a limited number of species clones were used in genetic improvement (Aitken et al. 2006), which varied in different countries. From the available reports, breeding programmes in India have been more successful and incorporated a relatively larger number of ancestral clones, such as 24 *S. officinarum*, 4 *S. barberi*, 2 *S. sinense*, and 5 *S. spontaneum* giving rise to commercial canes till the year 1986 (Sreenivasan et al. 1987). An assessment done from Indian sugarcane cultivars arising from 100 years of breeding (1918–2017) revealed that the improved commercial lines termed 'Co canes' were derived from 46 accessions of *S. officinarum*, 16 accessions of *S. robustum*, 11 accessions of *S. spontaneum*, 5 of *S. barberi*, and 4 of *S. sinense*, adding to 82 *Saccharum* species clones. Representatives from related genera including three accessions of *Erianthus arundinaceus* and one each of *S. bengalense, Narenga porphyrocoma, Sorghum durra,* and African sorghum were also introgressed into commercial varietal pool. This probably would be the most diverse varietal pool in the world with nine different genera incorporated into the commercial varietal pool (Hemaprabha et al. 2021).

Improving the sugar content in sugarcane breeding programmes was a major value addition initiative in sugar yield. Improving commercial cane sugar yield (t/ha) has to be targeted, for being the most important component of cane sugar productivity. However, there is difficulty in improving sugar content and cane yield simultaneously on account of moderate to low negative direct effects observed for commercial cane sugar (CCS) % and sucrose on cane yield. Heritability for juice brix is moderate to high (Skinner et al. 1987; Milligan et al. 1990). The genetic variation for brix could largely be attributed to the additive effects (Hogarth et al. 1981). The choice of appropriate parental clones coupled with the adoption of different selection strategies has been successful (Legendre 1995; Lo and Chen 1995). With the bottleneck of a few founder parents, the lack of genetic variability among the inter-varietal crosses was observed for juice sucrose in different agroclimatic zones in India (Ram and Hemaprabha 1997). To pool the available genetic variability for the trait, recurrent selection programmes were adopted. In Barbados, three cycles of recurrent selection within a closed population resulted in significant advancement in sugar yield (Kennedy 2005). Potential sugarcane families harbouring clones combining high juice sucrose and red rot resistance were identified from biparental crosses involving Indian and exotic hybrid clones (Shanthi and Alarmelu 2007). India is one of the countries where germplasm was used to diversify the base of sugarcane hybrids and develop novel interspecific and intergeneric hybrids, and this resulted in a steady improvement in the sucrose content of the varieties developed during 1991–2000 (Hemaprabha and Lavanya 2015). Adopting a simple recurrent selection scheme with four cycles of crossing and selection developed a source population for high sucrose content in sugarcane and achieved an overall improvement of 19.0% over the base population. Success in quality breeding has been enormous with the identification of sucrose-rich varieties as well as sugar-rich parents of high breeding value. Co 775, CoC 671, Co 86011, CoT 8201, and Co 11015 are the best examples of such parents in India. Among these, Co 11015 not only serves as a donor for sucrose content but also for early maturity which is essential to increase sugar productivity. Population improvement also has been practised in several countries using elite clones emanating from selection programmes, introduction of new foreign clones, and elimination of the unproductive parents (Cox et al. 2000).

Sugarcane improvement mainly involved *S. spontaneum*. Though *S. robustum* is in close kinship to *S. officinarum*, its successful incorporation in commercial cultivars remains scanty except from Hawaiian Breeding station where many popular commercial cultivars have *S. robustum* in their pedigree as male parent (Heinz 1965). While a few popular varieties of Hawaii immensely benefitted from this wild species, only the Port Moresby type could find a place in the ancestry. In India, several Co canes could be developed with the improved clones of *S. robustum* in their pedigree. *S. robustum* derived backcross hybrids combined high yield and high sucrose and intercrossed hybrids of backcross derivatives were better than the advanced backcrosses for these traits (Ram and Hemaprabha 1997). Several hybrids of commercial status were developed under the ISH series and the notable among them are ISH7, ISH 21, ISH 69, and ISH 100, which are promising general combiners. A recent variety Co 18009 has been developed from ISH 69 with exemplary performance in stress environments.

2.8 Jaggery Varieties

Sugarcane was grown from time immemorial for its juice and jaggery. The jaggery industry has been regarded as one of the oldest cottage industries in several countries including India. Jaggery is not only a sweetener but also used in the preparation of several traditional dishes and sweets for its taste and immense medicinal value, low cost, and easy availability (Singh et al. 2011; Nath et al. 2015). Jaggery contains iron (11%), calcium (0.4%), phosphorous (0.045%), glucose and fructose (10–15%), protein (0.25%), and fat (0.05%) (Kumbhar 2016). In India, about 21% of sugarcane produced is utilized in jaggery and khandsari manufacturing in the decentralized sector (Singh et al. 2013). Farmers prefer sugarcane varieties with high cane yield potential and have little concern about the juice quality components. Generally, the high-yielding varieties lack better sugar content and, hence, are not fit for jaggery production, particularly in the early part of crushing season, i.e., October and November, owing to poor jaggery setting. When the climate is ideal for juice accumulation, sugar millers start crushing and canes are diverted to sugar production. There are certainly other problems specific to sugarcane like climate, the economic situation of the mill, and pressures for planting other crops which are less labour-intensive and more remunerative.

Breeding varieties for jaggery production have been limited and are confined to countries with small land holdings. For instance, in India, any variety released for commercial cultivation for sugar production is invariably tested for its jaggery-making properties, jaggery quality, and shelf life. The advantage of jaggery farmers is the flexibility they enjoy in producing the commodity as these farmers themselves make jaggery or sell it to large jaggery manufacturers. Generally, early maturing high sucrose varieties are best suited for jaggery production for an extended production of jaggery starting with the early season, i.e., October and November, and for the same reason, high-yielding varieties late maturing varieties are not fit for the jaggery production.

Generally, selection among the elite clones was adopted to develop varieties suitable for jaggery production as the jaggery yield has a direct bearing on the sugar yield. There were several studies on the identification of suitable varieties for normal and organic jaggery manufacture. Guddadamath et al. (2014) reported a wide range and variation in jaggery recovery, yield, and juice quality parameters among the test clones and indicated the scope of identifying varieties combining good quality and desirable jaggery productivity. Good varieties are characterized by superior jaggery recovery and productivity over commercial varieties with highly acceptable light golden yellow colour, crystalline nature, and long shelf life without adding any chemical clarificants. Several varieties have been identified as suitable for jaggery from the initial years of breeding till recently. Not all sugarcane varieties are suitable for jaggery making, Co 62175, Co 8014, and Co 92005 are known as jaggery varieties for good colour and texture of the jaggery without adding chemical clarificants and are used for jaggery making on a large scale in India. In Kolhapur, known for its jaggery market and has a GI tag for jaggery, the jaggery prepared from Co 92005 fetched an additional rate of Rs.250 to 550/per quintal as compared with other released varieties in the region for its excellent jaggery properties. Anakapalle in Andhra Pradesh is the second place known for its jaggery market, which is moving in a downward trend. Sustaining this industry profitably is a challenge and depends on different value-added products from jaggery and their market availability.

Higher net rendement values (NRV) determine better quality and are positively associated with sucrose, purity, and colour intensity. Reducing sugars and ash content was detrimental to quality and had a significant positive association with ash. Sucrose is positively correlated with purity, colour intensity, and NRV and negatively correlated to ash content and reducing sugars. Higher NRV determines better jaggery quality, which was found positively associated with sucrose, purity, and colour intensity (Sharma et al. 1979). Jaggery, being rich in minerals has a direct bearing with the sugarcane variety. Many good jaggery varieties have been identified including Co 62175, Co 0238, Co 0118, CoC 671, CoSe 03234, CoS 08279, and CoS 08272, hence shows the scope of genetic improvement for jaggery. When selecting parents for jaggery improvement, it is essential to create genetic variability for jaggery colour and texture. This approach helps achieve sustainable genetic improvement and results in the release of improved cultivars suitable for jaggery production as well. There is also a need for continuous improvement of the populations to develop donor parents for good jaggery characteristics.

2.9 Energy Canes

Though sugarcane is a major industrial crop cultivated over centuries as a feed-stock for sugar production, recently the crop has attained status as an ethanol producer (Diniz et al. 2013). The global socio-political situations, dwindling petroleum sources, and climate change necessitates replacement of fossil fuels with an alternative source for energy, especially from renewable sources including green energy. The lignocellulosic biomass derived from sugarcane, primarily of bagasse and straw, is used not only to generate electricity and second-generation ethanol but also to create chemicals and bioproducts in a biorefinery (Kim and Day 2013). Many countries around the world have focused on sugarcane breeding programmes for producing energy cane with high fibre content and biomass yield. The concept of 'energy cane' was proposed by Alexander in 1985 and led to the energy cane concept to develop varieties with higher fibre content and higher biomass yield (Cursi et al. 2022). Energy cane improvement is envisaged to increase the total sugarcane biomass in terms of greater lignocellulosic biomass utilizing the whole plant including stalks, leaves, and straw for bioethanol production. This involves breeding programmes with wild germplasm to incorporate genetic diversity for sugar and fibre content. *Erianthus arundinaceus* and tall forms of *S. spontaneum* have huge biomass. Several species clones exhibit gigantic growth in the natural habitat, as well as in the germplasm gardens. Either modern hybrids of sugarcane are crossed with high biomass wild accessions or can be intercrossed or selfed. The main idea is to increase the energy conversion efficiency of primary sugarcane energy, agricultural productivity, or total bioenergy per hectare (de Souza Barbosa et al. 2020). Energy canes are characterized by its high tillering and a vigorous root system with rhizomes. These traits enable energy cane to capture natural resources efficiently and convert more energy into biomass than sugarcane (Santos et al. 2015). Such programmes are reported from many countries including Brazil, USA, India, Argentina, Thailand, and Japan, where new sugarcane or energy cane varieties are available. In Brazil, genetic improvement for energy cane is done to produce high-productivity varieties adapted to the different Brazilian biomes, as well as to develop customized cultivars for possible future demands. In the United States, a joint programme between Louisiana State University and USDA-ARS developed an energy cane cultivar with a substantial increase in biomass and fiber content (28.0%) and it continued to increase with each ratoon (Negri et al. 2010). The team released several energy cane varieties for use as bioenergy feedstocks.

The energy canes are hybrids between commercial canes and their wild ancestors with high bioenergy feedstock potential. Breeding programmes aimed at developing energy cane are achieved through the utilization of high fibre and high biomass producing *Erianthus arundinaceus* clones or *S. spontaneum* with *S. officinarum* or other commercial cultivars (Wang et al. 2008). Two types of energy canes are available, namely the Type I and Type II canes, derived from *S. spontaneum* and *E. arundinaceous* background with high biomass, fibre, and bioenergy potential. For Type I energy canes, *Saccharum spontaneum* provides high fibre content, higher number of canes, wider adaptability to adverse agroclimatic conditions, and better ratoonability for 7–8 seasons. *Erianthus arundinaceus* is widely used for developing Type II energy canes since it produces large amounts of biomass as well as has resistance to most pests and diseases. It has been observed that an increase in fibre content is typically accompanied by a reduction in sucrose levels since carbon allocation after photosynthesis is distributed between sucrose storage, respiration, and cell wall biosynthesis (Waclawovsky et al. 2010). These energy canes also vary for fibre content in cane as well as brix % in juice. The clones, with >15% juice brix and >20% cane fibre, were designated as Type I energy canes. While the juice can be used for direct fermentation in distilleries to produce alcohol, the bagasse is used in the co-generation unit for generating electricity. Type II energy canes are characterized by >25% cane fibre and the juice brix <15%. This type of energy canes can be harvested as whole canes with trash and tops and directly fed into the boilers for producing electricity. In India, about 21 energy canes were developed which are suitable for cultivation for the production of biomass under different agroclimatic conditions. The identified clones are not recommended for cultivation under normal agroclimatic conditional as these clones are not developed to compete with sugarcane for area. It should be planted where any other agricultural crop cannot be profitably cultivated owing to low rainfall, waterlogging, saline, and alkaline soils, etc.

Breeding for energy cane was attempted in India. Selection from open-pollinated fluff of *Erianthus arundinaceus* clone IK 76-75 led to the identification of energy canes. Six energy canes with high biomass were, viz., SBIEC 14001, SBIEC 14002, SBIEC 14003, SBIEC 14004, SBIEC 14005, and SBIEC 14006 and a registered genetic stock SBIEC 11002 exhibited high biomass under suboptimal management condition with limited irrigation and fertigation in plant and ratoon crops (Govindaraj 2020). Among these, SBIEC 14006 was identified as a promising clone based on high harvestable biomass and fibre content in cane and recorded mean harvestable biomass of 265.28 t/ha, which was 241.47 t/ha under limited irrigation conditions and 289.08 t/ha under normal irrigation condition. SBIEC 14006 recorded a fibre content of 27.54%, average juice brix of 8.94%, and sucrose of 4.72%, cane diameter of 2.18 cm, and single cane weight of 1.24 kg. All these factors make this an ideal energy cane characterized by more biomass yield per unit area and less cost of cultivation for its low nutrient requirement and multiple ratoonability of 7–8 times, which avoids the need for replanting every year. By virtue of the tall and non-lodging nature of the energy canes, these canes are amenable to mechanical harvesting, which is essential while taking up large plantations. To test their suitability in multiple environments,18 biomass types including 6 energy canes were evaluated under the subtropical conditions at Karnal, Haryana, India. The type I canes such as SBIEC 11008 (204.15 t/ha), SBIEC 11005 (192.93 t/ha), SBIEC 13008 (201.26 t/ha), SBIEC 13009 (196.58 t/ha), SBIEC 13002 (170.15 t/ha), and SBIEC 13007 (173.76 t/ha) consistently outperformed the check varieties, while among the type II canes, SBIEC 11004 (225.78 t/ha), SBIEC 11006 (184.89 t/ha), and SBIEC 14006 (184.73 t/ha) produced significantly higher biomass than commercial checks. Hence these forms were identified to be superior for co-generation with an estimated energy output from the energy canes, energy utilization efficiency in plants and ratoon crops for energy canes and higher fibre contents. In contrast to the commercial varieties, these energy cane present a vital option to meet the future demand of bioenergy and fibre for their versatility to flourish under marginal lands without competing with the food crops cultivated in the cultivable areas.

Since energy canes grow throughout the year, its plantation offers a continuous supply of raw materials for biofuel industries. A 10 MW/h power industry requires 3700 ha/year of energy cane crop, with an average harvestable biomass of 150 t/ha. Such power plants can cater to agriculture and domestic needs for a cluster of villages and to support local agro-based industries. Energy canes can be profitably utilized in the paper industries and facilitate reduced consumption of forest wood. It is estimated that 1 t of energy cane including foliage contains 4500 MJ energy, and ethanol yield is approximately equal to one barrel of oil. Integrated production of sugar, ethanol, and energy as electricity from sugarcane and energy canes can boost social and economic growth of rural areas to accelerate greening the growth of a nation.

2.10 Red-Fleshed Canes with Antioxidant Properties

S. robustum is in close kinship to *S. officinarum*, the red-fleshed '*Sanguineum*' ($2n = 60$) was not reported in the pedigree of any of the commercial cultivars or advanced breeding lines (Heinz 1965). The forma '*sanguineum*' exhibits high biomass and high fibre content and is adapted to a wide range of environmental conditions (Cheavegatti-Gianotto et al. 2011). Several clones of *S. robustum* are reported to be flood tolerant and naturally found on the riverbanks prone to impending flood (Moore 1987). The future of sugarcane rests on its manyfold applications. The studies on prospecting sugarcane germplasm for antioxidants (Rakkiyappan et al. 2012) confirmed that the red flesh clones of *S. robustum* had manyfold higher antioxidant compounds in their juice compared to the other sugarcane cultivars. Also, introgression of *sanguineum* group ($2n = 60$) has been reported in the evolution of *S. officinarum* (Grassl 1974; Daniels and Roach 1987). The field gene bank of ICAR-SBI Research Centre, Kannur, Kerala, has nine germplasm clones of red-fleshed *S. robustum* (28 NG 219, NG 77-73, NG 77-75, NG 77-76, NG 77-78, NG 77-84, NG 77-88, NG 77-90, and NG 77-132).

It is also interesting to note that the flesh colour variation in the *S. officinarum* accessions could identify pink to red-coloured flesh (NG 77-62, NG 77-66, NG 77-70, NG 77-130, NG 77-131), apart from their gigantic growth. The major hurdle in utilizing the red-fleshed clones in breeding programme was their rare flowering nature. Chandran et al. (2020) utilized three rare flowering forms (NG 77-84, NG 77-73, and NG 77-75) and identified these genotypes as potential sources for antioxidant compounds. They characterized interspecific progenies generated from polycrosses of two red-fleshed forms, and this was the first report of utilization of the red-fleshed *S. robustum* in sugarcane breeding programme. The progenies were not only highly variable for morphological and yield traits but also exhibited considerable improvement over the parental clones of *S. robustum*. The report has high applied values, not only to sugarcane improvement for breaking the yield barriers and red-coloured flesh with antioxidant properties (Suresha et al. 2017, 2023). Premachandran (1994) reported a leaf mutant with greenish purple leaf in *S. edule* and opined that to be the result of the activity of transposable elements on the gene responsible for anthocyanin synthesis. However, the purple-leaved clones were relatively poor in sucrose content (6.08–12.7%) in the juice.

2.11 Ornamental Sugarcane

S. officinarum is characterized by a wide variation of its attractive rind colour and stripes and some forms have purple-pigmented leaves, making these ideal ornamental chewing plants. According to Artschwager and Brandes (1958), the aborigines selected softer and sweeter forms with bright colours and stripes on the stem. The Papuan gardens thus had a collection of bright, multicoloured, and striped canes (Stevenson 1965a, b). Though, Hackel (1889) classified *S. officinarum* into three groups: 'genium' with yellow-green canes, 'violaceum' having purple-hued leaves, and 'litteratum' with yellow-green rind and red stripes, breeders did not concentrate on improving the ornamental properties for a long time. Recently, across the globe, growing of chewing canes as ornamental plant is gaining importance. These are *S. officinarum* clones like Florida red, Java green, green canes with red stripes, ribbon canes which form attractive landscapes (Chandran et al. 2017b). They reported 110 clones in the *S. officinarum* collection to have vertical rind stripes in varying combinations and intensities of yellow, red, green, purple, etc., and 10 with purple leaves.

Genetic improvement for ornamental types is possible due to their variability in cane yield traits, such as cane number, thickness, and cane weight, which are positively correlated with cane yield. The striped forms exhibited a range of 6–41 canes, 1.76–3.7 cm average cane diameter, 79–358 cm cane length, and 0.3–2.7 kg single cane weight. The brix value ranged from 11.1% to 21.6% at 7 months and from 11.8% to 21.9% at 10 months, while sucrose content at 10 months varied from 5.5% to 20.5%. The study highlighted the scope of breeding short-duration ornamental chewing canes, ready to use from 7 months onwards. Online sale of a few rare forms is in place, for instance, the rarest plant Hilo Buddha, a Hawaiian sugarcane plant is the most sought-after ornamental sugarcane for its beautifully shaped striped canes. Hilo Buddha exhibited a different flavour and texture from other varieties. Ballooning of the colourful nodes at maturity containing airy, cotton-candy-like sweet fibres is a preferred cane and grows up to 8 ft. *Saccharum officinarum* 'Pele's Smoke' is yet another ornamental and edible sugarcane, named after the Hawaiian volcano goddess Pele. The foliage is dark purple to black in colour. The canes quickly grow to a height of 7–10 ft, while the blades of grass grow 2–3 ft long, though the stalks are thinner than the normal canes. The plants are dark-coloured and ornamental resembling bamboo in habit (Chandran et al. 2017a, b). They attributed ornamental chewing nature to the clones 51 NG 131, 57 NG 149, 57 NG 172, Chittan, Creoula Rayada, Louisiana striped, Preanger striped, Striped Mauritius, etc., by virtue of their attractive stems and high brix (17.0–20.9%). Purple-leaved forms invariably were characterized by attractive dark purple colour rind and Brix ranging from 10.6% (IK 76-29) to 20.24% (51 NG 142). Nonflowering nature of such clones needs to be addressed to facilitate inheritance studies and the development of more attractive chewing types.

At this time, when sugarcane is considered a magic plant that can transform livelihoods, these ornamental chewing types hold value for additional income generation. Genetic improvement for ornamental sugarcane is attempted in India through selective crossing of ornamental canes and through induction of somaclonal variation and mutation (Chandran et al. 2023). For induction of variation in ornamental *S. officinarum* through hybridization, they selected 35 progenies of three crosses and found GUK 21-368 (striped cane green/yellow), GUK 21-383 (yellow rind and thin cane), GUK 21-408 (black purple rind), GUK 21-413 (green rind), GUK 21-479 (yellow rind), GUK 21-511 (dark purple/light purple striped rind with mild wax), and GUK 21-544 (purple rind with weather mark) as promising for ornamental purpose.

2.12 Induction of Variation Through In Vitro Callus Culture

Somaclonal variation was induced through chemical mutagenesis using EMS in selected ornamental plants by Indian workers. Field evaluation of the somaclones revealed distinct morphological variations among these and from their respective parental clones. The tumescent striped cane morphology of 51 NG 131 was completely transformed and all the regenerants showed close similarities among them but were quite distinct from the mother plant for their uniform yellow colour with ornamentation of red blotches and absence of deep internodal cracks. Variation in leaf colour was noticed in a regenerant of NG 77-314. From NG 77-92 with dark purple stripes on purple coloured rind, two morphotypes were observed among the somaclones, one with striped rind and another without stripes and uniform purple colour on the rind. Similarly, the two regenerants from NG 77-952 differed, one with stripes and another with uniform rind colour. The regenerants from IJ 76314 and IJ 76-559 also showed variation from mother plant. The somaclones TC1 of 51 NG 131, TC 6 of NG 77-92, and TC 8 of IJ 76-314 were identified as promising in the initial evaluation.

2.13 Sugarcane as a Vegetable

The species *S. edule* Hassk is characterized by aborted inflorescence, cultivated from New Guinea to Fiji for its edible inflorescence (Massal and Barrau 1956) and was mentioned as Ova piscium for the inflorescence resembling the egg-mass of fish (Daniels and Roach 1987). There are many popular names for the species as pitpit, coastal pitpit, Fiji asparagus, duruka, dule, naviso, etc. and occur widely in Melanesia (Warner and Grassl 1958; Berding and Koike 1980). According to Grassl (1967), the New Guinea type of *S. edule* originated through introgression of *S. robustum* × *Miscanthus* and the Fijian type from *S. officinarum* × *Miscanthus*, while Daniels and Roach (1987) regarded it as a product of introgression of other related genera in *S. officinarum* and *S. robustum*. Cytological studies of the species supported the latter view owing to the availability of polyploid series $2n = 60$, 70, and 80 with aneuploid forms. Premachandran (2006) referred to the edible inflorescence as 'cane flower' having a similar origin in cauliflower caused by mutation of flowering gene. The edible inflorescence inside the leaf sheath was largely creamy yellow in colour in the varieties of white duruka, which differ in stem size, inflorescence length, and time of harvest, whereas a variant had a deeper yellowish-orange colour, called red duruka or minimanu (Waqaniu-Rogers 1986; Chandran et al. 2017a, b). ICAR-SBI, Kannur germplasm bank housed 17 accessions of the species, among which 9 flowered regularly. Chandran et al. (2017a, b) evaluated these clones for 30 qualitative and 10 quantitative traits and found high dissimilarity coefficient in majority of the accessions to suggest these as independent units. Two accessions IJ 76-329 and IJ 76-312 exhibited more than 75% similarity and were found to have been collected from the same locality, whereas the accession 28 NG 82 was the most dissimilar. Saraswati et al. (2013)also made a similar observation in the *S. edule* collection from the lowland to highland in Papua, Indonesia. Natural hybridization led to sterile and aborted inflorescences, hence further improvement of the species did not happen and remained as discrete population in specific geographical locations. In this regard, Premachandran (2006) suggested that the aborted inflorescence of *S. edule* might have resulted in cal and ap1 mutants of the flowering gene as reported in Arabidopsis, cauliflower, maize, rice, etc., or its orthologs in *S. robustum*. The possibility of transferring this gene to sugarcane thus would facilitate adding one more economic product from this crop. In this era on genomics, this would look promising and the value addition in sugarcane could lead to a new dimension of stem used for sugar and inflorescences as vegetables.

The study revealed that the accessions IJ 76-368, IJ 76-360, and IJ 76-552 are better yielders for edible inflorescence on account of fresh weight of edible inflorescence, good tillering, good flowering percentage of stalks, and also for biomass yield. *S. edule* is suitable for intercropping with soybean and rotation with early flowering red duruka with late flowering white types for better commercial exploitation (Saraswati et al. 2013). They also suggested wider spacing (200 × 200 cm) for improving the growth and yield, with greater plant height, more number of suckers and flowers and increased flower weight, compared to the traditional planting with 150 × 150 and 100 × 150 cm spacing. The variety purple duruka flowered twice a year when introduced to Viti Levu, Fiji, and became popular and a canning unit was proposed in Fiji to sell this as Fijian asparagus (Waqaniu-Rogers 1986). *S. edule* is well adapted to marginal land and can check soil erosion (Waqaniu-Rogers 1986), hence is suitable for such areas where sugarcane is not commercially viable. Since it has a short flowering period of 9 weeks (from October second week to December second week), suitable genotypes with early flowering, late flowering, and that flower twice in a year are needed to be identified to have a continuous availability of fresh vegetable in the market. In order to overcome the quick perishable nature of the inflorescence, there is also a need to develop suitable packing/canning of this vegetable to ensure its availability in the export market.

2.14 New Traits Through Genetic Engineering

Though different sequencing projects of parental species and hybrids were carried out in sugarcane, none is comprehensive enough to understand the genomes. A recent report on the comprehensive reference genome for 'R570' (Healey et al. 2024) is considered a landmark advancement for agricultural biotechnology, particularly at a time when sugarcane is considered one of the most productive crops and a source of sugar and renewable carbon. At this time of genomic revolution, this study holds promise to make substantial improvements in the genetic understanding of agronomic traits through the exploration of allelic variation, copy number, and gene presence/absence variation. This information is critical in modern cultivar breeding programmes worldwide. Since the last decade of 20th century, the development of transgenic sugarcane to improve agronomic traits such as resistance to disease, pest, herbicide, and abiotic stresses and to improve industrial traits progressed well. Improving sugar, bioenergy, and biofuel production, alteration of the carbon-partitioning balance for biodiesel production, and production of bioplastics like polyhydroxy alkanoates and high-value therapeutic proteins have been reported (Zale et al. 2016; Petrasovits et al. 2012).

Sugarcane for biofarming has immense benefits such as large biomass, high growth rate, efficient carbon fixation with C4 pathway, and a well-developed storage tissue system in sugarcane. In sugarcane, the culm contributes about 70% dry weight, which is largely made of storage parenchyma cells and the vacuoles in the storage parenchyma cells store the juice. Therefore, it is logical to target any recombinant protein to the vacuole, so as to attain the maximum protein yield. A novel vacuolar-targeting determinant to localize recombinant proteins in the vacuole in high concentrations was identified and such targeted r-proteins could be purified from the juice in simple steps (Palaniswamy et al. 2016). This study opened up significant applications of sugarcane as a biofactory. In addition to the benefits mentioned above, sugarcane carries minimal transgene dispersal owing to its vegetative propagation through stem cuttings, about 70% of extractable juice having minimal protein content (0.01–0.02%), and a well-established downstream processing technology to mobilize a foreign protein into the vacuole (Appunu et al. 2017). In India, work in this line is in progress for targeting and accumulating glucocerebrosidase, insulin, and interferon in a public–private partnership mode.

2.15 Genome Editing

Sugarcane plays a central role in the sugar and energy sector. Genome editing opens up affordable modifications in its genome to control plant behaviour and develop designer sugarcane varieties for different needs. Genome editing for specialty sugars, biomass modification, and edible vaccines in sugarcane juice is interesting research in progress. One of the main advantages of using genome editing is that elite breeding lines or commercial varieties can be edited (Lowe et al. 2016), thus eliminating the need for backcrossing, which is not practical in sugarcane to realize the same genetic background. This way not only the time taken to develop a variety is reduced but also eliminates linkage drag caused by undesirable alleles from the donor parent.

The challenges of polyploidy cause problems in the number of copies to be edited in sugarcane. Despite this, genome editing techniques could improve yield attributes and plant metabolism, leading to enhanced yield on a sustainable basis (Jung and Altpeter 2016). Similarly, genes conferring resistance against diseases, yield, and sucrose recovery were edited (Wu et al. 2023). Different candidate genes for imparting tolerance against rust and smut diseases, abiotic stresses (drought, salinity, cold, or chilling stress, and oxidative stress) in sugarcane were used through precise genome editing using the CRISPR/Cas9 technique (Li and Iqbal 2024).

There are different gene editing tools employed to modify the genome, such as mitochondrial genome editing, anti-sense transcription, zinc-finger nuclease techniques, site-specific recombinase, base editing, transcription activator-like effector nucleases (TALENs) (Jung and Altpeter 2016; Kannan et al. 2018; Peng et al. 2015; Ma et al. 2016; Zaidi et al. 2017), but off-targeting was a stumbling block to perfect the technique. CRISPR technique is reported to be a more advanced genetic tool for editing the sugarcane genome (Hussin et al. 2022). More details are included in subsequent chapters. These techniques are effective in value addition in sugarcane.

2.16 Conclusions

Sugarcane is a complex polyploid, and major traits for sugarcane improvement are polymorphic and polygenic in inheritance, which complicates the breeding process. Interspecific hybridization has provided breakthroughs in yield and disease resistance and adaptation to stresses to a larger extent. Sugarcane, originally used for chewing and for juice-based sweeteners has transformed into a wonder plant that can produce even high-value molecules, through the advancement of research through conventional and omics approaches. The crop has also become increasingly resilient to stress. All these achievements are largely possible through systematic breeding for the past 120 years. However, time has come to a stage where incremental gains cannot meet the growing demands of sugarcane. Modern tools in genomics, phenomics, AI, big data analysis drone-based phenotyping, etc., are essential to use the germplasm effectively and to make quantum jumps in realizing multiproducts. For instance, genomic research has evolved to reveal that 8.7 billion decoded base pairs make up the full sugarcane genome (Healey et al. 2024) and the genetic information can precisely be used in trait enhancement/modifications. Precision breeding for yield optimization in specific agroclimatic regions and product diversification have to be placed in a cyclical way to make the crop competitive, and more relevant in the tropical countries and sustainable to the planet Earth. Breeding for value addition is expected to scale greater heights with precise knowledge guided by cutting-edge technologies including information technology and to develop designer crop varieties in a sustainable way.

References

Abbas SR, Sabir SM, Ahmad SD, Boligon AA, Athayde ML (2014) Phenolic profile, antioxidant potential and DNA damage protecting activity of sugarcane (*Saccharum officinarum*). Food Chem 147:10–16

Adhini SP, Durai AA, Govindaraj (2017) Co 205; the Coimbatore legacy. In: Hemaprabha G, Viswanathan R, Ramasubramanian T, Bhaskaran A, Mohanraj K, Ram B (eds) Proceedings of international symposium on sugarcane research since Co 205: 100 years and beyond (SucroSym 2017). ICAR-Sugarcane Breeding Institute, Coimbatore, pp 173–176

Aitken K, Jackson PA, McIntyre CL (2006) Quantitative trait loci identified for sugar related traits in a sugarcane (Saccharum spp.) cultivar x Saccharum officinarum population. Theor Appl Genet 112(7):1306–1317

Allen CJ et al (1997) New technologies for sugar milling and by-product modification. In: Keating BA, Wilson JR (eds) Intensive sugarcane production: meeting the challenges beyond 2000, proceedings of the sugar 2000 symposium. CABI, Wallingford, pp 267–285

Amalraj VA, Balasundaram N (2006) On the taxonomy of the members of "Saccharum complex". Genet Resour Crop Evol 53:35–41

Appunu C, Ram B, Subramonian N (2017) Sugarcane: an efficient platform for molecular farming. In: Mohan C (ed) Sugarcane biotechnology: challenges and prospects. Springer, Cham. https://doi.org/10.1007/978-3-319-58946-6_7

Artschwager E, Brandes EW (1958) Sugarcane (Saccharum officinarum), origin, characteristics and descriptions of representative clones. USDA agriculture handbook no. 122. USDA, Washington, DC, p 307

Berding N, Koike H (1980) Germplasm conservation of the Saccharum complex: a collection from the Indonesian Archipelago. Hawaiian Planters Rec 59:87–176

Brandes EW, Sartoris GB (1936) Sugarcane: its origin and improvement. Yearbook U.S. Department of Agriculture's 1936, pp 561–624

Bremer G (1923) A cytological investigation of some species and species hybrids within the genus Saccharum. Genetica 5:97–148, and 273–326

Brown ADH, Daniels J, Latter BDH (1969) Quantitative genetics of sugarcane. II. Correlation analysis of continuous characters in relation to hybrid sugar cane breeding. Theor Appl Genet 39:1–10

Calcino DV (1994) Australian sugarcane nutrition manual. SRDC/BSES, Indooroopilly

Chandran K, Nisha M, Mahesh P, Arunkumar R (2017a) Saccharum edule Hassk. an under-exploited germplasm resource of sugarcane. Indian J Plant Genet Resour 30(1):72–76. https://doi.org/10.5958/0976-1926.2017.00010.9

Chandran K, Nisha M, Gireesan PP (2017b) Ornamental chewing canes. J Sugarcane Res 7(2):169–173

Chandran K, Nisha M, Gireesan PP (2020) Characterization of progenies from polycrosses of *S. robustum* clones *f. sanguineum*. Sugar Tech 22:379–388. https://doi.org/10.1007/s12355-019-00785-y

Chandran K, Nisha M, Gopi R et al (2023) Sugarcane genetic resources for challenged agriculture. Sugar Tech 25:1285–1302. https://doi.org/10.1007/s12355-023-01313-9

Cheavegatti-Gianotto A, de Abreu HMC, Arruda P, Filho JCB, Burnquist WL, Creste S, di Ciero L, Ferro JA, de Oliveira Figueira AV, de Sousa Filgueiras T, Grossi-de-Sá MF, Guzzo EC, Hoffmann HP, de Andrade Landell MG, Macedo N, Matsuoka S, de Castro Reinach F, Romano E, da Silva WJ, de Castro Silva Filho M, César Ulian E (2011) Sugarcane (*Saccharum × officinarum*): a reference study for the regulation of genetically modified cultivars in Brazil. Trop Plant Biol 4:62–89

Conab (2011) Acompanhamento da SafraBrasileira: Cana-de-açúcar, report no. TerceiroLevantamento

Cox M, Hogarth M, Smith G (2000) Chapter 5: Cane breeding and improvement. In: Hogarth M, Allsopp P (eds) Manual of cane growing. Bureau of Sugar Experiment Stations, Indooroopilly, pp 91–108

Cursi DE, Hoffmann HP, Barbosa GVS, Bressiani JA, Gazaffi R, Chapola RG, Fernandes JAR, Balsalobre TWA, Diniz CA, Santos JM (2022) History and current status of sugarcane breeding, germplasm development and molecular genetics in Brazil. Sugar Tech 24:112–133

D'Hont A, Grivet L, Feldmann P, Glaszmann JC, Rao S, Berding N (1996) Characterisation of the double genome structure of modern sugarcane cultivars (Saccharum spp.) by molecular cytogenetics. Mol Gen Genet 250(4):405–413

D'Hont A, Ison D, Alix K, Roux C, Glaszmann JC (1998) Determination of basic chromosome numbers in the genus Saccharum by physical mapping of ribosomal RNA genes. Genome 41(2):221–225. https://doi.org/10.1139/g98-023

Daniels J, Roach BT (1987) Taxonomy and evolution (Chapter 2). In: Heinz DJ (ed) Sugarcane improvement through breeding, vol 11. Elsevier, Amsterdam, pp 7–84

de Matos AT et al (2003) Removal of Cu and Zn from swine raising wastewater using organic filters. Environ Technol 24(2):171–178

de Souza Barbosa GV, dos Santos JM, Diniz CA, Cursi DE, Hoffmann HP (2020) Energy cane breeding. In: Sugarcane biorefinery, technology and perspectives. Elsevier, Amsterdam, pp 103–116. https://doi.org/10.1016/B978-0-12-814236-3.00006-8

Diniz AL, Ferreira SS, ten-Caten F, Margarido GRA, dos Santos JM, Barbosa GVS, Carneiro MS, Souza GM (2013) Genomic resources for energy cane breeding in the post genomics era. Comput Struct Biotechnol J 17:1404–1414. https://doi.org/10.1016/j.indcrop.2013.12.004

Dunlap CE, Callihan CD (1969) Fermentative utilization of sugar cane bagasse. Am Soc Sugar Cane Technol 16:82–90. https://archive.org/stream/americansocietyo1617amer/americansocietyo1617amer_djvu.txt

Edme SJ, Comstock JC, Miller JD, Tai PYP (2005) Determination of DNA content and genome size in sugarcane. J Am Soc Sugar Cane Technol 25:1–16

Fauconnier R (1993) Sugar cane. Macmillan Press Ltd., London, pp 1–140

Glyn J (2004) Sugarcane. Blackwell Publishing, Oxford, p 224

Gopinathan MC, Sudhakaran R (2009) Biofuels: opportunities and challenges in India. In Vitro Cell Dev Biol Plant 45(3):350–370. https://doi.org/10.1007/s11627-009-9217-7

Govindaraj P (2020) SBIEC 14006. A high biomass energycane for power, alcohol and paper industries. J Sugarcane Res 10(1):100–106. https://doi.org/10.37580/JSR.2020.1.10.100-106

Grassl CO (1967) Introgression between Saccharum and Miscanthus in New Guinea and Pacific Area. Proc Int Soc Sugar Cane Technol 12:995–1003

Grassl CO (1974) The origin of sugarcane producing cultivars of Saccharum. Sugarcane Breed Newslett 34:10–18

Guddadamath SG, Patil SB, Khadi BM et al (2014) Genetic enhancement of sugarcane for the production of organic Jaggery. Sugar Tech 16:86–91. https://doi.org/10.1007/s12355-013-0257-2

Hackel E (1889) Andropogaoneae. In: de Candolle A, de Candolle C (eds) Monograph of phanerogamarum, vol 6, pp 1–716

Harrison J (1888) Seedlings of sugar cane at Barbados (Saccharum officinarum L.). Bull Miscellaneous Inf (Royal Gardens Kew) 1888:294–296

Healey AL, Garsmeur O, Lovell JT, Shengquiang S, Sreedasyam A, Jenkins J, Plott CB, Piperidis N, Pompidor N, Llaca V, Metcalfe CJ, Doležel J, Cápal P, Carlson JW, Hoarau JY, Hervouet C, Zini C, Dievart A, Lipzen A, Williams M, Boston LB, Webber J, Keymanesh K, Tejomurthula S, Rajasekar S, Suchecki R, Furtado A, May G, Parakkal P, Simmons BA, Barry K, Henry RJ, Grimwood J, Aitken KS, Schmutz J, D'Hont A (2024) The complex polyploid genome architecture of sugarcane. Nature 628:804–810. https://doi.org/10.1038/s41586-024-07231-4

Heinz DJ (1965) Wild Saccharum species for breeding in Hawaii. Paper no. 180 in the Journal Series of the Experiment Station, Hawaiian Sugar Planters' Association, Honolulu, Hawaii

Heinz DJ (1987) Sugarcane improvement through breeding. In: Heinz DJ (ed) Plant breeding reviews. Elsevier, Amsterdam, pp 15–118

Hemaprabha G, Lavanya DL (2015) Sucrose specific TRAP markers as genus and species specific markers in Saccharum and Erianthus spp. Indian J Genet Plant Breed 75(1):99–104

Hemaprabha G, Mohanraj K, Chandran K, Alarmelu S, Ram B (2021) Parental diversity in Indian commercial sugarcane varieties-past, present and future. In: International conference on sugarcane research: sugarcane for sugar and beyond, June 19–22, 2021. ICAR-Sugarcane Breeding Institute, Coimbatore, pp 41–47

Hogarth DM, Wu KK, Heinz DJ (1981) Estimating genetic variance in sugarcane using a factorial cross design. Crop Sci 21:21–25

Hussin SH, Liu X, Li C, Diaby M, Jatoi GH, Ahmed R et al (2022) An updated overview on insights into sugarcane genome editing via CRISPR/Cas9 for sustainable production. Sustainability 14:12285. https://doi.org/10.3390/su141912285

Janikula M (2002) Policosanol: a new treatment for cardiovascular disease? Altern Med Rev 7(3):203–217

Jeswiet J (1930) The development of selection and breeding of the sugar cane in Java. Proc Int Soc Sugar Cane Technol 3:44–57

Jung JH, Altpeter F (2016) TALEN mediated targeted mutagenesis of the caffeic acid O-methyltransferase in highly polyploid sugarcane improves cell wall composition for production of bioethanol. Plant Mol Biol 92:k131–k142

Kanawade SM et al (2010) Low cost sugarcane bagasse ash as an adsorbent for dye removal from dye effluent. Int J Chem Eng Appl 1(4):309–318

Kannan B, Jung JH, Moxley GW, Lee SM, Altpeter F (2018) TALEN-mediated targeted mutagenesis of more than 100 COMT copies/alleles in highly polyploid sugarcane improves saccharification efficiency without compromising biomass yield. Plant Biotechnol J 16:856–866. https://doi.org/10.1111/pbi.12833

Kansal S (1998) Factors determining Indian sugar production and its comparative advantage. In: Proceedings of the Fiji/FAO 1997 Asia Pacific sugar conference. Food and Agriculture Organization, Rome, pp 78–92. www.fao.org/docrep/005/x0513e/x0513e16.htm

Karuppaiyan R, Nair NV, Ram B, Mohanraj K, Anna Durai A, Amudh P (2020) Evaluation of intra-specific hybrids of *Saccharum officinarum* for yield, quality and incidence of Yellow leaf disease. J Sugarcane Res 10:54–62. https://doi.org/10.37580/JSR.2020.1.10.54-62

Karve P, Karve P, Salunkhe RM, Karve AD (2001) A chain of technologies for using sugarcane trash as a household fuel. Boil Point 47:16–18

Kennedy AJ (2005) Breeding improved cultivars for the Caribbean by utilization of total biomass production. Proc Int Soc Sugar Cane Tech 25:491–499

Kim M, Day DFL (2013) Composition of sugarcane, energy cane, and sweet sorghum suitable for ethanol production at Louisiana sugar mills. J Ind Microbiol Biotechnol 38:803–807. https://doi.org/10.1007/s10295-010-0812-8

Kumbhar YS (2016) Study on gur (Jaggery) industry in Kolhapur. Int Res J Eng Technol 3(2):590–594

Legendre BL (1995) Potential for increasing sucrose content of sugarcane; an assessment of recurrent selection in Louisiana. Sugarcane 3:4–9

Li C, Iqbal MA (2024) Leveraging the sugarcane CRISPR/Cas9 technique for genetic improvement of non-cultivated grasses. Front Plant Sci 15:1369416. https://doi.org/10.3389/fpls.2024.1369416

Lo CC, Chen YH (1995) Breeding for sucrose content improvement in sugar cane. Taiwan Sugar 42:10–16

Lowe K, Wu E, Wang N, Hoerster G, Hastings C, Cho MJ et al (2016) Morphogenic regulators baby boom and Wuschel improve monocot transformation. Plant Cell 28(9):1998–2015. https://doi.org/10.1105/tpc.16.00124

Ma X, Zhu Q, Chen Y, Liu YG (2016) CRISPR/cas9 platforms for genome editing in plants: developments and applications. Mol Plant 9:961–974. https://doi.org/10.1016/j.molp.2016.04.009

Mackintosh D (2000) Sugar milling (Chapter 17). In: Hogarth M, Allsopp P (eds) Manual of cane growing. Bureau of Sugar Experiment Stations, Indooroopilly, pp 369–377. www.sugarresearch.com.au/icms_docs/166965_Chapter_17_Sugar_Milling.pdf

Massal E, Barrau J (1956) Food plants of the South Sea islands. South Pac Comm Tech Paper, vol 94, 51p

Milligan SB, Gravois KA, Bischoff KP, Martin FA (1990) Crop effects on broadsense heritabilities and genetic variances on sugarcane yield components. Crop Sci 30:344–349

Moore PH (1987) Breeding for stress resistance. In: Heinz DJ (ed) Sugarcane improvement through breeding. Elsevier, Amsterdam, pp 503–542

Mukherjee SK (1957) Origin and distribution of Saccharum. Bot Gaz 119:18

Nair NV (2008) Sugarcane Breeding Institute, Coimbatore: a perspective. Sugar Tech 10:285–292. https://doi.org/10.1007/s12355-008-0051-8

Nair NV, Sreenivasan TV, William Jebadhas A (1998) Intraspecific improvement in Saccharum. In: Proceedings of the 1st National Plant Breeding Congress. Indian Society of Plant Breeders, Tamil Nadu Agricultural University, Coimbatore, pp 140–145

Nath A, Dutta D, Kumar P, Singh JP (2015) Review on recent advances in value addition of jaggery based products. J Food Process Technol 6:440. https://doi.org/10.4172/2157-7110.1000440

Negri MC, Dao A, Prueger J (2010) Miscanthus and Saccharum as energy crops for biofuel production in the United States: a comparative analysis. Biofuels 1(5):761–776. https://doi.org/10.4155/bfs.10.60

Orellana C, Neto RB (2006) Brazil and Japan give fuel to ethanol market. Nat Biotechnol 24:232. https://doi.org/10.1038/nbt0306-232

Palaniswamy H, Syamaladevi DP, Mohan C, Philip A, Petchiyappan A, Narayanan S (2016) Vacuolar targeting of r-proteins in sugarcane leads to higher levels of purifiable commercially equivalent recombinant proteins in cane juice. Plant Biotechnol J 14(2):791–807. https://doi.org/10.1111/pbi.12430. Epub 2015 Jul 16

Peng R, Lin G, Li J (2015) Potential pitfalls of CRISPR/Cas9-mediated genome editing. FEBS J 283:1218–1231. https://doi.org/10.1111/febs.13586

Petrasovits LA, Zhao L, McQualter RB, Snell KD, Somleva MN, Patterson NA et al (2012) Enhanced polyhydroxybutyrate production in transgenic sugarcane. Plant Biotechnol J 10:569–578. https://doi.org/10.1111/j.1467-7652.2012.00686.x

Premachandran MN (1994) A spontaneous leaf colour mutant in Saccharum. Indian J Genet Plant Breed 54:96–98

Premachandran MN (2006) Cauliflower gene in sugarcane? Curr Sci 91:750–751

Qudsieh Hanan Yassin M, Yusof S, Osman A, Rahman RA (2001) Physico-chemical changes in sugarcane (Saccharum officinarum var yellow cane) and the extracted juice at different portions of the stem during development and maturation. Food Chem 75(2):131–137

Qureshi ME et al (2000) Mill mud case study in Mackay. An economic study on recycling sugar by-products for the Mackay Region. CRC publication series, CRC sustainable sugar production. James Cook University, Townsville

Rakkiyappan P, Chiranjivi Rao K, Jayamani V (2003) Evaluation of Saccharum officinarum clones for chewing. Sugar Tech 5(3):155–160

Rakkiyappan P, Suresha GS, Rupa TR, Hari K, Venkatesan S (2012) Prospecting sugarcane germplasm for antioxidants. In: Proceedings of international symposium on new paradigms in sugarcane research 15–18 October 2012. SBI, Coimbatore, pp 315–316

Ram B, Hemaprabha G (1997) Correlation and path analysis in inter-specific progenies in sugarcane (Saccharum spp.). Indian. J Agric Res 31(1):23–27

Ram B, Karuppaiyan R, Hemaprabha G (2022) Sugarcane breeding. In: Yadava DK, Dikshit HK, Mishra GP, Tripathi S (eds) Fundamentals of field crop breeding. Springer, Singapore, pp 499–570. https://doi.org/10.1007/978-981-16-9257-4_9
Roach BT (1989) Origin and improvement of the genetic base of sugarcane. In: Proceedings of the Australian Society of Sugar Cane Technologists, Brisbane, pp 35–47
Sainz MB, Dale J (2009) Towards cellulosic ethanol from sugarcane bagasse. In: Proceedings of the Australian Society of Sugar Cane Technologists, vol 31, pp 18–23. www.assct.com.au/media/pdfs/2009-M-29-Sainz-Dale.pdf
Sales A, Lima SA (2010) Use of Brazilian sugarcane bagasse ash in concrete as sand replacement. Waste Manag 30(6):1114–1122. https://doi.org/10.1016/j.wasman.2010.01.026
Santos F, Borém A, Caldas C (2015) Sugarcane: agricultural production, bioenergy and ethanol. Academic Press, San Diego. https://doi.org/10.1016/C2014-0-02692-5
Saraswati P, Mawikere NL, Djuuna IAF, Pakiding F (2013) The diversity and cultivation system of Saccharum edule L. and its role as an edible plant source in Papua, Indonesia. Acta Hort (ISHS) 979:373–380
Shanmuganathan M, Gayathry G, Maheshwari P et al (2024) Identification of flavor producing compounds and multi elements from chewing cane (Saccharum officinarum L. cv. Badila). Sugar Tech 26:87–94. https://doi.org/10.1007/s12355-023-01322-8
Shanthi RM, Alarmelu S (2007) Assessment of the first cycle hybrid progenies for early high sugar content in sugarcane. Indian J Genet Plant Breed 67(4):384–387
Sharma SC, Johary PC, Rao GSC (1979) Some observation on the chemical composition of the cane juices of different promising varieties grown in North Bihar conditions and their clarification characteristics. Indian Sugar 29(3):141–147
Shen Y, Zhou G, Liang C, Tian Z (2022) Omics-based interdisciplinarity is accelerating plant breeding. Curr Opin Plant Biol 66:102167. https://doi.org/10.1016/j.pbi.2021.102167
Singh J, Singh RD, Anwar SI, Solomon S (2011) Alternative sweeteners production from sugarcane in India: lump sugar (jaggery) (report). Sugar Tech 13:366–371
Singh J, Solomon S, Kumar D (2013) Manufacturing Jaggery, a product of sugarcane, as health food. Agrotechnol S11:1–3
Siti R, Baharuddin B (1994) Tebukuning-Varietitebujus.TelcnologiPelbagaiTanaman, Jilid 10
Skinner JC, Hogarth DM, Wu KK (1987) Selection methods, criteria and indices. In: Heinz DJ (ed) Sugarcane improvement through breeding. Elsevier, Amsterdam, pp 409–453
Sreenivasan TV, Ahloowalia BS, Heinz DJ (1987) Cytogenetics. In: Heinz DJ (ed) Sugarcane improvement through breeding. Elsevier, Amsterdam, pp 211–253
Stevenson GC (1965a) Genetics and breeding of sugarcane. Tropical science series. Longmans, Green and Co Ltd, London, p 278
Stevenson GC (1965b) Genetics and breeding of sugarcane. Longmans, London, p 31
Sun J, He XM, Zhao MM, Li L, Li CB, Dong Y (2014) Antioxidant and nitrite-scavenging capacities of phenolic compounds from sugarcane (Saccharum officinarum L.) tops. Molecules 19(9):13147–13160. https://doi.org/10.3390/molecules190913147. PMID: 25162956; PMCID: PMC6270697
Suresha GS, Chandran K, Nisha M, Kumar A, Hari K (2017) Assessing the inheritance of redflesh colour and antioxidant activity from the polycross progenies of Saccharum robustum genotypes. In: Proceedings of international symposium on sugarcane research since Co 205: 100 years and beyond (Sucrosym2017), Coimbatore, 18–21 September 2017, pp 24–25
Suresha GS, Chandran K, Hari K, Sreenivasa V, Hemaprabha G (2023) Functional characterization of bioactive compounds from the stem extracts of red fleshed Saccharum robustum genotypes and their polycross progenies from Indian sugarcane germplasm. In: Proceedings of the national seminar on research imperatives for sustaining sugarcane, sugar and ethanol production (SugarSem-2023), 25 August 2023. ICAR-Sugarcane Breeding Institute, Coimbatore, p 146
Tew TL (1987) New varieties. In: Heinz DJ (ed) Sugarcane improvement through breeding. Elsevier, New York, pp 559–594
Thirumala Rao W (1954) Sugarcane varieties for chewing. In: Proceedings of second biennial conference of sugarcane research and development workers, Jullunder, p 115

Waclawovsky AJ, Sato PM, Lembke CG, Moore PH, Souza GM (2010) Sugarcane for bioenergy production: an assessment of yield and regulation of sucrose content. Plant Biotechnol J 8:263–276

Walker DIT (1987) Manipulating the genetic base of sugarcane. In: Proceedings of Copersucar's international sugarcane workshop, Sao Polo. Brazil, pp 321–334

Wang L, Jackson PA, Lu X, Fan YH, Foreman JW, Chen XK, Deng HH, Fu C, Ma L, Aitken KS (2008) Evaluation of sugarcane x *Saccharum spontaneum* progeny for biomass composition and yield components. Crop Sci 48:951–961

Waqaniu-Rogers A (1986) Some observations on duruka, Saccharum edule, in VitiLevu, Fiji. J Polynesian Soc 95:475–478

Warner JN, Grassl CO (1958) The 1957 sugarcane expedition to Melanesia. Hawaiian Planters Rec 55:209–236

Wu ZY, Raven PH (1994) Flora of China (English edition), vol 22, pp 571, 576, 589, Saccharum arundinaceum. http://www.efloras.org/florataxon.aspx.lora_id=2&taxon_id=200026230

Wu R, Lin X, He J, Min A, Pang L, Wang Y, Lin Y, Zhang Y, He W, Li M, Zhang Y, Luo Y, Wang X, Tang H, Chen Q (2023) Hexokinase1: a glucose sensor involved in drought stress response and sugar metabolism depending on its kinase activity in strawberry. Front Plant Sci 14:1069830. https://doi.org/10.3389/fpls.2023.1069830

Xu Y, Zhang X, Li H, Zheng H, Zhang J, Olsen MS et al (2022) Smart breeding driven by big data, artificial intelligence, and integrated genomic-enviromic prediction. Mol Plant 15:1664–1695. https://doi.org/10.1016/j.molp.2022.09.001

Zaidi SS, Mahfouz MM, Mansoor S (2017) CRISPR-Cpf1: a new tool for plant genome editing. Trends Plant Sci 22:550–553. https://doi.org/10.1016/j.tplants.2017.05.001

Zale J, Jung JH, Kim JY, Pathak B, Karan R, Liu H et al (2016) Metabolic engineering of sugarcane to accumulate energy-dense triacylglycerols in vegetative biomass. Plant Biotechnol J 14:661–669

Zhang C, Jiang S, Tian Y, Dong X, Xiao J, Lu Y et al (2023) Smart breeding driven by advances in sequencing technology. Mod Agric 1:43–56. https://doi.org/10.1002/moda.8

Technological Interventions in the Production of Sugarcane Juice Concentrates

3

Giriyapura Shivalingamurthy Suresha, Hari Kuppusamy, Murali Palanichamy, R. Lavanya, Kalai Selvi, Gopalareddy Krishnappa, R. T. Maruthi, Huskur Kumaraswamy Mahadeva Swamy, and Amaresh

Abstract

India is producing surplus amounts of sugarcane and sugar. Over the past few years, India has produced more than 300 million tonnes of sugarcane from about five million ha of land with an average yield of around 75 t/ha and a sugar recovery rate of around 10.0%. India requires 25.0 MT of sugar every year for its domestic consumption. In recent years, India has produced nearly 10.0 million tonnes of excess sugar, which needs to be diverted for the protection of the livelihood of sugarcane farmers, safety of sugar mills and stability of the sugar market. Therefore, diversifying cane juice is inevitable and product diversification is very much needed for efficient use of the produce and to generate more income. Sugarcane juice, as an energy drink with health benefits, offers tremendous opportunities for value addition and product diversification. Hence, it is highly important to explore ways of converting sugarcane juice into new value-added products by employing advanced juice concentration technologies. Therefore, this chapter aims to review the recent advances in sugarcane juice concentrates and the new processes for converting the juice into value-added products. Technologies for the production of new concentrated products from sugarcane juice, namely sugarcane juice powder using spray drying and freeze-drying technologies, jams, jellies, gummies, candies and jaggery and its formulations, and advances in membrane filtration research for concentrating sugarcane juice have

G. S. Suresha (✉) · H. Kuppusamy
Crop Production Division, ICAR-Sugarcane Breeding Institute, Coimbatore, Tamil Nadu, India

M. Palanichamy · R. Lavanya · K. Selvi
ICAR-Sugarcane Breeding Institute, Coimbatore, Tamil Nadu, India

G. Krishnappa · R. T. Maruthi · H. K. M. Swamy · Amaresh
Crop Improvement Division, ICAR-Sugarcane Breeding Institute, Coimbatore, Tamil Nadu, India

G. S. Suresha et al. (eds.), *Value Addition and Product Diversification in Sugarcane*, https://doi.org/10.1007/978-981-97-7228-5_3

been discussed in this chapter. This chapter also discusses optimization of the processing parameters for developing these products. Nutritional benefits and commercial applications of sugarcane juice-based value-added products have been thoroughly reviewed in this chapter. This would help the readers and stakeholders enrich their knowledge of sugarcane juice concentrates. This would also provide enormous opportunities for young entrepreneurs and the food processing sector to explore the potential of sugarcane juice for commercial applications.

Keywords

Value addition · Sugarcane juice · Nutrition · Concentrates · Food products

3.1 Introduction

Sugarcane is one of the most resourceful crops in terms of harnessing solar energy. It is the main source of sugar and an important cash crop around the world. India is the second-largest sugar producer in the world, after Brazil. Apart from sugar, sugarcane can produce various commercially important products like biofuels, fibres, manure and a slew of co-products that have ecological value. Sugarcane juice is used to make white sugar, brown sugar (*khandsari*) and *gur* (Shrivastava and Singh 2020). Sugarcane is known for its high sucrose content, which makes it a primary producer of sugar worldwide. Sucrose, a disaccharide, gives sugarcane its characteristic sweetness. The sucrose content in sugarcane juice ranges from 10% to 20% or more, depending on the variety of the crop and the growing conditions (Shanthi et al. 2022). Sugarcane is the direct source of sucrose, jaggery and a range of by-products, which provides wide opportunities for production of value-added products. Besides this, sugar industry by-products are capable of providing energy as a renewable source. The sugar industry therefore occupies a predominant place in many countries where sugarcane is under cultivation. Out of 300 million tonnes of sugarcane produced in our country, about 60% is processed by sugar factories for the production of white sugar and the rest goes for preparation of different forms of jaggery, feed, seed and chewing. Sugarcane varieties cultivated in India contain around 20% of total soluble solids in the juice at the age of about 12 months, of which a major part is sugar and the rest are minerals and health promoting constituents. Thus, sugarcane juice is a refreshing natural energy drink (Kaavya et al. 2019). The colour of the juice varies with the sugarcane cultivar and peel colour. In India, the sugar industry is the largest among the processing industries, next only to the cotton textile and agro-processing industries. In India, the livelihoods of nearly five million sugarcane growers and about 0.35 million workers are closely associated with the sugar industry and sugarcane crop. Therefore, in the present socio-economic context, it is imperative that the maximum benefit from sugarcane is accrued to all sectors of the sugar industry (Solomon 2016). This is possible through better utilization of sugarcane in the food and nutraceutical industries for value addition. Value-added products, especially modified, new and novel products from sugarcane, can

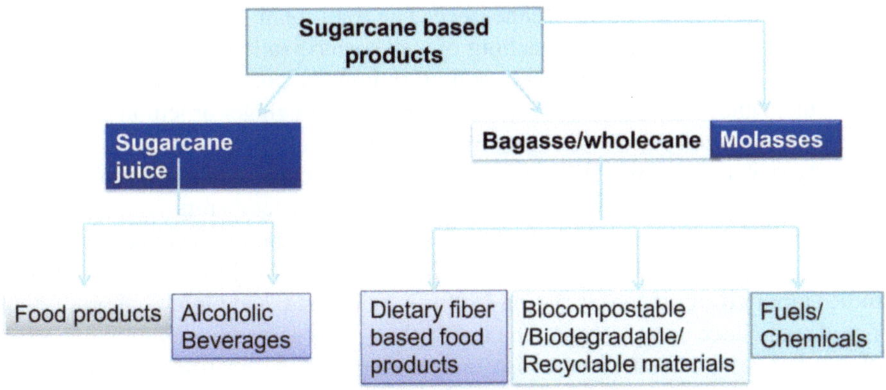

Fig. 3.1 An overview of product diversification in sugarcane

play a significant role in bridging the food security and nutritional safety gaps of the country. Product diversification from sugarcane through value addition would provide scope for sugarcane growers to increase their revenue in a sustainable way (Suresha et al. 2023). The excess cane produced can be utilized for the development of novel products with high nutritional value to cater to the nutritional needs of our people. This will also result in generation of employment in rural areas. In the sugarcane value chain system, sugarcane juice, bagasse and molasses are used as potential raw materials to produce value-added products (Fig. 3.1). Therefore, in this chapter, we have reviewed and discussed the recent research advances in sugarcane juice concentrates and their processes, technologies and value-added products.

3.2 Sugarcane Juice as Beverage

Sugarcane juice is a high-energy drink that is natural, sweet and a healthy alternative to refined sugar-added drinks (Arif et al. 2019). Besides sucrose, sugarcane juice contains suspended fine particles of bagasse, waxy substances and other elements (Kaavya et al. 2019). Sugarcane juice is acidic in nature with a pH range of 4.5–5.5. Another major component of this juice is colloids which are particles existing in a permanent state of dispersion imparting turbidity to the juice. These colloidal substances include waxes, proteins, pentosans, pectic substances, natural gums, starch and silica. In India, ginger and lemon are added to sugarcane juice to enhance its flavour and taste, and this is sold as an energizing drink in shops and street sideways. Cane is directly fed to crushers and the extracted juice is consumed fresh. However, if the juice is kept for a few hours, it tends to deteriorate quickly. This poses a major research problem in the area of sugarcane juice preservation, which is yet to be established. Also, sugarcane juice cannot be stored under refrigerated conditions for a long time. Therefore, it would be very difficult to maintain the fresh sugarcane juice without adding preservatives. Attempts have been made to preserve the sugarcane juice. The traditional sugarcane juice preservation methods followed are blanching of stalks, sterilization by heat followed by addition of chemical preservatives such as sodium or potassium metabisulphite, benzoates, vitamin C, etc. The juice is value added with ginger or lemon extract for enhancing taste and flavour. Bottling of juice is being done in glass containers followed by sterilization. When the juice was sterilized at 80–95 °C for variable duration, maximum shelf life was obtained at sterilization temperature of 90 °C for 5 minutes (Karmakar et al. 2011). Some reports used gamma irradiation to preserve the juice for extended periods (Oliveira et al. 2007; Mishra et al. 2011). Many studies reported significant variation in physico-chemical and biological properties of sugarcane juice upon storage. The major changes observed were fall in pH, total soluble solids and total sugars, rise in reducing sugars and titrable acidity and changes in sensory qualities on storage (Kapur et al. 1978; Mann and Singh 1988; Rao 1990; Bhupinder et al. 1991; Chauhan et al. 2002; Silva et al. 2006; Chun et al. 2007). In general, sterilized glass bottles are employed for packing juice because tinplate cans are inferior as juice sensory quality becomes unsatisfactory on storage (Alvarez et al. 1997). Recent reports on sugarcane juice preservation suggest that the juice can be suitably bottled and stored for more than 6–8 weeks appropriate for small scale businesses. For bottling, sugarcane juice is heated to 70 °C for 15 min, filtered, followed by the addition of sodium metabisulphate at 1 g/8 l juice, poured into sterilized bottles and sealed (Dhansu et al. 2023). Current developments in food technology have paved the way for preserving liquid food items with or without adding preservatives. Panigrahi et al. (2021) suggested five major categories of technologies: use of preservatives; use of clarifying agents; non-thermal processing techniques; thermal techniques; and hurdle technology. These technologies would help the processing industries to store and sell fresh sugarcane juice in bottles. One such technology is high pressure processing (HPP). Using this technology, liquid food products such as fruit juices are subjected to an extremely high pressure for a short time, resulting in juice with a high shelf life. This is a non-thermal process that helps preserve the colour, flavour and fresh attributes of the juice, resulting in a product that tastes similar to freshly squeezed juice. A recent study has revealed that the shelf life of sugarcane juice can be enhanced using HPP technology. Sreedevi and Rao (2018) tested the efficacy of HPP in sugarcane juice within a pressure range of 300–600 MPa at ambient temperature of around 30 °C for an incubation time of 1 s to 25 min. Their study revealed alterations in the various characteristics of the juice, such as pH, total soluble solids (TSSs) and titratable acidity. However, there was significant difference in the colour of the juice and reduction in the bacterial load with a rise in pressure and incubation period. The authors concluded that processing conditions such as 600 MPa pressure and a 20 min treatment duration were optimum to kill germs with negligible variations in the physical and chemical features of the juice. However, the cost of establishing such manufacturing units is high which would render the processed juice more expensive than fresh sugarcane juice. Nevertheless, such technologies will find an important place in juice preservation in future.

3.3 Nutritional and Nutraceutical Properties of Sugarcane Juice

Sugarcane juice is considered a medicinal drink due to its rich nutritional composition which was well mentioned in Ayurveda and Unani systems of medicine (Nadkarni 1976; Kirtikar and Basu 1991; Kaushik and Dhiman 1999; Sammbamurthi 2006; Warrier et al. 2010). Sugarcane contains various phenolic compounds, and its extracts have displayed a wide range of biological activities including antioxidant (Zamir et al. 2012), immune stimulation (El-Abasy et al. 2002), anti-inflammatory, antithrombosis and anti-stress effects (Barocci et al. 1999). Sugarcane juice is a high-energy drink that is natural, sweet and is a healthy alternative to refined sugar-added drinks. There are many studies reported on the composition of sugarcane juice which revealed that it contains around 70% of water and remaining are sugars, vitamins, minerals and organic acids (Longvah et al. 2017). Apart from these, sugarcane juice also contains carotenoids, fatty acids, flavonoids, phenolic acids and polyphenols (Panigrahi et al. 2021). Due its high mineral content, it helps the body to prevent dehydration (Yadav and Solomon 2006). Sugarcane juice contains the phytochemicals 5-hydroxymethylfurfural and cyclopropyl 4-methoxyphenyl ketone which help to combat various diseases like tumour, burning sensation during urination, symptoms of jaundice like irregular haem breakdown, liver problems, bile duct blockage, anuresis condition due to lack of urine production, damage of blood vessels resulting in haemorrhage and heart related problems (Karthikeyan and Samipillai 2010; Panigrahi et al. 2021; Rajendran et al. 2017). Functional role of bioactive compounds present in sugarcane juice was also studied to identify the various mechanisms of pharmacological actions like anti-tumour, anti-asthmatic, anti-arthritic, antioxidant, anti-cancer, anti-inflammatory, anti-microbial and anti-diabetic properties (Duke 2017). Sugarcane juice also helps diabetic patients due to its low glycaemic index (Khare et al. 2012).

Recently, interest towards nutraceuticals with the help of latest analytical techniques lead to identification of health-promoting substances and substances with medicinal properties such as antioxidants, polyphenols, pigments, flavonoids, omega 3 fatty acids, phytosterols and other high molecular weight fatty acids in plant products (Puri et al. 2022). Sugarcane juice is a healthy component of human diet, serves as natural hydrating agent due to its water content and electrolytes like potassium. According to studies, the plant polyphenols in sugarcane juice have a high antioxidant and anti-cancer activity (Prakash et al. 2021). During sugarcane crushing, large portion of powdery waxy substances are detached and mixed with the expressed juice. Although wax compositions of sugarcane plant parts varied significantly, C28 (about 81%) was the main component in all the sugarcane samples (Asikin 2008). Research studies indicated that policosanols in sugarcane wax lowers cholesterol levels in the human blood by inhibiting cholesterol biosynthesis and enhancing LDL decatabolism. Thus, these additives generally referred as phytosterols were used in functional foods for prevention of cardiovascular disease (Menendez et al. 1994; Varady et al. 2003; Irmak et al. 2006; Fernandes and Cabral 2007). The presence of flavonoids and antioxidant compounds in sugarcane juice has been well studied. Colombo et al. (2006) identified flavonoid group of polyphenolic compounds and reported that sugarcane juice had total flavonoids content of 0.6 mg/mL which was comparatively higher than other food products. Two novel phenylpropanoid glucosides, 3-O- (β-D-glucopyranosyl) -1-(3′,5′-dimethoxy −4′-hydroxyphenyl) -1-propanone and 3-O-(6”-O-α-D-glucopyranosyl-β-D-glucopyranosyl)-1-(3′,5′-dimethoxy-4′-hydroxyphenyl)-1-propanone and phenolic compounds, namely syringic acid-1-O-β-D-glucose ester and 6-β-D-glucopyranosyl-8-α-L-arabino pyranosyl apigenin were reported in sugarcane juice (Naranjo et al. 1999; Tamaki et al. 2001). Mauricio Duarte-Almeida et al. (2006) and Duarte-Almeida et al. (2011) identified flavones (apigenin, luteolin and tricin derivatives) and phenolic acids (hydroxycinnamic, caffeic and sinapic acids) from sugarcane juice. The phenolic compounds fraction extracted from sugarcane juice exhibited defensive role against methyl mercury poisoning and suppressed lipid peroxidation under in vivo conditions, which indicated a potential use for beneficial health effects and/or therapeutic applications (Kadam et al. 2008). Sugarcane juice, syrup and molasses had high antioxidant activity (Saska and Chou 2002; Guimaraes et al. 2005; Saska and Chou 2006). Vila et al. (2008) reported that the content of sugarcane juice flavonoids (0.241 ± 0.001 mg total flavonoids/ml juice) was comparable to other food sources of flavonoids. Thus, sugarcane juice has been suggested as potential dietary source of natural antioxidants. In India, sugarcane juice has been recommended to cure jaundice and liver-related disorders (Kadam et al. 2008). It is reported that flavonoids present in the sugarcane juice helps the body cells to overcome from degenerative processes and protect from cancer and cardiovascular diseases (Hudson et al. 2000; Hollman 2001).

3.4 Microbiological Quality of Sugarcane Juice

Sugarcane juice as such is highly nutritious and prone to spoilage due to yeast and bacteria. Even within few hours of extraction of the juice, fermentation sets in and the juice loses its natural refreshing properties. The juice quality is considerably altered during post-harvest storage of sugarcane (Uppal and Sharma 1999; Uppal 2003). Among different sugarcane varieties in cultivation, some varieties give good quality juice and hence are suitable as beverage (Yusof et al. 2000; Verruma-Bernardi et al. 2007). Fresh sugarcane juice deteriorated in a day when stored at 27 ± 1 °C. Microbial count, especially lactic acid bacteria and yeasts increased during storage of cane juice. However, juice stored at 10 °C did not suffer deterioration for up to a week and when stored at −20 °C in a frozen state, the juice lasted for 10–12 months without spoilage (Yusof et al. 2000). The most common way of extracting sugarcane juice is by crushing the canes through motor-powered crushers which are specifically manufactured for sugarcane juice extraction. However, this process is not hygienic and highly prone for contamination. This is a major problem with respect to road side vending of sugarcane juice. The sugarcane juice vended by the road side sellers in Brazil failed to meet the Brazilian sanitary norms and reported microbial and coliform contaminations (Hoffmann et al. 2006; Oliveira et al. 2008). Lopes et al. (2006) reported inadequate hygienic practices despite satisfactory microbial levels in sugarcane juice sold by street vendors. Evaluation of sugarcane juice sold by road side vendors in India showed the presence of high microbial load and coliforms indicating unhygienic handling. Hazard Analysis Critical Control Point (HACCP) analysis showed that use of contaminated cane, water, ice, hands of personnel, filters and roller drum crusher contributed to the microbial load in the juice (Nagalakshmi and Reddy 1999; Kowsalya and Shyny 2005; Subbannayya et al. 2006).

3.5 Concentrated Liquid Forms of Sugarcane Juice

Recently, various concentrated forms of sugarcane juice products, viz., value-added jaggery, juice powder, syrup, jam, jelly and spread are gaining importance in the sugar-based food sector. Concentrating the juice increases the sugar content and reducing the water activity of the product resulting in inhibition of microbial growth and the spoilage (Jebitta et al. 2012). In general, sugarcane juice is evaporated by heating till it reaches appropriate consistency. Singh et al. (2002) reported a method for preparation of concentrated sugarcane juice in which citric acid and sodium benzoate was used as a preservative. The product prepared by this method had good quality and shelf life. However, physico-chemical features of the juice products did not display significant variation excluding discolouration with elevation in reducing sugars content. Ghazal et al. (2003) reported physical and chemical properties of syrup produced from sugarcane cultivars (from Egypt). Syrup produced from different cultivars were significantly different with respect to specific gravity, viscosity, colour absorbency, pH, purity, non-sucrose substances, reducing sugar, starch, total N, total ash, titratable acidity total solids, total sugars and moisture content.

3.5.1 Jaggery and its Products

Jaggery is the most common conventional sugarcane juice concentrate having different formulations, namely powdered jaggery, liquid jaggery, bucket jaggery, cube jaggery, granulated jaggery, jaggery balls and other value-added or fortified jaggery products like candy / Toffee (Anwar 2017). Jaggery is a natural traditional sweetener prepared mostly in rural areas. The juice extracted from the sugarcane is filtered and boiled to evaporate the water and then syrup is added to moulds after cooling. This is highly nutritious sweetener used for making traditional sweets (Verma and Narain 1990). Jaggery is known to comprise high amount of vital minerals compared to refined cane sugar (Ghosh et al. 1998). Jaggery is also called as non-centrifugal sugar which is well-known among the consumers since from long period for the last 3000 years (Zidan and Azlan 2022). Jaggery is the most important part of the food especially in the villages. It is a healthy sweetener rich in vitamins and minerals. Jaggery is the most demanded commodity when compared to white sugar due to its nutrients composition and health advantages. It can be value added with health-promoting medicinal and herbal extracts through biofortification (Rao and Singh 2022). As per the study, about 30% percent of total sugarcane production in the world is being diverted towards jaggery production. India is the largest producer and consumer of jaggery. India alone contributes about 70% of total jaggery production of the world (Singh et al. 2011; Jagannadha Rao et al. 2007). In India, major contribution of around 80–90% of total jaggery production come from the states Karnataka, Maharashtra, Tamil Nadu, Uttar Pradesh and Andhra Pradesh. About 53% of total sugarcane produced in India (about 300 MT) is used for white sugar production, remaining 47% is diverted to jaggery (36%), seed (8%) and for juice (3%). It is estimated that about 54 MT of sweeteners is required by 2020 AD which is mainly supplied through jaggery (40%) (Hirpara et al. 2020). It is reported that people in rural area consumes more of jaggery which is about 30% when compared other population (Madan et al. 2004). About 4.3 million tonnes of sugarcane is crushed for jaggery production in delta district in Southern India, where Cauvery River is the main source of irrigation for growing sugarcane. Presently, In India, about 25% of the sugarcane harvested is being used for jaggery making. Jaggery is healthy, nutritious and readily obtainable in all parts of the country. It is prepared at a pH of around 5.8–6.4. Compared to white sugar, jaggery contains many vitamins (Pro-vitamin, vitamin A, vitamin B1, vitamin B2, vitamin B5, vitamin B6, vitamin C, vitamin D2 and vitamin E) and minerals, such as calcium, phosphorus, magnesium and potassium. It is reported that hundred grams of jaggery have around 70–85% sucrose, 7–10 g of reducing sugars, 0.5–0.75 g fats, 0.35–0.45 g proteins and 0.5–1.0 g ash contents. Jaggery is also rich source for minerals like calcium, magnesium, potassium, phosphorus, sodium, iron, manganese, zinc, copper and chloride, and vitamins like A, B1, B2, B5, B6, C, D2, E and nicotinic acid (Shrivastava and Singh 2020). Singh et al. (2013) reported the mineral composition of jaggery, viz. calcium content of 40–100 mg, magnesium of 70–90 mg, potassium of 1056 mg, phosphorus of 2090 mg, sodium of 19–30 mg, iron of 10–13 mg, manganese of 0.2–0.5 mg, zinc of 0.2–0.4 mg, copper of 0.1–0.9 mg and chloride of 5.3 mg per 100 g of jaggery. They have also reported the composition of vitamins in jaggery with a vitamin A of 3.8 mg, vitamin B1 of 0.01 mg, vitamin B2 of 0.06 mg, vitamin B5 of 0.01 mg, vitamin B6 of 0.01 mg, vitamin C of 7.00 mg, vitamin D2 of 6.50 mg, vitamin E of 111.30 mg, vitamin PP of 7.00 mg and protein of 280 mg per 100 g of jaggery (Table 3.1).

Constant consumption of white sugar led to many health problems like obesity, diabetes, stomach problems, etc. Therefore, jaggery is preferred as sweetener over white sugar due to its intrinsic composition in minerals, healthy sugars and antioxidants. Japanese jaggery is popularly called as Kokuto known to contain antioxidant compounds namely syringaresinol, medioresinol, coniferyl alcohol, sinapyl alcohol, and 3-hydroxy-1-(4-hydroxy-3,5 dimethoxy phenyl)-1 propanone (Nakasone et al. 1996). Jaggery prepared from sugarcane juice reported to have total phenolics

Table 3.1 Physico-chemical characteristics of jaggery and its formulations

Characteristics	Types of jaggery			Reference
	Solid	Liquid	Granular	
Physical characteristics				
• Moisture	3–10%	30–35%	1–2%	Singh (1998)
• Water activity (a_w)	0.45–0.59	0.733	0.36–0.70	Rajendran et al. (2021)
• Colour	Golden yellow to dark brown	Brown	Light to dark brown	Sahu et al. (2024)
• Texture	Firm	Liquid	Crystalline	Kumar et al. (2013)
Chemical characteristics				
• pH	6.75 ± 0.03	5.03 ± 0.15	6.60	Aralkar et al. (2023)
• Ash (%)	0.6–1.0	0.75	0.6–1.0	Said and Pradhan (2013)
• Energy (kcal)	383	300	383	Rao et al. (2007)
• Total sugar (%)	65–85	40–60	80–90	Rao et al. (2007)
• Reducing sugar (%)	10–15	10–15	5–9	Singh (1998); Rao et al. (2007)
Vitamins				
• Vitamin A	3.8 mg/100 g	3.8 mg/100 g	3.8 mg/100 g	Hirpara et al. (2020)
• Vitamin C	7.0 mg/100 g	7.0 mg/100 g	7.0 mg/100 g	Nisha et al. (2017)
• Vitamin E	111.3 mg/100 g	111.3 mg/100 g	111.3 mg/100 g	Shrivastava and Singh (2020); Singh et al. (2013)
Fat	0.1/100 g	0.1/100 g	0.1/100 g	Hirpara et al. (2020)
Protein	0.4 g/100 g	0.10/100 g	0.4 g/100 g	Kumar and Smita (2020)

content of 280–320 mg/100 g jaggery (Jaffe 2015). According to Indian system of Ayurveda and medicine, jaggery has many health-promoting activities such as blood cleansing, rheumatic illnesses prevention, bile associated ailments, and these benefits have been scripted (Sushruta Samhita, Chapter 45, Sloka 146; (Sharma et al. 2012). Jaggery has been principal source of sugar in traditional foods due to its taste and medicinal properties (Rao et al. 2007). Among the countries in the world, India, Srilanka, Pakistan, Myanmar, Africa, America and Caribbean are the major contributors for jaggery production. Different countries have different traditional names for jaggery, namely "Gur" in India, "Desi" in Pakistan, "Naam Taan Oi" in Thailand and "Hakura" in Srilanka (Kumar and Singh 2020; Nath et al. 2015). It is known that magnesium helps to improve our nerve problems, relieve strain in muscles, reduce fatigue problem and keep healthy blood vessels. Selenium present in the jaggery plays an important role as antioxidant and helps to overcome stress conditions through scavenging free radicals from our body. Presence of high concentration of potassium and sodium in the jaggery benefits to stabilize pH of cells and maintain redox potential thereby helped to control blood pressure. It was also reported that jaggery is rich in iron and helps to prevent anaemia. Jaggery was also reported to control migraine and also helped pregnant women to relieve from weakness (Sahu and Paul 1998; Ghosh et al. 1998; Singh et al. 2008, 2013).

The functional role of bioactive compounds like essential amino acids, minerals and vitamins present in the jaggery has given much scope for its use in functional foods and nutraceutical industry (Shahi 1999). Presence of high concentration of calcium in the jaggery helps for strengthening bones through jaggery-based sweets. Jaggery also comprised iron which improves anaemic condition especially in vegetarians and women. Jaggery has been studied to have anti-allergy properties, which supported to combat against stress and controled asthma-related health issues (Singh et al. 2008; Hirpara et al. 2020).

During the chemical processing of sugarcane juice for white sugar production, major nutrients present in the juice will be lost. Since the jaggery processing retains these nutrients, this would be the nutritive source of sugar over chemically processed refined sugar. The production of jaggery and refined sugar are completely separate processing technologies. In refined sugar production, the process involves high amount of nutrients loss due to the use of chemicals, whereas in jaggery production, it is not so. Because, processing condition is entirely natural with no addition of chemicals. This helps the product to retain most of the nutrients present in the sugarcane juice. In the ancient Indian Ayurvedic system, jaggery was cited to treat most of the diseases affecting the human body (Pandey 2010). It was reported that the presence of high amount of potassium in jaggery helps human body for digestion, weight loss, detoxification and relief from constipation (Tal et al. 2019; Ravisankar et al. 2016; Umate et al. 2021). It was also studied that due to the addition of ferrous salts from the iron vessel during processing, jaggery can be the source for iron and prevent anaemia (Resmi and Latheef 2016; Samson et al. 2022). People with regular health problems like cough, cold and asthma are practising household remedy by using jaggery which gives relief for throat infections (Ogawa et al. 2013). It was also studied that jaggery could be beneficial for lungs for those who are working in an air polluted area (Sahu and Saxena 1994; Sahu and Paul 1998). Magnesium present in the jaggery helped to relieve from muscle ache, nerve problems and obesity (Kumar and Singh 2020). Jaggery has been a good source of antioxidant and a blood purifier (Harish Nayaka et al. 2009; Singh and Singh 2008). Jaggery is being used as potential ingredient and source of sugar in ayurvedic tonics to cure many diseases (Sekar and Mariappan 2008; Singh et al. 2013). Jaggery was also recommended for improving immunity against common cold and flu (Deotale et al. 2019) (Table 3.2).

Hussain et al. (2012) evaluated jaggery characteristics such as colour, flavour and shelf life. During jaggery processing, sulphur dioxide (SO_2) is added as juice clarificant to get golden colour which is dangerous for health. Therefore, it has been studied that light colour jaggery was an adulterated one when compared to dark colour

Table 3.2 Health benefits of jaggery and its products

Health benefits	Effect	Reference
Antibacterial	Antibacterial, antioxidant and anti-microbial property	Barrera et al. (2020)
Antioxidant	Free radical scavenging activity and protection of DNA damage	Harish Nayaka et al. (2009)
Cytoprotective	Cytoprotective activity	Harish Nayaka et al. (2009)
Prevent anaemia	Improves anaemic condition	Resmi and Latheef (2016), Samson et al. (2022)
Cardioprotective effect	Cardioprotective effect, reducing total fat levels, low-density lipoprotein cholesterol and triglycerides	Hudson et al. (2000), Hollman (2001)

jaggery (Uppal et al. 2005). Another study reported that granular jaggery had high amount of minerals (0.6–1%) with a calcium content of 9 mg, phosphorous of 4 mg and iron of 12 mg (Singh et al. 1978). Jaggery is known to have complex nutrients than sugar. Jaggery contains nutrients such as minerals and fibre which helps in slower digestion and gradual release of sugar to the blood (Patil et al. 1985). Jaggery is most commonly used for making many traditional sweet dishes, such as payasam, obbattu (holige), Hurakki Holige stuffing, Karadantu, Madeli, Godhi huggi, Gugagana huggi, Allittu and unday (Inamdar et al. 2005). The most common bakery product is peanut and chickpea chikki which is made from jaggery (Tidke et al. 2017). There are many value-added products are being made using jaggery as sweetener namely chocolate vinegar, gazak, reori, chikki and ramdana (Shahi 1999). There are reports which studied the role of jaggery in the prevention of lung damage (Sahu and Saxena 1994). Since the Sucrose is the major sugar in jaggery, it is advantageous for diabetic patients for using it as sweetener over starch-based sugar sources. Due to its lower glycaemic index than glucose, it adds better processing properties for consistency, colour, palatability and shelf life (De Souza et al. 2013).

The physico-chemical characteristics such as specific gravity, foaming, binding, coagulation, softness, beating temperature and highest thickness of muffins were improved with the addition of jaggery along with sugar (Sahi and Alava 2003; Agrahar-Murugkar et al. 2016; Lamdande et al. 2018). Another study on the physico-chemical characteristics of jaggery muffins revealed higher moisture content due to the inversion and higher moisture content in the jaggery (Mandal et al. 2006). Sugar content in the jaggery muffins was lower when compared to white sugar-added muffins. They also observed increase in ash content due to the presence of minerals in the jaggery added muffins over sugar-added muffins (Jagannadha Rao et al. 2007).

In fact, jaggery has a mineral content approximately 50 times greater than refined sugar and highly nutritious sweetener compared to refined sugar (Gopalan et al. 1991). Therefore, attractive products can be produced and popularized for higher consumption of jaggery-based products. This will improve the health of children rather than consuming white sugar-based products. In order to attract children and adults, jaggery has been produced in various shapes and sizes flavoring with different spices and other natural flavors. Many attractive traditional sweets, chocolates, toffee, confectionery and beverages can also be prepared using jaggery (Baboo et al. 1988). Most popular commercially available product is peanut chikki made by mixing fried ground nuts with jaggery. High-value chikki prepared using dry nuts, dry fruits and chocolate powder is gaining momentum. Preparation of variants by fortification with mineral, vitamins and medicinal substances can be prepared to address specific target areas to mitigate deficiency problems. Granulated version of jaggery can be prepared by dry extrusion and spheronization, which is less hygroscopic, attractive, free flowing, stable spoon able and has better shelf life than sugarcane syrup / liquid jaggery (Kuruba et al. 2020). However, liquid jaggery is having its own advantageous over solid jaggery. A modified process in which sugarcane juice is evaporated and brought into honey like consistency and used as liquid jaggery. At commercial level, sugarcane juice is concentrated using vacuum evaporators combined with heating to produce syrup like product. Liquid jaggery / syrup contain vitamins and minerals such as calcium, phosphorus, magnesium and potassium. Though, liquid jaggery is used as a regular sweetener in our country, but it is not being consumed by large population. However, value-added and fortified products can be prepared from jaggery instead of using white sugar and promoted to increase the level of consumption to tackle the nutritional security. Liquid jaggery or syrup can be directly used as spread on Indian tradition foods or on bread instead of synthetic jams and jellies available in the market (Rajendran et al. 2021).

3.6 New Approaches for Producing Concentrated Sugarcane Juice

Sugarcane juice is always susceptible to microbial deterioration and decomposition. To avoid this, preservatives are added in the sugarcane juice products. This will decrease the normal taste and is harmful to health. Sugarcane juice concentrate prepared by prolonged heating combined with clarification using chemicals eliminates essential nutrients and nutraceuticals. To preserve all the goodness, dehydration is the quick remedy. Also concentrating the juice is advantageous in terms of packing, storage and transport costs. If drying whole juice is carried out with less exposure to high temperatures, there is a possibility of obtaining powder/concentrated form of product retaining the natural qualities (Hari et al. 2013). Most recently, researchers have come with new processes to concentrate sugarcane juice, namely spray and freeze drying, reverse osmosis and vacuum concentration (Fig. 3.2). By employing drying processes, viz., freeze and spray drying, sugarcane juice in the form of powder or granulated product was prepared (Hari et al. 2013). Variants can be prepared by the addition of natural colour, flavour, spices, etc. The dried product preserved all the natural nutritional and nutraceutical qualities of sugarcane juice. All these processes require high investment costs, but it is possible to prepare dried product of sugarcane juice which has longer shelf life as the dehydrated product prevents deterioration due to microbial activity (Hari et al. 2013).

This dried product is similar to milk powder available in the market and can be reconstituted whenever required. Since these methods use whole sugarcane juice to produce a powder, these products are expected to retain all the nutrition and nutraceutical substances of juice which are essential for human health.

3.6.1 Spray Drying

Spray drying is the most common process being employed in food industry sector (Cano-Chauca et al. 2005). The processing parameters like drying air temperature and the feed rate are important in drying process to obtain desired colour, particle size, bulk density, moisture content, nutrient retention, average time of wettability and number of insoluble solids (Greenwald and King 1981; Welti and Lafuente 1983; Chegini and Ghobadian 2005; Fazaeli et al. 2012). Sugarcane juice contains about 70% water. To retain inherent physico-chemical characteristics like taste, flavour and other nutritional composition, it is very much required to reduce moisture and water activity of the juice to a maximum extent. Spray drying is one of the most commonly used methods in nutraceutical and pharmaceutical industries to produce dried products from various sources (Sagar and Suresh Kumar 2010). In this process, juice is subjected to less heating

Fig. 3.2 Physico-chemical characteristics and uses of novel sugarcane juice-based concentrates

environment under vacuum and the final dried powder produced is in the form of granules with maximum retention of original juice nutrients (Adhikari et al. 2004; Jebitta et al. 2012). In case of spray drying of fruit juices, stickiness was the major problem due to more hygroscopic effect. This was mainly due to the less glass transition temperature of the products and low molecular weight compounds like sugars and organic acids (Bhandari et al. 1993; Tonon et al. 2011). Similarly, process of spray drying of sugarcane juice had many challenges due to its composition of low molecular weight compounds like sugars and organic acids which are known to have low Tg. It is known that spray drying process is more efficient under higher drying temperatures (>100 °C). But it leads to low powder yield due to high adhesiveness of the product to inner wall of collecting chamber which is made up of glass. In order to increase the Tg of juice, high molecular weight compounds like maltodextrins, cellulose, arabica gum and waxy starch was added (Bhandari and Howes 1999; Truong et al. 2005; Shrestha et al. 2007). Similar problems exist with spray drying of sugarcane juice also. In general, spray dried products with proper packing are not prone for growth of microorganisms which would enhance the shelf life of the powdered product. Spray and drum drying techniques were suggested for production of dried sugarcane juice (Nigam and Raha 1982; Campo et al. 2004). Studies conducted at Sugarcane Breeding Institute, Coimbatore showed that sugarcane juice can be successfully dried to powder. Effect of spray drying parameters like inlet temperature and feed volume of juice was studied using laboratory capacity spray drying machine. The results from this study indicated that inlet temperature and feed volume were significantly affect powder yield. Other conditions used in the spray dryer were vacuum: 90 mmHg; spray pressure: 1.5 kg cm^2; juice Brix: 17.8–19.50; total soluble solids: 20.2–22.6% and maltodextrin: 20%. The obtained powder yield was ranged from 0 to 65.4. The highest powder yield of 187.75 g/l juice was attained at the inlet temperature of 140° or 160°C and at a feed rate of 100 ml/h. The minimum yield was obtained at 100°C–120°C temperature when the feed rate was at 100, 150, 200, 250 and 300 ml/h. The highest moisture content of 2.5% was obtained when the inlet temperature was at 100 °C. The lowest moisture content of 1% was obtained at 220°C and 100 ml/h pressure. But the powder formed was dark coloured with a jaggery flavour (Jebitta et al. 2012; Hari et al. 2013). It was reported that powder production was not efficient in spray drying raw sugarcane juice but was improved considerably by addition of bulking agents. Diaz et al. (2007) reported the production of dried juice powder from sugarcane juice was similar to raw sugar using Niro-atomizer spray drying technique by aspersion. The low efficiency was generally associated with low glass transition substances such as simple sugars and organic acids (Bhandari and Howes 1999). Bhandari et al. (1997) reported that adhesiveness was a main cause that restricts the concentration of various sweet food products through spray drying due to high sugar content. More hygroscopicity of powdered products, easily soluble nature of sugars due to increase in temperature and lesser melting point and glass transition temperature contributed to the stickiness problem. This problem was addressed by use of bulking agents. Bhandari et al.

(1993) reported spray drying of blackcurrant, apricot and raspberry juice into powder production using maltodextrin DE6 as additive. Similarly, orange juice powder was reported by Chegini and Ghobadian (2005). Though sugarcane juice contains high amount of low Tg substances, similar approaches can be tried to produce a desirable powder.

Unlike fruit juices, conversion of sugarcane juice to powdered form is really a challenging due its inherent composition. The powdered product obtained has many advantageous in terms of shelf life and also has huge commercial potential to use it as natural sweetener. This product can be used as ready to dilute powder and as a natural sweetener and flavouring agent in many bakery products. Unlike refined sugar, sugarcane juice powder retains natural flavour, colour, minerals and vitamins (Oliveira et al. 2007). When compared various fruit juice powdered products available in the market, not much information is available on the drying technology of sugarcane juice to powder. There are many challenges during the processing of fruits juice to powdered form. As per the published literature, bringing back to original juice consistency after the powder is reconstituted with water is a big challenge. Due to high sugar content especially sucrose, glucose and fructose in fruit juices make the drying process more difficult as the process poses problems of thermoplasticity and hygroscopicity at high temperatures (Bhandari et al. 1997; Adhikari et al. 2004; Cano-Chauca et al. 2005; Gabas et al. 2007; Bhandari et al. 1997; Cheuyglintase and Morison 2009). Optimization of physico-chemical characteristics like glass transition temperature (Tg), hygroscopy, melting point and water solubility is very much required to get desired product without stickiness after drying (Adhikari et al. 2003; Cheuyglintase and Morison 2009). These problems were fixed by using drying carriers like maltodextrin and gum arabica during drying process (Bhandari et al. 1993; Cano-Chauca et al. 2005; Gabas et al. 2007). The main reason for adding these drying carriers was to higher the Tg of powder which was achieved from these high Tg compounds (Shrestha et al. 2007). Maltodextrin is the well-known drying carrier being widely used in drying of fruit juices which has high water solubility (Cano-Chauca et al. 2005; Langrish et al. 2007).

Spray drying technique can be applied to sugarcane juice to convert into powder. However, it needs optimization of processing parameters and more importantly the benefit-cost ratio need to take into consideration. Drying carriers like maltodextrin and gum arabica need to be added in higher quantity to get desired powder for its commercial application. Hari et al. 2013 optimized the process for the production of spray dried sugarcane juice powder (SJP) using bench top spray dryer under laboratory conditions. Juice from 12-month-old sugarcane variety Co 99,006 was processed for spray drying and analysed for physico-chemical characteristics. At the beginning of the experiment, the yield of the powder was very low due to its adhesion to the dryer chamber wall. Further optimization of processing parameters like temperature, aspirator level, pressure and feed rate improved the yield of the SJP with <5.0% moisture content. Next level of optimization by using different bulking agents has significantly increased the SJP yield and moisture content of <4%, water activity of <0.34, bulk density of >0.5 g cm^{-3} and improved flow ability of <47° than SJP without bulking agents. Microbial analysis of the SJP revealed little bacterial load and complete non-appearance of other microbial contamination. Use of bulking agents in this process enhanced the glass transition (Tg) and decomposition temperature. However, addition of sucrose as bulking agent showed undesirable result on Tg and decomposition temperature (Hari et al. 2013). They have also analysed particle size, shape and surface characteristics through scanning electron microscope and noticed differences in all these characteristics. SJP formed spherical particles when they use maltodextrin and soluble starch as bulking agents but composition of the product unchanged by retaining significant amount of nutrients (Hari et al. 2013). It is estimated that the cost of production of sugarcane juice powder varied from 0.77 USD and 2.06 USD per kg of powder. This cost of production can be lowered by using preconcentrated juice through evaporator before drying process which has increased the powder yield (Khuenpet et al. 2016).

3.6.2 Freeze Drying

Evaporation is considered to be the most economical and widely used method of concentration. As compared to the conventional evaporation processes, concentration by freezing is potentially a superior and economic process for aroma-rich liquid foods. Freeze-drying is one of the best methods for producing high quality dried food products as compared to other drying technologies. Freeze-drying process reduces the moisture content in food products with low process temperature thereby reducing the deterioration and microbial reactions in foods. Therefore, both preservation and value addition of high-value foods are achieved by using this technology (Toledo 1999; Chen et al. 2000). The vacuum freeze-drying process involves two important systems, namely freezing system and freeze-drying system. Ideally, in two-component liquid systems, water and solute in any food will crystallize by freezing through removal of heat (Becker and Fricke 1999). In freeze drying system, water must be removed from the product by sublimation process which also involves mass and heat transfer (Sanaye and Malekmohammadi 2004). Deshpande et al. (1984a, b) reviewed several aspects of the theoretical considerations behind freeze concentration of fruit juices in pharmaceutical and high-value foods. Though this method is expensive, it is highly suited for aroma rich substances. Freeze drying of orange juice resulted hygroscopic product which tend to agglomerate into large solid lumps (Kramer et al. 1988). Various freeze-dried powdered products were reported, namely guava, mango puree, passion fruit and pineapple juice (Moy 2006); lime juice (Theansuwan et al. 2008); Garcinia fruit extract (Gurak et al. 2008); dragon fruit (Liaotrakoon et al. 2012); mushroom *Inonotus obliquus* polysaccharides (Ma LiShuai et al. 2013); red wine polyphenols (Sanchez et al. 2013) and yoghurt (Moghbel et al. 2013).

Due to its high sensitivity for enzymatic deterioration and very low shelf life, sugarcane juice as such cannot be stored in liquid form (Sankhla et al. 2012; Singh et al. 2014). Therefore, production of sugarcane juice powder from the juice is the best approach to use as fresh drink. Comparative analysis of sugarcane juice and powder for nutritional composition revealed 0.16% and 0.059% of acidity, 4.89 and

5.8 of pH, 5.39 mg/100 g and 4.29 mg/100 g vitamin C content, 15.17 mg/100 g and 11.06 mg/100 g calcium content, 25.36 mg/100 g and 22.57 mg/100 g potassium content, 152.44 mg/100 g and 127.66 mg/100 g phosphorus content, 112.97 mg/100 g and 9.44 mg/100 g iron content, respectively (Nishad et al. 2019). Freeze drying of sugarcane juice involves a sublimation process where freezing the juice followed by dehydration. In this process, ice is getting converted directly in to vapour form without passing through the liquid state (Harguindeguy and Fissore 2020). By following this technique, the product will have very less moisture content. Freeze drying of sugarcane juice retained maximum nutrients along with flavour and colour. The shelf life of the juice powder also increased due to less moisture content and low microbial activity (Acevedo et al. 2014). Dal Magro et al. 2019 reported the process of making sugarcane juice powder using freeze drying and evaluated the various physico-chemical characteristics of the powder. They have also reported that powdered product had good solubility with a water activity of 0.49 which reduced the microbial contamination significantly and increased the shelf life of the product.

Advantageous of using freeze-dried sugarcane juice powder is due to its longer shelf life than fresh sugarcane juice or even regular sugarcane juice powder. It is also having added benefit of being in powdered form, making it easier to measure and incorporate into recipes. Overall, freeze-dried sugarcane juice powder is a cost-effective and long-lasting way to incorporate the natural sweetness and flavour of sugarcane juice into a variety of culinary creation (Dal Magro et al. 2019).

3.6.3 Reverse Osmosis

Evaporation is considered to be the most economical and most widely used method of concentration. However, it is not suited for food products with delicate flavours. To overcome this problem, reverse osmosis using membranes could be employed. Membrane technology for the processing of fruit juices and beverages has been applied mainly for clarification using ultrafiltration and microfiltration and for concentration using reverse osmosis (RO). Though commercial processes for the concentration of such products by membrane separation techniques are not yet available, recent developments in membrane technology offer a variety of membranes for concentrating liquid substances. Due to this, many attempts have been reported recently in concentrating juices using RO membranes. Girarda and Fukumotoa (2000) reported that for processing fruit juices, product preparation, membrane selection and operating parameters are important factors influencing filtration rate and product quality. Akin et al. (2012) have suggested that membrane separation technologies provide unique advantages offering great potential for the nutraceutical and functional food industry. Medina and Garcia 1988 reported a successful RO process for orange juice concentration up to 50% hydraulic recovery with a polyamide membrane. Permeate fluxes and solute recoveries were determined at transmembrane pressures of 6.21 and 4.14 MPa. Pectinase treatment was required to prevent fouling and to allow cleaning in place procedures. Enzymatic treatment did not affect permeate flux or solute recoveries. Overall recovery of sugars, organic acids and flavour-volatile components was approximately 93%. Similar studies were reported for pineapple juice (Bowden and Isaccs 1989; Couto et al. 2011), apple juice (Sheu and Wiley 1984; Chua et al. 1987), orange juice (Jesus et al. 2007), mandarin juice (Fukutani and Ogawa 1983), tomato juice (Yildiz et al. 1993), grape fruit and lemon juices (Braddock et al. 1988), grape juice (Gurak et al. 2008), carrot juice (Di Giacomo and Taglieri 2009), passion fruit (Sipoli and Barros 2011) and peach, pear and mandarin (Echavarria et al. 2012). Major problem in this process is blocking of the RO membranes due to the high molecular substances present in fruit juices and hence prefiltering is required. Fouling of RO membranes is a common problem when fruit juices are used which require frequent cleaning for efficient RO process (Madaeni et al. 2011).

Similar to fruit juices, sugarcane juice was also concentrated using various membrane technologies. However, evaporation method by heating poses problems. Heating the sugarcane juice decomposes the sugar and other substances resulting in low-quality and dark-coloured product. To overcome this, Madaeni et al. (2004) have reported a novel method of making syrup without heating the juice by employing low-energy-consuming reverse osmosis process. In this study, different commercial reverse osmosis membranes were used and successfully concentrated sugarcane and sugar beet juice into sugar syrup. Jegatheesan et al. (2009) explored different pore sizes for concentrating the sugarcane juice using lime. The results from this study revealed that membrane with pore size of 0.05 μm was better and given higher yield with good quality filtered juice and maintained average flux. Clarifying the cane juice using membranes significantly decreased the turbidity levels, colour and starch content in the juice but purity of the juice has increased by 1.4% (Sim et al. 2009). Shalini and Nayak (2016) compared the sugarcane juice which was concentrated by thermal and forward osmosis membrane processes. At 0.79 L/m^2 h transmembrane flux, they achieved two-fold concentration of sugarcane juice from 17.6 to 31.7 °Brix. From this study, it is more evident that forward osmosis process is the most energy efficient process to clarify sugarcane juice to syrup. Akhtar et al. (2021) studied the concentration of sugarcane juice through aquapor in hollow fibre forward membrane and achieved 53–60% sugar concentration in 12 min time with specific reverse solute flux of 0.684 g/l after 12 min reflux.

3.6.4 Jam and Jellies

Jam is a well-known food product widely available in the market. Most of the jam products which are sold in the retail shops are basically made from the different fruits. But process for preparation of cane jam is completely a different method and involves many unique steps to overcome the difficulties to get desired product. The major problem of using sugarcane juice for preparation of jam is its vulnerability to quick deterioration and short shelf life. This process is called as browning due to the activity of two major enzymes, namely invertases and polyphenol oxidases present in the juice (Suresha et al. 2017). Due to the high activity of these enzymes upon crushing of canes, colour of the juice gets turned to brown within the 30 min. This type deterioration is called as inversion where sucrose is enzymatically converted into glucose and fructose which in turn undergoes oxidative reactions to convert the juice colour and quality (Shivalingamurthy et al. 2018). Polyphenol oxidase of sugarcane juice involved in wound inducible oxidation of phenolic compounds and responsible for browning in sugarcane juice (Suresha et al. 2017). Therefore, production of sugarcane juice-based novel value-added products is challenging process in order to keep the quality of the product unchanged with longer shelf life. For the first time, ICAR-Sugarcane Breeding Institute has come up with a novel process for preparation of sugarcane juice-based jam. This process has many inventive steps like process for removing the impurities using indigenous filtration step and prevention of juice browning, precise regulation of boiling temperature, optimizing the brix concentration at each stage of process, concentration of thickeners and process for prevention of clump formation, optimizing threshold point for final setting point, process for stopping crystallization and jaggery flavour in the final product (Suresha et al. 2021; Lavanya et al. 2023). Sugarcane juice jam has been trademark registered as CANEJAM® by ICAR-SBI for commercial production. The product is exclusively made from the sugarcane juice which has natural characteristic of cane flavour. Sugarcane juice jam can be readily mixed with other natural flavours or plant extracts to produce various fortified and flavoured jam products without adding refined sugar. The pulp of fruits like mango, pineapple, apple, watermelon, strawberry, papaya, grapes, etc., and spices like cinnamon, ginger, cardamom, chilli, etc., can be added to cane jam to produce different flavoured products. Sugarcane juice jam is a distinct product when compared to fruits jam. This can be used readily in our regular food. Currently, jam prepared exclusively from sugarcane juice does not exist in food market. This product has enormous applications in food industries similar to other jams. Comparative analysis of sugarcane juice jam with fruit pulp-based jam revealed that cane jam has many advantageous over fruit jam due to its nutritional composition. Sugarcane juice jam was reported to contain less sugars (45%) when compared to fruits jams which are having >50%. This is mainly due to the non-addition of refined sugar during the process (Suresha et al. 2021). Sugarcane juice is rich in minerals, which is carried forward to the products prepared from the juice. Our report on the mineral composition of cane jam samples revealed high concentration of minerals such as potassium, sodium, calcium and magnesium. It also contains high concentration of vitamins, polyphenols and flavonoids. Routine jam making process using fruit pulp involves use of equal proportion of fruit pulp and table sugar (Awulachew 2021). But sugarcane juice-based jam production does not use any chemically processed or synthesized sugars and sweeteners except natural source of sugar from sugarcane juice. Study of economics of cane jam making revealed that jam production from sugarcane juice is having vast potential for alternate revenue generation. If a farmer divert sugarcane for jam production, he may get eight times more income which is nearly Rs. 25,000/-per tonne when compared to minimum support price fixed by the government which is nearly Rs 3000/-per tonne of sugarcane. This is the best example for Government of India's vision for doubling farmer's income. Sugarcane juice jam has wide applications as natural sweeteners. It can be used in daily diet starting from morning breakfast with breads and dosa and meals with chapati and roti during lunch and dinner. Cane jam can also be used in topping of cakes, donuts and other bakery products. It can also be used in sauce, fruit salads and other various food products as sweetener (Suresha et al. 2021).

Jellies and gummies made from cane juice with the addition of certain ingredients makes it a value-added product. Since jelly known to have high amount of water, it helps for digestion and easy movement of food along the digestive tract. Gelatine also has the similar function in ensuring the smooth movements in the intestinal muscles, which lead to advance the digestion process and retain nutrients in the body. The addition of gelatin in the right proportion and the quantity of sugarcane juice contributes to the final product. The study revealed that jelly prepared with 40 g corn starch, 0.5 g citric acid and 15 g agar and 0.5 g citric acid in 1000 ml of sugarcane juice and 10 g of agar along with 0.25 g citric acid in 500 ml of sugarcane juice was found to be more stable and good for consumption. The standardized jelly and gummy products were subjected to various physio-chemical, microbial and sensory analysis and found to have an excellent shelf life and stability (Liju et al. 2023).

3.7 Conclusions

Although sugarcane is known as commercial crop, area under sugarcane is gradually decreasing. Farmers are apprehensive that cultivating sugarcane may not be profitable in the near future. Experience with sugarcane famers indicated that they are desperately in the need of other mitigating avenues for getting an interim source of income rather than waiting for a full year for sugarcane harvest. Therefore, product diversification from sugarcane through value addition would provide a great opportunity for sugarcane growers to increase their revenue in sustainable way for doubling farmers' income. In order to create awareness and educate end users about the importance of value addition and its new technologies and products, this review has meticulously discussed on the new niche products of sugarcane juice concentrates, nutritional composition, health benefits and provide new avenues for generating alternate source of income through value-added products of sugarcane. This would also help create significant amount of employment in secondary and service sector of urban and rural communities. Therefore, value addition is a new way for farmers, rural youth and rural women not only to increase their income but also compete with value-added agricultural products in the food basket.

References

Acevedo BA, Chaves MG, Avanza MV, Dellacassa ES (2014) Freeze-drying concentration of Rangpur lime juice. Int J Food Sci Technol 49:423–428

Adhikari B, Bhandari BR, Howes T, Troung V (2003) Characterization of the surface stickiness of fructoseemaltodextrin solutions during drying. Dry Technol 21:17–34

Adhikari B, Howes T, Bhandari BR, Troung V (2004) Effect of addition of maltodextrin on drying kinetics and stickiness of sugar and acid—rich foods during convective drying: experiments and modelling. J Food Eng 62:53–68

Agrahar-Murugkar D, Zaidi A, Kotwaliwale N, Gupta C (2016) Effect of egg-replacer and composite flour on physical properties, color, texture and rheology, nutritional and sensory profile of cakes. J Food Qual 39:425–435

Akhtar A, Singh M, Subbiah S, Mohanty K (2021) Sugarcane juice concentration using a novel aquaporin hollow fiber forward osmosis membrane. Food Bioprod Process 126:195–206

Akin O, Temelli F, Koseoglu S (2012) Membrane applications in functional foods and nutraceuticals. Crit Rev Food Sci Nutr 52(4):347–371. https://doi.org/10.1080/10408398.2010.500240

Alvarez IL, Marti BS, Utset E, Hernandez GM (1997) Analysis of the sensory characteristics of natural cane juice in three types of packages. Alimentaria 35:73–76

Anwar SI (2017) Development of modified jaggery moulding framefor cubical shape jaggery. Agric Eng Today 41(1):44–47

Aralkar S, Kshirsagar R, Lande V, Agarkar B (2023) Exploring the nutritive sweetener liquid jaggery: physicochemical characteristics and its application. Pharma Innov J SP-12(12):388–391

Arif S, Batool A, Nazir W, Khan RS, Khalid N (2019) 8—Physiochemical characteristics nutritional properties and health benefits of sugarcane juice. In: Grumezescu AM, Holban AM (eds) Non-alcoholic beverages. Woodhead Publishing, pp 227–257., ISBN 9780128152706. https://doi.org/10.1016/B978-0-12-815270-6.00008-6

Asikin Y (2008) Waxes, policosanols and aldehydes in sugarcane (Saccharum officinarum L.) and okinawan brown sugar (kokuto) Bogor Agricultural University, Master of Science Thesis Department of Food Science and Technology

Awulachew M (2021) Fruit jam production. Int J Food Sci Nutr Dietetics 10(4):532–537

Baboo B, Zahoor MA, Garg SK (1988) Design and development of gur moulding frame. Indian J Sugarcane Technol 5(2):89–92

Barocci S, Re L, Capotani C, Vivani C, Ricci M, Rinaldi L (1999) Effects of some extracts on the acetyl-choline release at the mouse neuromuscular joint. Pharmacol Res 39:239–245

Barrera C, Betoret N, Seguí L (2020) Phenolic profile of cane sugar derivatives exhibiting antioxidant and antibacterial properties. Sugar Tech 22:798–811

Becker RB, Fricke AB (1999) Freezing times of regularly shaped food items. Int Comm Heat Mass Transfer 26(5):617–626

Bhandari BR, Howes H (1999) Implication of glass transition for the drying and stability of dried foods. J Food Eng 40:71–79

Bhandari BR, Senoussi A, Dumoulin ED, Lebert A (1993) Spray drying of concentrated fruit juices. Dry Technol 11:1081–1092

Bhandari BR, Dattaa N, Howesb T (1997) Problems associated with spray drying of sugar-rich foods. Dry Technol Int J 15:671–684

Bhupinder K, Sharma KP, Harinder K (1991) Studies on the development and storage stability of ready-to-serve bottled sugarcane juice. Int J Trop Agric 9:128–134

Bowden RP, Isaccs AR (1989) Concentration of pineapple juice by reverse osmosis. In: The Australian Institute of Food Science and Technology. Convention Papers, 30th April to 5th May, Perth W.A, 30th April to 5th May 1989, Perth WA

Braddock RJ, Nikdel S, Nagy S (1988) Composition of some organic and inorganic compounds in reverse osmosis-concentrated citrus juices. J Food Sci 53(2):508–512

Campo F, Santo Tomas JF, Bermello A, Leal C, Diaz A, Reyes I, Hernandez Y, Gallardo R (2004) Cane juice drying—a preliminary study in continuous and batch conventional dryers. Revista ICIDCA Sobre los Derivados de la Cafia de Azucar 38:54–59

Cano-Chauca M, Cal-Vidal J, Ramos AM, Stringheta PC (2005) Effect of the carriers on the microstructure of mango powder obtained by spray drying and its functional characterization. Innov Food Sci Emerg Technol 6:420–428

Chauhan OP, Singh DS, Tyagi M, Balyan DK (2002) Studies on preservation of sugarcane juice. Int J Food Prop 5:217–229

Chegini GR, Ghobadian B (2005) Effect of spray-drying conditions on physical properties of orange juice powder. Dry Technol 23:657–668

Chen W, Gast BKL, Smithey S (2000) The effects of different freeze-drying processes on the moisture content, color and physical strength of roses and carnations. Sci Hortic 84:321–332

Cheuyglintase K, Morison KR (2009) Crystal induction by dietary fiber in spray drying of fruit juice. In: Proceedings of the 6th Asia-Pacific drying conference (ADC2009), Bangkok, Thailand, pp. 606:613

Chua HT, Rao MA, Acree TE, Cunningham DG (1987) Reverse osmosis concentration of apple juice: flux and flavor retention by cellulose acetate and polyamide membranes. J Food Process Eng 9(3):231–245

Chun ML, Quan XY, Fei Q (2007) Maintaining the quality of sugarcane juice with blanching andascorbic acid. Food Chem 104:740–745

Colombo R, Fernando M, Lanças Janete H, Yariwake. (2006) Determination of flavonoids in cultivated sugarcane leaves, bagasse, juice and in transgenic sugarcane by liquid chromatography-UV detection. J Chromatogr A 1103(1):118–124., ISSN 0021-9673,. https://doi.org/10.1016/j.chroma.2005.11.007

Couto DS, Cabral LM, Da MVM, Deliza RFDG (2011) Concentration of pineapple juice by reverse osmosis: physicochemical characteristics and consumer acceptance. Brazil, Ciéncia e Tecnologia de Alimentos 31:905–910

Dal Magro C, Bulegon R, de Almeida Gomes G (2019) Development of freeze dried sugarcane juice powder. In Revista do Congresso Sul Brasileiro de Engenharia de Alimentos (Vol. 4, No. 1)

De Souza VR, Pereira PAP, Pinheiro ACM, Bolini HMA, Borges SV, Queiroz F (2013) Analysis of various sweeteners in low-sugar mixed fruit jam: equivalent sweetness, time intensity analysis and acceptance test. Int J Food Sci Technol 48:1541–1548

Deotale S, Bhotmange M, Halde P, Chitale M (2019) Study of traditional Indian sweetener 'Jaggery' and its storage behavior. Int J Chem Stud 7:410–416

Deshpande SS, Cheryan M, Sathe SK, Salunkhe DK, Luh BS (1984a) Freeze concentration of fruitjuices. J C R C Crit Rev Food Sci Nutr 20:31–88

Deshpande SS, Cheryanb M, Sathec SK, Salunkhed DK, Luhe BS (1984b) Freeze-drying has been widely used in the production of pharmaceuticals, vaccines and high-valued foods. CRC Crit Rev Food Sci Nutr 20:173–248

Dhansu P, Ram B, Singh AK, Tomar SK, Karuppaiyan R, Kumar R, Chhabra ML, Singh A, Raja AK, Kaushik P, Pandey SK (2023) Different treatments for sugarcane juice preservation. Food Secur 12:311. https://doi.org/10.3390/foods12020311

Di Giacomo G, Taglieri L (2009) A New High-Yield Process for the Industrial Production of Carrot Juice. Food Bioprocess Technol 2: 441–446. https://doi.org/10.1007/s11947-009-0207-x

Diaz A, Campos F, Diaz AO, Hernandez Y, Lopez P (2007) Evaluation of drying parameter to obtain dried sugarcane juice by using an aspersion dryer ICIDCA. Sobre Ios Derivados de la Cana de Azucar 41:48–51

Duarte-Almeida JM, Salatino A, Genovese MI, Lajolo FM (2011) Phenolic composition and antioxidant activity of culms and sugarcane (Saccharum officinarum L.) products. Food Chem 125(2):660–664, ISSN 0308-8146,. https://doi.org/10.1016/j.foodchem.2010.09.059

Duke JA (2017) Handbook of phytochemical constituents of GRAS herbs and other economic plants. CRC Press, New York

Echavarria AP, Falguera V, Torras C, Berdt'm C, Pagan J, Lbarz A (2012) Ultra filtration and reverse osmosis for clarification and concentration of fruit juices at pilot plant scale. LWT Food Sci Technol 46:189–195

El-Abasy M, Motobu M, Shimura K, Na KJ, Kang CB, Koge K, Onodera T, Hirota Y (2002) Immunostimulating and growth-promoting effects of sugar cane extract (SCE) in chickens. J Vet Med Sci 64(11):1061–1063

Fazaeli M, Emam-Djomeh Z, Kalbasi-Ashtari A, Omid M (2012) Effect of process conditions and carrier concentration for improving drying yield and other quality attributes of spray dried black mulberry (Morus nigra) juice. Int J Food Eng 8:1–20

Fernandes P, Cabral JM (2007) Phytosterols: applications and recovery methods. Bioresour Technol 98(12):2335–2350. https://doi.org/10.1016/j.biortech.2006.10.006. Epub 2006 Nov 22
Fukutani K, Ogawa H (1983) A comparison of membrane's suitability and effect of operating pressure for juice concentration by reverse osmosis. Nippon Shokuhin Kogyo Gakkaishi 30(11):636–641
Gabas AL, Sobral PJA, Telis VRN, Telis-Romero J (2007) Effect of maltodextrin and Arabic gum in water vapor sorption thermodynamic properties of vacuum dried pineapple pulp powder. J Food Eng 82:246–252
Ghazal GA, El-Mansy HA, Shady AK, Gaber A (2003) Physical, chemical and sensory properties of juice and syrup produced from sugarcane cultivars. Ann Agric Sci 41:1177–1193
Ghosh AK, Shrivastava AK, Agnihotri VP (1998) Production Technology of Lump Sugar-Jaggery/ Jaggery. Daya Publishing House, Delhi, India, p 287. (Reprinted 2017)
Girarda B, Fukumotoa LR (2000) Membrane processing of fruitjuices and beverages: a review. Crit Rev Food Sci Nutr 40:91–157
Gopalan C, Shastri BVR, Balasubramanian SC (1991) Nutritive value of Indian foods. National Institute of Nutrition Hyderabad, ICMR, India
Greenwald CG, King CJ (1981) The effects of design and operating conditions on particle morphology for spray dried foods. J Food Process Eng 4:171–187
Guimaraes C, Martinez SS, Pintado AE, Giao M, Pintado ME, Malcata FX, Bento LS. Antioxidant activity and phenolic compounds in extracts of sugar molasses (poster) 2nd Intern. Congress on Antioxidant Methods, June 22-24, 2005, Orlando, FL
Gurak PD, Matta VM, Cabral LMC, Rocha-Leao MHM (2008) Grape juice encapsulated obtained by lyophilization after concentration by reverse osmosis; International Commission of Agricultural Engineering (CIGR), Institut fur Landtechnik, Bonn, Germany, Central theme, technology for all: sharing the knowledge for development. Proceedings of the International Conference of Agricultural Engineering, XXXVII Brazilian Congress of Agricultural Engineering, International Livestock Environment Symposium—ILES VIII, Iguassu Falls City, Brazil, 31st August to 4″1 September, 2008
Harguindeguy M, Fissore D (2020) On the effects of freeze-drying processes on the nutritional properties of foodstuff: a review. Dry Technol 38(7):846–868. https://doi.org/10.1080/07373937.2019.1599905
Hari K, Reginold Jebitta S, Sivaraman K (2013) Production and characterization of sugarcane juice powder. J Sugarcane Res 3(1):20–34
Harish Nayaka MA, Sathisha UV, Manohar MP, Chandrashekar KB, Dharmesh SM (2009) Cytoprotective and antioxidant activity studies of jaggery sugar. Food Chem 115:113–118
Hirpara P, Thakare NU, Kele VD, Patel D (2020) Jaggery: a natural sweetener. J Pharmacogn Phytochem 9:3145–3148
Hoffmann P, Reis JA, dos Castro LP, de Hoffmann FL (2006) Microbiological quality of samples ofsugarcane juices marketed in the city of 850 José in Rio Preto, SP, Higiene Alimentar 20:79–83
Hollman PCH (2001) Evidence for health benefits of plant phenols: local or systemic effect? J Sci Food Agr 81:842–852
Hudson EA, Dinh PA, Kokubu T, Simmonds MSJ, Gescher A (2000) Characterization of potentially chemopreventive phenols in extracts of brown rice that inhibit the growth of human breast and colon cancer cells. Cancer Epidemiol Biomarkers Prev 9(11):1163–1170
Hussain Z, Islam M, Mohammad Z, Khan KM, Perveen S, Afzal M (2012) The effect of pretreatment of juice on the properties and composition of jaggery. Sugar Techol 14:291–294
Inamdar V, Chimmad BV, Naik R (2005) Nutrient composition of traditional festival foods of North Karnataka. J Hum Ecol 18:43–48
Irmak S, Dunford NT, Milligan J (2006) Policosanol contents of beeswax, sugarcane and wheat extracts. Food Chem 95(2):312–318, ISSN 0308-8146,. https://doi.org/10.1016/j.foodchem.2005.01.009
Jaffe WR (2015) Nutritional and functional components of non -centrifugal cane sugar: a compilation of the data from theanalytical literature. J Food Compos Anal 43:194–202

Jagannadha Rao PVK, Madhusweta D, Das SK (2007) Jaggery—a traditional Indian sweetener. Indian J Tradit Knowl 6:95–102

Jebitta RS, Hari K, Sivaraman K (2012) Characterization of sugarcane juice powder from different varieties. In: Viswanathan R, Hemaprabha G, Bhaskaran A, Mohanraj K, Jayakumar V, Ramasubramanian T, Nair NV (eds) Proc. of International Symposium on New Paradigms in Sugarcane Research, October 15–18, 2012. ICAR-SBI, Coimbatore, India., p 323

Jegatheesan V, Phong DD, Shu L, Aim RB (2009) Performance of ceramic micro-and ultrafiltration membranes treating limed and partially clarified sugarcane juice. J Membr Sci 327(1–2):69–77

Jesus DF, Leite MF, Silva LFM, Modesta RD, Matta VM, Cabral LMC (2007) Orange (Citrus sinensis) juice concentration by reverse osmosis. J Food Eng 81(2):287–291. https://doi.org/10.1016/j.jfoodeng.2006.06.014

Kaavya R, Pandiselvam R, Kothakota A, Banuu Priya EP, Arun PV (2019) Sugarcane juice preservation: a critical review of the state of the art and way forward. Sugar Tech 21(1):9–19. https://doi.org/10.1007/s12355-018-0622-2

Kadam US, Ghosh SB, De S, Suprasanna P, Devasagayam TPA, Bapat VA (2008) Antioxidant activity in sugarcane juice and its protective role against radiation induced DNA damage. Food Chem 106(3):1154–1160, ISSN 0308-8146. https://doi.org/10.1016/j.foodchem.2007.07.066

Kapur KL, Singh VP, Garg RG (1978) Preliminary studies on preservation of sugarcane juice. Indianfood Packer 32:32–33

Karmakar R, Ghosh AK, Gangopadhyay H (2011) Effect of pretreatments on Physico-chemical characteristics of sugarcane juice. Sugar Tech 13:47–50

Karthikeyan J, Samipillai SS (2010) Sugarcane in therapeutics. J Herb Med Toxicol 4:9–14

Kaushik P, Dhiman AK (1999) Medicinal plants and raw drugs of India. Doon Phototype Printers, Dehradun, pp 560–564

Khare A, Behari Lal A, Singh A, Pratap Singh A (2012) Shelflife enhancement of sugarcane juice. Hrvatski Časopis Za Prehram benu Tehnologiju, Biotehnologiju i Nutricionizam 7(3):179–183

Khuenpet K, Charoenjarasrerk N, Jaijit S, Arayapoonpong S, Jittanit W (2016) Investigation of suitable spray drying conditions for sugarcane juice powder production with an energy consumption study. Agric Nat Res 50(2):139–145

Kirtikar KR, Basu BD (1991) In: Blatter E, Caius JF, Mhaskar KS (eds) Indian medicinal plants, vol 5, pp 2661–2669

Kowsalya S, Shyny PA (2005) Microbial contamination and adulteration of selected beverages of Malappuram District, Kerala. Indian J Nutr Diet 42:492–502

Kramer GN, Angelis LD, Pauwels J, Ooghe W, Belliardo JJA (1988) The freeze-drying of orange juice for the preparation of reference materials. Fresenius' Zeitschrift'fflr analytische Chemie 332:694–697

Kumar A, Singh S (2020) The benefit of Indian jaggery over sugar on human health. In: Preuss HG, Bagchi D (eds) Dietary sugar, salt and fat in human health. Academic Press. (Ch. 16)

Kumar A, Smita S (2020) The benefit of Indian jaggery over sugar on human health. In Book: Dietary Sugar, Salt and Fat in Human Health. pp. 347–359

Kumar D, Singh J, Rai D, Kumar RM, Bhatia S (2013) Effect of modified atmosphere packaging on keeping quality of jaggery. Sugar Tech 15:203–208. https://doi.org/10.1007/s12355-012-0197-2

Kuruba EK, Rao PV, Khokhar D, Patel S (2020) Technologies for preparation of solid and granular jaggery: a review. Curr J Appl Sci Technol 39(30):105–113

Lamdande AG, Khabeer ST, Kulathooran R, Dasappa I (2018) Effect of replacement of sugar with jaggery on pasting properties of wheat flour, physico-sensory and storage characteristics of muffins. J Food Sci Technol 55:3144–3153

Langrish TAG, Chan WC, Kota K (2007) Comparison of maltodextrin and skim milk wall deposition rates in a pilot-scale spray dryer. Powder Technol 179:84–89

Lavanya R, Suresha GS, Hari K, Murali P, Ram B (2023) Cane Jam: an innovative value-added product of sugarcane for entrepreneurship. Proceedings of the National Seminar on Research Imperatives for Sustaining Sugarcane, Sugar and Ethanol Production (SugarSem-2023), 25 August 2023, ICAR-sugarcane breeding institute Coimbatore, Tamil Nadu, India. p. 160

Liaotrakoon W, Clercq N, de Lewille B, Dewettinck K (2012) Physicochemical properties, glass transition state diagram ancrcolour stability of pulp and peel of two dragon fruit varieties (Hylocereus spp.) as affected by freeze-drying. Int Food Res J 19:743–750

Liju LS, Leo A, Suresha GS, Murali P Hari K, Kalaiselvi K, Malini P, Sreenoob S (2023) Development of process for preparation of Jellies and Gummies from sugarcane juice. Proceedings of the National Seminar on Research Imperatives for Sustaining Sugarcane, Sugar and Ethanol Production (SugarSem-2023), 25 August 2023, ICAR-sugarcane breeding institute Coimbatore, Tamil Nadu, India. p. 161

Longvah T, Anantan I, Bhaskarachary K, Venkaiah K (2017) Indian food composition tables. National Institute of Nutrition, Indian Council of Medical Research, Hyderabad

Lopes G, Cresto R, Carraro CNM (2006) Microbiological analysis of sugarcane juice sold on the streets of Curitiba, PR, Higiene Alimentar 20:40–44

Ma LiShuai, HaiXia C, WenChai Z, ZhaoShuai W (2013) Effect of different drying methods on physicochemical properties and antioxidant activities of polysaccharides extracted from mushroom Inonotus obliquus. Food Res Int 50L:633–640

Madaeni SS, Tahmasebi K, Kerendi SH (2004) Sugar syrup concentration using reverse osmosis-membranes. Eng Life Sci 4:187–190

Madaeni SS, Sasanihoma A, Zereshki S (2011) Chemical cleaning of reverse osmosis membrane fouled by apple juice. J Food Process Eng 34:1535–1557

Madan HK, Jaiswal UK, Kumar JS, Khanna SK (2004) Improvement in gur(jaggery) making plant for rural areas. J Rural Technol 1:194–196

Mandal D, Tudu S, Mitra SR, De GC (2006) Effect of common packing materials on keeping quality of sugarcane jaggery during monsoon season. Sugar Tech 8:137–142

Mann RS, Singh S (1988) Processing, packaging and preservation of sugarcane, juice. Indian Farm 37:15–17

Mauricio Duarte-Almeida J, Novoa AV, Linares AF, Lajolo FM, Ines Genovese M. (2006) Antioxidant activity of phenolics compounds from sugar cane (Saccharum officinarum L.) juice. Plant Foods for Human Nutrition. 61:187–92.

Medina BG, Garcia A (1988) Concentration of orange juice by reverse osmosis. J Food Process Eng 10:217–230

Menendez R, Fernandez SI, Del Rio A, Gonzalez RM, Fraga V, Amor AM, Mas RM (1994) Policosanol inhibits cholesterol biosynthesis and enhances low density lipoprotein processing in cultured human fibroblasts. Biol Res 27(3–4):199–203. PMID: 8728831

Mishra BB, Gautam S, Sharma A (2011) Shelf-life extension of sugarcane juice using preservatives and gamma radiation processing. J Food Sci 76:573–578

Moghbel A, Ghiasvand M, Salimi E, Karami M, Abbaspour H (2013) Preparation of yogurt lyophilized powder to achieve a complementary tablet. Sci Med J 12:1–12

Moy JH (2006) Vacuum-puff freeze drying of tropical fruit juices. J Food Sci 36:906–910

Nadkarni KM (1976) Indian Materia Medica. Bombay Popular Prakashan, Mumbai, pp 1085–1089

Nagalakshmi AVD, Reddy MU (1999) Quality analysis of selected fruit juices sold by street vendors in Hyderabad city. Indian J Nutr Diet 36:78–83

Nakasone Y, Takara K, Wada K, Tanaka J, Yogi S, Nakatani N (1996) Antioxidative compounds isolated from Kukuto, non-centrifugal cane sugar. Biosci Biotech Bioch 60:1714–1716

Naranjo S, Armas R, de Rodriguez CW, Martinez M, Legaz ME, Vicente C (1999) Identification of some phenolics and hydroxycinnamic acids in sugarcane juice. Revista Biologia (Habana) 13:7–11

Nath A, Dutta D, Kumar P, Singh J (2015) Review on recent advances in value addition of jaggery based products. J Food Process Technol 6:1–4

Nigam, Raha (1982) Spray drying of sugarcane juice. Proceeding of 37th Annual Convention of Sugar Technologies Association of India, pp 206–215

Nisha M, Chandran K, Gopi R, Krishnapriya V, Mahendran B (2017) Nutritional and therapeutic benefits of sugarcane and its products. Journal of Sugarcane Research 7(1):01-10.

Nishad J, Mir SA, Walia K (2019) Optimization of spray drying technology for sugarcane juice using natural and synthetic encapsulating agents. Sugar tech 21:749–755

Ogawa H, Fujimura M, Ohkura N, Makimura K (2013) Chronic cough management: dealing with a sensation of irritation in the throat. Respirology 18:1278–1279

Oliveira ACGD, Spoto MHF, Canniatti-Brazaca SG, Sousa CP, de Gallo CR (2007) Effects of heat treatment and gamma radiation on the characteristics of pure sugarcane juice and mixed with fruit juices. Ciéncia e Tecnologia de Alimentos 27:863–873

Oliveira TS, de Ribeiro DS, Paulo EM (2008) Microbiological analysis of sugarcane juice (with orwithout ice), sold in the streets of Feira de Santana, BA, Higiene Alimentar 22:56–60

Pandey J (2010) Harit Samhita, Chaukhambha Vishvabhararti, Varanasi, India, pp. 544

Panigrahi C, Shaikh AEY, Bag BB, Mishra HN, De SA (2021) Technological review on processing of sugarcane juice: spoilage, preservation, storage, and packaging aspects. J Food Process Eng 44:e13706. https://doi.org/10.1111/jfpe.13706

Patil JR, Wandre SS, Gojar DG (1985) Kakavi quality in relation to sterilization and preservatives. In: Proceedings of National Seminar-cum group discussion on jaggery manufacture and storage, Lucknow, p 84

Prakash MD, Stojanovska L, Feehan J, Nurgali K, Donald EL, Plebanski M, Flavel M, Kitchen B, Apostolopoulos V (2021) Anti-cancer effects of polyphenol-rich sugarcane extract. PLoS One 16(3):e0247492. https://doi.org/10.1371/journal.pone.0247492

Puri V, Nagpal M, Singh I, Singh M, Dhingra GA, Huanbutta K, Dheer D, Sharma A, Sangnim T (2022) A comprehensive review on nutraceuticals: therapy support and formulation challenges. Nutrients 14(21):4637. https://doi.org/10.3390/nu14214637

Rajendran P, Bharathidasan R, Sureshkumar K (2017) GC-MS analysis of Phyto-components in raw and treated sugarcane juice. Int J Curr Microbiol App Sci 5(1):78–94. https://doi.org/10.20546/ijcmas.2016.501.007

Rajendran I, Palaniswami C, Vennila A (2021) Improved method of liquid jaggery preparation. J Sugarcane Res 10:2. https://doi.org/10.37580/JSR.2020.2.10.107-112

Rao KC (1990) Bottling of sugarcane juice—a profitable venture (small scale or on large scale). In: Proceedings of 40th Annual Convention—Deccan Sugar Technologists' Association, pp. A207-A212

Rao GP, Singh P (2022) Value addition and fortification in non-centrifugal sugar (jaggery): a potential source of functional and nutraceutical foods. Sugar Tech 24:387–396. https://doi.org/10.1007/s12355-021-01020-3

Rao J, Das M, Das S (2007) Jaggery—a traditional Indian sweetener. Indian J Tradit Knowl 6(1):95–102

Ravisankar P, Koushik O, Reddy A, Kumaru AP, Pragna P (2016) A detailed analysis on acidity and ulcers in esophagus, gastric and duodenal ulcers and management. IOSR J Dent Med Sci 15:94–114

Resmi S, Latheef FRV (2016) Effectiveness of herbal extract in enhancing the level of HB among adolescent girls with iron deficiency anaemia at selected higher secondary schools at Bangalore. Int J Health Sci Res 6:174–179

Sagar VR, Suresh Kumar P (2010) Recent advances in drying and dehydration of fruits and vegetables: a review. J Food Sci Technol 47:15–26

Sahi SS, Alava JM (2003) Functionality of emulsifiers in sponge cake production. J Sci Food Agric 83:1419–1429

Sahu AP, Paul BN (1998) The role of dietary whole sugarjaggery in prevention of respiratory toxicity of air toxics and in lung cancer. Toxicol Lett 95:154

Sahu AP, Saxena AK (1994) Enhanced translocation of particles from lungs by jaggery. Environ Health Perspect 102:211–214

Sahu T, Kumar D, Patel S, Sahu S, Painkra AK (2024) Physico-chemical properties of solid and molten Jaggery from CoLk94184. Int J Adv Biochem Res 8(1S):381–384. https://doi.org/10.33545/26174693.2024.v8.i1Sf.340

Said PP, Pradhan RC (2013) Preservation and value addition of jaggery. 569–574

Sammbamurthi AVSS (2006) Dictionary of medicinal plants. CBS publishers and distributors, New Delhi, pp 263–264

Samson KLI, Fischer JAJ, Roche ML (2022) Iron status, anemia, and iron interventions and their associations with cognitive and academic performance in adolescents: a systematic review. Nutrients 14:224

Sanaye S, Malekmohammadi HR (2004) Thermal and economical optimization of air conditioning units with vapor compression refrigeration system. Appl Therm Eng 24:1807–1825

Sanchez V, Baeza R, Galmarini MV, Zamora MC, Chirife J (2013) Freeze-drying encapsulation of red wine polyphenols in an amorphous matrix of maltodextrin. Food Bioprocess Technol 6:1350–1354

Sankhla S, Chaturvedi A, Kuna A, Dhanlakshmi K (2012) Preservation of sugarcane juice using hurdle technology. Sugar Tech 14:26–39

Saska M, Chou CC (2002) Antioxidant properties of sugarcane extracts. Proc. First Biannual World Conf. on Recent Developments in Sugar Technology, Delray Beach, FL, May 16–17

Saska M, Chou CC (2006) Antioxidants: an excellent phytochemical functional food from sugarcane. Meeting of Sugar Industry Technologists, Inc., LaBaule, France, May 14–17, 2006Paper # 898

Sekar S, Mariappan S (2008) Traditionally fermented biomedicines, arishtas and asavas from ayurveda. Indian J Tradit Knowl 7:548–556

Shahi HN (1999) Sustainability of jaggery and khandsari industry in India. In: Singh J (ed) Status, problems and prospects of jaggery and khandsari industry in India. Indian Institute of Sugarcane Research, Lucknow, pp 19–27

Shalini HN, Nayak CA (2016) Forward osmosis membrane concentration of raw sugarcane juice. In Recent advances in chemical engineering: select proceedings of ICACE 2015 (pp. 81-88). Springer, Singapore

Shanthi RM, Alarmelu S, Mahadeva Swamy HK, Lakshmi Pathy T (2022) Impact of climate change on sucrose synthesis in sugarcane varieties. In: Verma KK, Song XP, Rajput VD, Solomon S, Li YR, Rao GP (eds) Agro-industrial perspectives on sugarcane production under environmental stress. Springer, Singapore. https://doi.org/10.1007/978-981-19-3955-6_2

Sharma H, Sharma H, Sharma H (2012) Sushruta-samhitA—A critical review Part-1: historical glimpse. Ayu 33:167–173

Sheu MJ, Wiley RC (1984) Influence of reverse osmosis on sugar retention in apple juice. J Food Sci 49(1):304–305. https://doi.org/10.1111/j.1365-2621.1984.tb13738.x. First published: January 1984

Shivalingamurthy SG, Anangi R, Kalaipandian S, Glassop D, King GF, Rae AL (2018) Identification and functional characterization of sugarcane invertase inhibitor (ShINH1): a potential candidate for reducing pre-and post-harvest loss of sucrose in sugarcane. Front Plant Sci 3(9):598

Shrestha AK, Howes T, Adhikari SP, Bhandari BR (2007) Water sorption and glass transition properties of spray dried lactose hydrolyzed skim milk powder. LWT Food Sci Technol 40:1593–1600

Shrivastava AK, Singh P (2020) Jaggery (Gur): the Ancient Indian open-pan non-centrifugal sugar. In: Mohan N, Singh P (eds) Sugar and sugarderivatives: changing consumer preferences. Springer, Singapore, pp 283–307. https://doi.org/10.1007/978-981-15-6663-9_19

Silva S, da Faria J, De AF (2006) Quality of sugarcane juice packed by hot fill and asepticprocesses. Ciḗncia e Tecnologia de Alimentos 26:754–758

Sim L, Shu L, Jegatheesan V, Phong DD (2009) Effect of operating parameters and cleaning on the performance of ceramic membranes treating partially clarified sugarcane juice. Sep Sci Technol 44(15):3506–3537

Singh J (1998) Jaggery and Khandsari research digest, 140. Indian Institute of Sugarcane Research, Lucknow

Singh J, Singh R (2008) Nutritive and eco-friendly jaggery, processing, handling and storage of sugarcane jaggery. Indian Institute of Sugarcane Research, pp 3–10

Singh M, Bharadwaj KM, Agrawal ML (1978) Storage of jaggery. Co-oper Sugar 3:14–17

Singh P, Shahi HN, Suman A, Jain RC (2002) Sugarcane juice concentrate—preparation, preservation and storage. J Food Sci Technol (Mysore) 39:96–98

Singh N, Kumar D, Raisuddin S, Sahu AP (2008) Genotoxic effects of arsenic: prevention by functional food-jaggery. Cancer Lett 268:325–330

Singh J, Singh RD, Anwar SL, Solomon S (2011) Alternative sweeteners production from sugarcane in India: lump sugar (jaggery). Sugar Tech 13:366–371

Singh J, Solomon S, Kumar D (2013) Manufacturing Jaggery: a product of sugarcane, as health food. Agrotechnology 11:1–3

Singh S, Omre PK, Gaikwad K (2014) Sensory evaluation of optimized and stabilized sugarcane juice. IJERGS 2:637–648

Sipoli CC, Barros STD (2011) Concentration of passion fruit juice by reverse osmosis. Revista Brasileira de Produtos Agroindustriais, Campina Grande, Brazil, Revista Brasileira de Produtos Agroindustriais 13:179–187

Solomon S (2016) Sugarcane production and development of sugar industry in India. Sugar Tech 18:588–602. https://doi.org/10.1007/s12355-016-0494-2

Sreedevi P, Rao PS (2018) Microbial destruction kinetics of high-pressure-processed sugarcane juice (Saccharum officinarum). J Food Process Eng 41(7):e12850. https://doi.org/10.1111/jfpe.12850

Subbannayya K, Bhat GK, Shetty S, Junu VG (2006) How safe is sugarcane juice? Indian J Med Microbiol 25:73–74

Suresha GS, Prasath J, Naveenarani M, Appunu C, Vasantha S, Hari K (2017) Assessing the effect of soluble acid invertase activity in post-harvest sugarcane under tropical conditions. J Sugarcane Res 7(2):93–99

Suresha GS, Murali P, Hari K, Puthira Prathap D, Antony Leo C, Lavanya R, Hemaprabha G (2021) Cane jam: an innovative product for entrepreneurship, Pamphlet No. SBI-ABI/2022/E/01, ICAR-SBI, Coimbatore

Suresha GS, Hari K and Murali P (2023) Technological developments in sugarcane for value addition and product diversification. In: Book of Abstracts, XVI Agricultural Science Congress and ASC Expo, Transformation of Agri-food systems for achieving sustainable development goals, National Academy of Agricultural Sciences, New Delhi and ICAR-Central Marine Fisheries Research Institute, Kochi,10-13 October 2023, Kochi, India, 127

Tal B, Sack J, Yaron M, Shefer G, Buch A, Ben Haim L, Marcus Y, Shenkerman G, Sofer Y, Shefer L, Margaliot M, Stern N (2019) Increment in dietary potassium predicts weight loss in the treatment of the metabolic syndrome. Nutrients 11:1256

Tamaki H, Takaichi S, Yoshida A, Okuno M (2001) Polyphenols in sugarcane juice. Proceedings of the Research Society of Japan Sugar Refineries' Technologists 49: 41–46

Theansuwan W, Triratanasirichai K, Tangchaichit K (2008) Continuous production of lime juice by vacuum freeze drying. Am J Appl Sci 5(8):959–962

Tidke B, Sharma HK, Kumar N (2017) Development of peanut and chickpea nut brittle (Chikki) from the incorporation of sugar, jaggery and corn syrup. Int Food Res J 24:657–663

Toledo TR (1999) Fundamentals of food process engineering, 2nd edn. Mari yard, Gaithersburg

Tonon RV, Freitas SS, Hubinger MD (2011) Spray drying of açai (Euterpe oleraceae Mart.) juice: effect of inlet air temperature and type of carrier agent. J Food Process Preserv 35(5):691–700

Truong V, Bhandari BR, Howes T (2005) Moisture and glass optimization of co-current spray drying process of sugar rich foods: transition temperature profile during drying. J Food Eng 71:55–65

Umate P, Khobragade P, Thanvi H, Zilate S (2021) A review on the clinical aspect of guda (jaggery) in brihatrayi and laghutrayi. Int J Curr Res Rev 13:51

Uppal SK (2003) Post harvest losses in sugarcane. Sugar Tech 5:93–94

Uppal SK, Sharma S (1999) Relative performance of genotypes to post harvest inversions in subtropical region. Indian Sugar 49:345–348

Uppal SK, Thind KS, Gill RS (2005) Relative performance of sugarcane genotypes for jaggery (gur) quality and storage. Sugar Tech 7(4):136–140

Varady KA, Wang Y, Jones PJH (2003) Role of policosanol in the prevention and treatment of cardiovascular disease. Nutr Rev 61:376–383

Verma VK, Narain M (1990) Moisture absorption isotherms of jaggery. J Stored Prod Res 26:6146–6147

Verruma-Bernardi MR, Borges MTM, Lopes RCH, Della-Modesta RC, Ceccato-Antonini SR (2007) Avaliação microbiológica, físico-química e sensorial de açúcares mascavos comercializados na cidade de São Carlos-SP. Braz J Food Technol 10(3):205–211. Retrieved from http://www.ital.sp.gov.br/bj/artigos/html/busca/PDF/v10n3293a.pdf

Vila FC, Colombo R, Lira TOD, Yariwake JH (2008) HPLC microfractionation of flavones and antioxidant (radical scavenging) activity of Saccharum officinarum L. J Braz Chem Soc 19:903–908

Warrier PK, Nambiaar VPK, Ramankutti C (2010) Indian medicinal plants a compendium of 500 species -Vaidyaratnam PS warriers Arya Vaidya sala universities press (india private limited) vol 5: 26–44

Welti JS, Lafuente B (1983) Spray drying of comminuted orange products, I. Influence of air temperature and feed rate on product quality. Chem Eng Prog 79:80–85

Yadav RL, Solomon S (2006) Potential of developing sugarcane by-product based industries in India. Sugar Tech 8:104–111. https://doi.org/10.1007/BF02943642

Yildiz F, Bozkurt M, Wiley RC (1993) Comparison of reverse osmosis and evaporation processing on the quality characteristics of tomato concentrates. Food Control 4(3):155–158., ISSN 0956-7135. https://doi.org/10.1016/0956-7135(93)90303-6

Yusof S, Shian LS, Osman A (2000) Changes in quality of sugar-cane juice upon delayed extractionand storage. Food Chem 68:395–401

Zamir R, Khalil SA, Shah ST, Ahmad N, Saima S (2012) Antioxidant activity influenced by in vivo and in vitro mutagenesis in sugarcane (Saccharum officinarum L.). Afr J Biotechnol 11(54):11686–11692

Zidan D, Azlan A (2022) Non-centrifugal sugar (NCS) and health: a review on functional components and health benefits. Appl Sci 12(1):460

Biotechnological Manipulations in Sugarcane for Bioenergy Applications

4

Baskaran Kannan, Rajesh Yarra, Thakku R. Ramkumar, and Kandhalu Sagadevan Dinesh Babu

Abstract

Sugarcane is the most preferred feedstock for bioenergy productions owing to highly efficient C4 photosynthetic machinery, high sugar content, relatively high cellulosic biomass, and a favorable energy input and output ratio per hectare relative to other bioenergy-producing plants. In addition to table sugar production, sugarcane is being used as a feedstock for sugar derived and lignocellulose-derived bioethanol, and recently advanced biofuels like biodiesel are also obtained from metabolically engineered sugarcane also known as oilcane. Bioethanol is generally derived from sucrose by means of enzymatic hydrolysis. However, biotechnological manipulations help to enhance the accumulation of sugar in the sugarcane stalks and improve saccharification efficiency which in turn increases the bioethanol production. Sugar content and its concentration have been enhanced by altering the expression of sucrose synthase, trehalose phosphate, or soluble acid invertase genes. Altering lignin content or cell wall composition, particularly downregulation of caffeic acid *O*-methyltransferase

B. Kannan (✉)
Agronomy Department, Institute of Food and Agricultural Sciences, University of Florida, Gainesville, FL, USA
e-mail: kbaskaran@ufl.edu

R. Yarra
Department of Plant and Agroecosytem Sciences, University of Wisconsin-Madison, Madison, WI, USA

T. R. Ramkumar
Department of Plant and Soil Sciences, Institute of Genomics for Crop Abiotic Stress Tolerance, Texas Tech University, Lubbock, TX, USA

K. S. Dinesh Babu
Department of Botany and Microbiology, Hemvati Nandan Bahuguna Garhwal University, Chauras Campus, Tehri Garhwal, Uttarakhand, India

G. S. Suresha et al. (eds.), *Value Addition and Product Diversification in Sugarcane*, https://doi.org/10.1007/978-981-97-7228-5_4

(COMT) and 4-coumarate:coenzyme A ligase (Sh4-CL) or knock out of lignin biosynthesis genes by gene editing has also been demonstrated to improve lignocellulose to bioethanol conversions in sugarcane. Recently, metabolic engineering approach has been used to hyper accumulate neutral lipids such as triacyl glycerol (TAG) in the vegetative tissues of sugarcane by mostly co-expressing Wrinkled (WRI1), diacylglycerol acyltransferase1-2 (DGAT1-2), and oleosin1 (OLE1). Stored lipids or TAGs can be converted into biodiesel by transesterification process. Therefore, coproduction of bioethanol and biodiesel from sugarcane feedstock can be feasible through genetic manipulations using advanced biotechnological tools like genetic and metabolic engineering and genome editing.

Keywords

Sugarcane · Oilcane · Triacyl glycerol (TAG) · COMT · Bioenergy · Gene editing

4.1 Introduction

Sugarcane is a highly efficient C4 biomass crop and converts 6–7% of sunlight into biomass and grown in more than 80 countries primarily for table sugar production (Tew and Cobill 2008). However, Brazil is the highest bioethanol-producing country from sugarcane feedstock. Of the 610 million tons of sugarcane produced in Brazil, 321 million tons (more than 50%) were used for bioethanol production in 2020–2022, followed by India, where 11.5 million tons of sugarcane used for ethanol production (OECD/FAO 2023). In the USA, corn is the major feedstock for bioethanol production with 15.4 billion gallons, whereas soybean used for the 3.1 billion gallons of biodiesel or renewal diesel produced in 2022 (www.ers.usda.gov/data-products/u-s-bioenergy-statistics/). Therefore, due to its high biomass and sugar-yielding ability, sugarcane is considered as the prime feedstock for the second-generation biofuels such as lignocellulose-derived bioethanol and advanced biofuels like biodiesel. Recent biotechnological advances provided opportunities for genetic manipulation of sugarcane and makes it sugar-producing crop to biofactory for sustainable bioethanol and advanced biofuels productions (Dinesh Babu et al. 2022). Sugar crops have widely been used for bioethanol production by enzymatic hydrolysis. However, lignocellulosic biomass of sugarcane has also been used for bioethanol owing to improved bioconversion efficiency by reduction of lignin, which acts as a barrier to cellulase enzyme access, and altering lignin composition through biotechnological manipulations of sugarcane feedstock (Jung et al. 2013; Ko et al. 2018). Recently, techno-economic analysis of lipid extraction and biodiesel derived from metabolically engineered sugarcane, which accumulates triacylglycerol (TAG) in the vegetive tissues, were demonstrated (Arora and Singh 2020; Kumar et al. 2021; Maitra et al. 2022). Hence, the coproduction of bioethanol and biodiesel from sugarcane feedstock can be feasible through genetic manipulations using advanced biotechnological tools like genetic and metabolic engineering and genome editing.

4.2 Improving Sugar Yield and Saccharification Efficiency for Biofuel Production

Sugarcane varieties across its geographical range display diverse nature in their agricultural traits. A higher accumulation of sucrose in the whole plant and reduced lignin content are considered as the important traits in sugarcane cultivation for their selection in biofuel production (Jackson 2005). The conventional breeding programs in sugarcane consume more than a decade to release an improved variety, and being a polyploid plant, the effect of homeo-alleles with undesirable phenotype occasionally emerges in the later generations (Khan et al. 2019). Furthermore, choosing the right parental stocks with better trait heritability, high cellulose content, better disease resistance, higher sucrose accumulation, and yield characteristics is a real challenge (Meena et al. 2022). These challenges drive scientists to utilize transgenic technology for introducing desired traits into selected breeds for agronomic enhancement, especially for reduced lignin content by introducing lignin-modifying gene(s), and higher sugar content (Scortecci et al. 2012; Murugan et al. 2021). The advantage of transgene technology is that it does not interfere with or alter the previously selected breed traits, which is not the case with conventional breeding practices (Ramkumar et al. 2020).

4.2.1 Enhancing Sugar Content by Modulating the Expression of Target Genes

The Sucrose Phosphate Synthase (SPS) is an enzyme that plays an important role in the conversion of uridine diphosphate-glucose and fructose-6-phosphate into sucrose. Transgene expression of maize Sucrose Phosphate Synthase (ZmSPS) in tomatoes has increased the sucrose: starch ratio as well as increased the sucrose unloading in tomato fruits (Worrell et al. 1991; Nguyen-Quoc et al. 1999). The overexpression of ZmSPS gene in tobacco increased the sucrose synthesis in older leaves, and further accelerated the growth of the whole plant (Baxter 2003; Park et al. 2008). In sugarcane, Anur et al. (2020) overexpressed the SoSPS and reported an increase in SPS enzyme activity and total sugar content in transgenic sugarcane leaves. Additionally, they have also observed an increase in glucose and fructose levels in the leaves and a significant increase in plant height and stalk number in transgenic sugarcane when compared with non-transgenic controls.

In a similar study, Rahman et al. (2021) showed that overexpression of a Tomato ethylene responsive factor gene (TERF1) in sugarcane has improved drought tolerance and increased accumulation of soluble sugar, proline, and betaine, possibly by activating stress-related genes and increasing solutes as a defense activation by TERF1 in plants exposed to stress. Chemical treatment of sugarcane also improves sugar content. For instance, downregulation of Soluble Acid Invertase (SAI) gene expression by foliar chemical treatment resulted in improved sugar content (Jain et al. 2017). Similarly, the effect of sugarcane invertase inhibitors on the preservation of sucrose during pre- and postharvest was identified and validated that gene in *in vitro* conditions (Shivalingamurthy et al. 2018).

A study conducted by Gabriel et al. (2021) showed that manipulation of trehalose 6-phosphate (Tre6P) metabolism can alter the soluble sugars in a matured sugarcane stem. They observed that an overexpression of *E. coli* otsA (trehalose 6-phosphate synthase; TPS) gene has lowered total sucrose levels in the mature canes. Whereas, overexpression of *E. coli* otsB (trehalose 6-phosphate phosphatase; TPP) in sugarcane has increased total sucrose levels in the mature canes as compared to wild-type plant canes. There were about 15-fold increased sucrose levels in the mature internodal tissues than in the immature internodal tissues. However, in the case of immature internodal tissues, inverted sugars such as glucose and fructose levels were higher than the mature internodes. Therefore, the manipulation of Tre6P/trehalose metabolism has the potential to modify a profile of soluble sugars accumulated in mature sugarcane stems (Gabriel et al. 2021). In another study, stem-specific overexpression of Trehalulose synthase gene II in sugarcane has resulted in a higher level of sucrose concentration in mature sugarcane culms than leaf tissues. Moreover, these transgenic sugarcane lines have also yielded higher sugar recovery (14.9%) when compared to control lines (8.5%) (Awan et al. 2022).

Reduction of Pyrophosphate: D-fructose-6 phosphate 1-phosphotransferase (PFP) in sugarcane can lead to increased sucrose content in transgenic sugarcane lines (van der Merwe et al. 2010). They confirmed that an increase in sugar concentrations went up three to six-fold in PFP down-regulated transgenic sugarcane lines. The high sucrose accumulation is easily detectable in the immature internodes compared to their wild-type counterparts. Also, an increase in the hexose phosphate concentrations results from a restriction in the conversion of hexose phosphates to triose phosphates, which drives sucrose synthesis in the young internodes. However, these effects are less pronounced in mature tissues. The lower levels of pyrophosphate in older internodes might also contribute to the reduced sucrose content in the mature tissues of transgenic sugarcane containing lower levels of PFP expression (van der Merwe et al. 2010).

In general, the cellular sugar concentration has been regulated by various pathways and mechanisms that limit the manipulation of sucrose concentration beyond a certain level. To avoid these sucrose-based cellular limitations, researchers have tried to concentrate other forms of sugar molecules in sugarcane so that the total sugar concentration can be increased. For instance, successful expression of sucrose isomerase (SI) gene in vacuolar compartmentation resulted in a better increase of Iso-Maltose (IM) concentration in cane without affecting the concentration of sucrose.

4.2.2 Altering Lignin Compositions for Improving Sugar to Bioethanol Conversion Efficiency

Apart from first-generation biofuel from sugarcane, recent research involving engineering of lignocellulosic biomass of sugarcane to reduce lignin, which is recalcitrance to saccharification process. This technology proved that efficient production of second-generation biofuels by utilizing abundant lignocellulosic biomass generated in sugarcane industries. Because the second-generation biofuel production derived from lignocellulosic biomaterials has multiple barriers such as various pretreatment methods and utilization of expensive cellulose-degrading enzymes. Moreover, during the process of lignocellulosic biomass digestion many sugar molecules are degraded, the lignin content of a biomass are interfere with the digestion process, and the presence of inhibitor molecules was also observed in the case of conventional digestion processes. Lignin present in the plant cell is a hydrophobic molecule that contains a complex heterogenous mixture of polymers. Therefore, lignin acts as a barrier and poses a challenge to remove it from cellulosic fibrils without any harsh chemical treatment. The lignin embedded in a complex with cell wall polysaccharides prevents their degradation. Even after the removal of lignin, the residual lignin molecules act as inhibitors for the degrading cellulase enzymes which in turn reduce the efficiency of saccharification. Hence, in addition to developing sugarcane varieties with high sucrose content, increased biomass, or higher fiber content, attempts were made to develop sugarcane with reduced lignin content and altered cell wall components. Multiple research groups have already tried to reduce the total lignin content in sugarcane by down-regulate specific target genes that are involved in the lignin biosynthetic pathway (Khan et al. 2019).

The plant cell wall lignins are mainly made up of polymers of monolignols, hydroxyl cinnamic acids, and ferulate molecules as structural components. The lignin biosynthetic process is complex and requires the involvement of approximately 10 different enzymes and more than 25 unigenes for monolignol biosynthesis. Specific enzymatic genes such as caffeic acid *O*-methyltransferase (COMT) and cinnamyl alcohol dehydrogenase (CAD) are considered as best targets for reducing total lignin content, without hindering the plants growth and development. The COMT suppression mutants have shown reduced lignin content (Fu et al. 2011), and RNAi-mediated downregulation of cinnamyl alcohol dehydrogenase can enhance the glucose production and cellulase pretreatment efficiency in switchgrass (Saathoff et al. 2011). In sugarcane, an experiment involving RNAi-mediated downregulation of the COMT displayed a reduction in lignin levels by 3.9–13.7% (Jung et al. 2012). The transgenic lines showed lignin content as low as 156.6 compared to 181.4 mg g^{-1} of the wild-type plants. RNAi-mediated downregulation of the COMT gene has resulted in a moderate reduction of up to 8.4%, which was recommended to limit the recalcitrance of cane biomass without compromising the plant performance (Jung et al. 2012). In field experiments, COMT-suppressed transgenic lines showed 91% of transcript reduction, 12% reduction in lignin content, 19–23% increment in saccharification efficiency, reduction in one-third of hydrolysis time, and three to four-fold less requirement of hydrolytic enzymes for yielding an equal amount of fermentable sugars than the wild-type plants (Jung et al. 2013). In addition to targeting COMT gene, suppression of 4-coumarate: coenzyme A ligase (Sh4-CL), which involved in lignin biosynthesis, resulted in 16.5% reduction of lignin content and 52–76% improved saccharification efficiency (Jung et al. 2016). These results suggest that the altered lignin content and modified lignin monomers will enhance the saccharification efficiency of sugarcane lignocellulosic biomass and make it second-generation bioethanol production from sugarcane an economically viable alternative to fossil fuels.

4.2.3 Genetic Manipulation of Sugarcane to Improve Biomass Traits

In the current scenario, sugarcane has been identified as the prime feedstock for the efficient production of both first-generation and second-generation biofuels. Therefore, increasing the total biomass yield of sugarcane can increase the total amount of biofuel produced from the unit area of land. Previously, Garcia Tavares et al. (2018) identified that ScGAI, gibberellin-related gene, mediated the regulation of growth and development of sugarcane culm via modulation of the ethylene signaling pathway. They also showed that silencing of the ScGAI gene increases plant height and intermodal length and increases carbon allocation to the stem. Downregulation of ScGAI gene has suggested that increasing the biomass potential is another promising strategy for producing higher amounts of biofuels from sugarcane (Garcia Tavares et al. 2018).

4.3 Expression of Transgenes for Triacyl Glycerol Production

Sugarcane is a potential production platform as value-added fuels, and chemicals are synthesized and stored on the vegetive tissues (Altpeter and Oraby 2010). It has been demonstrated that, in contrast to oilseed crops, neutral lipids such as triacyl glycerol (TAG) or oil can be stored in the leaves and stems of sugarcane (Zale et al. 2016; Kannan et al. 2022) and energycane, a cellulosic biomass-yielding energy crop derived from sugarcane (Luo et al. 2022; Cao et al. 2023a). Elevated TAG accumulation in vegetative tissues was made possible through metabolic engineering of multigenes, which includes at least three simultaneous actions of fatty acid synthesis, conversion of diacyl glycerol (DGAT) to TAG, and protection of TAG from degradation by "PUSH-PULL-PROTECT" strategies, respectively (Vanhercke et al. 2014a).

4.3.1 Fatty Acid Synthesis

Fatty acids are carboxylic acids and building blocks of fats. It is made up of a hydrocarbon chain. Fat or plant oils' energy content is twice (9 kcal) as that of carbohydrates (Ohlrogge and Chapman 2011) and has 25% more energy than ethanol. Fatty acids are synthesized in plastids de novo by diverting carbon flux from sugars, which accumulated during photosynthesis. Then, fatty acids are exported to endoplasmic reticulum (ER) and assembled into neutral storage lipid or TAG. The TAG molecules are packed into an oil droplet with phospholipid monolayer and coated with structural protein Oleosin (Ole) to increase the stability of synthesized TAGs and oil droplets then released into cytoplasm of plant cells. TAG molecules can be extracted from vegetative tissues and converted to biodiesel by transesterification process (Lourinho and Brito 2015). In nature, a small amount of TAG about 0.05% synthesized in plant vegetative cells and used for glycerolipid cell membranes synthesis (Xu and Shanklin 2016). Therefore, overexpression of TAG synthesis's enzymatic machinery through metabolic engineering would enhance the hyperaccumulation of oil bodies in sugarcane vegetative biomass. In sugarcane, metabolic engineering for co-expressing Wrinkled1 (WRI1), diacylglycerol acyltransferase1-2 (DGAT1-2), oleosin1 (OLE1), and suppressing peroxisomal ATP-binding cassette transporter 1 (PXA1) contributed to a total fatty acid content elevated up to 4.7% or 1.7% in leaves or stems dry weight, respectively (Zale et al. 2016). In case of energycane, total fatty acid content increased up to 4.4% of its leaf dry weight (Luo et al. 2022). Further elevation of total fatty acid content (12.9% of its leaf dry weight) has been reached in energycane in which intron-mediated enhancement of DGAT1 combined with other lipogenic factors (Cao et al. 2023b).

The studies on the model plant *Arabidopsis* showed that transcription factors such as WRI1 (*Wrinkled1*) and Lec2 (*Leafy cotyledons2*) or committed enzymes like ACCase (Acetyl-CoA carboxylase) were involved in the fatty acid biosynthesis (Vanhercke et al. 2014b). WRI1 is a positive and master regulator of fatty acid biosynthesis and belongs to APETALA2 (AP2)/ethylene-responsive element-binding protein (EREBP) subfamily. For instance, overexpression of the WRI1 in *A. thaliana* yielded a 2.8-fold increase in TAG content (Sanjaya et al. 2011). Overexpression of ZmWRI1 in maize also increased seed oil content (Shen et al. 2010). However, in sugarcane, WRI1 has been introduced as a part of multigenic construct since synergistic effects of multigenic construct on the enhanced accumulation of TAG is superior to a single gene construct (Zale et al. 2016; Parajuli et al. 2020). Because, expression of WRI1, DGAT1-2, or Ole1 yielded 1.5- to 9.5-fold increase of TAG content in sugarcane, whereas co-expression of those genes resulted in 95-fold increase of leaf TAG content (Zale et al. 2016). Synergistic effects of WRI1 and DGAT1 for oil accumulation in *Nicotiana benthamiana* exceed the additive effects of those genes individually expressed (Vanhercke et al. 2013). Sorghum, a closely related species of sugarcane, was also transformed with ZmWRI1, UrDGAT2a. and SiOle genes and led to 3–8.4% increased TAG content on the leaves (Vanhercke et al. 2019). Lec2 is also a transcription factor and acts on the upstream of WRI1 and induces the expression of WRI1. Expression of Lec2 displayed increased TAG content in Arabidopsis (Kim et al. 2015; Alameldin et al. 2017). However, constitutive expression of Lec2 affects plants' growth and development in Arabidopsis (Kim et al. 2015). There have been no reports of Lec2 expression in sugarcane for TAG synthesis. ACCase is the first committed enzyme involved in plastidial fatty acid synthesis and it converts Acetyl-CoA to malonyl-CoA in plastids. This conversion is a potentially rate-limiting step in fatty acid synthesis. Plant contains ACCase in both plastids and cytosol of the cell, and light signal strongly modulates the plastidial ACCase activity. Therefore, during light, de novo fatty acid synthesis is higher than dark condition (Sasaki et al. 1997). Except WRI1, Lec2 or ACCase expressions in sugarcane for TAG synthesis have not yet been reported.

4.3.2 Conversion of DGAT to TAG

Fatty acids synthesized in the plastids and those free fatty acids are transported to ER and sequentially acylated to form diacyl glycerol (DAG) in Kennedy pathway. DGAT is involved in the conversion of DAG to TAGs or it can be used as a substrate for membrane lipid biosynthesis. DGAT is the final rate-limiting enzyme for the TAG assembly in ER. In case of an acyl-CoA-independent pathway, phosphatidylcholine (PC) pool is formed from free fatty acids in ER, and TAG is assembled directly from the PC pool by phospholipid:diacylglycerol acyltransferase 1 (PDAT1) (Xu and Shanklin 2016; Xu et al. 2018). It appears that the amount of PDAT1 and DGAT1 transcript levels varies between developmental stages of the plant. For instance, developing and matured leaves have higher level of PDAT1, whereas senescence leaf contains higher levels of DGAT1 (Fan et al. 2013; Xu and Shanklin 2016). Though overexpression of PDAT1 in sugarcane for TAG synthesis has not been reported, overexpression of DGAT1 as a part multigene for TAG synthesis has been demonstrated in sugarcane and energycane. Intron-mediated enhancement of DGAT1 expression in energycane indicates that DGAT1 is a rate-limiting lipogenic factor for TAG assembly in ER (Cao et al. 2023b).

4.3.3 Protection of Lipid Droplets

Assembled TAG molecules are initially covered by phospholipid monolayer and then coated with structural protein Oleosin and form oil bodies or lipid droplets. As TAGs, Oleosin and phospholipids are also synthesized in ER and provides stability to TAGs and prevent it from coalescing and hydrolysis by lipases (Hsieh and Huang 2004). Further, engineered oleosins with strategically placed multiple cysteine residues were developed to increase the stability of Oleosin and used for metabolic engineering in plants. Moreover, recently, mutant oleosin containing one or more amino acid substitutions (arginine substitution for lysine including deletion of few amino acids) were developed and that the mutant version of oleosin provides higher TAGs compared to wild-type version of Oleosin (Shanklin and Anaokar 2021). In addition to Oleosin, genes involved in lipid catabolism such as peroxisomal β-oxidation (PXA1), lipid hydrolysis (sugar-dependent 1, SDP1), or reimport of phospholipids into the plastid by trigalactosyl diacylglycerol 1 (TGD1), were also suppressed by RNA interference (RNAi) approach to increase TAG content in the vegetative tissues of sugarcane (Zale et al. 2016; Parajuli et al. 2020; Luo et al. 2022).

Metabolically engineered sugarcane plants were developed by co-expression of WRI1, DGAT1-2, Ole1, and co-suppression of ADP-glucose pyrophosphorylase (AGPase) and PXA1 and evaluated in greenhouse condition displayed that 95-fold (1.9% of its dry weight) increased TAG content in leaves (Zale et al. 2016). These plants were evaluated under field conditions over a 2-year period to study the stable expression of transgenes and lipogenic factors. TAG accumulation in field-grown engineered plants was 70-fold higher (4.4% of leaf dry weight) than non-transgenic control and about two times higher TAG content than that grown in greenhouse (Kannan et al. 2022). In a similar approach, Parajuli et al. (2020) introduced large multigenic construct containing WRI1 from Sorghum, engineered version of CysOle1, which harbors 6 cysteine residues, DGAT1-2 and RNAi suppression of SDP1 and TGD1 was introduced into sugarcane and reported that TAG content in stem reached to 4.3% and leaf TAG content reached up to 8.0% of its dry weight, which is equivalent to 400-fold increase as compared to non-transgenic sugarcane. The dramatic increase of 400-fold change in TAG accumulation was attributed to the alternative version of gene variants and level of transgene expressions than reported in previous studies (Parajuli et al. 2020). For instance, Cao et al. (2023b) were compared variant of codon-optimized TmDGAT1 from *Tropaeolum majus* with and without intron along with co-expression of WRI1 and Ole1 in greenhouse experiment, and engineered energycane plants displayed 192-fold increased TAG content than non-transformed energycane and 3.8-fold increase than its intron less counterpart. In another study, intron-mediated enhanced expression of *Tropaeolum majus* DGAT1 combined with wildtype version of WRI1 and Ole1 displayed 245-fold (4.8%) increased TAG content and 330-fold (6.5%) increased fatty acid content in energycane leaves in field conditions. Moreover, some sections of energycane leaves displayed up to 9.9% TAG and 12.9% fatty acids in its leaf dry weight (Cao et al. 2023a). A remarkable increase of TAG content in the vegetative tissues of sugarcane and energycane through metabolic engineering proves their potential as advanced biofuel feedstocks for bioenergy applications.

4.3.4 Conversion of TAG into Biodiesel

Several reports were published describing the conversion of sugar and lignocellulosic biomass of sugarcane into bioethanol (Naik et al. 2010; Dias et al. 2014; Jung et al. 2016; Kannan et al. 2018; Ko et al. 2018; Vieira et al. 2020). However, biodiesel production from metabolically engineered sugarcane is in its early stages unlike biodiesel conversion from oilseed crops. Recently, a comprehensive techno-economic analysis for the feasibility of co-producing biodiesel and bioethanol from metabolically engineered sugarcane and energycane was assessed (Huang et al. 2016; Kumar et al. 2021). Pilot scale processing for co-production of lipids and fermentable sugars from engineered oil-producing sugarcane (oilcane) feedstock was performed. Lipid yield of 0.20 tons/ha was recovered from vegetative tissues of oilcane with 55% recovery efficiency and 3.3%, 3.6%, and 0.4% lipid content in leaves, bagasse, and juice, respectively. With 100% recovery efficiency, lipid yield is projected to 0.35 tons/ha, which is closer to soyabean lipid yield of 0.44 tons/ha (Maitra et al. 2022). Even with 55% recovery efficiency, enhanced TAG content that recently achieved in engineered energycane could results in 35% more oil yield than that of 0.20 tons/ha reported (Cao et al. 2023a). Sugarcane can accumulate sucrose in the stems about 20% of its dry weight and modeling suggests that this sucrose can be diverted into lipid biosynthesis pathway by refining metabolic engineering strategies and can increase of TAG content up to 20% of its dry weight (Huang et al. 2016). Therefore, enhanced accumulation of TAG content in sugarcane vegetative biomass combined with improved lipid extraction efficiency, lipid yield of engineered sugarcane will surpass the soybean lipid yield per hectare of land area.

4.4 Genome Editing Approaches in Sugarcane for Biofuel Production

Genome editing has revolutionized as a possible alternative for accelerated improvement of polyploid crops (May et al. 2023; Mall et al. 2024). Genome-editing technologies for targeted mutagenesis in plants using engineered site-specific nucleases such as mega nucleases, zinc-finger nucleases (ZFNs), transcription activator-like effector nucleases (TALENs), and clustered regularly interspaced short palindromic repeats and CRISPR-associated protein (CRISPR/Cas) have been successfully utilized for crop improvement (May et al. 2023). Despite their utilization in various crop improvement approaches, genome editing technologies have yet to be reached their full potential for sugarcane crop improvement. The highly complex, polyploid and heterozygous nature of sugarcane genome poses significant challenges for targeted gene editing. Among these specific nucleases, TALENs and CRISPR/Cas9 have been employed for targeted mutagenesis in sugarcane (Jung and Altpeter 2016; Kannan et al. 2018; Eid et al. 2021; Oz et al. 2021; Laksana et al. 2024). The genome editing methodologies have opened the doors for biofuel production from vegetative tissues of sugarcane.

4.4.1 TALEN-Mediated Genome Editing for Biofuel Production

As the rise in energy resource challenges and their environmental impacts, it is vital to utilize sustainable energy obtained from lignocellulosic biomass (Wang et al. 2021). The bioethanol production from lignocellulosic biomass has a profound and feasible choice to petroleum fuels for overcoming energy crisis around the globe

(Wang et al. 2021). However, the recalcitrant nature of sugarcane lignocellulosic biomass limits their bioconversion into simple sugars and bioethanol (Vieira et al. 2020). High concentration of lignin and its monomer syringyl to guaiacyl ratio in the bagasse is the key component of saccharification in sugarcane (Jung et al. 2012, 2016; Jung and Altpeter 2016). Therefore, it is necessary to alter the constituents of lignin to enable cellulase enzymes to access lignocellulose molecules for the production of ethanol. Genome editing methodologies (TALENs, ZFNs, and CRISPR/Cas) could be useful in altering the lignin monomers by editing their genes (Caffeic acid *O*-methyltransferase (*COMT*), caffeoyl-CoA *O*-methyltransferase (*CCoAOMT*), and *Ferulate 5-hydroxylase* (F5H)) involved in the lignin biosynthetic pathway. The sequence-specific nucleases (SSNs) such as TALENs contain a nonspecific DNA-cleaving nuclease (Fok1) fused to a DNA-binding domain of 12–28 repeats, a nuclear localization signal (NLS), and an acidic domain for target gene transcription activation that can be easily engineered to make site-specific gene modifications. TALEN-mediated genome editing has been reported to create site-specific changes in sugarcane genome to improve cell wall composition for biofuel production and enhanced saccharification efficiency (Jung and Altpeter 2016; Kannan et al. 2018). TALEN-mediated site-specific changes of the lignin biosynthetic gene, *caffeic acid O-methyltransferase* (*COMT*) to reduce the lignin content was reported in sugarcane (Jung and Altpeter 2016). They targeted a conserved region of the first exon of sugarcane *COMT* gene using TALEN to edit multiple alleles of the sugarcane, and TALEN cassette were introduced into sugarcane genome via agrobacterium-mediated transformation or minimal TALEN expression cassette was introduced via biolistic-mediated transformation to target *COMT* gene. To enable screening of large number of *COMT* gene edited lines, capillary electrophoresis method has also been used and found up to 74% target mutation in *COMT* gene edited lines. Lignin reduction was reported from 29% to 32% in edited lines and it was positively correlated with mutation frequency of *COMT* gene. Further, TALEN-mediated mutagenesis targeted more than 100 *COMT* copies/alleles in sugarcane and results in improved saccharification efficiency without compromising biomass yield (Kannan et al. 2018). It appears that the conserved domain of *COMT* gene in sugarcane plants was targeted and that led to mutation of more than 100 *COMT* copies/alleles. Field-grown *COMT* mutants exhibited improved saccharification efficiency of up to 43.8% with 19.7% of lignin reduction compared to wild-type sugarcane plants (Kannan et al. 2018). Moreover, *COMT* mutants did not show any significant reduction in biomass yield in the field conditions compared to wild-type plants. *COMT* mutants without compromising biomass yield will serve as an efficient feedstock for bioethanol production. The combination of TALEN-mediated *COMT* edited line and engineered yeast strain was used to co-ferment glucose and xylose and resulted in 148% more ethanol production than non-edited wild-type sugarcane plants (Ko et al. 2018). This result indicates the economic feasibility of bioethanol production from genome-edited sugarcane plants.

4.4.2 CRISPR/Cas9-Mediated Genome Editing

CRISPR/Cas technology has revolutionized genetic engineering by enabling precise and efficient modification of DNA sequences in several model plants including annual, biennial, and perennial plants (Wada et al. 2020). Due to large genome size and highly polyploid nature of sugarcane genome with introgression of *Saccharum spontaneum* genome makes it difficult to edit all the alleles/copies of the target gene. Moreover, polygenic nature of yield-related traits, limited availability of genomic resources, and redundancy of gene copies in sugarcane also cause challenges in identifying desirable target regions for CRISPR/Cas editing. After successful application of *TALEN*-mediated genome editing in sugarcane to target multiple copies/alleles of a gene, Fredy Altpeter's group from University of Florida reported the proof of concept of CRISPR/Cas9-mediated editing of multiple alleles of *magnesium chelatase subunit I,* a key enzyme for chlorophyll biosynthesis (Eid et al. 2021). This study provided a rapidly scorable phenotype in polyploid sugarcane for multiallelic editing and further expanded the genome editing scope for the optimization of editing reagents and delivery methods for efficient co-editing of large number of alleles. Same group from University of Florida reported the efficient co-editing of three copies of acetolactate synthase (*ALS*) gene via template-mediated and homology-directed repair (HDR) of DNA double strand breaks induced by CRISPR/Cas9 nuclease (Oz et al. 2021). The edited sugarcane lines showed a targeted nucleotide substitution in both W574L and S653I in *ALS*-targeted region and displayed tolerance to herbicide. This study shows the conversion of inferior alleles to superior alleles through targeted nucleotide substitutions for sugarcane crop improvement.

4.4.3 CRISPR/Cas9 Technology for Biofuel Production

Recently, a research group from Thailand reported the lignin reduction in sugarcane stalks by targeted mutagenesis of *SoLIM* gene (LIM domain transcription factor) via CRISPR/Cas9 technology (Laksana et al. 2024). Three single gRNAs targeting exon 2 and exon 5 of the *SoLIM* gene were introduced into the calli of sugarcane via *Agrobacterium*-mediated transformation. The edited lines with single nucleotide insertions and deletions displayed the downregulation of lignin biosynthesis genes (*SoPAL*, *SoC4H*, and *SoCAD*). Consequently, the gene edited lines showed a reduction of lignin content of 9.7–51.5% compared to the wild-type sugarcane plants. The mutated sugarcane lines with low lignin content will provide a reliable feed stock for biofuel production from sugarcane plants. Although CRISPR/Cas9-mediated genome editing has not widely been used in sugarcane to target biofuel traits, above reports indicate the CRISPR/Cas9 has enormous potential to improve biofuel traits through precise gene modifications in sugarcane.

4.5 Conclusions and Future Prospects

The theoretical maximum biomass yield of sugarcane can be reached at 177 tons/ha year in the tropical and subtropical regions of the world (Waclawovsky et al. 2010). Given the sugarcane's superior biomass yield, developing sugarcane feedstock with elevated TAG content and improved lipid extraction method will make the metabolically engineered sugarcane as a viable feedstock for sustainable biodiesel production. Recent reports confirm these strategies are feasible in sugarcane or energycane owing to advances in biotechnological manipulations (Maitra et al. 2022; Cao et al. 2023a). Currently, significant yield reductions were reported in the engineered sugarcane with elevated lipid accumulations (Parajuli et al. 2020; Kannan et al. 2022; Cao et al. 2023a). Therefore, use of genomic resources and advanced biotechnological tools to overcome this yield drag by directing the TAG or fatty acids accumulation to later stages of plant growth or at the maturity stage of plants would be critical to increase lipid yield without compromising biomass yield. This strategy can be achieved by the use of stem tissue-specific, chemical or temperature inducible or developmentally regulated promoters of lipid genes. Further refinement of metabolic engineering strategies like intron-mediated enhancement of all lipid pathway genes, use of engineered lipid gene variants, gene staking of additional lipogenic factors, and knocking out lipid droplet catabolizing genes using precise genome editing tools can boost TAG accumulation in the vegetative tissues of sugarcane biomass. In addition, improvement in the bioprocessing of oilcane feedstock by various pretreatments and crop management practices can also increase the lipid yield of oilcanes.

The use of targeted gene editing technology in sugarcane is still in its infancy in terms of finding specific targets due to highly polyploidy genome and applicability in order to become an efficient method for precision sugarcane genome editing. However, recent efforts in targeting multiple alleles/copies of a gene in the sugarcane genome demonstrated that genome editing is highly feasible in sugarcane. Considering the interest in developing a sugarcane cultivar with improved biofuel traits, CRISPR/Cas9 mediated genome editing would serve as the precise and accelerated approach. Specifically, in polyploid crops like sugarcane, it is necessary to edit all alleles/copies of genes with the same function due to its functional redundancy between para homologous genes. By considering this, optimization of Cas9 expression and identifying a target region could impact the editing efficiency in sugarcane. The mutation analysis is also the critical task in genome editing due to the lack of high-throughput screening methods, which is important for implementing the CRISPR/Cas technology in highly polyploid sugarcane. However, it has been reported that targeted mutations in sugarcane genome would help in biofuel production by increasing the saccharification efficiency through altered lignocellulose constituents. Further, CRISPR/Cas technology will serve as the important tool for the precise editing of biofuel traits.

References

Alameldin H, Izadi-Darbandi A, Smith SA, Balan V, Jones AD, Sticklen M (2017) Production of seed-like storage lipids and increase in oil bodies in corn (Maize; Zea mays L.) vegetative biomass. Ind Crops Prod 108:526–534. https://doi.org/10.1016/j.indcrop.2017.07.021

Altpeter F, Oraby H (2010) Sugarcane. In: Biotechnology in agriculture and forestry. Springer International Publishing, Berlin, pp 453–472

Anur RM, Mufithah N, Sawitri WD, Sakakibara H, Sugiharto B (2020) Overexpression of sucrose phosphate synthase enhanced sucrose content and biomass production in transgenic sugarcane. Plants 9(2):200. https://doi.org/10.3390/plants9020200

Arora A, Singh V (2020) Biodiesel production from engineered sugarcane lipids under uncertain feedstock compositions: process design and techno-economic analysis. Appl Energy 280:115933. https://doi.org/10.1016/j.apenergy.2020.115933

Awan MF, Ali S, Iqbal MS, Sharif MN, Ali Q, Nasir IA (2022) Enhancement of healthful novel sugar contents in genetically engineered sugarcane juice integrated with molecularly characterized ThSyGII (CEMB-SIG2). Sci Rep 12(1):18621. https://doi.org/10.1038/s41598-022-23130-y

Baxter CJ (2003) Elevated sucrose-phosphate synthase activity in transgenic tobacco sustains photosynthesis in older leaves and alters development. J Exp Bot 54(389):1813–1820. https://doi.org/10.1093/jxb/erg196

Cao VD, Kannan B, Luo G, Liu H, Shanklin J, Altpeter F (2023a) Triacylglycerol, total fatty acid, and biomass accumulation of metabolically engineered energycane grown under field conditions confirms its potential as feedstock for drop in fuel production. GCB Bioenergy 15(12):1450–1464. https://doi.org/10.1111/gcbb.13107

Cao VD, Luo G, Korynta S, Liu H, Liang Y, Shanklin J, Altpeter F (2023b) Intron-mediated enhancement of DIACYLGLYCEROL ACYLTRANSFERASE1 expression in energycane promotes a step change for lipid accumulation in vegetative tissues. Biotechnol Biofuels Bioprod 16(1):153. https://doi.org/10.1186/s13068-023-02393-1

Dias MOS, Cavalett O, Filhob RM, Bonomi A (2014) Integrated first and second generation ethanol production from sugarcane. Chem Eng Trans 37:445–450. https://doi.org/10.3303/CET1437075

Dinesh Babu KS, Janakiraman V, Palaniswamy H, Kasirajan L, Gomathi R, Ramkumar TR (2022) A short review on sugarcane: its domestication, molecular manipulations and future perspectives. Genet Resour Crop Evol 69(8):2623–2643. https://doi.org/10.1007/s10722-022-01430-6

Eid A, Mohan C, Sanchez S, Wang D, Altpeter F (2021) Multiallelic, targeted mutagenesis of magnesium chelatase with CRISPR/Cas9 provides a rapidly scorable phenotype in highly polyploid sugarcane. Front Genome Ed 3:654996. https://doi.org/10.3389/fgeed.2021.654996

Fan J, Yan C, Zhang X, Xu C (2013) Dual role for phospholipid:diacylglycerol acyltransferase: enhancing fatty acid synthesis and diverting fatty acids from membrane lipids to triacylglycerol in arabidopsis leaves. Plant Cell 25(9):3506–3518. https://doi.org/10.1105/tpc.113.117358

Fu C, Mielenz JR, Xiao X, Ge Y, Hamilton CY, Rodriguez M, Chen F, Foston M, Ragauskas A, Bouton J, Dixon RA, Wang Z-Y (2011) Genetic manipulation of lignin reduces recalcitrance and improves ethanol production from switchgrass. Proc Natl Acad Sci 108(9):3803–3808. https://doi.org/10.1073/pnas.1100310108

Gabriel C, Fernhout J, Fichtner F, Feil R, Lunn JE, Kossmann J, Lloyd JR, van der Vyver C (2021) Genetic manipulation of trehalose-6-phosphate synthase results in changes in the soluble sugar profile in transgenic sugarcane stems. Plant Direct 5(11):e358. https://doi.org/10.1002/pld3.358

Garcia Tavares R, Lakshmanan P, Peiter E, O'Connell A, Caldana C, Vicentini R, Soares JS, Menossi M (2018) ScGAI is a key regulator of culm development in sugarcane. J Exp Bot 69(16):3823–3837. https://doi.org/10.1093/jxb/ery180

Hsieh K, Huang AHC (2004) Endoplasmic reticulum, oleosins, and oils in seeds and tapetum cells. Plant Physiol 136(3):3427–3434. https://doi.org/10.1104/pp.104.051060

Huang H, Long S, Singh V (2016) Techno-economic analysis of biodiesel and ethanol co-production from lipid-producing sugarcane. Biofuels Bioprod Biorefin 10(3):299–315. https://doi.org/10.1002/bbb.1640

Jackson PA (2005) Breeding for improved sugar content in sugarcane. Field Crop Res 92(2–3):277–290. https://doi.org/10.1016/j.fcr.2005.01.024

Jain R, Singh SP, Singh A, Singh S, Kishor R, Singh RK, Chandra A, Solomon S (2017) Soluble acid invertase (SAI) activity and gene expression controlling sugar composition in sugarcane. Sugar Tech 19(6):669–674. https://doi.org/10.1007/s12355-017-0511-0

Jung JH, Altpeter F (2016) TALEN mediated targeted mutagenesis of the caffeic acid O-methyltransferase in highly polyploid sugarcane improves cell wall composition for production of bioethanol. Plant Mol Biol 92(1–2):131–142. https://doi.org/10.1007/s11103-016-0499-y

Jung JH, Fouad WM, Vermerris W, Gallo M, Altpeter F (2012) RNAi suppression of lignin biosynthesis in sugarcane reduces recalcitrance for biofuel production from lignocellulosic biomass. Plant Biotechnol J 10(9):1067–1076. https://doi.org/10.1111/j.1467-7652.2012.00734.x

Jung JH, Vermerris W, Gallo M, Fedenko JR, Erickson JE, Altpeter F (2013) RNA interference suppression of lignin biosynthesis increases fermentable sugar yields for biofuel production from field-grown sugarcane. Plant Biotechnol J 11(6):709–716. https://doi.org/10.1111/pbi.12061

Jung JH, Kannan B, Dermawan H, Moxley GW, Altpeter F (2016) Precision breeding for RNAi suppression of a major 4-coumarate:coenzyme A ligase gene improves cell wall saccharification from field grown sugarcane. Plant Mol Biol 92(4–5):505–517. https://doi.org/10.1007/s11103-016-0527-y

Kannan B, Jung JH, Moxley GW, Lee SM, Altpeter F (2018) TALEN-mediated targeted mutagenesis of more than 100 COMT copies/alleles in highly polyploid sugarcane improves saccharification efficiency without compromising biomass yield. Plant Biotechnol J 16(4):856–866. https://doi.org/10.1111/pbi.12833

Kannan B, Liu H, Shanklin J, Altpeter F (2022) Towards oilcane: preliminary field evaluation of metabolically engineered sugarcane with hyper-accumulation of triacylglycerol in vegetative tissues. Mol Breed 42(10):64. https://doi.org/10.1007/s11032-022-01333-5

Khan MT, Khan IA, Yasmeen S (2019) Genetically modified sugarcane for biofuels production: status and perspectives of conventional transgenic approaches, RNA interference, and genome editing for improving sugarcane for biofuels. In: Khan MT, Khan IA (eds) Sugarcane biofuels. Springer International Publishing, Cham, pp 67–96

Kim HU, Lee K, Jung S, Shin HA, Go YS, Suh M, Kim JB (2015) Senescence-inducible LEC2 enhances triacylglycerol accumulation in leaves without negatively affecting plant growth. Plant Biotechnol J 13(9):1346–1359. https://doi.org/10.1111/pbi.12354

Ko JK, Jung JH, Altpeter F, Kannan B, Kim HE, Kim KH, Alper HS, Um Y, Lee SM (2018) Largely enhanced bioethanol production through the combined use of lignin-modified sugarcane and xylose fermenting yeast strain. Bioresour Technol 256:312–320. https://doi.org/10.1016/j.biortech.2018.01.123

Kumar D, Long SP, Arora A, Singh V (2021) Techno-economic feasibility analysis of engineered energycane-based biorefinery co-producing biodiesel and ethanol. GCB Bioenergy 13(9):1498–1514. https://doi.org/10.1111/gcbb.12871

Laksana C, Sophiphun O, Chanprame S (2024) Lignin reduction in sugarcane by performing CRISPR/Cas9 site-direct mutation of SoLIM transcription factor. Plant Sci 340:111987. https://doi.org/10.1016/j.plantsci.2024.111987

Lourinho G, Brito P (2015) Advanced biodiesel production technologies: novel developments. Rev Environ Sci Biotechnol 14(2):287–316. https://doi.org/10.1007/s11157-014-9359-x

Luo G, Cao VD, Kannan B, Liu H, Shanklin J, Altpeter F (2022) Metabolic engineering of energycane to hyperaccumulate lipids in vegetative biomass. BMC Biotechnol 22(1):24. https://doi.org/10.1186/s12896-022-00753-7

Maitra S, Viswanathan MB, Park K, Kannan B, Alfanar SC, McCoy SM, Cahoon EB, Altpeter F, Leakey ADB, Singh V (2022) Bioprocessing, recovery, and mass balance of vegetative lipids from metabolically engineered "oilcane" demonstrates its potential as an alternative feedstock for drop-in fuel production. ACS Sustain Chem Eng 10(50):16833–16844. https://doi.org/10.1021/acssuschemeng.2c05327

Mall AK, Manimekalai R, Misra V, Pandey H, Srivastava S, Sharma A (2024) CRISPR/Cas-Mediated Genome Editing for Sugarcane Improvement. Sugar Tech. 1–13. https://doi.org/10.1007/s12355-023-01352-2

May D, Paldi K, Altpeter F (2023) Targeted mutagenesis with sequence specific nucleases for accelerated improvement of polyploid crops: progress, challenges, and prospects. Plant Genome 16(2):e20298. https://doi.org/10.1002/tpg2.20298

Meena MR, Appunu C, Arun Kumar R, Manimekalai R, Vasantha S, Krishnappa G, Kumar R, Pandey SK, Hemaprabha G (2022) Recent advances in sugarcane genomics, physiology, and phenomics for superior agronomic traits. Front Genet 13:854936. https://doi.org/10.3389/fgene.2022.854936

Murugan N, Mohan C, Kannan B (2021) RNAi-based gene silencing in sugarcane for production of biofuel. In: Basu C (ed) Biofuels and biodiesel. Springer, New York, pp 141–155

Naik SN, Goud VV, Rout PK, Dalai AK (2010) Production of first and second generation biofuels: a comprehensive review. Renew Sust Energ Rev 14(2):578–597. https://doi.org/10.1016/j.rser.2009.10.003

Nguyen-Quoc B, N'Tchobo H, Foyer CH, Yelle S (1999) Overexpression of sucrose phosphate synthase increases sucrose unloading in transformed tomato fruit. J Exp Bot 50(335):785–791. https://doi.org/10.1093/jxb/50.335.785

OECD/FAO (2023) OECD-FAO agricultural outlook 2023-2032. OECD Agriculture statistics (database). https://doi.org/10.1787/agr-outl-data-en

Ohlrogge J, Chapman K (2011) The seeds of green energy. Bioechem Soc 33(2):34–38. https://doi.org/10.1042/BIO03302034

Oz MT, Altpeter A, Karan R, Merotto A, Altpeter F (2021) CRISPR/Cas9-mediated multi-allelic gene targeting in sugarcane confers herbicide tolerance. Front Genome Ed 3:673566. https://doi.org/10.3389/fgeed.2021.673566

Parajuli S, Kannan B, Karan R, Sanahuja G, Liu H, Garcia-Ruiz E, Kumar D, Singh V, Zhao H, Long S, Shanklin J, Altpeter F (2020) Towards oilcane: engineering hyperaccumulation of triacylglycerol into sugarcane stems. GCB Bioenergy 12(7):476–490. https://doi.org/10.1111/gcbb.12684

Park J-Y, Canam T, Kang K-Y, Ellis DD, Mansfield SD (2008) Over-expression of an arabidopsis family A sucrose phosphate synthase (SPS) gene alters plant growth and fibre development. Transgenic Res 17(2):181–192. https://doi.org/10.1007/s11248-007-9090-2

Rahman MA, Wu W, Yan Y, Bhuiyan SA (2021) Overexpression of TERF1 in sugarcane improves tolerance to drought stress. Crop Pasture Sci 72(4):268–279. https://doi.org/10.1071/CP20161

Ramkumar TR, Lenka SK, Arya SS, Bansal KC (2020) A short history and perspectives on plant genetic transformation. In: Rustgi S, Luo H (eds) Biolistic DNA delivery in plants: methods and protocols. Springer, New York, pp 39–68

Saathoff AJ, Sarath G, Chow EK, Dien BS, Tobias CM (2011) Downregulation of cinnamyl-alcohol dehydrogenase in switchgrass by RNA silencing results in enhanced glucose release after cellulase treatment. PLoS One 6(1):e16416. https://doi.org/10.1371/journal.pone.0016416

Sanjaya DTP, Weise SE, Benning C (2011) Increasing the energy density of vegetative tissues by diverting carbon from starch to oil biosynthesis in transgenic Arabidopsis. Plant Biotechnol J 9(8):874–883. https://doi.org/10.1111/j.1467-7652.2011.00599.x

Sasaki Y, Kozaki A, Hatano M (1997) Link between light and fatty acid synthesis: thioredoxin-linked reductive activation of plastidic acetyl-CoA carboxylase. Proc Natl Acad Sci 94(20):11096–11101. https://doi.org/10.1073/pnas.94.20.11096

Scortecci KC, Creste S, Calsa T Jr, Xavier MA, Landell MGA, Figueira A, Benedito VA (2012) Challenges, opportunities and recent advances in sugarcane breeding. In: Plant breeding. IntechOpen, London, pp 267–296

Shanklin J, Anaokar SP (2021) Modified form of oleosin that when expressed in plants leads to increased triacylgycerol (oil) accumulation. https://patents.google.com/patent/US20210261632A1/en

Shen B, Allen WB, Zheng P, Li C, Glassman K, Ranch J, Nubel D, Tarczynski MC (2010) Expression of ZmLEC1 and ZmWRI1 increases seed oil production in maize. Plant Physiol 153(3):980–987. https://doi.org/10.1104/pp.110.157537

Shivalingamurthy SG, Anangi R, Kalaipandian S, Glassop D, King GF, Rae AL (2018) Identification and functional characterization of sugarcane invertase inhibitor (ShINH1): a potential candidate for reducing pre- and post-harvest loss of sucrose in sugarcane. Front Plant Sci 9:598. https://doi.org/10.3389/fpls.2018.00598

Tew TL, Cobill RM (2008) Genetic improvement of sugarcane (Saccharum spp.) as an energy crop. In: Vermerris W (ed) Genetic improvement of bioenergy crops. Springer, New York, pp 273–294

van der Merwe MJ, Groenewald J-H, Stitt M, Kossmann J, Botha FC (2010) Downregulation of pyrophosphate: d-fructose-6-phosphate 1-phosphotransferase activity in sugarcane culms enhances sucrose accumulation due to elevated hexose-phosphate levels. Planta 231(3):595–608. https://doi.org/10.1007/s00425-009-1069-1

Vanhercke T, El Tahchy A, Shrestha P, Zhou XR, Singh SP, Petrie JR (2013) Synergistic effect of WRI1 and DGAT1 coexpression on triacylglycerol biosynthesis in plants. FEBS Lett 587(4):364–369. https://doi.org/10.1016/j.febslet.2012.12.018

Vanhercke T, El Tahchy A, Liu Q, Zhou X, Shrestha P, Divi UK, Ral J, Mansour MP, Nichols PD, James CN, Horn PJ, Chapman KD, Beaudoin F, Ruiz-López N, Larkin PJ, Feyter RC, Singh SP, Petrie JR (2014a) Metabolic engineering of biomass for high energy density: oilseed-like triacylglycerol yields from plant leaves. Plant Biotechnol J 12(2):231–239. https://doi.org/10.1111/pbi.12131

Vanhercke T, Petrie JR, Singh SP (2014b) Energy densification in vegetative biomass through metabolic engineering. Biocatal Agric Biotechnol 3(1):75–80. https://doi.org/10.1016/j.bcab.2013.11.010

Vanhercke T, Belide S, Taylor MC, El Tahchy A, Okada S, Rolland V, Liu Q, Mitchell M, Shrestha P, Venables I, Ma L, Blundell C, Mathew A, Ziolkowski L, Niesner N, Hussain D, Dong B, Liu G, Godwin ID, Lee J, Rug M, Zhou X, Singh SP, Petrie JR (2019) Up-regulation of lipid biosynthesis increases the oil content in leaves of Sorghum bicolor. Plant Biotechnol J 17(1):220–232. https://doi.org/10.1111/pbi.12959

Vieira S, Barros MV, Sydney ACN, Piekarski CM, de Francisco AC, de Souza Vandenberghe LP, Sydney EB (2020) Sustainability of sugarcane lignocellulosic biomass pretreatment for the production of bioethanol. Bioresour Technol 299:122635. https://doi.org/10.1016/j.biortech.2019.122635

Waclawovsky AJ, Sato PM, Lembke CG, Moore PH, Souza GM (2010) Sugarcane for bioenergy production: an assessment of yield and regulation of sucrose content. Plant Biotechnol J 8(3):263–276. https://doi.org/10.1111/j.1467-7652.2009.00491.x

Wada N, Ueta R, Osakabe Y, Osakabe K (2020) Precision genome editing in plants: state-of-the-art in CRISPR/Cas9-based genome engineering. BMC Plant Biol 20(1):234. https://doi.org/10.1186/s12870-020-02385-5

Wang F, Ouyang D, Zhou Z, Page SJ, Liu D, Zhao X (2021) Lignocellulosic biomass as sustainable feedstock and materials for power generation and energy storage. J Energy Chem 57:247–280. https://doi.org/10.1016/j.jechem.2020.08.060

Worrell AC, Bruneau JM, Summerfelt K, Boersig M, Voelker TA (1991) Expression of a maize sucrose phosphate synthase in tomato alters leaf carbohydrate partitioning. Plant Cell 3(10):1121–1130. https://doi.org/10.1105/tpc.3.10.1121

Xu C, Shanklin J (2016) Triacylglycerol metabolism, function, and accumulation in plant vegetative tissues. Annu Rev Plant Biol 67:179–206. https://doi.org/10.1146/annurev-arplant-043015-111641

Xu X-Y, Yang H-K, Singh SP, Sharp PJ, Liu Q (2018) Genetic manipulation of non-classic oilseed plants for enhancement of their potential as a biofactory for triacylglycerol production. Engineering 4(4):523–533. https://doi.org/10.1016/j.eng.2018.07.002

Zale J, Jung JH, Kim JY, Pathak B, Karan R, Liu H, Chen X, Wu H, Candreva J, Zhai Z, Shanklin J, Altpeter F (2016) Metabolic engineering of sugarcane to accumulate energy-dense triacylglycerols in vegetative biomass. Plant Biotechnol J 14(2):661–669. https://doi.org/10.1111/pbi.12411

Sugarcane Metabolomics and Its Applications in Value Addition

5

V. Krishnapriya, S. Anusha, V. Vinu, M. Nisha, Adhini S. Pazhany, P. T. Prathima, R. Arunkumar, M. Alagupalamuthirsolai, and R. Gomathi

Abstract

Metabolomics is a dynamic and reliable database of plant functionality as metabolites surpass the complexity of genes, transcripts, and proteins to emerge as ultimate receptors of genetic information. In the context of nutrition and health-conscious attitudes prevalent today, sugarcane metabolomics is gaining considerable progress in terms of technical capability as well as financial support. With the ever-rising cases of lifestyle diseases such as obesity, diabetes, and high blood pressure, a majority of the global population is turning towards natural and sustainable options to combat health issues. Integration of state-of-the-art *-omics* approaches bring with them the competence to uncover novel and therapeutic compounds ignored either due to insufficient analytical capability or low abundance in the metabolome. Several metabolites including carbohydrates, phenolic compounds, and terpenoids with potential health benefits and amenable for industrial value addition have been identified in different organs (leaf, cane, root), products (juice, jaggery-based products, wax), and by-products (molasses,

V. Krishnapriya (✉) · S. Anusha · R. Arunkumar · M. Alagupalamuthirsolai · R. Gomathi
Division of Crop Production, ICAR-Sugarcane Breeding Institute, Coimbatore, Tamil Nadu, India

V. Vinu
Division of Crop Improvement, ICAR-Sugarcane Breeding Institute, Coimbatore, Tamil Nadu, India

ICAR-Indian Institute of Spices Research, Kozhikode, Kerala, India

M. Nisha
ICAR-SBI Research Centre, Kannur, Kerala, India

A. S. Pazhany · P. T. Prathima
Division of Crop Improvement, ICAR-Sugarcane Breeding Institute, Coimbatore, Tamil Nadu, India

G. S. Suresha et al. (eds.), *Value Addition and Product Diversification in Sugarcane*, https://doi.org/10.1007/978-981-97-7228-5_5

bagasse, press mud) of sugarcane. This chapter synthesises the available information on methodologies employed for metabolome development and cutting-edge technologies in sugarcane and other crops, metabolite profiles of sugarcane organs, products, and by-products, and their applications in value addition.

Keywords

Saccharum spp. · Bioactive compounds · Foodomics · Molecular pharming · LC-MS · GC-MS · HPLC · Nutrigenomics

5.1 Introduction

Surpassing the complexity of genes, transcripts, and proteins, metabolites are the ultimate receptors of genetic information; hence *metabolomics* emerges as the most dynamic and reliable database in the context of plant functionality. Sugarcane metabolomics has made considerable progress in the last two decades in terms of identifying molecules and pathways determining functional traits (Molina-Cortes et al. 2023). Sucrose accumulation and stalk development were some of the earliest areas of interest in sugarcane metabolomics (Bosch et al. 2003; Glassop et al. 2007). Data generated from metabolomics combined along with functional genomics aid in deciphering signal cascades operating during the different phenophases, and in response to biotic and abiotic stresses. Differential metabolite profiles of sugarcane genotypes have been reported (Coutinho et al. 2016), highlighting the significance of phenolic acids, flavones, and glucoside derivates. Metabolites showing potential for industrial application have been documented in different organs (leaf, cane, root), products (juice, jaggery-based products, wax), and by-products (molasses, bagasse, press mud) of sugarcane. Some of these metabolites possess nutraceutical properties exhibiting anti-oxidant, anti-bacterial, anti-diabetic, and other beneficial effects on human health. This chapter focuses on methodologies or approaches employed for metabolome development in sugarcane, scope for improvisation by taking cues from other cutting-edge technologies, metabolite profiles of sugarcane organs, products and by-products, and their applications in value addition.

5.2 Methodologies for Metabolome Development

Methodologies employed for developing metabolome profiles in sugarcane across different organs, developmental stages, and diverse genotypes are presented in Table 5.1. The estimation of the steady-state levels of metabolites aids in the identification of metabolic control points of specific pathways functioning within the plant system. In this context, Bosch et al. (2003) employed Gas Chromatography coupled with Mass Spectroscopy (GC-MS) to identify the differential accumulation of amino acids, organic acids, and sugars (trehalose) between low and high sucrose-storing sugarcane varieties, as these metabolites were less abundant and difficult to detect otherwise. Derivatisation with *N*-methyl-*N*-(trimethylsilyl)trifluoracetamide

Table 5.1 Approaches employed for metabolome profiling in sugarcane

Methodology	Objectives	References
GC-MS	Differentiate between low and high sucrose-storing varieties	Bosch et al. (2003)
GC-MS	Stalk developmental stages	Glassop et al. (2007)
GC-MS	Culm developmental stages	Perlo et al. (2022)
GC-TOF/MS	Effect of plant growth-promoting bacteria and humic acid on sugarcane leaf metabolome	Aguiar et al. (2018)
^{1}H-NMR ^{1}H-^{13}C-NMR	Differentiate between embryogenic and non-embryogenic callus tissues	Mahmud et al. (2014)
^{1}H-NMR	Metabolomic alterations to sugarcane juice during processing	Date et al. (2022)
^{1}H-NMR LC-DAD-MS	Screening of genotypes through metabolite profiling of leaves	Coutinho et al. (2016)
^{1}H-NMR UPLC-UV-ESI-MS GC-MS	Metabolite fingerprinting and quantification in molasses	Ali et al. (2019)
HPLC	Screening of genotypes through sugar profiling in sugarcane juice	Pereira et al. (2017)
HPAEC-PAD	Developmental stages of leaf, culm, and root	De Souza et al. (2018)
UHPLC-QTOF-MS	Metabolomic alterations to sugar from different extraction methods	Azlan et al. (2023)

(TMS) or *N*-methyl-*N*-(*tert*-butyldimethylsilyl)trifluoracetamide (TBS) helped increase the metabolites being detected through GC-MS in a study undertaken to decipher the stalk developmental stages in sugarcane (Glassop et al. 2007). Following GC-MS, Automated Mass Spectral Deconvolution and Identification System (AMDIS) was employed to analyze the generated chromatograms. In a similar study, metabolic alterations during culm development observed through GC-MS in diverse sugarcane genotypes revealed an increasing trend in the quantum of lignin, sucrose, carbohydrates, and amino acids including aspartic acid, serine, alanine, methionine, threonine, 3-cyano-L-alanine, and citric acid (Perlo et al. 2022). The effect of plant growth-promoting bacteria (PGPB) and humic acid (HA) on sugarcane leaf metabolome was determined by GC coupled with Time of Flight and MS (GC-TOF/MS), indicating significant changes to the plant metabolism following individual and co-inoculation of PGPB and HA (Aguiar et al. 2018). One- (^{1}H) and two-dimensional (^{1}H-^{13}C) Nuclear Magnetic Resonance (NMR) spectroscopy were used to distinguish the metabolic profiles of embryogenic and non-embryogenic callus tissues (Mahmud et al. 2014). The concentration of sugars such as glucose, fructose, sucrose, and maltose, amino acids such as valine, leucine, alanine, threonine, asparagine, glutamine, and organic acid derivates such as lactate, 2-hydroxy isobutyrate, 4-aminobutyrate, malonate and choline were higher in embryogenic callus as compared to non-embryogenic callus tissues of sugarcane. Date et al.

(2022) used an NMR-based metabolomic approach to evaluate the effect of centrifugation, filtration, and thermal pretreatments on juice metabolomic variability, hence NMR was designated as a potential tool to improve sugarcane quality and productivity. Coutinho et al. (2016) employed Preparative High-Performance Liquid Chromatography (HPLC) to fractionate sugarcane leaf extracts of different genotypes and subjected the latter to ^{1}H-NMR and Liquid Chromatography coupled with Diode Array Detection and Mass Spectroscopy (LC-DAD-MS) for the development of metabolite markers. It was observed that isomeric flavone C-glycosides were abundant in the leaves across genotypes, which were clustered according to the concentration of sugars, phenolic acids, and flavones. Ali et al. (2019) employed a multiplex metabolomics approach wherein ^{1}H-NMR was used for metabolite fingerprinting and quantification in molasses; the secondary metabolites mainly polyphenols such as flavonoids, fatty acids, and sterols were detected through Ultra HPLC coupled with UV-Visual spectra and high-resolution Electrospray Ionisation MS (UPLC-UV-ESI-MS), whereas primary metabolites such as sugars, organic acids, and amino acids were identified through GC-MS. Differential contents of fructose, glucose, and sucrose were detected in the juice of sugarcane genotypes (Pereira et al. 2017), while High-Performance Anion-Exchange Chromatography with Pulsed Amperometric Detection (HPAEC-PAD) was the method of choice to quantify non-structural carbohydrates (glucose, fructose, sucrose and raffinose) in leaf, culm, and root during different phenophases (De Souza et al. 2018). Ultra HPLC coupled with a Mass Q-TOF mass spectrometer with a dual ESI source spectrometer (UHPLC-QTOF-MS) was employed to assess the sugar obtained through different extraction methods (Azlan et al. 2023). Sensitivity for detecting sporadic compounds was higher in LC-MS as compared to HPLC, wherein it was observed that less refined sugar and molasses-added sugar were good sources of phenolic acids and flavonoids.

Destailleur et al. (2021) highlighted the significance of organelle-level metabolomics, wherein the subcellular components of tomato leaves were separated by non-aqueous fractionation and quantified using either UHPLC coupled to a hybrid quadrupole/ion mass spectrometer (UHPLC-QTRAP) or UHPLC-DAD-MS. Results revealed significant metabolic alterations in the vacuole, being the major storage organelle for sucrose at the initial stage of leaf development, alongside secondary metabolites and sugars, whilst organic acid accumulation increased during maturity. Reversed-phase Liquid Chromatography coupled with High-Resolution Mass Spectrometry (LC-HRMS) was employed to identify adulteration markers in cocoa including flavonoids, fatty acids, terpenoids, lysophospholipids, and galatolipids (Greno et al. 2023), while near-infrared spectroscopy (NIR) (Quelal-Vasconez et al. 2018) and LC interfaced to ultraviolet detection and MS (LC-UV-MS) (Yang et al. 2015) also detected adulterants in food samples. Apart from the above, stable isotope-assisted metabolomics (Flasch et al. 2022), supercritical fluid chromatography coupled with MS (SFE-MS) (Segers et al. 2019), and capillary electrophoresis (CE) (Mandal et al. 2023) are less explored approaches to develop metabolome profiles in sugarcane. Sugarcane being a sucrose accumulator, metabolomics to decipher subcellular level changes across developmental stages and organ types might reveal novel insights for value addition and molecular pharming.

Some of the common methods for statistical analysis of metabolomic datasets include principal component analysis (PCA), hierarchical cluster analysis (HCA), partial least squares discriminant analysis (PLS-DA), and orthogonal projections to latent structures discriminant analysis (OPLS-DA) (Mandal et al. 2023; Sarkar et al. 2023). The use of artificial intelligence helps to identify rare and novel metabolites with immense advantages that are likely to be missed due to overrepresentation of common metabolites. Machine learning algorithms combined with curated databases such as Kyoto Encyclopedia of Genes and Genomes (KEGG), Human Metabolome Database (HMDB), Yeast Metabolome Database (YMDB), LipidMaps, FoodDB database, and Tandem Mass Spectral Library of Standards (METLIN) would aid in-depth analysis (Greno et al. 2023; Sarkar et al. 2023), as in the case of Bueschl et al. (2022) who reported the usefulness of PeakBot, a machine learning approach to optimise and distinguish chromatographic peaks from background noise in LC-HRMS datasets. Multivariate analysis of ^{1}H-NMR metabolomics data was employed to determine the quality and origin of sparkling wines and to monitor their aging; results further revealed the superiority of OPLS-DA over PCA and PLS-DA methods (Mao et al. 2023; Charnock et al. 2024).

5.3 Metabolome Generated from Sugarcane Tissues

Metabolites in sugarcane include carbohydrates, phenolic compounds, and terpenoids which are predominant in different organs of the plant, products, and by-products of sugarcane, apart from melanoidins, synthesised via the Maillard reaction during post-harvest processing. Among these, bioactive compounds show potential for application in value-addition due to their multifaceted roles (Molina-Cortes et al. 2023).

5.3.1 Leaf

The levels of glucose, fructose, and sucrose peaked at 3rd and 12th months, whereas starch content in leaves was constant till 9th month, but doubled during the 12th month in sugarcane (De Souza et al. 2018). Along with soluble sugars, raffinose was detected in leaves, prevalent during the third and sixth months of plant growth. Further, the diurnal variation in soluble sugars was reported in leaves of sugarcane and energy cane, indicating their differential growth pattern and carbon allocation (Da Silva et al. 2023). Micro fractionation of methanolic extracts of sugarcane leaves revealed the presence of phenolic compounds such as flavones and glycosides (Singh et al. 2015). Rao et al. (2021) reported differential metabolome profiles in leaves of diverse sugarcane varieties, wherein the proportion of amino acids, glucose, and carbohydrates was generally high in young leaves, while secondary metabolites such as phenolics and flavonoids were higher in mature leaves.

Significant differences observed in secondary metabolites indicated the differential tolerance of sugarcane varieties to stress conditions and acclimatization potential to unfavourable environments. Metabolites such as shikimic acid, caffeic acid, hydroxycinnamic acid, putrescine, behenic acid, quinoline xylulose, galactose, lactose, proline, oxyproline, and valeric acid increased in response to stress, while adenine and adenosine derivatives, ribose, ribonic acid, and citric acid were associated with cellular growth in sugarcane leaf tissues (Aguiar et al. 2018). Co-inoculation of HA and PGPB in 45-days-old sugarcane stalks elevated the levels of 110 metabolites, with major role in glucose and pentose pathways, nucleoside biosynthesis, shikimic acid pathway, mevalonate pathway and tricarboxylic acid (TCA) cycle; stimulating the synthesis of phytosterols, vitamins, glucoronate, amino sugars, lipids, and amino acids (Aguiar et al. 2018). Sugarcane leaf polysaccharide identified from leaf tissues (Sun et al. 2023) exhibited manifold effects including anti-oxidant, anti-tumor, hypoglycaemic, and therapeutic effects, reiterating the potential of sugarcane metabolites for nutraceutical value addition.

5.3.2 Cane

Metabolite profiling during the cane developmental process is of utmost interest in sugarcane, owing to its commercial value and utility. Glucose and fructose in canes were greater during the first 3 months, while sucrose increased sevenfold in sixth as compared to the first month, after which the sucrose levels did not show significant variation (De Souza et al. 2018). Metabolic pathways related to biomass production were more active in energy canes as compared to sugarcanes (Da Silva et al. 2023). Sugarcanes had a small proportion of starch in the canes (De Souza et al. 2018), whilst mature internodes of energy canes accumulated a high concentration of starch (Da Silva et al. 2023). TCA cycle intermediates and amino acids were prevalent in the actively growing tissues from the apical meristem to the second internode, with a concomitant reduction in the mature internodes. Trehalose and raffinose showed a positive association with sucrose concentration in sugarcane stalks (Glassop et al. 2007). Perlo et al. (2022) reported a significant increase in the concentration of sucrose, trehalose, tryptophan, and phenylalanine during the developmental stages of sugarcane, with a concomitant increase in the abundance of their biosynthetic enzymes. Metabolome and transcriptome correlation network analysis showed a significant positive association between the metabolites and their regulatory genes. Immature internodes exhibited higher levels of amino acids and organic acids, especially aconitic acid; essential for growth and respiration (Bosch et al. 2003). Further, trehalose levels were higher in immature compared to mature internodes. Sugarcane rinds were rich in phytosterols, *p*-coumaric acid, gallic acid, chlorogenic acid, ferulic acid, and anthocyanin (Molina-Cortes et al. 2023). Moreover, the abundance of unidentified metabolites, labelled as mass spectral tags (MSTs) contributed substantially to the overall metabolite profile (Glassop et al. 2007), which may prove to be useful once identified.

5.3.3 Root

Sugarcane roots showed high levels of glucose, fructose, and sucrose at third month stage, while the starch content was almost negligible (De Souza et al. 2018). Reports on metabolite profiles of sugarcane roots are scarce, nevertheless, roots of *Saccharum officinarum* and *S. spontaneum* are known to exhibit diuretic and demulcent effects, useful to treat skin and eye diseases, anaemia, constipation, debility, joint pain, and whooping cough (Nisha et al. 2017).

5.4 Metabolomics in Products and By-Products of Sugarcane

5.4.1 Juice

Sugarcane juice is mainly composed of water, sucrose, fibre, and other reducing sugars, apart from amino acids, organic acids, phenolic acids, flavones, and fatty acids (Nisha et al. 2017; Ali et al. 2019). Phenolic compounds and flavones impart the juice colour. Phenolic acids such as hydroxycinnamic acid, sinapic acid, caffeic acid, and flavones including apigenin, luteolin, and tricin were predominant in sugarcane juice (Singh et al. 2015). Some novel acylated flavone glycosides isolated from juice (tricin-7-*O*-β-(6′-methoxycinnamic)-glucoside, luteolin-8-C-rhamnosyl glucoside, tricin-4′-*O*-(erthroguaicylglyceryl)-ether, orientin), and flavones (swertisin, tricin-7-*O*-neohesperoside-4′-*O*-rhamnoside, tricin-7-*O*-methylglucuronate-4′-*O*-rhamnoside and tricin-7-*O*-methylglucuronide) are of particular interest with greater scope for utility in value addition. Weather conditions prevailing during the harvest of canes influenced the metabolome of sugarcane juice, especially in the case of trans aconitic acid, threonine, and alanine (Date et al. 2022). Further, polycyclic aromatic hydrocarbons (PAHs) formed during post-harvest processing included benz(a)anthracene, benzo(b)fluranthene, benzo(k)fluoranthene, and benzo(a)pyrene, which were reported to be carcinogenic in nature (Silvia et al. 2009). The quantum of phenolic acids and bioactive compounds in sugarcane juice was on par with that of orange juice or black tea (Molina-Cortes et al. 2023).

5.4.2 Centrifugal and Non-centrifugal Sugar

Sugar extracted with acidified ethanol resulted in the highest yield of phenolic acids amongst other extraction methods, with correspondingly higher anti-oxidant activity (Azlan et al. 2023). Further, less refined sugars showed highest abundance in phenolic acids, followed by brown sugars and refined sugars. Various phenolic compounds including benzoic acid derivates (*p*-hydroxybenzoic acid, vanillic acid, syringic acid), cinnamic acid derivates (ferulic acid, *p*-coumaric acid), and volatile compounds (1-methyl-2-pyrrolidinone, 2,3-butanediol, 4-hydroxybenzaldehyde, benzyl alcohol, syringaldehyde, dimethylsulphoxide, benzophenone) were reported in brown sugar (Payet et al. 2005; Singh et al. 2015). Non-centrifugal sugar (NCS) or jaggery (in Asia) is one of the most preferred products of sugarcane, rich in carbohydrates and minerals. Since time immemorial, jaggery was used in traditional medicine as a blood purifier or base material for syrups (Nisha et al. 2017). Owing to the abundance of amino acids, minerals, vitamins, and antioxidants, NCS assumes an indispensable role in nutraceuticals and other biological applications. Iron or ferrous salts are essential constituents of NCS, making it a suitable medicine for anaemic patients (Sharifi-Rad et al. 2023). Apart from this, bioactive compounds including phenols such as chlorogenic acid, coumaric acid, caffeic acid, gentisic acid, gallic acid, protocatechuic acid, ferulic acid, 4-hydroxyphenyl acetic acid, vanillic acid, flavones, tricin, and apigenin were reported (Nayaka et al. 2009). Aromatic compounds like furfuryl alcohol, carboxylic acids, pyrazines, esters, and heterocyclic compounds were also detected in NCS (Chen et al. 2020). These bioactive compounds in NCS impart anti-oxidant, anti-microbial, and protective functions against DNA damage, whereas except for the presence of flavonoids and phenolic acids in brown sugars, other types of centrifugal or refined sugars exhibit negligible amounts of bioactive compounds (Molina-Cortes et al. 2023).

5.4.3 Wax

Wax in sugarcane is an important adaptive response to abiotic stress particularly high temperature or high light incidence. Depending on variety, the wax content ranges from 0.1% to 0.3% (Granja et al. 1999). Sugarcane wax consists of a mixture of long chain aliphatic fatty acids (D-003), policosanols (tetracosanol, hexacosanol, heptacosanol, octacosanol, nonacosanol, triacontanol, dotriacontanol, tetratriacontanol), phytosterols (brassicasterol, campesterol, betasitosterol, stigmasterol), steroids (24-methylcholest-3,6-dione, 24-ethylcholest-3,6-dione, 24-ethylcholest-22-en-3,6-dione, 6-hydroxycampest-4-en-3-one, 6-hydroxystigmast-4-en-3-one, 6-hydroxystigmast-4,22-dien-3-one), and terpenoids (arundoin, saw amilletin) (Singh et al. 2015).

5.4.4 Molasses

Molasses is a rich in phenolic acids, flavonoids, and lignin in comparison to sugarcane juice or syrup. The abundance of these bioactive compounds imparts antibacterial, anti-diabetic, and anti-oxidant properties to molasses (Molina-Cortes et al. 2023). Although the disaccharides like sucrose were abundant amongst the metabolites identified in molasses, tricin-7-*O*-deoxyhexosyl glucuronide, a C/O flavone was detected in sugarcane juice and molasses by Ali et al. (2019). The anti-hyperglycemic effect of extracts from molasses was determined from in vitro antioxidant activity and free radical scavenging assays, suggesting the potential for employing molasses in the preparation of functional foods with health benefits (Ali et al. 2019). Other novel *O*-glycosides detected in molasses are isoorientin-7,3′-*O*-dimethyl ether and dehydroconiferylalcohol-9′-*O*-β-D-glucopyranoside (Takara et al. 2007), both exhibiting anti-bacterial effects.

5.4.5 Bagasse

Sugarcane bagasse, constituting about 30% of the stalk biomass, is mainly composed of cellulose, hemicellulose, and lignin (Taneja et al. 2023). Quantification of flavonoids in bagasse revealed on par concentration with fruits like peach and tomato (Slimestad et al. 2008). Terpenoids such as tocopherol and phytosterols (sitosterol, campesterol, campestanol, stigmasterol, stigmastanol, 7-oxositosterol) have been identified in bagasse. Similar to molasses, the phenolic compounds in bagasse showed anti-microbial, anti-oxidant, and inhibitory properties (Patil et al. 2023).

5.4.6 Press Mud

Filter mud or press mud accounts for about 30 kg per ton of milled sugarcane, with traditional utility as fertiliser (Li et al. 2011). Further, the higher proportion of bioactive compounds was retained in the juice and not in the press mud. Molina-Cortes et al. (2023) reiterate the industrial valorisation of press mud to increase its appeal as a fertiliser fortified with bioactive compounds.

5.5 Metabolites with Potential for Value Addition

Integrating several *omics* approaches such as transcriptomics, proteomics, and genomics along with metabolomics and food science has led to new branches like *foodomics* and *nutrigenomics*, which are important in evaluating the quality of harvested produce, processing, and storage methods and detection of food adulterants. Whole genome sequencing along with state-of-the-art metabolomic techniques helps to carry out trait-specific association studies for designing therapeutic strategies. Metabolites identified in sugarcane organs, products, and/by-products possess widespread industrial application and value-addition in sugarcane (Fig. 5.1). This section highlights the utility of such novel metabolites.

5.5.1 Glucans

Bioactive compounds such as α-glucans and β-glucans are composed of D-glucose monomers with α-1,4 and β-glycosidic bonds, respectively (Venkatachalam et al. 2020). Linear molecules formed with α-1,4-D-glucan monomeric units are referred to as resistant starch, which is the portion of starch or its products exhibiting

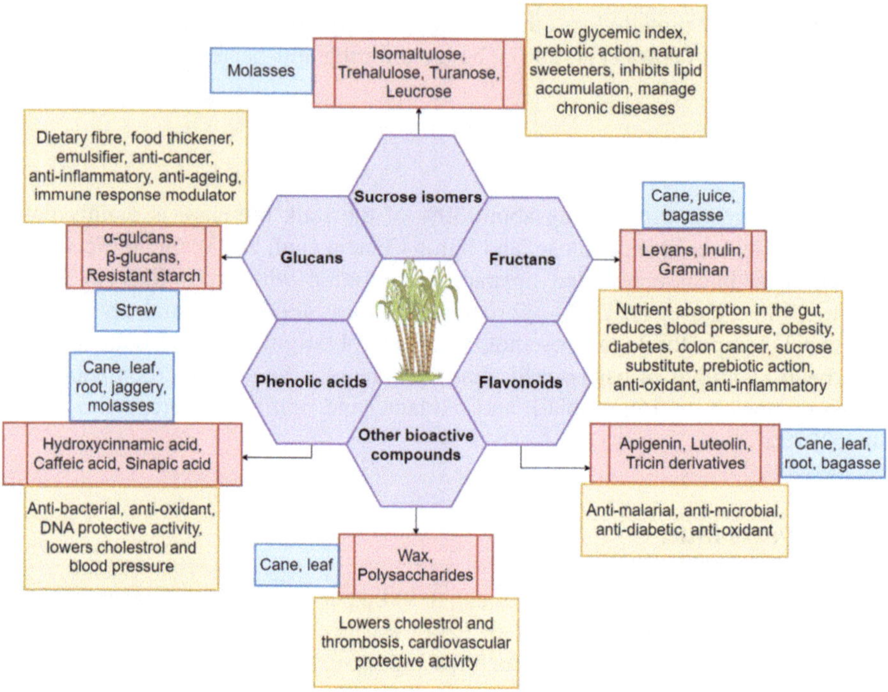

Fig. 5.1 Major metabolites present in sugarcane, its products and by-products, potential benefits to human health and application in value addition

resistance to digestion in the small intestine/gastrointestinal tract (Tabibloghmany and Ehsandoost 2014). Resistant starch was recently identified as a rich source of fibre with several beneficial effects on gut health, as the digestive enzymes like amylase and amyloglucosidases are ineffective against its hydrolysis (Homayouni et al. 2014). β-glucans are used as dietary fibres, food thickeners and emulsifiers, exhibiting anti-cancer, anti-inflammatory, anti-aging, and immune response-modulating effects (Zhu et al. 2016). Sugarcane straw was employed for production of β-glucans through saccharification using cellulase enzymes from novel strains of *Lasiodiplodia theobromae* (Abdeshahian et al. 2020).

5.5.2 Fructans

Fructans and/or Fructooligosaccharides are branched polymers of β-fructofuranosyl-fructose monomers, and are categorised as inulin, levans, and graminan, based on their β-linkages (Shi et al. 2023). Fructans are carbohydrate reserves in plants, implicated with multiple health benefits such as increasing the guts' ability to absorb nutrients and reduce the risk of high blood pressure, obesity, diabetes, and colon cancer (Apolinario et al. 2014; Verma et al. 2021). Fructooligosaccharides have been used successfully as sucrose substitute and prebiotic constituent (Correa et al. 2024), hence forming one of the potential ingredients for functional foods. Industries specialising in ready-to-eat foods, bakery, dairy, and meat products are exploiting the potential benefits of fructans and/or fructo-oligosaccharides in their value chain (Sangeetha et al. 2005; Dominguez et al. 2014). The desirable properties of levans include biodegradability, biocompatibility, anti-oxidant, and anti-inflammatory effects (Srikanth et al. 2015). Among other substrates, sugarcane bagasse was the most promising for the production of fructooligosaccharides through solid-state fermentation (de la Rosa et al. 2019; Iram et al. 2023). A composite medium containing sugarcane juice and chicken feather peptone was reported to be 20% lower in cost as compared to refined sugar for the production of levans (Veerapandian et al. 2022).

5.5.3 Sucrose Isomers

Sucrose isomers including isomaltulose, trehalulose, turanose, and leucrose are naturally occurring compounds prevalent in honey, pollen, and sugarcane (Souza et al. 2022). Isomaltulose with low glycemic index and prebiotic activity is superior to sucrose as a natural sweetener and is implicated in management of chronic diseases (Shyam et al. 2018). Turanose and leucrose help to control obesity due to their inhibitory effect on lipid accumulation, with potential applications in increasing the elasticity of rice flour, stability of rice dough, and cooking efficiency of noodles (Park et al. 2019). Metabolic engineering to produce trehalulose in sugarcane vacuoles (Hamerli and Birch 2010) offers much promise as trehalulose is sweeter than isomaltulose, but difficult to produce. Sugarcane molasses are good substrates for the production of these sucrose isomers in a safe, efficient, and cost-effective process (Tian et al. 2019).

5.5.4 Polyphenols

The abundant presence of polyphenols, flavonoids, and anti-oxidants enabled the use of sugarcane stem, root, and leaf in traditional systems of Indian medicine since time immemorial. Phenolic acids like hydroxycinnamic acid, caffeic acid, sinapic acid, and polyphenols, and flavonoids such as apigenin, luteolin, and tricin derivatives present in sugarcane extracts impart anti-bacterial and anti-oxidant characteristics (Azlan et al. 2023). Computational analysis of the structure of sugarcane polyphenol oxidase (PPO) enzyme (Patil et al. 2023) provided the basic data and information to design novel PPO inhibitors and additives to improve the shelf like of produce prone to enzymatic browning. Among sugarcane products, NCS or jaggery exhibited the highest anti-oxidant potential (Sharifi-Rad et al. 2023) and high DNA protective activity (Nayaka et al. 2009). Anti-bacterial properties of NCS (Barrera et al. 2020) and molasses (Takara et al. 2007) make these sugarcane products amenable to various pharmaceutical and industrial applications. Several reports suggest the cardioprotective effects of NCS by lowering the cholesterol and blood pressure thereby reducing the risk of associated lifestyle diseases (Taneja et al. 2020; Zidan and Azlan 2022). Coutinho et al. (2016) identified sugarcane genotypes with significantly high quantities of flavones, especially luteolin and tricin derivatives, which imparted anti-malarial, anti-microbial, anti-diabetic, and anti-oxidant properties.

5.5.5 Other Bioactive Compounds

Wax derived from sugarcane juice is a plausible substitute for carnauba wax, as the latter is expensive. The lipophilic components present in sugarcane wax make it effective in the treatment of cholesterol and thrombosis (Singh et al. 2015). Sun et al. (2023) reported the cardiovascular protective effects of polysaccharides present in sugarcane leaves. This was deciphered by in vitro and in vivo anti-oxidant activities; however, the underlying mechanism remains unexplored.

5.6 Conclusions and Way Forward

With the ever-rising cases of lifestyle diseases such as obesity, diabetes, and high blood pressure, a majority of the global population is turning toward natural and sustainable options to combat health issues. Integration of state of the art *-omics* approaches brings with them the competence to uncover novel and therapeutic compounds ignored either due to insufficient analytical capability to low abundance in the metabolome. Bioactive compounds present in by-products such as bagasse and molasses offer much scope for devising strategies to capitalise on the waste generated from sugarcane industries. Apart from bioactive compounds identified in different organs, products, and by-products, sugarcane itself can act as a biological factory for the production of metabolites and vaccines. Molecular pharming is a viable platform, which may drastically reduce the cost of production of therapeutic compounds conventionally manufactured through enzymatic or industrial processes. Metabolomic approaches are already an indispensable part of the food inspection protocols, thus ensuring robust and foolproof methodologies to warrant the safety of human and animal health.

References

Abdeshahian P, Ascencio JJ, Philippini RR, Antunes FAF, Santos JC, Silva SS (2020) Utilization of sugarcane straw from production of β-glucan biopolymer by *Lasiodiplodia theobromae* CCT 3966 in batch fermentation process. Bioresour Technol 314:123716. https://doi.org/10.1016/j.biortech.2020.123716

Aguiar NO, Olivares FL, Novotny EH, Canellas LP (2018) Changes in metabolic profiling of sugarcane leaves induced by endophytic diazotrophic bacteria and humic acids. PeerJ 6:e5445. https://doi.org/10.7717/peerj.5445

Ali SE, El Gedaily RA, Mocan A, Farag MA, El-Seedi HR (2019) Profiling metabolites and biological activities of sugarcane (*Saccharum officinarum* Linn.) juice and its product molasses via a multiplex. Molecules 24:934. https://doi.org/10.3390/molecules24050934

Apolinario AC, Damasceno BPGL, Beltrao NEM, Pessoa A, Converti A, Silva JA (2014) Inulin-type fructans: a review on different aspects of biochemical and pharmaceutical technology. Carbohydr Polym 101:368–378. https://doi.org/10.1016/j.carbpol.2013.09.081

Azlan A, Sultana S, Mahmod II (2023) Effect of different extraction methods on the total phenolics of sugar cane products. Molecules 28:4403. https://doi.org/10.3390/molecules28114403

Barrera C, Betoret N, Segui L (2020) Phenolic profile of cane sugar derivatives exhibiting antioxidant and antibacterial properties. Sugar Tech 22:798–811. https://doi.org/10.1007/s12355-020-00817-y

Bosch S, Rohwer JM, Botha FC (2003) The sugarcane metabolome. In: Proceedings of the South African Sugar Technologists Association, vol 7, pp 129–133

Bueschl C, Doppler M, Varga E, Seidl B, Flasch M, Warth B, Zanghellini J (2022) Peakbot: machine-learning-based chromatographic peak picking. Bioinformatics 38:3422–3428. https://doi.org/10.1093/bioinformatics/btac344

Charnock HM, Pickering GJ, Kemp BS (2024) Application of ^{1}H NMR metabolomics analysis to sparkling wines aged with different types of sugarcane added to the *liqueur de dosage*. J Food Compos Anal 125:105834. https://doi.org/10.1016/j.jfca.2023.105834

Chen E, Song H, Li Y, Chen H, Wang B, Che X, Zhang Y, Zhao S (2020) Analysis of aroma components from sugarcane to non-centrifugal cane sugar using GC-O-MS. RSC Adv 10:32276–32289. https://doi.org/10.1039/D0RA05963C

Correa AC, Lopes MS, Perna RF, Silva EK (2024) Fructan-type prebiotic dietary fibers: clinical studies reporting health impacts and recent advances in their technological application in bakery, dairy, meat products and beverages. Carbohydr Polym 323:121396. https://doi.org/10.1016/j.carbpol.2023.121396

Coutinho ID, Baker JM, Ward JL, Beale MH, Creste S, Cavalheiro AJ (2016) Metabolite profiling of sugarcane genotypes and identification of flavonoid glycosides and phenolic acids. J Agric Food Chem 64:4198–4206. https://doi.org/10.1021/acs.jafc.6b01210

Da Silva JJB, Abreu LGF, Ferrari AJR, De Carvalho LM, Grandis A, Buckeridge MS, Fill TP, Pereira GAG, Carazzolle MF (2023) Diurnal metabolism of energy-cane and sugarcane: a metabolomic and non-structural carbohydrate analysis. Ind Crop Prod 202:117056. https://doi.org/10.1016/j.indcrop.2023.117056

Date Y, Ishikawa C, Umeda M, Tarumoto Y, Okubo M, Ta Y, Ono H (2022) Sugarcane metabolome compositional stability in pretreatment processes for NMR measurements. Metabolites 12:862. https://doi.org/10.3390/metabo12090862

de la Rosa O, Flores-Gallegos AC, Muniz-Marquez D, Nobre C, Contreras-Esquivel JC, Aguilar CN (2019) Fructooligosaccharides production from agro-wastes as alternative low-cost source. Trends Food Sci Technol 91:139–146. https://doi.org/10.1016/j.tifs.2019.06.013

De Souza AP, Grandis A, Arenque-Musa BC, Buckeridge MS (2018) Diurnal variation in gas exchange and nonstructural carbohydrates throughout sugarcane development. Funct Plant Biol 45:865–876. https://doi.org/10.1071/FP17268

Destailleur A, Poucet T, Cabasson C, Alonso AP, Cocuron JC, Larbat R, Vercambre G, Colombie S, Petriacq P, Andrieu MH, Beauvoit B, Gibon Y, Dieuaide-Noubhani M (2021) The evolution of leaf function during development is reflected in profound changes in the metabolic composition of the vacuole. Metabolites 11:848. https://doi.org/10.3390/metabo11120848

Dominguez AL, Rodrigues LR, Lima NM, Teixeira JA (2014) An overview of the recent developments on fructooligosaccharide production and applications. Food Bioprocess Technol 7:324–337. https://doi.org/10.1007/s11947-013-1221-6

Flasch M, Bueschl C, Del Favero G, Adam G, Schuhmacher R, Marko D, Warth B (2022) Elucidation of xenoestrogen metabolism by non-targeted, stable isotope-assisted mass spectrometry in breast cancer cells. Environ Int 158:106940. https://doi.org/10.1016/j.envint.2021.106940

Glassop D, Roessner U, Bacic A, Bonnett GD (2007) Changes in the sugarcane metabolome with stem development. Are they related to sucrose accumulation? Plant Cell Physiol 48:573–584. https://doi.org/10.1093/pcp/pcm027

Granja LA, Hernandez JM, Quintana DC, Valmana LA, Ferreiro RM, Mesa MG (1999) Mixture of higher primary aliphatic alcohols, its obtention from sugar cane wax and its pharmaceutical uses. US patent no 5856316

Greno M, Plaza M, Marina ML, Puyana MC (2023) Untargeted HPLC-MS-based metabolomics approach to reveal cocoa powder adulterations. Food Chem 402:134209. https://doi.org/10.1016/j.foodchem.2022.134209

Hamerli D, Birch RG (2010) Transgenic expression of trehalulose synthase results in high concentrations of the sucrose isomer trehalulose in mature stems of field-grown sugarcane. Plant Biotechnol J 9:32–37. https://doi.org/10.1111/j.1467-7652.2010.00528.x

Homayouni A, Amini A, Keshtiban AK, Mortazavian AM, Esazadeh K, Pourmoradian S (2014) Resistant starch in food industry: a changing outlook for consumer and producer. Starch 66:102–114. https://doi.org/10.1002/star.201300110

Iram A, Ozcan A, Turhan I, Demirci A (2023) Production of value-added products as food ingredients via microbial fermentation. Processes 11:1715. https://doi.org/10.3390/pr11061715

Li X, Song H, Yao S, Jia C, Yang Y, Zhu W (2011) Quantitative analysis and recovery optimisation of flavonoids and anthocyanins in sugar-making process of sugarcane industry. Food Chem 125(2011):150–157

Mahmud I, Thapaliya M, Boroujerdi A, Chowdhury K (2014) NMR-based metabolomics study of the biochemical relationship between sugarcane callus tissues and their respective nutrient culture media. Anal Bioanal Chem 406:5997–6005. https://doi.org/10.1007/s00216-014-8002-6

Mandal P, Lanaridi O, Warth B, Ansari KM (2023) Metabolomics as an emerging approach for deciphering the biological impact and toxicity of food contaminants: the case of mycotoxins. Crit Rev Food Sci Nutr 7:1–25. https://doi.org/10.1080/10408398.2023.2217451

Mao IL, Costa GD, Bautista C, Revel G, Richard T (2023) Application of ^{1}H NMR metabolomics analysis to French sparkling wines. Food Control 145:109423. https://doi.org/10.1016/j.foodcont.2022.109423

Molina-Cortes A, Quimbaya M, Toro-Gomez A, Tobar-Tosse F (2023) Bioactive compounds as an alternative for the sugarcane industry: towards an integrative approach. Heliyon 9:e13276. https://doi.org/10.1016/j.heliyon.2023.e13276

Nayaka HMA, Sathisha UV, Manohar MP, Chandrashekar KB, Shylaja MD (2009) Cytoprotective and antioxidant activity studies of jaggery sugar. Food Chem 115:113–118. https://doi.org/10.1016/j.foodchem.2008.11.067

Nisha M, Chandran K, Gopi R, Krishnapriya V, Mahendran B (2017) Nutritional and therapeutic benefits of sugarcane and its products. J Sugarcane Res 7:1–10

Park Y, Oh IK, Park SW, Ryu K, Lee S (2019) Elucidation of rheological, microstructural, water mobility, and noodle-making properties of rice flour affected by turanose. Food Chem 276:9–14. https://doi.org/10.1016/j.foodchem.2018.09.168

Patil SA, Akki AJ, Raghu AV, Kulkarni RV, Akamanchi KG (2023) Sugarcane polyphenol oxidase: structural elucidation using molecular modelling and docking analyses. Process Biochem 134:243–249. https://doi.org/10.1016/j.procbio.2023.09.013

Payet B, Shum Cheong Sing A, Smadja J (2005) Assessment of antioxidant activity of cane brown sugars by ABTS and DPPH radical scavenging assays: determination of their polyphenolic and volatile constituents. J Agric Food Chem 53:10074–10079

Pereira LFM, Ferreira VM, De Oliveira NG, Sarmento PLVS, Endres L, Teodoro I (2017) Sugars levels of four sugarcane genotypes in different stem portions during the maturation phase. Ann Braz Acad Sci 89:1231–1242. https://doi.org/10.1590/0001-3765201720160594

Perlo V, Furtado A, Botha FC, Margarido GRA, Hodgson-Kratky K, Choudhary H, Gladden J, Simmons B, Henry RJ (2022) Transcriptome and metabolome integration in sugarcane through culm development. Food Energy Secur 11:e421. https://doi.org/10.1002/fes3.421

Quelal-Vasconez MA, Perez-Esteve E, Arnau-Bonachera A, Barat JM, Talens P (2018) Rapid fraud detection of cocoa powder with carob flour using near infrared spectroscopy. Food Control 92:183–189. https://doi.org/10.1016/j.foodcont.2018.05.001

Rao MJ, Ahmed U, Ahmed MH, Duan M, Wang J, Wang Y, Wang L (2021) Comparison and quantification of metabolites and their antioxidant activities in young and mature leaves of sugarcane. ACS Food Sci Technol 1:362–373. https://doi.org/10.1021/acsfoodscitech.0c00133

Sangeetha PT, Ramesh MN, Prapulla SG (2005) Recent trends in the microbial production, analysis and application of fructooligosaccharides. Trends Food Sci Technol 16:442–457. https://doi.org/10.1016/j.tifs.2005.05.003

Sarkar S, Sha SP, Ghatani K (2023) Metabolomics of ethnic fermented foods and beverages: understanding new aspects through omic techniques. Front Sustain Food Syst 7:1040567. https://doi.org/10.3389/fsufs.2023.1040567

Segers K, Declerck S, Mangelings D, Heyden YV, Eeckhaut AV (2019) Analytical techniques for metabolomic studies: a review. Bioanalysis 11:2297–2318. https://doi.org/10.4155/bio-2019-0014

Sharifi-Rad J, Painuli S, Sener B, Kilic M, Kumar NVA, Semwal P, Docea AO, Suleria HAR, Calina D (2023) Revisiting the nutraceutical profile, chemical composition, and health benefits of jaggery: updates from recent decade. eFood 4:e75. https://doi.org/10.1002/efd2.75

Shi Y, Si D, Zhang X, Chen S, Han Z (2023) Plant fructans: recent advances in metabolism, evolution aspects and applications for human health. Curr Res Food Sci 7:100595. https://doi.org/10.1016/j.crfs.2023.100595

Shyam S, Ramadas A, Chang SK (2018) Isomaltulose: recent evidence for health benefits. J Funct Foods 48:173–178. https://doi.org/10.1016/j.jff.2018.07.002

Silvia AV, Natali GS, Milton BN (2009) Polycyclic aromatic hydrocarbons in sugarcane juice. Food Chem 116:391–394

Singh A, Lal UR, Mukhtar HM, Singh PS, Shas G, Dhawan R (2015) Phyto chemical profile of sugarcane and its potential health aspects. Pharm Rev 9:45–54

Slimestad R, Fossen T, Verheu MJ (2008) The flavonoids of tomatoes. J Agric Food Chem 56:2436–2441

Souza WFC, Almeida FLC, Castro RJS, Sato HH (2022) Isomaltulose: from origin to application and its beneficial properties—a bibliometric approach. Food Res Int 155:111061. https://doi.org/10.1016/j.foodres.2022.111061

Srikanth R, Reddy CHSSS, Siddartha G, Ramaiah MJ, Uppuluri KB (2015) Review on production, characterization and applications of microbial levan. Carbohydr Polym 120:102–114. https://doi.org/10.1016/j.carbpol.2014.12.003

Sun K, Yuan R, He J, Zhuo Y, Yang M, Hao E, Hou X, Yao C, Yang S, Gao H (2023) Sugarcane leaf polysaccharide exerts a therapeutic effect on cardiovascular diseases through necroptosis. Heliyon 9:e21889. https://doi.org/10.1016/j.heliyon.2023.e21889

Tabibloghmany FS, Ehsandoost E (2014) Investigation of nutritional and functional properties of resistant starch in food industry: a review. Int J Rec Res Rev 7:27–44

Takara K, Ushijima K, Wada K, Iwasaki H, Yamashita M (2007) Phenolic compounds from sugarcane molasses possessing antibacterial activity against cariogenic bacteria. J Oleo Sci 56:611–614

Taneja D, Rai S, Yadav K (2020) Evaluation of promotion of iron-rich foods for the prevention of nutritional anemia in India. Indian J Public Health 64:236–241

Taneja A, Sharma R, Khetrapal S, Sharma A, Nagraik R, Venkidasamy B, Ghate MN, Azizov S, Sharma S, Kumar D (2023) Value addition employing waste bio-materials in environmental remedies and food sector. Metabolites 13:624. https://doi.org/10.3390/metabo13050624

Tian Y, Deng Y, Zhang W, Mu W (2019) Sucrose isomers as alternative sweeteners: properties, production, and applications. Appl Microbiol Biotechnol 103:8677–8687. https://doi.org/10.1007/s00253-019-10132-6

Veerapandian B, Shanmugam SR, Sivaraman S, Sriariyanun M, Karuppiah S, Venkatachalam P (2022) Production and characterization of microbial levan using sugarcane (*Saccharum* spp.)

juice and chicken feather peptone as a low-cost alternate medium. Heliyon 9:e17424. https://doi.org/10.1016/j.heliyon.2023.e17424

Venkatachalam G, Arumugam S, Doble M (2020) Industrial production and applications of α/β linear and branched glucans. Indian Chem Eng 63:533–547. https://doi.org/10.1080/00194506.2020.1798820

Verma DK, Patel AR, Thakur M, Singh S, Tripathy S, Srivastav PP, Chavez-Gonzalez ML, Gupta AK, Aguilar CN (2021) A review of the composition and toxiciology of fructans, and their applications in foods and health. J Food Compos Anal 99:103884. https://doi.org/10.1016/j.jfca.2021.103884

Yang WI, Hu MH, Chen SW, Wang Q, Zhu S, Dai J, Li XZ (2015) Identification of adulterated cocoa powder using chromatographic fingerprints of polysaccharides coupled with principal component analysis. Food Anal Methods 8:2360–2367. https://doi.org/10.1007/s12161-015-0126-6

Zhu F, Du B, Xu B (2016) A critical review on production and industrial applications of beta-glucans. Food Hydrocoll 52:275–288. https://doi.org/10.1016/j.foodhyd.2015.07.003

Zidan D, Azlan A (2022) Non-centrifugal sugar (NCS) and health: a review on functional components and health benefits. Appl Sci 12:460. https://doi.org/10.3390/app12010460

Sugarcane-Based Chemicals and Its Applications

6

Kingsley O. Iwuozor, Kehinde Temitope Micheal, Taiwo Temitayo Micheal, Ebuka Chizitere Emenike, Cletus Chinedu Onwuegbusi, Linus Anietie Umoren, and Adewale George Adeniyi

Abstract

The growing demand for sustainable chemical alternatives in various industries and the environmental impact of conventional practices in the past decades have increased, prompting an urgent need to investigate potential of sugarcane in reshaping chemical production. The primary aim of this chapter is to examine both the synthesis and applications of sugarcane-based chemicals across diverse fields. The study revealed that the possible chemicals produced from sugarcane can cover beyond alcohols such as ethanol, butanol, xylitol, sorbitol, and glycerol; organic acids like lactic, levulinic, and succinic acids; furan-based chemicals such as furfural and furfuryl alcohol; and bio-plastics like polyethylene and polyethylene terephthalate. These chemicals prove to be valuable in industries spanning food, pharmaceuticals, textiles, cosmetics, medicine, and agriculture. The study also presents recommendations based on challenges in the field, offer-

K. O. Iwuozor (✉)
Nigeria Sugar Institute, Ilorin, Kwara State, Nigeria

Department of Pure and Industrial Chemistry, Nnamdi Azikiwe University, Awka, Nigeria

K. T. Micheal
Department of Mechanical Engineering, Ladoke Akintola University of Technology, Ogbomosho, Nigeria

T. T. Micheal
Department of Agricultural Engineering, Ladoke Akintola University of Technology, Ogbomosho, Nigeria

E. C. Emenike · C. C. Onwuegbusi · L. A. Umoren
Department of Pure and Industrial Chemistry, Nnamdi Azikiwe University, Awka, Nigeria

A. G. Adeniyi
Department of Chemical Engineering, Landmark University, Omu-Aran, Nigeria

Department of Chemical Engineering, University of Ilorin, Ilorin, Nigeria

G. S. Suresha et al. (eds.), *Value Addition and Product Diversification in Sugarcane*, https://doi.org/10.1007/978-981-97-7228-5_6

ing insights for future research. In conclusion, the utilization of sugarcane for chemical production represents a harmonious blend of economic viability, environmental sustainability, and technological innovation. As industries worldwide seek greener alternatives, sugarcane emerges as a beacon of hope, offering solutions that not only meet current needs but also pave the way for a more sustainable and resilient future.

Keywords

Biomass utilization · Bio-refinery · Circular economy · Green chemistry · Sustainability

6.1 Introduction

The contemporary chemical industry is at a crossroads, grappling with a critical challenge rooted in the over-reliance on finite fossil fuel resources. This reliance not only poses a threat to the industry's long-term viability but also contributes significantly to environmental degradation, exacerbating concerns about sustainability on a global scale (Ahmad et al. 2022). The extraction, processing, and utilization of fossil fuels in chemical production release substantial greenhouse gas emissions, further exacerbating climate change (Adeola et al. 2023; Emenike et al. 2023b). As the demand for chemicals continues to surge worldwide, driven by industrial processes, agriculture, and consumer products, the urgency to identify alternative, renewable sources becomes imperative. The finite nature of fossil fuels and their environmental impact highlights the need for a strategic shift in the industry's approach to raw material acquisition and production processes (Conteratto et al. 2021). This shift involves recognizing the inadequacy of current resource utilization practices, acknowledging the ecological consequences associated with conventional chemical production methods, and actively seeking sustainable alternatives. The pursuit of sustainable alternatives is not merely a matter of environmental responsibility; it is an economic necessity and a means to ensure the industry's resilience in the face of resource scarcity and regulatory changes. The paradigm shift towards sustainable alternatives encompasses a multifaceted approach. Researchers and industry leaders are exploring diverse avenues, such as bio-based feedstocks, circular economy principles, and green chemistry practices (Ighalo et al. 2021). Bio-based feedstocks, derived from renewable sources like plant biomass or agricultural residues, hold promise as they offer a potential inexhaustible supply while mitigating environmental impacts (Adeniyi et al. 2023d; Emenike et al. 2023c).

In the pursuit of sustainable alternatives for the chemical industry, sugarcane emerges as a compelling solution. As a tropical grass with high biomass productivity, sugarcane offers a renewable resource that can address the challenges posed by the overreliance on finite fossil fuels (Emenike et al. 2023a; Iwuozor et al. 2023b). The cultivation of sugarcane for its sucrose content is a well-established practice in the production of sugar and ethanol (Iwuozor et al. 2022a, d, e, 2023f; Mohammed et al. 2024), but its potential extends far beyond these traditional applications.

Sugarcane-based chemicals present a promising avenue for mitigating the environmental impacts associated with conventional chemical production (Iwuozor et al. 2023c). The promise of sugarcane-based chemicals lies in their potential to bridge the gap between sustainable agriculture and renewable resources (Iwuozor et al. 2022f, 2023d). Through innovative processes, sugarcane can be transformed into a myriad of valuable products, including biofuels like ethanol, bio-based plastics, and a variety of bio-chemicals (Iwuozor et al. 2022b). These products not only offer economic opportunities but also provide renewable alternatives to their petrochemical counterparts. One of the most significant advantages of sugarcane-based chemicals is their reduced environmental impact. Unlike traditional petrochemical-based production, the production of chemicals from sugarcane can significantly lower greenhouse gas emissions, energy consumption, and pollution. Additionally, sugarcane cultivation contributes to carbon sequestration and soil health improvement, fostering a virtuous cycle of sustainability (Lee et al. 2019). Beyond their environmental benefits, sugarcane-based chemicals can have far-reaching economic and social implications. By diversifying agricultural production and offering additional revenue streams, they can contribute to the economic well-being of farming communities.

Various reviews have emerged in the past on the utilization of a part or by-product of sugarcane for the production of chemicals. For example, Rajamani et al. (2021) reviewed the modification of sugarcane bagasse using ionic liquids for the production of value-added chemicals. Bezerra and Ragauskas (2016) reviewed the utilization of sugarcane bagasse for the generation of power and the production of ethanol. Sindhu et al. (2016) reviewed the conversion of sugarcane residue to value-added products. Even though these reviews exist, there is yet to be a study that is primarily centred on chemicals produced from sugarcane and its by-products. This is an important knowledge gap that this chapter hopes to fill. The purpose of this chapter is centred on exploring sugarcane as a potential feedstock for chemical production by examining the complexities of its production processes and various applications. The novelty of this study comes from reviewing the innovative ways undertaken by various studies to produce chemicals from sugarcane and its by-products, which are typically sourced from finite and environmentally burdensome petrochemical origins. The importance of the study is reinforced by its focus on revolutionizing the chemical industry and offering a sustainable alternative that tackles environmental concerns associated with conventional practices. By emphasizing environmental and sustainability considerations, the study positions itself as a catalyst for change within an industry wrestling with the necessity to embrace eco-friendly practices.

6.2 Sugarcane as a Feedstock for Chemical Production

The use of renewable resources for purposes other than fuels, including chemicals, paper, and textiles, generally yields greater value (Tsiropoulos et al. 2015). Sugarcane is a notable example in this context. Sugarcane, scientifically known as *Saccharum officinarum*, is a versatile tropical and subtropical crop that has been cultivated for centuries primarily for its high sugar content (Afiomah and Iwuozor 2020; Iwuozor et al. 2023a). Beyond its historical significance as a sweetening agent, sugarcane possesses a range of properties that make it an exceptional feedstock for chemical production, playing a pivotal role in the transition toward sustainable and renewable resources. Utilizing sugarcane as a feedstock for the production of chemicals undoubtedly promotes sustainable chemistry. Its utilization in sustainable chemistry is a result of its capacity to yield a diverse range of compounds, chemicals, and materials, including sugars such as sucrose, glucose, and xylose; macromolecules present in bagasse and straw that can be converted into hexose, pentose, and phenol sources; vinasse; and additional compounds such as higher alcohols (Vaz 2020). Today, the sugarcane sector is recognized as a multifaceted industry with involvement in agro-food, chemicals, energy, and the extensive domain of synthetic biology (Arruda 2011). This is because sugarcane serves as a valuable source material for the production of bioenergy and various bio-products (Hiloidhari et al. 2021). Sugarcane, a crucial primary raw material for sugar and ethanol production in numerous countries (Candido and Goncalves 2019), holds considerable promise as a favoured raw material for implementing advanced techniques in the production of diverse chemicals essential for meeting modern societal needs (Robins and Speight 2016).

Various materials containing abundant carbohydrates, designed for utilization in either animal feed or human consumption, such as corn, sugarcane, wheat, rice, maize, and cassava, beetroot, have been employed in the production of chemicals like bioethanol (Afiomah and Iwuozor 2020). Among these, sugarcane stands out as a superior feedstock due to its elevated levels of easily fermentable sugars, valuable minerals, and organic nutrients (De Almeida and Colombo 2023). It is considered to be a more sustainable option compared to other similar sources of feedstock, such as corn (Villela Filho et al. 2011). Additionally, the energy balance of ethanol derived from sources other than sugarcane, determined by the ratio of fossil energy input to ethanol energy output, does not match that of sugarcane (Arruda 2011; Jun et al. 2023). Moreover, sugarcane has exceptional productivity, excellent conversion yields, and cost-effectiveness (Castañeda-Ayarza and Cortez 2017), and exhibits notable productivity and superior performance, primarily due to its efficient biochemical photosynthesis pathway known as the C4 route (Hiloidhari et al. 2021; Villela Filho et al. 2011). Hence, these properties render sugarcane an extensively utilized carbon source for the chemical industry and the most fitting raw material for chemical and bioethanol productions (De Almeida and Colombo 2023; Villela Filho et al. 2011). Therefore, sugarcane retains its significance as a key crop and biomass provider for industrial biotechnology processes, both in the present and the foreseeable future (Robins and Speight 2016).

Sugarcane is renowned for its exceptionally high sugar content, primarily in the form of sucrose (Raza et al. 2019). This sugar content, which can range from 12% to 18% by weight in mature cane, serves as the foundational raw material for a wide array of chemical compounds. The abundance of sucrose makes sugarcane an attractive and energy-rich source for chemical production. Sugarcane is an annual crop that can be cultivated year after year, making it a renewable resource (De Miranda and Fonseca 2020). Its rapid growth and ability to thrive in tropical and subtropical climates ensure a consistent and abundant supply of feedstock for chemical production. This renewability distinguishes sugarcane from non-renewable resources typically used in chemical production. Another critical factor is the comprehensive utilization of the entire crop (Villela Filho et al. 2011). Sugarcane presents various promising feedstock options for industrial biotechnology (Robins and Speight 2016). This can be attributed to the fact that, in the process of producing chemicals such as bioethanol, alcohol, and organic acid from sugarcane, various wastes are generated. These include agricultural residues resulting from the burning or crushing of sugarcane during harvesting (Parsaee et al. 2019), sugarcane straw, also referred to as sugarcane trash, which is a by-product of mechanized harvesting and sugar production procedures (Sukphun et al. 2023), sugarcane tops that emerge during the harvesting of the sugarcane stem (Khaire et al. 2021), and industrial by-products like ash, bagasse, filter cake, molasses, and vinasse (Iwuozor et al. 2024b, d; Parsaee et al. 2019), which are by-products produced from the extraction of juice from crushed sugarcane stems for the production of sugar and first-generation ethanol (Torgbo et al. 2021).

The intermediate substances and by-products obtained from sugarcane processing, as they undergo the transformation into white sugar, can be subjected to fermentation to yield chemicals and bio-products (Robins and Speight 2016). This means that much of the waste and by-products obtained from sugarcane processing can also be used as primary feedstock, with great potential to be bio-converted into commercially valuable products (Torgbo et al. 2021). For instance, the exploration of sugarcane straw processing as a viable feedstock for acquiring bio-products and chemicals such as ethanol, enzymes, and xylitol has garnered significant interest (Aguiar et al. 2021). Sugarcane bagasse finds application in the manufacturing of cellulosic ethanol and furfural (Santos et al. 2020). Although the characteristics and potential uses of the different intermediates and by-products produced in the processing of sugarcane vary based on their cost and market dynamics, the unique cost and purity profiles associated with each feedstock play a crucial role in determining how seamlessly it can be integrated into a bioprocess and also dictate the need to assess the displacement of current applications (Robins and Speight 2016).

6.3 Production of Sugarcane-Based Chemicals

Sugarcane plantations have a substantial impact on global sugar and ethanol manufacturing (Schmitt et al. 2020). It is a versatile plant that serves as a valuable provider of various resources, including food in the form of sucrose, jaggery, and syrups. It also yields cellulose, is found in fibres, and serves as fodder with its green leaves and cane tops. Furthermore, it contributes to fuel and chemicals through products like bagasse, molasses, and alcohol (Iwuozor et al. 2024a; Solomon 2011). Chemicals such as ethanol, xylitol, sorbitol, and glycerol in the alcohol category, as well as organic acids like levulinic acid and succinic acid, along with furans such as furfural, furan, and methylfuran, are prominent chemicals derivable from sugarcane

(Fig. 6.1) (Rabelo et al. 2020). The advancement of biomass-derived chemicals as an alternative to petrochemicals has propelled the adoption of renewable chemistry. This approach prioritizes the use of renewable raw materials over traditional oil-based sources (Vaz 2020). During the raw sugar manufacturing process, the sugarcane preparation and milling equipment dismantle the organized structure of fibres in the sugarcane plant. This breakdown is essential to disrupt the parenchyma cells, facilitating the extraction of sucrose (De Almeida and Colombo 2023). As per the research conducted by American scientists, approximately 12 chemicals can be derived from carbohydrates through either chemical or biological conversion (Rabelo et al. 2020). However, it's worth noting that this information was revised in 2010 to reflect advancements in technology. These compounds fall into three primary functional groups—alcohols, organic acids, and furanics (Rabelo et al. 2020).

Sugarcane-based chemicals are produced through two distinct pathways: alcohol chemistry and fermentative processes (Villela Filho et al. 2011). The alcohol chemistry pathway merges two effective technologies: alcohol fermentation and catalytic processes. Numerous bio-based products can be derived from ethanol using this route. Chemical pathways for alcohol commence with the ethanol oxidation process. This process produces acetaldehyde, a crucial intermediate for manufacturing various chemicals, including butanol, acetic acid, acetic esters, peroxyacetic acid, acetic anhydride, and pyridine (Villela Filho et al. 2011). Microbial fermentation of sugars sourced from sugarcane, along with the associated agricultural and agro-processing residues, has been studied for the production of diverse substances. These include organic acids such as lactic, polymalic, succinic, itaconic, and gluconic acids, along with higher alcohols like butanol and xylitol (Rabelo et al. 2020). The fermentative pathway involves the transformation of carbohydrates into the desired compound by microorganisms. A prominent illustration of this process is ethanol fermentation (Villela Filho et al. 2011). The production of some of the major chemicals in the aforementioned groups is briefly discussed below.

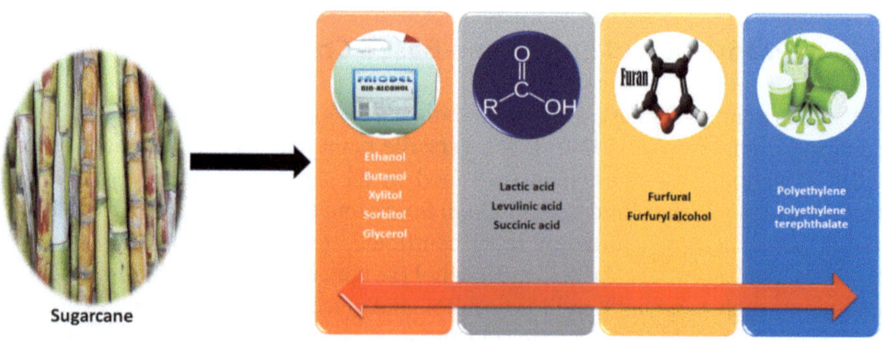

Fig. 6.1 Overview of sugarcane-based chemicals

6.4 Alcohols

Ethanol, butanol, and polyols such as xylitol, sorbitol, and glycerol are notable examples of the categories of alcohols that can be derived from sugarcane, its waste, and by-products.

6.4.1 Production of Ethanol

Ethanol, among the widely favoured biofuels, is primarily derived from crops such as sugarcane, beets, and grains (Lisboa et al. 2011). There are two methods for producing ethanol from sugarcane: 1G ethanol production technology and 2G ethanol production technology. The former involves the use of sugarcane juice, molasses, or a combination of both (Rabelo et al. 2020). The latter, on the other hand, is derived from sugarcane waste and by-products such as sugarcane straw (Aguiar et al. 2021; Rabelo et al. 2020), sugarcane tops (Khaire et al. 2021), and sugarcane bagasse (Das Bhowmik et al. 2016). The production of 1G ethanol is relatively uncomplicated in comparison to the intricate procedure involved in generating 2G ethanol. This complex process entails the initial separation from the lignin fraction, followed by hydrolysis, and ultimately, the fermentation of monomers into ethanol (Schmitt et al. 2020).

In the first generation (1G), the initial steps closely mirror those of sugar production (Santos et al. 2020). The first step is the harvesting of cane from the field. Sugarcane can be gathered from the field using either manual cutting or mechanized methods (Das Bhowmik et al. 2016). During this process, the leaves and upper portions of the sugarcane plant are detached from the stalk. This separation occurs either through burning or as part of the mechanical harvesting procedure, during which the stalk is cut into smaller billets for convenient transportation to the factory (Rackemann et al. 2016). In manual cutting, the harvesting process involves either cutting the sugarcane after it has been burned or leaving it in its natural green state without burning. Following the harvest, the sugarcane is transported to a processing facility (De Almeida and Colombo 2023), which is not always far from the areas where sugarcane is cultivated because the amount of sugar obtained from sugarcane diminishes as the duration between harvesting and processing increases (Das Bhowmik et al. 2016). The processing stage at the facility includes several steps. Upon the arrival of sugarcane at the production facility, it undergoes a washing process before being sent to the broth preparation and extraction system (Santos et al. 2020). After washing or cleaning with water, pricking, defibrillation, grinding, and soaking are done, followed by the extraction and treatment of the raw broth, fermentation of the treated juice, as well as the distillation and dehydration of bioethanol (De Almeida and Colombo 2023). The extraction process separates the broth from the fibre (bagasse) (Santos et al. 2020). In the fermentation step, the purified broth is combined with yeast, which plays a crucial role in fermenting glucose to generate ethanol and carbon dioxide. This fermentation process typically lasts around 36 h, resulting in a wine mixture (De Almeida and Colombo 2023;

Igwegbe et al. 2022), containing unfermented sugar, yeast, and a fermented blend containing 7–10% alcohol concentration (De Almeida and Colombo 2023; Santos et al. 2020). After the fermentation process, the yeasts are collected and processed for subsequent utilization, while the wine is transported to the distillation columns, where it is heated to induce evaporation, followed by condensation to separate and obtain hydrated bioethanol with a 96% alcohol content. The final step is dehydration, which can be accomplished through various methods, such as introducing cyclohexane, utilizing molecular sieves for adsorption, or employing extractive distillation with mono-ethyleneglycol (Santos et al. 2020), aiming to separate water molecules from the hydrated bioethanol. After dehydration, the hydrated bioethanol is transformed into anhydrous ethanol, with an alcohol content of 99.5% (De Almeida and Colombo 2023).

The second-generation (2G) technology of producing ethanol from sugarcane involves the use of wastes and by-products derived from sugarcane, which are often referred to as lignocellulosic plant material. Examples of such materials include sugarcane bagasse, leaves or trash of sugarcane (Das Bhowmik et al. 2016), filter cake (FC), vinasse, and molasses (Sukphun et al. 2023), which remain after the harvesting and crushing processes (Das Bhowmik et al. 2016). Sugarcane bagasse refers to the solid and fibrous portion (lignocellulose) that remains after the traditional milling process of sugarcane (Iwuozor et al. 2022c; Santos et al. 2016). FC is the by-product obtained by removing impurities from sugarcane juice in the decanter tank during the initial clarification stage before the production of sugar (Sukphun et al. 2023). Vinasse, also known as stillage or distilled water, is the ultimate product obtained by distilling fermented sugarcane juice to acquire ethanol (Sukphun et al. 2023). Molasses is the final output in raw sugar factories (Torgbo et al. 2021). Sugarcane bagasse as a feedstock is the most used for 2G production of ethanol, and this comes from its composition. Similar to other lignocellulosic materials, SCB possesses limited nutritional value, alleviating concerns related to the food versus fuel conflict (Reena et al. 2022). However, the production of ethanol or bioethanol from many of these sugarcane wastes or by-products mirrors each other. For instance, sugarcane bagasse, like sugarcane straw, can be employed in the second-generation bioethanol production process to generate bioethanol. The production procedure mirrors that of straw, encompassing pre-treatment steps, enzymatic hydrolysis, and fermentation (De Almeida and Colombo 2023). The process of generating bioethanol from sugarcane bagasse (SCB) involves five key phases, including biomass pre-treatment, cellulose hydrolysis, fermentation, separation, and effluent treatment (Reena et al. 2022). SCB, composed of lignocellulosic elements comprising approximately 50% cellulose, 25% hemicellulose, and 25% lignin (Devadiga et al. 2020), possesses a complex inherent structure and lacks accessible monosaccharides for microorganisms during fermentation (Reena et al. 2022). Instead, they comprise polysaccharides like cellulose and hemicelluloses, which need to undergo hydrolysis using acids or enzymes to transform into fermentable sugars (Cardona et al. 2010). The intertwining of cellulose microfibrils with hemicellulose and lignin causes the formation of a resistant structure in lignocellulosic biomass, making it challenging for direct biological degradation. Consequently, a pre-treatment process becomes crucial to disrupt the resistant structures of biomass, enhancing its susceptibility to acid or enzyme catalysis (Jin et al. 2020). It has been established that there is always a low yield of glucose from lignocellulosic materials when enzymatic hydrolysis is conducted without implementing a prior pre-treatment (De Araujo Guilherme et al. 2019). Hence, the pre-treatment step is necessary to facilitate material breakdown and increase the glucose yield.

Various pre-treatment methods have been created to facilitate cellulose hydrolysis, encompassing physical, chemical, physicochemical, and biological approaches (Cardona et al. 2010; Reena et al. 2022). Chemical pre-treatment methods encompass acid, alkali, ozone, and solvent-based treatments. Physical preparation involves comminution and hydro-thermolysis, while steam explosion and ammonia fibre explosion are examples of physiochemical treatment. Biological treatment involves the use of microorganisms and/or microbial enzyme systems (Reena et al. 2022). Following pre-treatment, cellulose obtained from the process undergoes hydrolysis to break down into glucose (saccharification) using acids or enzymes (Cardona et al. 2010). When acid is utilized, it is termed acid hydrolysis, whereas the use of enzymes is commonly known as enzymatic hydrolysis. Acid hydrolysis involves the generation of glucose through the breakdown of cellulose in an acidic environment. On the other hand, enzymatic hydrolysis relies on enzymes, and these enzymes contribute to the degradation of hemicellulose, ultimately yielding reducing sugars (Vaz 2020). For achieving high product yields, which is crucial for economic success, the enzymatic hydrolysis pathway stands out as the most promising and favourable method (Cardona et al. 2010; Vaz 2020). It is important to pre-treat biomass in the production of ethanol for greater yield. Uppal et al. (2011) researched improving the processes of chemical pre-treatment and acid saccharification for the ethanol conversion of sugarcane bagasse. The findings indicated that the most efficient ethanol production through fermentation occurs in shorter timeframes when using bagasse that undergoes pre-treatment followed by saccharification, as opposed to utilizing directly pre-treated bagasse.

After the breaking down of cellulose into glucose by an enzyme such as cellulases, the next step is the fermentation of the glucose (De Araujo Guilherme et al. 2019). The selection of the fermentation mode for bioethanol production hinges on the specific properties of the fermenting microbe and the enzyme involved. The fermentation methods employed in bioethanol production include separate hydrolysis and fermentation (SHF), simultaneous saccharification and fermentation (SSF), simultaneous saccharification and co-fermentation (SSCF), and consolidated bioprocessing (CBP) (Khaire et al. 2021). De Araujo Guilherme et al. (2019) performed an experiment on ethanol production from sugarcane bagasse using different fermentation strategies. The SHF process involves individual cellulose hydrolysis and fermentation steps. Although this process has been reported to be effective, it is lengthy and requires additional steps. SSF achieves high ethanol productivity but faces challenges with differing optimal temperatures for hydrolysis and fermentation. SSCF, using a mix of yeasts, enhances sugar conversion but demands careful temperature management. Sindhu et al. (2016) reported this method for ethanol production from sugarcane bagasse by *Zymomonasmobilis*. CBP integrates enzyme production, biomass hydrolysis, and sugar fermentation in a single reactor, simplifying the process, albeit requiring a microorganism capable of simultaneous C6 and C5 sugar fermentation. SHF and SSF are the main approaches in the literature, but CBP and SSCF are gaining recognition. Many authors have resorted to combining SHF and SSF in a pre-saccharification step called PSSF (Aguiar et al. 2021), potentially leading to decreased processing time and energy usage (De Araujo Guilherme et al. 2019). Finally, the separation is done to recover the bioethanol, followed by treatment.

6.4.2 Production of Polyols

In addition to ethanol, sugarcane is employed in the production of polyols such as sorbitol and xylitol. D-xylitol is a five-carbon sugar alcohol derived through microbial fermentation (Das Bhowmik et al. 2016). It is a naturally occurring polyol with significant sweetness, present in a variety of fruits and vegetables (Sindhu et al. 2016). Xylitol possesses attributes such as anticancer effects, reduced calorie content, and the capability to stimulate the re-hardening of teeth (Aguiar et al. 2021; Das Bhowmik et al. 2016). Polyols such as xylitol and sorbitol are primarily produced through the catalytic hydrogenation of sugars, but they can also be generated through biochemical processes. However, the chemical method is more technically advanced (Rabelo et al. 2020; Rackemann et al. 2016). The biotechnological synthesis of xylitol from lignocellulosic biomass has been studied extensively as a potential substitute for traditional techniques. This method has an advantage over conventional procedures because it purifies xylose without requiring high pressures or temperatures, and it is a biologically selective process that does not require the use of a chemical catalyst. As such, it offers a more advantageous choice concerning the energy intensity of xylitol production and helps to lower process costs overall. The bioprocess uses NAD(P)H-dependent xylose reductase as the first step in xylose metabolism, taking advantage of the ability of pentose-assimilating yeasts to convert xylose into xylitol (Aguiar et al. 2021). Similar to how ethanol is made from sugarcane waste, xylitol is also produced by hydrolyzing biomass and fermenting it to produce glucose and xylose, which are then fermented. However, for this biorefinery, pre-treatments with steam and diluted acid are frequently used techniques. During the dilute acid pre-treatment step, xylan is hydrolyzed to create xylose, which is then fermented to yield xylitol (Raj and Krishnan 2020).

6.4.3 Production of Butanol

Chemical processes such as the Oxo process, which uses propylene, and the Aldol process, which starts with acetaldehyde, are used to produce butanol. The high cost of the substrate and the detrimental effects of butanol toxicity on microorganisms that ferment food pose a major obstacle (Aguiar et al. 2021). Butanol is also among the various bio-products attainable via the biochemical conversion processes of sugarcane and its remnants (Santos et al. 2020). The use of lignocellulosic biomass as a feedstock for butanol production—such as sugarcane—has received increased attention lately. In an effort to lessen or completely eradicate butanol toxicity from the culture medium, acetone-butanol-ethanol (ABE) fermentation has been re-evaluated in response to this renewed interest. Manipulating the culture conditions is the goal in order to improve product yield and specificity (Aguiar et al. 2021). However, there are drawbacks to using sugar from lignocellulosic biomass in the ABE fermentation process to produce butanol, including low productivity and decreased concentration of *n*-butanol as a result of product inhibition. Thus, the ethanol chemistry route, which uses ethanol as the feedstock, is an alternate method frequently used to produce *n*-butanol from lignocellulosic biomass (Sindhu et al. 2016). Moving the harvested sugarcane to an ethanol processing facility is the first step in the butanol production process. Here, sugarcane juice is made by cleaning and crushing the cane. Afterwards, the ethanol plant supplies the sugarcane juice to the butanol plant. The juice is then treated with screens and hydrocyclones to remove fibre and sand. The mixture is then heated and allowed to settle after the addition of lime and phosphoric acid, which effectively removes a variety of impurities (Mariano et al. 2013). The interesting thing about this is that it becomes a more economically viable process when *n*-butanol production is added to sugarcane biorefineries. This yields a biofuel that can be used in the chemical industry in addition to accounting for economic considerations (Sindhu et al. 2016).

6.5 Organic Acids

6.5.1 Lactic Acid (LA)

The production of lactic acid is possible either through chemical synthesis or fermentation (Sindhu et al. 2016). Production through microbial fermentation is primarily for commercial purposes (Rackemann et al. 2016). The current focus on fermentative lactic acid production is driven by its potential to provide an eco-friendly alternative to the environmental issues associated with the petrochemical industry and the finite supply of petrochemical resources (Sindhu et al. 2016). The techno-economic production of lactic acid faces several obstacles when using 2G sugars, even though 90% of commercially produced lactic acid is derived from the microbial route. Currently, a significant limitation of LA fermentation is its requirement for a pH-controlled environment to sustain continuous production of lactic acid (Agrawal and Kumar 2023). Microbes like lactic bacteria and fungi have been extensively employed in the generation of lactic acid, with examples such as *Lactobacillus plantarum* utilizing sugarcane bagasse hemi-cellulosic hydrolysate and molasses for this purpose (Rabelo et al. 2020).

It is important to highlight that when sugar substances like molasses and sugarcane juice serve as raw materials, there is no need for milling, pre-treatment, hydrolysis, or detoxification. Conversely, when dealing with starchy materials, the usual processes involve milling, liquefaction, and saccharification to produce fermentable sugar (Castañeda-Ayarza and Cortez 2017). In a standard second-generation biorefinery, the initial step involves creating a sugar platform through enzymatic saccharification. Subsequently, the hydrolysate, rich in sugars, undergoes fermentation. There are limited instances in which these fermentation setups were employed to produce lactic acid using 2 g of sugars obtained from sugarcane bagasse (SCB) or pre-treated SCB, resulting in a final product concentration exceeding 50 g/L (Agrawal and Kumar 2023). But for the lactic acid that is produced through the fermentation of molasses, initially, the molasses sourced from a sugar mill are diluted to a 12% concentration before being subjected to fermentation by *Lactobacillus delbrueckii* bacteria. This fermentation process occurs within a fermenter tank, maintaining specific conditions such as a temperature of 46 °C, atmospheric pressure, and a pH range of 5–6. Nutrients, including malt sprouts, are added, and $CaCO_3$ serves as a pH stabilizer. The entire fermentation process spans approximately 21 h, resulting in a lactic acid yield of 90% (Zahara 2018). Lunelli et al. (2010) conducted an experiment in which the authors reported the fermentative synthesis of lactic acid utilizing sucrose derived from sugarcane molasses.

6.5.2 Levulinic Acid

Levulinic acid can be generated by subjecting hexose and pentose sugars to acid-catalyzed dehydration and hydrolysis processes (Rackemann et al. 2016). In the production process of levulinic acid, bagasse is first fed into a reactor, where it reacts with sulphuric acid. Subsequently, the resulting mixture progresses to a second reactor, specifically a continuous stirred-tank reactor (CSTR). Upon completion of the reaction in the second reactor, the combination of lignin, sulphuric acid, and levulinic acid is discharged from the system. The separation of lignin from levulinic acid is accomplished through centrifugation, while the recovery of sulphuric acid involves a decanter, with the reclaimed acid being reintroduced into the initial reactive extruder. Finally, the purification of levulinic acid occurs through a reaction with lime, resulting in the sedimentation of calcium sulphate (Zahara 2018).

6.5.3 Succinic Acid

Succinic acid is a dicarboxylic acid with the molecular formula $C_4H_6O_4$, which is also known as amber acid or butanedioic acid. It is a starting point for the production of numerous chemicals with a wide range of uses in the food processing, pharmaceutical, and agricultural industries (Borges and Pereira 2011). Thus, it is among the top 12 bulk chemicals according to the US Department of Energy's ranking (Shaji et al. 2021). It is presently manufactured from crude oil through either the catalytic hydrogenation of maleic anhydride to succinic anhydride followed by hydration or by directly catalytically hydrogenating maleic acid (Borges and Pereira 2011; Rabelo et al. 2020). Nevertheless, sustainability considerations have generated enthusiasm for the bio-based method of producing succinic acid, particularly utilizing lignocellulosic biomass as the primary raw material (Shaji et al. 2021). Succinic acid can be transformed into 1,4-butanediol, which, in turn, can be converted into tetrahydrofuran or utilized in the synthesis of polymers like polybutylene terephthalate and polybutylene succinate (Adeniyi et al. 2023c; Rabelo et al. 2020). Similar to ethanol, the fermentative production pathway is utilized for the synthesis of succinic acid (Borges and Pereira 2011). In the process of producing succinic acid, sugarcane bagasse undergoes an initial acid pre-treatment. This pre-treatment aims to break down the xylan in hemicellulose into xylose while preserving the majority of cellulose and lignin. Following pre-treatment, the mixture containing xylose and residue is directed to washer and screw press units to separate liquid xylose from the solid cellulose and lignin residue. Subsequently, the isolated xylose is directed to either an evaporator or a membrane, depending on the needs of the fermentation reactor. Within the fermentation reactor, xylose undergoes conversion into succinic acid (SA) and acetic acid (Shaji et al. 2021).

Moreover, sugarcane can be used to produce organic acids such as citric and amino acids. Amino acids are essential building blocks of proteins and can be produced via the fermentation process. A few amino acids are useful feed additives. These amino acids are used in the pharmaceutical industry to formulate parenteral nutrition since they are protein constituents. Certain amino acids also have important functions in other food-related applications (Villela Filho et al. 2011). Studies on using agricultural residues to produce citric acid have been carried out recently by a number of researchers (Sindhu et al. 2016). The Taguchi design was used, according to the Sindhu et al. (2016) report, to determine the best circumstances for turning sugarcane bagasse into citric acid. The study proved that sugarcane bagasse is a financially feasible citric substrate.

6.6 Furans

Furanics can be generated via acid-catalysed dehydration of hexose and pentose sugars, utilizing either homogeneous or heterogeneous processes. These compounds have the potential to be transformed into biofuels within the diesel range through condensation reactions. Additionally, furans exhibit versatility in their chemical applications (Rackemann et al. 2016) and can undergo reduction via hydrogenation, resulting in the production of various solvents, fuel additives (Rabelo et al. 2020), acrylate and polyester monomers, pharmaceuticals, and agrichemicals. These heterocyclic organic compounds are characterized by a five-membered aromatic ring comprising four carbon atoms and one oxygen atom (Rackemann et al. 2016).

Furfural stands out as a prominent illustration of a compound within the furan family. Furfural is a multipurpose industrial solvent that is produced in large quantities worldwide by dehydrating pentose sugars under the influence of acid. Furfuryl alcohol is produced from most of the furfural that is produced. Furfuryl alcohol is used primarily in the manufacturing of foundry resins, adhesives, and wetting agents, though it is also utilized as a solvent (Rackemann et al. 2016). It is important as a chemical, and chemical conversion techniques can be used to obtain it from sugarcane and its by-products (Santos et al. 2020). Because of its many uses, it also demonstrates great potential as a bio-based chemical and broad utility across a variety of industries (Zahara 2018). The process of producing formic acid and furfural follows comparable steps to that of levulinic acid. In this production method, bagasse is fed into a first reactor containing sulphuric acid. Subsequently, the mixture advances to a second reactor, where the resulting combination includes formic acid, furfural, and steam. The steam is then isolated from the chemical mixture in the second reactor through distillation column separation, yielding distinct formic acid and furfural components (Zahara 2018).

6.7 Bio-plastics

Replacing petroleum in the production of plastic has become dependent on the use of ethanol derived from sugarcane. The term "bio-plastics" refers to these materials that retain full recyclability while having chemical and physical characteristics similar to those of standard plastics like PET or polyethylene (Adeniyi et al. 2023a, b; Solomon 2016). The catalytic dehydration of ethanol results in ethylene, a C_2 building block used in the manufacturing of polymers such as high-density polyethylene, low-density polyethylene, polyvinyl chloride, and polyethylene terephthalate (Rabelo et al. 2020). The production of bio-based plastics is currently dominated by sugarcane ethanol-based bio-polyethylene (bio-PE) and bio-polyethylene terephthalate (bio-PET), and this dominance is expected to continue in the short- and medium term. Current sources of first-generation ethanol include sugarcane and other food crops. These sources also provide the raw materials for bio-PE and bio-PET (Tsiropoulos et al. 2015). One of the most widely utilized biopolymers globally is Braskem's green PE derived from sugarcane ethanol, which holds a prominent position (Santos et al. 2019).

6.7.1 Production of Bio-polyethylene

The process of producing biopolyethylene, whether LDPE (low-density polyethylene) or HDPE (high-density polyethylene), from sugarcane comprises four main stages. Initially, sugarcane is cultivated in the field, involving soil preparation, cane planting, fertilizer application, harvesting after 12–18 months, and subsequent transportation to an ethanol mill. The second stage entails the fermentation and distillation of ethanol. During this phase, sugarcane undergoes pre-treatment, including washing, crunching, and juice extraction at the mill. Following yeast fermentation to produce ethanol, distillation is employed to separate and purify the ethanol, resulting in a 95% concentration. The third stage involves the conversion of ethanol to ethylene. This process includes heating ethanol with a reactor product using a heat exchanger, further heating in a furnace, dehydration of ethanol on an aluminium catalyst to yield ethylene, and subsequent purification of the ethylene (Liptow and Tillman 2012). Depending on the intended use, ethylene undergoes conversion into LDPE using high-pressure tubular or autoclave reactors. HDPE, on the other hand, is predominantly manufactured in low-pressure reactors through suspension or gas-phase polymerization (Tsiropoulos et al. 2015). The final phase of LDPE production involves converting ethylene into LDPE through polymerization at pressures between 81 and 276 MPa and temperatures between 130 and 330 °C (Liptow and Tillman 2012).

6.7.2 Production of Bio-polyethylene Terephthalate (PET)

In the production of bio-polyethylene terephthalate, the process commences with the blending of ethylene, oxygen, argon, CO_2, and either nitrogen or methane. This gas mixture is introduced into a tubular catalytic reactor, where the highly exothermic reaction transforms ethylene into ethylene oxide (EO). To regulate the reaction temperature, steam is generated, and the pressure in the steam drum is meticulously controlled. The resulting ethylene oxide is treated with water through a scrubbing process, eliminating any accompanying by-product CO_2, which is then reintroduced into the reactor loop. The ethylene oxide is steam-stripped and recovered as a concentrated aqueous solution. This ethylene oxide stream is directed to a glycol reactor, where it reacts with water to generate ethylene glycols. Subsequently, a multi-effect evaporator system is employed to remove water from the glycols. The glycols undergo a series of processes, including drying and cooling, and are then sent to a distillation train for separation and purification. During this distillation process, monoethylene glycol (MEG) is separated from the heavier diethylene glycol (DEG) and triethylene glycol (TEG). Bio-MEG, constituting 27.7% of the mass inputs necessary for bio-PET production, is a bio-based monoethylene glycol. The other essential monomer for Bio-PET is purified terephthalic acid (PTA), which is the result of the combination of paraxylene and acetic acid. The PTA monomer is polymerized with MEG to create amorphous PET through either a direct esterification or melt polymerization process (Tsiropoulos et al. 2015).

6.8 Applications of Sugarcane-Based Chemicals

Sugarcane plays a crucial role in the economies of numerous countries, serving as a vital crop for both sustenance and bioenergy (Das Bhowmik et al. 2016). For example, the main source of Brazilian biofuel has been sugarcane, especially 1G bioethanol made from sugarcane, which has been produced and sold widely in Brazil for at least 40 years (Gnansounou et al. 2015). Beyond its role as a food source, virtually every component of sugarcane can be employed in the creation of various valuable products (Das Bhowmik et al. 2016). Industrial chemicals derived from sugarcane could play a significant role in producing the chemicals and materials required by society in the future (Robins and Speight 2016). Hence, various cutting-edge technologies are still under examination to transform sugarcane into essential chemicals like additives, resins, solvents, and inks, which are crucial components in the production of major products such as cosmetics, plastics, and pharmaceuticals (Rabelo et al. 2020). Below is a discussion of the application of chemicals derived from sugarcane.

6.8.1 Lactic Acid

The US Department of Energy anticipates lactic acid (LA) to emerge as a leading platform chemical derived from biomass, given its wide-ranging applications and substantial market value in the multi-billion-dollar range (Agrawal and Kumar 2023). Lactic acid, identified as 2-hydroxypropanoic acid ($CH_3CHOHCOOH$) (Zahara 2018), finds extensive use in diverse industries, such as textiles, polymers, cosmetics, and packaging (Agrawal and Kumar 2023). Additionally, it serves as a preservative and pH-adjusting agent in the food and beverage sectors (Rabelo et al. 2020; Zahara 2018) while playing a crucial role in the pharmaceutical and chemical industries as a solvent and serving as a foundational component for the synthesis of lactate esters (Rabelo et al. 2020; Vaz 2020). Lactic acid can serve as a precursor for various chemicals, including propylene glycol, 2,3-pentadione, and propanoic acid (Zahara 2018). In fact, presently, there has been a significant surge in the utilization of lactic acid, primarily attributed to its function as a monomer in the synthesis of the environmentally friendly polymer known as polylactic acid (PLA) (Sindhu et al. 2016). Sugarcane-derived PLA is manufactured through a fermentation process in which sugar undergoes initial conversion into lactic acid before progressing to the polymerization stage to form PLA (Silalertruksa and Gheewala 2020). This PLA can be utilized in medical settings for purposes like sutures, wound closure clips, and prosthetic devices (Laopaiboon et al. 2010).

6.8.2 Succinic Acid

Succinic acid can undergo transformation into 1,4-butanediol, which in turn can be further converted into tetrahydrofuran or utilized in the synthesis of polymers such as polybutylene terephthalate and polybutylene succinate (Rabelo et al. 2020). Its versatile uses span various industries, including those engaged in the manufacture of food, eco-friendly solvents, biodegradable plastics, and components employed to enhance plant growth. Succinic acid finds extensive application as a foaming agent, detergent, additive, and surfactant. Additionally, it serves as an ion chelator, playing a crucial role in preventing corrosion and pitting within the metal industry. Furthermore, it is employed as an antimicrobial and flavouring agent, and it serves as an additive in the manufacturing of vitamins, amino acids, and antibiotics (Sindhu et al. 2016). Succinic acid can be converted to 1,4-butanediol, which can then be used to synthesize polymers like polybutylene terephthalate and polybutylene succinate, or it can be further converted into tetrahydrofuran (Rabelo et al. 2020). Its diverse applications cover a wide range of industries, such as those producing food, environmentally friendly solvents, biodegradable plastics, and parts used to promote plant growth. Succinic acid is widely used as a surfactant, detergent, foaming agent, and additive. It also acts as an ion chelator, which is extremely important in the metal industry as it prevents corrosion and pitting. In addition, it is used as an additive in the production of vitamins, amino acids, and antibiotics, as well as an antimicrobial and flavouring agent (Sindhu et al. 2016).

6.8.3 Levulinic Acid

Levulinic acid serves as a versatile base chemical suitable for generating various bio-chemicals such as succinic acid, resins, polymers, herbicides, flavouring agents, surfactant solvents, plasticizers, biofuels, and oxygenated fuel additives (Rackemann et al. 2016). It is also capable of undergoing chemical transformation to produce pyrrolidones, lactones, and levulinate esters. These esters find application in the food industry, where they are utilized as solvents and plasticizers (Rabelo et al. 2020). γ-Valerolactone (GVL) can also be obtained from levulinic via catalytic hydrogenation, and it is employed in the production of perfumes, food additives, and eco-friendly solvents (Rabelo et al. 2020). Levulinic acid finds application in diverse industries such as cosmetics, food additives, pharmaceuticals, and agriculture. It serves as a plasticizer in the cosmetics industry. Furthermore, it is utilized in the synthesis of gamma-valerolactone and amino-levulinic acid, which are important ingredients in herbicides (Rajamani et al. 2021).

Other organic acids that can be produced from sugarcane include citric acid and amino acids, both of which have practical uses in a variety of fields. Proteins require amino acids as essential building blocks. A few amino acids are useful feed additives. These amino acids are used in the pharmaceutical industry to formulate parenteral nutrition since they are protein constituents. Certain amino acids also have important functions in other food-related applications (Villela Filho et al. 2011). Citric acid finds widespread application in various industries, including cosmetics, food, beverages, pharmaceuticals, chelating and buffering agents, as well as antioxidants, flavour enhancers, and preservatives (Sindhu et al. 2016). Citric acid, derived from sugarcane, is widely used as an acidulant in food and beverage products. Organic acids are widely used in the pharmaceutical industry as excipients in drug formulations.

6.8.4 Ethanol

The chemical transformation ability of ethanol enables it to enter a wide range of markets for a variety of uses, including food and packaging, agriculture, coatings and inks, building, cosmetics, and pharmaceuticals. Ethanol is widely used as an eco-friendly alternative to gasoline in flex-fuel vehicles and blends like E10 and E85. It reduces greenhouse gas emissions and enhances fuel efficiency. It serves as a solvent in various industries, including pharmaceuticals, cosmetics, and perfumery. Sugarcane-derived ethanol is also a key ingredient in the production of alcoholic beverages, including rum and other spirits. It is possible to replace methyl *tert*-butyl ether with ethyl *tert*-butyl ether, which is produced from ethanol and used as an oxygenate additive in gasoline. The mineral oil industry prefers this material over alternative biorenewable fuels because it can be used as a renewable fuel component that complies with European regulations (Villela Filho et al. 2011). It is also used in various capacities as an automotive fuel (Solomon 2011), which can be in its concentrated form or in a diluted state as an additive to gasoline. Its purpose is to enhance the octane level of gasoline and regulate emissions from vehicles (De Almeida and Colombo 2023). At 5% concentration, it acts as an oxygenator; at 5–20% concentration, it blends with gasoline (Gasohol); and at higher blends, up to F-95%, it can be used as a fuel extender or replacement. The well-established technical viability of ethanol integration into gasoline enables its use in gasoline blends ranging from 5% to 20% without requiring engine modifications (Solomon 2011). When compared to burning fossil fuels, the combustion of bioethanol made from sugarcane produces noticeably less carbon dioxide emissions (Das Bhowmik et al. 2016). Moreover, it can be utilized to produce diethyl ether, which is an essential solvent in the chemical industry. This material is essential to the synthesis of cellulose-based plastics, such as cellulose acetate (Villela Filho et al. 2011).

6.8.5 Butanol

Compared to ethanol, butanol is a primary alcohol with four carbon atoms that has a higher energy density and less volatility. With only minor engine modifications, it can be used as fuel in gasoline-powered engines currently in use (Sindhu et al. 2016). Additionally, it is used as a solvent and dilution agent in the production of hormones, cosmetics, brake fluid, and antibiotics (Cui et al. 2020). It is recognized for its "drop-in" characteristics as an automotive fuel and works well as a blend ingredient for ethanol, diesel, gasoline, and biodiesel (Mariano et al. 2013). Furthermore, it is an essential raw material for the chemical industry, contributing significantly to the production of paints, solvents, plasticizers, and chemical stabilizers (Mariano et al. 2013; Sindhu et al. 2016). Butanol and its derivatives play a crucial role as chemical constituents employed in the production of different polymers and plastics (Mariano et al. 2013).

6.8.6 Polyols

Xylitol is utilized as a sweetener (Rabelo et al. 2020) and has gained widespread acceptance as a sugar substitute within the health and wellness industry (Aguiar et al. 2021; Iwuozor et al. 2024c). Functioning as an alternative to sugar, it is incorporated into dietary products tailored for individuals with insulin deficiency (Sindhu et al. 2016). Its unique properties, such as anti-cavity characteristics, low-energy content, and the ability to promote tooth re-hardening, make it a foundational element for synthesizing valuable chemicals (Aguiar et al. 2021). Additionally, xylitol demonstrates anticancer effects and contributes to the re-hardening and re-mineralization of teeth. Its applications extend to the pharmaceutical sector, dental formulations, and the food industry (Das Bhowmik et al. 2016). Sorbitol functions as a foundational material for the production of polyurethanes, surfactants, and ascorbic acid (Rabelo et al. 2020).

6.8.7 Furfural

Furfural is a clear and fragrant oil that is also referred to as furfuraldehyde. It is only found in highly industrialized countries, where it is mostly used in the production of butadiene and nylon 55 (Solomon 2011). Furfural is needed as an intermediate in the synthesis of numerous chemical products, such as nylons, adhesives, lubricants, solvents, medications, and plastics. Furfuryl alcohol, which is produced by reducing furfural, can be mixed with urea, acetone, or phenol to form solid resins (Uppal et al. 2008). Furthermore, it has a wide range of uses; in the food and beverage sector, it is used as a transportation fuel, lubricant, nematicide, and fungicide (Rajamani et al. 2021). Furfural is used as a purification solvent for vegetable oils, wood resins, and lubricating oils. Furfuryl alcohol also serves as an essential component in the synthesis of furanic polymers, urea-based polymers, anticorrosive agents, fragrances, modified formaldehydes, and resins and dyes (Santos et al. 2020).

6.8.8 Bioplastics

The utilization of sugarcane for the production of bioplastics within a biorefinery has demonstrated its viability as a means to decrease carbon emissions (Villela Filho et al. 2011). The increasing public interest in technological advancements has led to a rise in the use of bio-based plastics in packaging. Noteworthy examples include bio-based polyethylene, polylactic acid, and partially bio-based PET (Tsiropoulos et al. 2015). Poly(hydroxy-butyrate) (PHB) is a biopolymer derived from sugarcane through a fully renewable process that exhibits intriguing properties like thermoplastic behaviour and the ability to undergo biodegradation (Kurusu et al. 2015; Villela Filho et al. 2011). PHB's biodegradability opens up novel applications; for example, it can be used to make seedling-nurturing tubes that decompose after the seedling is planted (Villela Filho et al. 2011). Sugarcane-based bio-plastics are used to create a wide range of packaging materials, including bottles, bags, and films. These products offer a sustainable alternative to traditional plastics and reduce plastic pollution. They are employed in agricultural applications, such as mulch films and nursery pots, to promote sustainable farming practices. Biodegradable tableware, such as utensils and plates, can be produced from sugarcane-based bio-plastics.

6.9 Future Perspectives

In envisioning the future of sugarcane-based chemicals and their applications, a series of transformative developments, as shown in Fig. 6.2, are anticipated to help shape the landscape of this industry. The optimization of sugarcane processing technologies should witness a paradigm shift towards fully automated and intelligent systems. Fermentation and distillation processes should be streamlined, leveraging nanotechnology and advanced catalysis for unprecedented efficiency. Waste utilization should transcend traditional methods with the emergence of bioconversion techniques that turn by-products into high-value bio-chemicals, fostering a closed-loop and zero-waste production cycle (Omoleye et al. 2023).

Circular economy integration is poised to redefine the concept of waste (Ighalo et al. 2022; Iwuozor et al. 2023e). Sugarcane by-products, such as bagasse and molasses, should be seamlessly integrated into a regenerative system where recycling and up-cycling are the norm. The industry's commitment to environmental stewardship should be manifested in the prolonged life cycles of products, contributing to a circular economy that minimizes resource depletion and environmental impact. In addition, market expansion should be driven by the discovery of novel applications for sugarcane-based chemicals. Bio-chemicals should not only dominate conventional industries but also find unprecedented utility in cutting-edge sectors like space exploration and advanced medical technologies. The versatility of

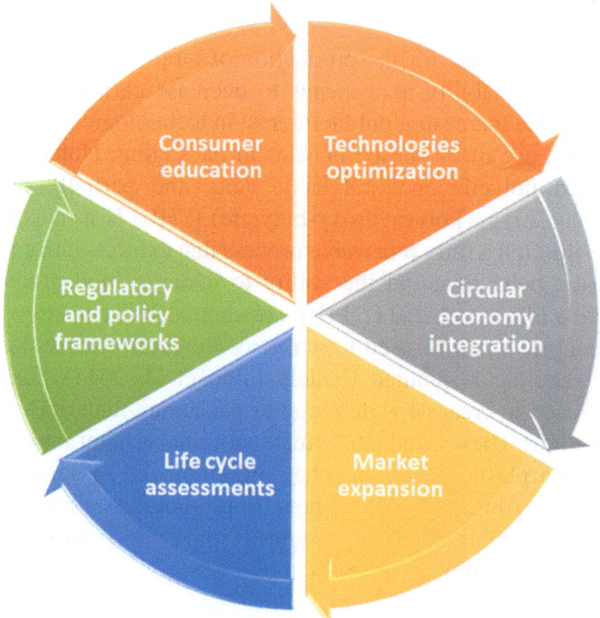

Fig. 6.2 Some future perspectives on sugarcane-based chemicals

sugarcane-derived compounds should be a catalyst for diversification, ensuring resilience in the face of evolving market demands.

Life cycle assessments should evolve into predictive models powered by advanced artificial intelligence (Iwuozor et al. 2023g). The entire life cycle of sugarcane-based chemicals should be meticulously simulated, allowing for real-time adjustments and continuous optimization. This foresight-driven approach will not only enhance environmental performance but will also provide dynamic insights into areas for improvement, ensuring sustainability at every stage of production. In addition, regulatory and policy frameworks will transform to accommodate the dynamic nature of sugarcane-based chemical industries. Governments and international bodies should collaborate to create agile regulatory structures that balance innovation with environmental responsibility. Incentives, standards, and sustainability certifications should be designed to encourage not just compliance but proactive participation in the industry's journey towards a sustainable future. Finally, consumer education and awareness should evolve into immersive experiences, utilizing virtual reality and interactive platforms to educate the public on the environmental benefits of sugarcane-based products. Social media and augmented reality applications will play an important role in shaping consumer perceptions and driving demand for sustainable alternatives. As consumers become more conscious of their choices, they will actively contribute to the industry's mission of sustainability.

6.10 Conclusion

The examination of sugarcane-based chemicals and their diverse applications discussed in this chapter emphasizes the crucial role of this versatile crop across various industries. Sugarcane transcends its conventional uses as a sweetener and bioenergy source, emerging as a sustainable powerhouse poised to reshape chemical production and contribute significantly to an environmentally friendly future. Chemicals that can be produced from sugarcane are not limited to alcohols such as ethanol, butanol, xylitol, sorbitol, and glycerol; organic acids such as lactic, levulinic, and succinic acids; furan-based chemicals such as furfural and furfuryl alcohol; and bio-plastics such as polyethylene and polyethylene terephthalate. Lactic acid is used in a variety of industries, such as textiles, cosmetics, and even medical settings. Succinic acid has multifaceted roles, such as a foaming agent, detergent, and antimicrobial. Furthermore, the base chemical levulinic acid serves as a foundation for a myriad of bio-chemicals, contributing to industries such as cosmetics, pharmaceuticals, and agriculture. The potential of sugarcane-based chemicals extends to the production of bio-based plastics, with examples like polylactic acid and poly(hydroxy-butyrate) showcasing the industry's commitment to sustainability. The environmental benefits of sugarcane-derived bioethanol are particularly noteworthy, producing significantly fewer carbon dioxide emissions compared to traditional fossil fuels. However, as with any emerging technology, challenges and considerations must be acknowledged. Further research is needed to optimize production processes, enhance yield, and address potential socio-economic impacts. Additionally, collaboration between researchers, industry stakeholders, and policymakers is crucial to fostering the widespread adoption of sugarcane-based chemicals. In essence, the utilization of sugarcane for chemical production represents a harmonious blend of economic viability, environmental sustainability, and technological innovation. As global industries seek greener alternatives, sugarcane emerges as a beacon of hope, offering solutions that not only meet current needs but also pave the way for a more sustainable and resilient future.

References

Adeniyi AG, Abdulkareem SA, Iwuozor KO, Emenike EC, Adeyanju CA, Abdulkareem MT, Omisore MO (2023a) Preparation and properties of biocomposite prepared from waste polystyrene and Prospopis africana biochar. Rec Innov Chem Eng 16:350–361
Adeniyi AG, Amusa VT, Emenike EC, Iwuozor KO (2023b) Hybrid biochar production from biomass and pigmented plastic for sustainable waste-to-energy. Emerg Mater 6:1481–1490
Adeniyi AG, Amusa VT, Iwuozor KO, Emenike EC (2023c) Valorization of waste biaxially-oriented polypropylene (BOPP) plastic films by its co-carbonization with almond leaves. Environ Prog Sustain Energy 42:e14064
Adeniyi AG, Iwuozor KO, Emenike EC, Sagboye PA, Micheal KT, Micheal TT, Saliu OD, James R (2023d) Biomass-derived activated carbon monoliths: a review of production routes, performance, and commercialization potential. J Clean Prod 423:138711
Adeola AO, Iwuozor KO, Akpomie KG, Adegoke KA, Oyedotun KO, Ighalo JO, Amaku JF, Olisah C, Conradie J (2023) Advances in the management of radioactive wastes and radionuclide contamination in environmental compartments: a review. Environ Geochem Health 45(6):2663–2689
Afiomah CS, Iwuozor KO (2020) Nutritional and phytochemical properties of Beta vulgaris Linnaeus (Chenopodiaceae)—a review. Nigerian J Pharm Appl Sci Res 9:38–44
Agrawal D, Kumar V (2023) Recent progress on sugarcane-bagasse based lactic acid production: technical advancements, potential and limitations. Ind Crop Prod 193:116132
Aguiar A, Milessi TS, Mulinari DR, Lopes MS, Da Costa SM, Candido RG (2021) Sugarcane straw as a potential second generation feedstock for biorefinery and white biotechnology applications. Biomass Bioenergy 144:105896
Ahmad F, Saeed Q, Shah SMU, Gondal MA, Mumtaz S (2022) Environmental sustainability: challenges and approaches. In: Natural resources conservation and advances for sustainability. Elsevier, Amsterdam, pp 243–270
Arruda P (2011) Perspective of the sugarcane industry in Brazil. Trop Plant Biol 4:3–8
Bezerra TL, Ragauskas AJ (2016) A review of sugarcane bagasse for second-generation bioethanol and biopower production. Biofuels Bioprod Biorefin 10:634–647
Borges ER, Pereira N Jr (2011) Succinic acid production from sugarcane bagasse hemicellulose hydrolysate by Actinobacillus succinogenes. J Ind Microbiol Biotechnol 38:1001–1011
Candido R, Goncalves AR (2019) Evaluation of two different applications for cellulose isolated from sugarcane bagasse in a biorefinery concept. Ind Crop Prod 142:111616
Cardona C, Quintero J, Paz I (2010) Production of bioethanol from sugarcane bagasse: status and perspectives. Bioresour Technol 101:4754–4766
Castañeda-Ayarza JA, Cortez LAB (2017) Final and B molasses for fuel ethanol production and some market implications. Renew Sust Energ Rev 70:1059–1065
Conteratto C, Artuzo FD, Santos OIB, Talamini E (2021) Biorefinery: a comprehensive concept for the sociotechnical transition toward bioeconomy. Renew Sust Energ Rev 151:111527

Cui Y, He J, Yang K-L, Zhou K (2020) Aerobic acetone-butanol-isopropanol (ABI) fermentation through a co-culture of Clostridium beijerinckii G117 and recombinant Bacillus subtilis 1A1. Metab Eng Commun 11:e00137
Das Bhowmik SS, Brinin AK, Williams B, Mundree SG (2016) Sugarcane biotechnology: tapping unlimited potential. In: Sugarcane-based biofuels and bioproducts. Wiley, New York, pp 23–52
De Almeida MA, Colombo R (2023) Production chain of first-generation sugarcane bioethanol: characterization and value-added application of wastes. Bioenergy Res 16:924–939
De Araujo Guilherme A, Dantas PVF, de Araújo Padilha CE, Dos Santos ES, de Macedo GR (2019) Ethanol production from sugarcane bagasse: use of different fermentation strategies to enhance an environmental-friendly process. J Environ Manag 234:44–51
De Miranda EE, Fonseca MF (2020) Sugarcane: food production, energy, and environment. In: Sugarcane biorefinery, technology and perspectives. Elsevier, Amsterdam, pp 67–88
Devadiga DG, Bhat KS, Mahesha G (2020) Sugarcane bagasse fiber reinforced composites: recent advances and applications. Cogent Eng 7:1823159
Emenike EC, Amusa VT, Iwuozor KO, Ojeyemi T, Micheal TT, Micheal KT, Adeniyi AG (2023a) Enhancing biochar properties through doping: a comparative study of sugarcane bagasse and chicken feather. Biofuels 15:627
Emenike EC, Iwuozor KO, Ighalo JO, Bamigbola JO, Omonayin EO, Ojo HT, Adeleke J, Adeniyi AG (2023b) Advancing the circular economy through the thermochemical conversion of waste to biochar: a review on sawdust waste-derived fuel. Biofuels 15:433–447
Emenike EC, Iwuozor KO, Saliu OD, Ramontja J, Adeniyi AG (2023c) Advances in the extraction, classification, modification, emerging and advanced applications of crystalline cellulose: a review. Carbohydr Polymer Technol Appl 6:100337
Gnansounou E, Vaskan P, Pachón ER (2015) Comparative techno-economic assessment and LCA of selected integrated sugarcane-based biorefineries. Bioresour Technol 196:364–375
Hiloidhari M, Haran S, Banerjee R, Rao AB (2021) Life cycle energy–carbon–water footprints of sugar, ethanol and electricity from sugarcane. Bioresour Technol 330:125012
Ighalo JO, Iwuozor KO, Ogunfowora LA, Abdulsalam A, Iwuchukwu FU, Itabana B, Bright OC, Igwegbe CA (2021) Regenerative desulphurisation of pyrolysis oil: a paradigm for the circular economy initiative. J Environ Chem Eng 9:106864
Ighalo J, Rangabhashiyam S, Dulta K, Umeh CT, Iwuozor KO, Aniagor CO, Eshiemogie S, Iwuchukwu FU, Igwegbe C (2022) Recent advances in hydrochar application for the adsorptive removal of wastewater pollutants. Chem Eng Res Des 184:38
Igwegbe CA, Obiora-Okafo IA, Iwuozor KO, Ghosh S, Kurniawan SB, Rangabhashiyam S, Kanaoujiya R, Ighalo JO (2022) Treatment technologies for bakers' yeast production wastewater. Environ Sci Pollut Res 29:1–23
Iwuozor KO, Anyanwu VU, Olaniyi BO, Mbamalu PS, Adeniyi AG (2022a) Adulteration of sugar: a growing global menace. Sugar Tech 24:914–919
Iwuozor KO, Emenike EC, Ighalo JO, Eshiemogie S, Omuku PE, Adeniyi AG (2022b) Valorization of sugar industry's by-products: a perspective. Sugar Tech 24:1052–1078
Iwuozor KO, Emenike EC, Ighalo JO, Omoarukhe FO, Omuku PE, Adeniyi AG (2022c) A review on the thermochemical conversion of sugarcane bagasse into biochar. Clean Mater 6:100162
Iwuozor KO, Mbamalu PS, Olaniyi BO, Anyanwu VU, Emenike EC, Adeniyi AG (2022d) Fortification of sugar: a call for action. Sugar Tech 24:1284–1294
Iwuozor KO, Ogunfowora LA, Oyekunle IP (2022e) Review on sugarcane-mediated nanoparticle synthesis: a green approach. Sugar Tech 24:1186–1197
Iwuozor KO, Oyekunle IP, Oladunjoye IO, Ibitogbe EM, Olorunfemi TS (2022f) A review on the mitigation of heavy metals from aqueous solution using sugarcane bagasse. Sugar Tech 24:1167–1185
Iwuozor KO, Adeniyi AG, Emenike EC, Adepoju MI, Ahmed MO (2023a) Sugarcane juice powder produced from spray drying technology: a review of properties and operating parameters. Sugar Tech 25:497–507

Iwuozor KO, Adeniyi AG, Emenike EC, Ojeyemi T, Egbemhenghe AU, Okorie CJ, Ayoku BD, Saliu OD (2023b) Prospects and challenges of utilizing sugarcane bagasse as a bio-coagulant precursor for water treatment. Biotechnol Rep 39:e00805
Iwuozor KO, Adeniyi AG, Emenike EC, Olaniyi BO, Anyanwu VU, Bamigbola JO, Ojo HT (2023c) Adsorption technology in the sugar industry: current status and future perspectives. Sugar Tech 25:1005–1013
Iwuozor KO, Emenike EC, Abdulkadir M, Ogunniyi S, Adeniyi AG (2023d) Effect of salt modification on biochar obtained from the thermochemical conversion of sugarcane bagasse. Sugar Tech 25:223–233
Iwuozor KO, Emenike EC, Omonayin EO, Bamigbola JO, Ojo HT, Awoyale AA, Eletta OA, Adeniyi AG (2023e) Unlocking the hidden value of pods: a review of thermochemical conversion processes for biochar production. Bioresour Technol Rep 22:101488
Iwuozor KO, Ojeyemi T, Emenike EC, Umeh CT, Egbemhenghe A, Ayoku BD, Ogunsanya TI, Ogunniyi S, Ighalo JO, Adeniyi AG (2023f) Management of sugar dust in the sugar industry. Heliyon 10:e23158
Iwuozor KO, Olaniyi BO, Anyanwu VU, Suleiman MA, Omoleye WS, Enahoro-Ofagbe FE, Katagum AA, Moronkola IA, Opeyemi AM (2023g) The effect of ChatGPT on sugar industry research. Sugar Tech 25:1–7
Iwuozor KO, Abdulkareem SA, Amoloye MA, Emenike EC, Ezzat AO, Mustapha JA, Egbemhenghe AU, Adeniyi AG (2024a) Eco-friendly composite materials: enhancing sustainability with sugarcane bagasse biochar and polystyrene resin. Sugar Tech 26:408–421
Iwuozor KO, Amoloye MA, Owolabi OO, Egbemhenghe AU, Emenike EC, Ezzat AO, Adeniyi AG (2024b) Sustainable production and comparative liquid phase exfoliation of onion peel-doped sugarcane bagasse hybrid biochar. Sugar Tech 26:502–512
Iwuozor KO, Emmanuel SS, Ahmed MO, Idris AM, Emenike EC, Saliu OD, Qudus AH, Adeniyi AG (2024c) Technologies for the extraction and post-extraction of Stevia rebaudiana leaves. Chem Afr 7:539–563
Iwuozor KO, Okoro HK, Adeniyi AG, Zvinowanda C, Ngila JC, Emenike EC (2024d) Sugarcane bagasse adsorbents: bibliometric insights and the influence of chemical treatment on adsorption performance in aqueous solution. Sugar Tech 26:333–351
Jin Y, Shi Z, Xu G, Yang H, Yang J (2020) A stepwise pretreatment of sugarcane bagasse by alkaline and hydroxymethyl reagent for bioethanol production. Ind Crop Prod 145:112136
Jun Z, Franca UA, Iwuozor KO (2023) Relationship between agricultural production, energy consumption, and climate change in Nigeria. In: Climate change impacts on Nigeria: environment and sustainable development. Springer, Singapore, pp 563–580
Khaire KC, Moholkar VS, Goyal A (2021) Bioconversion of sugarcane tops to bioethanol and other value added products: an overview. Mater Sci Energy Technol 4:54–68
Kurusu RS, Siliki CA, David E, Demarquette NR, Gauthier C, Chenal J-M (2015) Incorporation of plasticizers in sugarcane-based poly(3-hydroxybutyrate) (PHB): changes in microstructure and properties through ageing and annealing. Ind Crop Prod 72:166–174
Laopaiboon P, Thani A, Leelavatcharamas V, Laopaiboon L (2010) Acid hydrolysis of sugarcane bagasse for lactic acid production. Bioresour Technol 101:1036–1043
Lee SY, Kim HU, Chae TU, Cho JS, Kim JW, Shin JH, Kim DI, Ko Y-S, Jang WD, Jang Y-S (2019) A comprehensive metabolic map for production of bio-based chemicals. Nat Catal 2:18–33
Liptow C, Tillman AM (2012) A comparative life cycle assessment study of polyethylene based on sugarcane and crude oil. J Ind Ecol 16:420–435
Lisboa CC, Butterbach-Bahl K, Mauder M, Kiese R (2011) Bioethanol production from sugarcane and emissions of greenhouse gases—known and unknowns. GCB Bioenergy 3:277–292
Lunelli BH, Andrade RR, Atala DI, Wolf Maciel MR, Maugeri Filho F, Maciel Filho R (2010) Production of lactic acid from sucrose: strain selection, fermentation, and kinetic modeling. Appl Biochem Biotechnol 161:227–237
Mariano AP, Dias MO, Junqueira TL, Cunha MP, Bonomi A, Maciel Filho R (2013) Butanol production in a first-generation Brazilian sugarcane biorefinery: technical aspects and economics of greenfield projects. Bioresour Technol 135:316–323

Mohammed K, Iwuozor KO, Anyanwu VU, Olaniyi BO (2024) Sugar dust explosion in the sugar industry: case studies and prevention strategies. Sugar Tech 26:12–19

Omoleye WS, Fawole OB, Affinnih K, Aborode AT, Emenike EC, Iwuozor KO (2023) Bioremediation of Asa River sediment using agricultural by-products. In: Land remediation and management: bioengineering strategies. Springer, Singapore, pp 295–330

Parsaee M, Kiani MKD, Karimi K (2019) A review of biogas production from sugarcane vinasse. Biomass Bioenergy 122:117–125

Rabelo SC, De Paiva LBB, Pin TC, Pinto LFR, Tovar LP, Nakasu PYS (2020) Chemical and energy potential of sugarcane. In: Sugarcane biorefinery, technology and perspectives. Elsevier, Amsterdam, pp 141–163

Rackemann DW, Zhang Z, Doherty WO (2016) Conversion of sugarcane carbohydrates into platform chemicals. In: Sugarcane-based biofuels and bioproducts. Wiley, New York, pp 207–236

Raj K, Krishnan C (2020) Improved co-production of ethanol and xylitol from low-temperature aqueous ammonia pretreated sugarcane bagasse using two-stage high solids enzymatic hydrolysis and Candida tropicalis. Renew Energy 153:392–403

Rajamani S, Santhosh R, Raghunath R, Jadhav SA (2021) Value-added chemicals from sugarcane bagasse using ionic liquids. Chem Pap 75:5605–5622

Raza G, Ali K, Hassan MA, Ashraf M, Khan MT, Khan IA (2019) Sugarcane as a bioenergy source. In: Sugarcane biofuels: status, potential, and prospects of the sweet crop to fuel the world. Dokumen Pub, Dorval, pp 3–19

Reena R, Alphy MP, Reshmy R, Thomas D, Madhavan A, Chaturvedi P, Pugazhendhi A, Awasthi MK, Ruiz H, Kumar V (2022) Sustainable valorization of sugarcane residues: efficient deconstruction strategies for fuels and chemicals production. Bioresour Technol 361:127759

Robins KT, Speight RE (2016) Chemicals manufacture from fermentation of sugarcane products. In: Sugarcane-based biofuels and bioproducts. Wiley, New York, pp 111–136

Santos V, Ely R, Szklo A, Magrini A (2016) Chemicals, electricity and fuels from biorefineries processing Brazil's sugarcane bagasse: production recipes and minimum selling prices. Renew Sust Energ Rev 53:1443–1458

Santos JF, Macario KD, Jou RM, Oliveira FM, Cardoso RP, Diaz M, Anjos RM, Alves EQ (2019) Monitoring the biogenic fraction of sugarcane-based plastic bags. J Clean Prod 233:348–352

Santos F, Eichler P, Machado G, De Mattia J, De Souza G (2020) By-products of the sugarcane industry. In: Sugarcane biorefinery, technology and perspectives. Elsevier, Amsterdam, pp 21–48

Schmitt CC, Moreira R, Neves RC, Richter D, Funke A, Raffelt K, Grunwaldt J-D, Dahmen N (2020) From agriculture residue to upgraded product: the thermochemical conversion of sugarcane bagasse for fuel and chemical products. Fuel Process Technol 197:106199

Shaji A, Shastri Y, Kumar V, Ranade VV, Hindle N (2021) Economic and environmental assessment of succinic acid production from sugarcane bagasse. ACS Sustain Chem Eng 9:12738–12746

Silalertruksa T, Gheewala SH (2020) Competitive use of sugarcane for food, fuel, and biochemical through the environmental and economic factors. Int J Life Cycle Assess 25:1343–1355

Sindhu R, Gnansounou E, Binod P, Pandey A (2016) Bioconversion of sugarcane crop residue for value added products—an overview. Renew Energy 98:203–215

Solomon S (2011) Sugarcane by-products based industries in India. Sugar Tech 13:408–416

Solomon S (2016) Sugarcane production and development of sugar industry in India. Sugar Tech 18:588–602

Sukphun P, Wongarmat W, Imai T, Sittijunda S, Chaiprapat S, Reungsang A (2023) Two-stage biohydrogen and methane production from sugarcane-based sugar and ethanol industrial wastes: a comprehensive review. Bioresour Technol 386:129519

Torgbo S, Quan VM, Sukyai P (2021) Cellulosic value-added products from sugarcane bagasse. Cellulose 28:5219–5240

Tsiropoulos I, Faaij AP, Lundquist L, Schenker U, Briois JF, Patel MK (2015) Life cycle impact assessment of bio-based plastics from sugarcane ethanol. J Clean Prod 90:114–127

Uppal S, Gupta R, Dhillon R, Bhatia S (2008) Potential of sugarcane bagasse for production of furfural and its derivatives. Sugar Tech 10:298–301

Uppal S, Kaur R, Sharma P (2011) Optimization of chemical pretreatment and acid saccharification for conversion of sugarcane bagasse to ethanol. Sugar Tech 13:214–219

Vaz S (2020) Sugarcane for renewable chemistry. In: Sugarcane biorefinery, technology and perspectives. Elsevier, Amsterdam, pp 165–178

Villela Filho M, Araujo C, Bonfa A, Porto W (2011) Chemistry based on renewable raw materials: perspectives for a sugar cane-based biorefinery. Enzyme Res 2011:654596. https://doi.org/10.4061/2011/654596. PMID: 21637329; PMCID: PMC3102444

Zahara ZF (2018) Economic Assessment of the Sugarcane-based Bio-refinery in Indonesia. Evergreen 5(2):67–77. https://doi.org/10.5109/1936219

Lignin-Derived Aromatics in Sugarcane Biorefinery

7

Santhoshkumar Subramaniam, Oviya Govindaraj, Devi Priya Arumugam, Sugitha Thankappan, and Sivakumar Uthandi

Abstract

Sugarcane produces significant amounts of lignocellulosic waste, including sugarcane bagasse and straw, which can be used as cost-effective sources of lignocellulosic biomass (LCB). Lignin, the second most abundant biopolymer in these waste streams, can be converted into valuable lignin-derived aromatics for various purposes. However, the complex and heterogeneous structure of lignin makes it challenging to extract efficiently. This chapter provides an overview of the current status and future prospects of producing lignin-derived aromatics from sugarcane biorefinery. It covers the characteristics and extraction methods of sugarcane lignin, the depolymerization methods for breaking it down, and strategies for converting obtained aromatics into value-added bioproducts including approaches through microbial and metabolic engineering. Successful integration of lignin-derived aromatics can enhance the economic viability and environmental sustainability of sugarcane processing. Further research and innovation are needed to develop efficient and scalable processes for the production and utilization of these valuable lignin-derived compounds.

Santhoshkumar Subramaniam, Oviya Govindaraj, Devi Priya Arumugam, Sugitha Thankappan and Sivakumar Uthandi contributed equally with all other contributors.

S. Subramaniam · O. Govindaraj · D. P. Arumugam · S. Uthandi (✉)
Biocatalysts Laboratory, Department of Agricultural Microbiology, Tamil Nadu Agricultural University, Coimbatore, India
e-mail: usiva@tnau.ac.in

S. Thankappan
Karunya Institute of Technology and Sciences, Coimbatore, India

G. S. Suresha et al. (eds.), *Value Addition and Product Diversification in Sugarcane*, https://doi.org/10.1007/978-981-97-7228-5_7

Keywords

Lignin · Bio-plastics · Valorization · Sugarcane · LDAs

7.1 Introduction

Sugarcane is a large tropical grass that belongs to the Gramineae family and *Saccharum* genus. It is a primary crop for sugar production worldwide due to its notable characteristics like high biomass yield, sucrose content, and efficiency in solar energy accumulation (Godshall and Legendre 2003; Sindhu et al. 2016). After the harvest, the sugarcane stalks are transported to the sugar mills where the juice is extracted, while the leaves and tops are left in the field (Benjamin et al. 2013). Two significant by-products, harvest residue (straw) and fibrous fraction (bagasse), are considered abundant and cost-effective sources of Lignocellulosic biomass (LCB) (Sindhu et al. 2016). Sugarcane bagasse (SCB) and sugarcane straw (SCS) primarily consist of cellulose, hemicellulose, and lignin, with lower amounts of extractives and ash (del Río et al. 2015). SCB is extensively used as a fuel by the sugar industry, while SCS finds application as animal fodder or is burned in the field (Pandey et al. 2000; Sindhu et al. 2013).

Lignocellulosic biomass, known for its potential in chemical and material production due to its cost-effectiveness and availability, features lignin as the second most abundant biopolymer in nature (Gil-Chavez et al. 2021; Watkins et al. 2015). Lignin is a complex and abundant biopolymer that constitutes about 15–25% of the dry weight of sugarcane biomass. It is mainly composed of aromatic units linked by various types of bonds, which confer lignin with high chemical and thermal stabilities. Lignin is often considered a waste or a low-value by-product of sugarcane biorefinery, which aims to produce biofuels and bioproducts from sugarcane biomass. In sugarcane, lignin plays a crucial role in providing elasticity and mechanical strength. Sugarcane lignin, specifically, is highly acylated (p-coumaroylation) at its side chains, incorporates tricin flavonoid units, and includes ferulate residues cross-coupled between arabinoxylan and lignin (del Río et al. 2015). Structural differences between SCB and SCS lignins have been identified, with SCB being syringyl-rich, featuring alkyl–aryl ether substructures, and SCS exhibiting guaiacyl-rich lignin with higher concentrations of condensed structures. Additionally, there is a distinction in the degree of acetylation, with SCB having a higher acetylation level. However, lignin has great potential to be converted into lignin-derived aromatics, which are valuable organic compounds that can be used as fuels, chemicals, and materials. Lignin-derived aromatics can also contribute to the reduction of greenhouse gas emissions and fossil fuel dependence, as they can replace petroleum-based aromatics in various applications. The production of lignin-derived aromatics from sugarcane biorefinery involves three main steps: (1) the extraction of lignin from sugarcane biomass, (2) the depolymerization of lignin into lignin-derived aromatics, and (3) the conversion of lignin-derived aromatics into value-added bioproducts. Each step poses significant challenges and opportunities for the development of efficient and sustainable processes.

This chapter provides a comprehensive overview of the process of producing aromatics using lignin derived from sugarcane biorefinery. It covers the various methods used for extracting sugarcane lignin, as well as the biological and chemical techniques employed to depolymerize it into aromatics. Additionally, the passage delves into the strategies used to transform these aromatics into valuable bioproducts.

7.2 Sugarcane's Crucial Role in Advanced Biorefinery Practices

Sugarcane has been recognized as the main feedstock in modern biorefineries, opening up a wide range of opportunities for the production of diverse and sustainable bioproducts. This includes enzymes, organic acids, amino acids, pigments, animal feed, composites, chelating agents, alkaloids, bioethanol, biodiesel, biobutanol, 2, 3-butanediol, biohydrogen, bioelectricity, and biopolymers that can be produced using agro by-products and sugarcane by-products like bagasse, sugarcane tops, and molasses. The major significant sugarcane wastes like SCB, press mud cake (PMC), fly ash from SCB, and sugarcane trash are generated during the production of sugars from sugar manufacturing industries. SCB, a residual waste generated from the sugar processing industry, can be used as a raw material in agro-residue-based pulp and paper mills and also used to produce the solid fuel that has a significant calorific value of approximately 28.2 MJ kg^{-1} using torrefaction technology (Ahmad et al. 2021). During the torrefaction process, 70% of the biochar is produced. After being ground to a 300 μm sand size, the biochar can be compressed at 2–8 MPa using molasses in a 70:30 ratio to create bio-coal. This bio-coal exhibits a composition of 6.3% moisture content, 74.6% fixed carbon, and 1.4% ash content. Notably, sugarcane bio-coal boasts a prolonged burning period of 30 min and a short ignition time of 10 seconds (Ahmad et al. 2021).

Bagasse, a by-product of the sugar processing industry consisting of lignocellulose, possesses the potential to produce eco-friendly bio-coal. Bio-coal serves as a substitute for conventional coal due to its high heating value and higher fixed carbon content of 76% (Ahmad et al. 2021; Ribul et al. 2021). The exploration of producing high calorific value bio-coal from SCB represents a promising alternative to coal within the sugar industry (Buller et al. 2021). Due to minimal pollutants emitted during combustion, bio-coal stands as an environmentally friendly fuel. While cogeneration for steam and electricity remains the predominant industrial method for sugarcane bagasse (SCB) valorization, there's increasing interest in extracting high value co-products and biofuels from SCB (Meghana and Shastri 2020; Solomon 2011). Additionally, value-added products derived from SCB include animal feed and furfural, whereas PMC possesses high organic content, phosphate, calcium, and magnesium content and can be used as a fertilizer following bio-composting with biomethanated spent wash (distillery wastewater) (Korasapati et al. 2023).

Traditionally, sugarcane trash is disposed off by burning in the fields. SCB fly ash is commonly disposed off in pits, although it can also be used for soil improvement in certain regions or used for brickmaking. Apart from solid wastes generated from sugar production, a considerable amount of liquid waste known as molasses is left after the crystallization of sugars from juice. Molasses is usually employed in ethanol production through fermentation processes. Numerous sugar mills operate distilleries as their subsidiaries to maximize their profit margins by manufacturing extra-neutral alcohol and fuel ethanol. These surplus earnings are crucial, especially in recent years, as India has faced an accumulation of sugar inventory due to excess sugarcane production, resulting in a crisis. Additionally, the declining global sugar prices have posed challenges in ensuring fair and remunerative prices (FRP) for the farmers. The Government of India (GoI) decided to execute the ethanol blending program (EBP), aiming for a 10% blending (in petrol) nationwide, and reduce oil imports by 20% by 2030 in response to the sugar crisis. Furthermore, to enhance profits through molasses-based ethanol production, effective waste management is necessary to valorize the waste streams into value-added products (Korasapati et al. 2023).

7.3 Lignin Unveiled: Structure, Properties, and Natural Sources

Lignin is a complex organic polymer found in the cell walls of plants, particularly in wood and bark. It provides structural support and rigidity to the plant cell walls. The structure of lignin is intricate and varies among plant species, but it generally consists of phenylpropanoid units, primarily coniferyl, sinapyl, and p-coumaryl alcohols, which are linked together through various types of chemical bonds (Tobimatsu and Schuetz 2019). The basic building blocks of lignin are phenylpropane units, which are derived from the amino acid phenylalanine. These units polymerize to form a three-dimensional network (Lu et al. 2020). The main types of chemical bonds in lignin are ether bonds (β-O-4, α-O-4, 4-O-5, etc.), carbon–carbon bonds (β-5, 5–5, β-β, etc.), and occasionally carbon–oxygen bonds (1-O-4). These bonds create a highly cross-linked structure that contributes to lignin's strength and resistance to degradation. The exact arrangement and bonding patterns of lignin components can vary widely depending on factors such as plant species, tissue type, and developmental stage (Sun 2020).

Additionally, lignin can be found in non-woody plant materials such as grasses, agricultural residues, and herbaceous plants, albeit in lower concentrations compared to woody species. Furthermore, lignin is present in various agricultural byproducts and residues, including straw, sugarcane bagasse, and corn stover (Zakzeski et al. 2010). These diverse natural sources of lignin serve as valuable renewable feedstocks for the production of a wide range of bioproducts, including biofuels, bioplastics, and chemicals, thereby contributing to the sustainable utilization of plant biomass and the development of a bio-based economy.

The hydrophobic nature of lignin poses as a barrier to water penetration, providing resistance against decay and microbial attack (Vanholme et al. 2010). Furthermore, lignin's chemical resistance ensures durability and preservation of plant materials, such as wood, over extended periods. Its capacity to absorb and dissipate ultraviolet (UV) radiation contributes to UV protection, crucial for plant defense against photodamage (Boerjan et al. 2003). Lignin demonstrates high thermal stability, making it suitable for applications requiring resistance to elevated temperatures. Moreover, its adhesive properties enable lignin to form bonds with other molecules, serving as a natural adhesive in various industrial contexts (Sun et al. 2018). Despite its resistance to degradation under normal conditions, lignin remains biodegradable by specific microorganisms, facilitating its recycling in nature (Zakzeski et al. 2010). These diverse properties underscore lignin's significance in both natural ecosystems and industrial processes, driving ongoing research into its utilization and value-added applications.

7.4 Lignin Valorization in Sugarcane Biorefineries

A sustainable circular economy focuses heavily on transforming by-products of the sugarcane industry into value-added products. Moreover, the surge in the production of second-generation bioethanol has encouraged the development of technologies for transforming residuals into valuable bioproducts, thereby strengthening the overall process' sustainability. The production of valuable compounds derived from lignin remains a significant challenge because of its intricate structure, polydispersity, and recalcitrant nature. Cellulose is a straight-chain homopolysaccharide polymer made up of repeated units of D-anhydroglucose linked together via 1, 4-β-D-glycosidic bonds (Klemm et al. 1998). Hemicellulose is a diverse polymer made up of C5 sugars (xylose, arabinose), C6 sugars (mannose, glucose, and galactose), and uronic acids (Saha 2003). Each hemicellulose constituent differs depending on the source, and bagasse's hemicellulose has a comparatively high percentage (>90%) of xylan (Hendriks and Zeeman 2009). Hemicellulose forms hydrogen bonds with cellulose microfibrils in the cell wall, facilitating their linkage into microfibrils, thus contributing to lignin valorization in sugarcane biorefineries.

Moreover, isolating, fractionating, modifying, and characterization of lignin remain formidable tasks. Typically, the pretreatment process facilitates the separation of lignocellulosic biomass into its primary components, offering an efficient means of mitigating the inherent recalcitrance of lignocellulosic materials predominantly found in the cell wall (Benjamin et al. 2013). An effective pretreatment technique should effectively remove lignin from the lignocellulosic material and produce a high-quality, pure lignin fraction that meets the specifications needed for the following conversion processes. Following separation, lignin can be modified, fractionated, or depolymerized, offering viable pathways for its sustainable utilization. Among the constituents of lignocellulose, lignin is the most resistant to chemical and biological destruction. Moreover, it's a component of the cellular wall and provides structural support, hydrophobicity, and resistance against oxidative stress and microbial attack (Karp et al. 2013). Lignin serves as an adhesive and binds cellulose and hemicellulose, thereby enhancing the structural integrity and becoming more resistant to moisture and resistant to chemical and biological deterioration (Benjamin

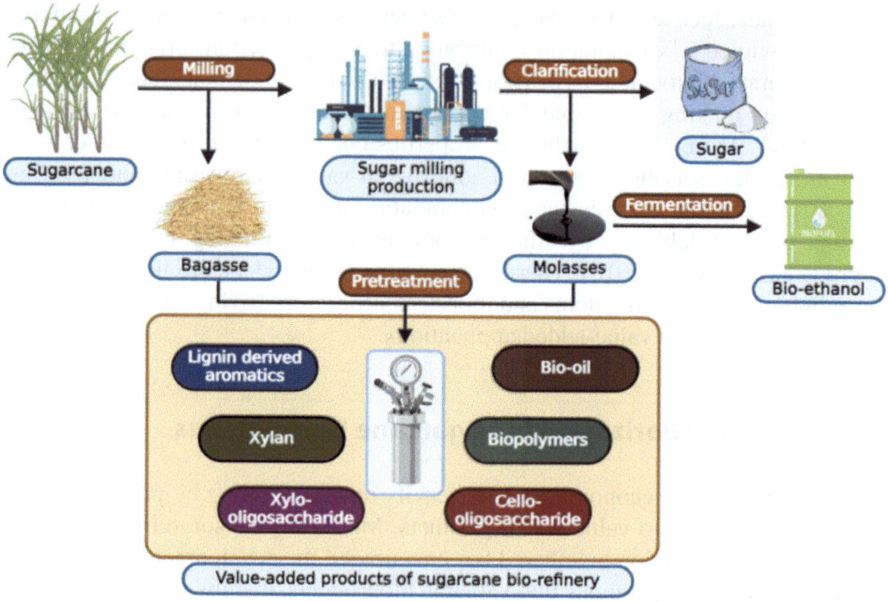

Fig. 7.1 Schematic representation showing valorization of sugarcane refinery wastes

et al. 2013). Investigations have focused on developing techniques for extracting and isolating lignin from sugarcane (Arni 2018; Karp et al. 2013) along with exploring its potential uses (Sarker et al. 2017).

Waste streams like SCB and SCS might serve as a significant lignocellulosic resource in nations with substantial sugarcane production, such as Brazil, India, and China (Orduña Ortega et al. 2020). Both SCB and SCS contain significant amounts of cellulose (ranging from 30% to 40% by weight), offering potential for the production of second-generation (2G) ethanol. Various chemical, physical, or biological pretreatment methods can be employed to convert them into fermentable sugars while simultaneously separating and valorizing lignin. However, recycling of SCS to generate higher value-added products has not been widely practiced. In the future, sugarcane biorefineries will facilitate the implementation of existing and emerging lignocellulosic biomass processing technologies at a commercial scale and broaden their product range by utilizing the potential for electricity and steam generation on-site (Fig. 7.1).

7.5 Methods for Isolating Lignin-Derived Aromatic Compounds (LDAs)

Lignin is a highly recalcitrant component of lignocellulosic biomass, comprising various linkages such as aryglycerol-β-ether dimer (β-O-4), aryglycerol-ether dimer (α-O-4), diaryl ether (4-O-5), resinol (β-5), diphenylethane (β-1), phenylcoumaran (β–β'), phenylcoumaran (β–β), and biphenyl (5–5) (Rawal et al. 2020). Therefore, lignin depolymerization is a challenging task, and numerous chemical and catalytic methods have been attempted. Pretreatment methods are very crucial in the extraction process of lignin. Currently, pretreatment techniques are categorized into chemical (acid/alkali treatment, supercritical fluids, ionic liquids, hydrogenolyis, oxidation, acid hydrotropic fractionation, solvolysis using deep eutectic solvents), physicochemical (steam and ammonia fiber explosion), biological (enzymatic methods) and Organosolv (ethanol) (Zhou et al. 2016). A combination of treatments is also employed for delignification and to increase lignin recovery. Lignin can be converted into various high-value commodity chemicals such as vanillin, furfural, phenol, and other aromatic compounds. However, the resulting products depend on the method of extraction, besides the nature of plant biomass. In this section, the extraction and isolation of lignin from sugarcane are briefly discussed.

7.5.1 Alkaline Methods

7.5.1.1 Alkaline Delignification

Alkaline delignification comprises two steps: Pretreatment and Pulping. After pretreatment (mechanical or hot water at 100 °C for 2 h with a solid-to-liquid ratio of 1:10 w/v), the biomass is treated with an alkaline sodium hydroxide solution. The solubilized hemicellulose and cellulose fraction are removed by centrifugation at 2000 rpm for 10 min. The separated solid phase is treated with 15% NaOH (w/v) at 90 °C for 1.5–4 h (1:10–1:18 solid to liquid ratio), under constant stirring. The black liquor obtained is filtered followed by acidification with 5N sulfuric acid (pH 2.0). The acidified lignin is separated by centrifugation and air-dried. This method has reported 15% lignin recovery from sugarcane bagasse (Wahba et al. 2015).

7.5.1.2 Soda Pulping

Lignin extraction by soda pulping is proposed by Costa et al. (2017) to solubilize the lignin present in the biomass. Different types of lignin namely Lig-Alk, Lig-Ox, and Lig-Ac are obtained by this method from sugarcane. The steps in the soda pulping technique are briefly discussed.

7.5.1.3 Lig-Alk

The sugarcane bagasse is treated with hot water at 80 °C and then dried at 40 °C for four consecutive days to achieve a moisture content of 10% by weight. The milled (powdered) bagasse is subjected to an autoclave process, where it is immersed in a 0.5 M sodium hydroxide (NaOH) solution at a ratio of 1 part bagasse to 10 parts solution by volume. After autoclaving, the liquids are filtered, and the resulting filtrate (called "black liquor") is adjusted to a pH of 5.5 using 1N hydrochloric acid (HCl). Three volumes of 95% ethanol are added to the filtrate to precipitate hemicellulose and other compounds. The ethanol is removed, and the pH of the remaining solution is lowered to 1.5 to precipitate lignin. The precipitated lignin is filtered, washed, and dried. This isolated lignin is referred to as "Lig-Alk."

7.5.1.4 Lig-Ox

The black liquor, which is the filtrate obtained after the initial alkaline extraction, is subjected to oxidation using hydrogen peroxide (14% w/v, which means 14 g hydrogen peroxide per 100 mL of solution). The oxidation process is carried out under reflux conditions, which means the solution is heated to its boiling point, and the vapors are condensed and returned to the reaction mixture. The reflux is performed at a pH of 13 and a temperature of 98 °C for 2 h. After the oxidation step, the precipitate is obtained and purified following the same procedure as described earlier, which involves precipitation with ethanol, acid precipitation of lignin, filtration, washing, and drying. The purified lignin obtained from this oxidation process is referred to as "Lig-Ox."

7.5.1.5 Acetylation of Lig-Alk

The lignin obtained from the initial alkaline extraction is dissolved in a mixture of anhydrous pyridine and anhydrous acetic anhydride in a ratio of 1:10:20 (Lig-Alk: pyridine: acetic anhydride). This mixture is agitated (stirred or shaken) at a temperature of 28 °C for 2 days. After agitation, a solution of dichloromethane and methanol (in a 1:1 volume ratio) is added to the mixture. The resulting solution is separated from the pyridine by repeatedly washing with 2N hydrochloric acid (HCl) until all the pyridine is removed. The remaining solution is washed with water and then dried to obtain the final product, which is called Lig-Ac. Costa et al. (2017) reported the following yields of lignin from sugarcane bagasse using the soda pulping technique:

- Lig-Alk (from the initial alkaline extraction): 7.2%.
- Lig-Ox (from the oxidation step): 5.5%.
- Lig-Ac (from the acetylation step): 6.2%.

7.5.1.6 Sequential Extraction by Alkaline Dioxane

In this method, lignin is extracted from sugarcane bagasse in three sequential steps using 96%, 50%, and 80% dioxane containing 1% NaOH at boiling temperature (Zhang et al. 2021). A detailed procedure is available (Arni 2018). The total lignin yield in this method is 7.2%, of which 70.8% is recovered in the first step itself.

7.5.2 Ionic Liquids

Ionic liquids (IL) have recently been used for lignocellulosic pretreatment, lignin fractionation, and removal from the biomass. The ILs mostly used for lignin removal include 1-ethyl-3-methylimidazolium acetate ([EMIM]OAc), 1-allylmethylimidazolium chloride ([AMIM]Cl), and 1-butyl-3-metylimidazolium chloride ([BMIM]Cl (Espinoza-Acosta et al. 2014). Using the IL method, sugarcane bagasse was fractionated into cellulose, hemicellulose, and lignin. Lan et al. (2011) used 1-butyl-3-metylimidazolium chloride followed by a precipitation step in 9:1 acetone/water (v/v) and extraction with 3% NaOH. This method recorded

36.78% cellulose. 26.04% hemicellulose and 10.51% of lignin which is considerably more compared to alkaline methods. In a study using 1-ethyl-3-methylimidazolium acetate ([EMIM]oAc) to extract lignin from sugarcane bagasse, at 20:1 (w/w) of IL to bagasse ratio, the presence of guiacyl and syringyl rings as well as p-coumaric and ferulic acid contents were confirmed through FTIR and UV-VIS spectrophotometries, respectively (Saha et al. 2017). Although ILs are more efficient in delignification, concerns about their toxicity, biodegradability, and high operating costs limit their application in biorefineries.

7.5.3 Organosolv Method

The Organosolv process is one of the most efficient processes for lignocellulosic biomass pretreatment in which the feedstock is treated with low boiling organic solvents such as alcohols, ketones, and organic acids. The Organosolv process offers good solubility and ease of recovery of lignin (Zhao et al. 2017). The operating conditions such as solvent-biomass ratio (S/L), reaction time, temperature, catalyst, and particle size influence the process efficiency. A higher solvent biomass ratio increases the delignification rate which is between 3 and 10 (Vallejos et al. 2015).

Pretreatment strategies often depend on the nature of the biomass and its morphological structure. In an experiment on kinetic modeling of hydrothermal and Organosolv pretreated sugarcane bagasse, Moreira Neto et al. (2023) used water/ethanol solution (1:2 *v*/*v*) for 150 min at 190 °C, followed by washing with 1% (*w*/*v*) sodium hydroxide solution to solubilize the residual lignin from the fibers. Considering the cellulase/cellulose system, the kinetic model represented the inhibition of sugars including cellobiose, glucose, and xylose as well as enzymatic adsorption. However, the kinetic parameters are observed to be significantly different for both hydrothermally and Organosolv pretreated sugarcane bagasse which indicates that the pretreatment process is highly influenced by the biomass composition. The enzymatic digestibility of sugarcane bagasse was enhanced and enriched in p-hydroxyphenyl (H) unit after ethanol-based Organosolv pretreatment. Further, apart from lignin structure, the predominant factors that influence enzymatic hydrolysis are the surface area of cellulose, crystallinity, and lateral order index as well as the specific surface area (Li et al. 2023).

7.5.3.1 Clean Fractionation Process

The Clean Fractionation (CF) process is a promising technology for synthesizing green chemicals, bioenergy, and biofuels. Nowadays, a one-step clean fractionation process utilizing a ternary solvent mixture (i.e., organic solvent, water, and alcohol) is being attempted to fractionate lignin from several agricultural residues. The ternary organic solvent systems improve the selectivity for different components of lignocellulosic biomass. A ternary system of acetone/butanol/ethanol achieved 98.5% of hemicellulose removal and 58.2% delignification in *Salix schwerinii* at 200 °C (Yang et al. 2022). A clean fractionation method developed by NREL using a ternary Organosolv system (acetone/MIBK/H_2O) resulted in 77.2% lignin and 74% hemicellulose from corn stover at 120 °C for 40 min. The cross-contamination of biopolymers was very low (Katahira et al. 2014). However, application of this method was limited due to lower lignin removal. Henceforth effective biomass fractionation could be addressed through a solvent system with monophasic deconstruction and biphasic separation. Accordingly, Li et al. (2023) developed a ternary solvent system comprising acetone, phenoxyethanol, and water (APW), which effectively fractionated sugarcane bagasse under optimal conditions of 130 °C, 70 min in the presence of 0.15 M sulfuric acid. At a liquid-solid ratio (LSR) of 20, 95.6% delignification and 98.39% hemicellulose removal were achieved, and 80.48% of cellulose was retained in the solid fraction.

Further, addition of antioxidants and surfactants such as tert-butylhydroquinone, 3-tert-butyl-4-hydroxyanisole, methyl 3,4,5-trihydroxybenzoate; surfactants Tween 20, Tween 80, and dimethyl sulfoxide (DMSO) collaborated with lignin and yielded 71% lignin in sugarcane bagasse (Schmatz et al. 2022). The process also improved enzymatic hydrolysis resulting in 98.9% cellulose conversion into glucose. High-purity lignin with <1% cross-contamination was achieved in a mild one-step Organosolv process with formic acid from sugarcane bagasse at 159 °C and 40 min residence (Suriyachai et al. 2018). The different Organosolv- based pretreatment studies in sugarcane are given in Table 7.1.

7.5.4 Enzymatic Process

Ligninolytic enzymes, mainly laccases and peroxidases, have been successfully employed for the biological pretreatment processes. Among them, laccases (EC 1.10.3.2) are promising enzymes for green and industrial processes as they are highly versatile with a catalytic action toward phenolic and non-phenolic substrates (in the presence of redox mediators). They are oxidative enzymes that reduce atmospheric oxygen to water. Lignin removal is directly correlated with enzymatic saccharification in any biomass. For instance, laccases from *Pycnoporus cinnabarinus* removed 27% of lignin from sugarcane bagasse in the presence of 1-HBT (1-hydroxybenzotriazole) as a mediator, thereby increasing the glucose yield to 39%. The 2D-NMR spectra revealed a significant reduction in the number of aliphatic side chains and the removal of H, G, and S lignin units as well as ferulates and coumarates (Rencoret et al. 2017). A study on laccase-mediated delignification on sugarcane tops as optimized by central composite design showed 79.1% delignification efficiency under optimal conditions of 21% solid loading rate, pH 7, enzyme titer of 430.3 IU mL^{-1}, temperature of 40 °C, and incubation of 6 h. Subsequent saccharification of sugarcane tops enhanced the reducing sugar yield (Sherpa et al. 2018). Enzyme-mediated systems for delignification of LCBs in sugarcane are given in Table 7.2.

Table 7.1 Recent studies on Organosolv-based pretreatment with emphasis on sugarcane bagasse

Raw material	Pretreatment	Delignification	Cellulose to glucose	Reference
Sugarcane bagasse	Ethanol-based Organosolv	p-hydroxy phenyl units (H)	39.96 g	Zhang and Wu (2015)
Sugarcane bagasse	Acetone/phenoxyethanol/water (APW)	95.6%	80.48%	Li et al. (2023)
Sugarcane bagasse	γ-Valerolactone (GVL)/water systems and a low concentration of sulfuric acid (SA)	75–96%	–	González et al. (2023)
Salix schwerinii	Acetone/butanol/ethanol	58.2%	12.1%	Yang et al. (2022)
Sugarcane bagasse	Antioxidants/surfactants applied to the Organosolv	71%	98.9%	Schmatz et al. (2022)
Corn stover	Acetone/MIBK/H_2O	77.2%	–	Katahira et al. (2014)
Sugarcane bagasse	One-step formic acid-catalyzed Organosolv process	80.4%	84.5% glucose recovery	Suriyachai et al. (2018)
Melia dubia	EnZolv process	48%	–	Muniraj et al. (2023)
Cotton stalk	EnZolv process	68.68%	–	Subramaniam et al. (2023)
Blowroom droppings	EnZolv process	61.34%	–	Subramaniam et al. (2024)
Flat strips	EnZolv process	61.64%	–	Subramaniam et al. (2024)
Lickerin fly	EnZolv process	41.85%	–	Subramaniam et al. (2024)
Microdust	EnZolv process	35.34%	–	Subramaniam et al. (2024)
Comber noils	EnZolv process	35.83%	–	Subramaniam et al. (2024)
Ginning mill waste	EnZolv process	65.51%	–	Subramaniam et al. (2023)
Banana fiber	EnZolv process	58.16%	–	Govindaraj et al. (2023)
Corn cob	Ultrasonic pretreatment	38.18%	–	Thangavelu et al. (2020)
Corn cob	Hydrodynamic cavitation and enzymatic process	47.4%	–	Thangavelu et al. (2018)
Corn cob	Combined lime and hydrodynamic cavitation process	38.1%		Thangavelu et al. (2022)

Table 7.2 Enzyme-mediated systems for delignification of LCBs

Enzyme	Microorganism	Substrate	Mediator	Delignification	Compounds	Reference
Laccase	*Pycnoporus sanguineus*	Sugarcane	Carminic acid	32.67% in 16 days	Low-molecular weight aromatic compounds	Hernández et al. (2017)
Laccase	*Pycnoporus* sp. P6 and *Trametes* sp. T3.	Sugarcane Vinasse	–	–	–	Ahmed et al. (2018)
Laccase	*Trametes vilosa*	Sugarcane bagasse	Tween 80 and 4-hydroxybenzoic acid, hydroxybenzotriazole (HBT),	–	–	Matei et al. (2020)
Laccase	*Trametes maxima*	Jute sticks	HBT	21.8%	Abundance of S lignin unit	Suman et al. (2022)
Laccase	*Chaetomium cupreum*	Sugarcane bagasse	–	–	96.1% more H_2 and 22.5% more organic acids	Dionizio et al. (2022)
Laccases/ ferroxidases	*Chrysoporthe cubensis*	Sugarcane bagasse	–	Oxidized 91% of the phenolic compound	–	Tavares et al. (2022)
Laccase (LccH)	*Hexagonia hirta* MSF2	Coirpith waste (CPW)	–	–	–	Arumugam and Uthandi (2024)

7.5.5 Purification and Modification of Lignin

The isolation and purification of lignin depend on the pretreatment methods used. Lignin from the paper and pulp industry (black liquor) is isolated by acid precipitation, whereas lignin from steam-exploded substrates is extracted by an alkaline method at a series of pH from 2 to 12 in the presence of sulphuric acid (Norgren and Edlund 2014). The fractions obtained at pH < 4 have low-polydispersity values and are usually low-molecular weight aromatic compounds, whereas at pH > 4, the fractions contained more nanoparticles and lignin aggregates of larger size. The optimal pH lies between 2 and 4 with low-molecular weight aromatic clusters (Brenelli et al. 2016). Flocculation is another method of lignin extraction besides the addition of ethanol or water and organic solvent fractionation (Oriez et al. 2020).

The most promising strategy studied is the membrane technology to enhance lignin purity. It does not employ any solvents, and the merits of membrane technology lie in low-energy consumption. Membrane technology separates the compounds based on their molecular weight. The ultrafiltration membrane process was successfully executed for fractionating black liquor in a paper and pulp plant in Finland back in 1985. Followed by this, several advancements in the technology have been made in order to achieve lignin with higher purity and more yield (Abdelaziz et al. 2019; Sultan et al. 2019).

Lignin modification can be done through depolymerization/fragmentation and functional group modification, more preferably the hydroxyl groups. Depolymerization is characterized using high temperatures and pressures in the presence of strong catalysts. According to Li et al. (2017), a co-operative catalytic action of an IL liquid pair exhibited significant catalytic activity and char inhibition ability with 66.7% lignin conversion, yielding 14.5% (wt) phenolic monomers and other volatile aromatics at 250 °C for 30 min. The other lignin depolymerization processes include hydrolysis, hydrogenolysis, pyrolysis, solvolysis, and oxidation. The catalyst monomeric anthraquinone acid is used to depolymerize sugarcane bagasse lignin, and an increase of carboxylic acid functional groups was detected in the FTIR spectrum (da Silva et al. 2013). For oxidation–hydrogenation catalytic process, methanol or ethanol is used along with Pd/C as catalysts in the presence of oxygen and they obtained 31.7% (w/w) of ferulic and coumaric acid derivatives optimal conditions (Cheng et al. 2020).

7.5.5.1 Enzymatic Depolymerization

Apart from laccases and peroxidases, β-etherases depolymerize lignin fragments wherein the β-aryl ether and biphenyl linkages within the lignin molecules are broken. The cleavage of β-O-4 ether bond in Cα-dehydrogenase LigD oxidizes the hydroxyl group at Cα position with the consumption of NAD^+. Subsequently, β-etherase, LigE, or LigF cleaves β-O-4 ether bond to produce the intermediates α-glutathionyl-β-hydroxypropiovanillone (GS-HPV) and vanillin. Further oxidation of GS-HPV with LigG and glutathione-S-transferase releases the final product, β-hydroxyproppiovanillone (HPV) (Chio et al. 2019). Lignin is depolymerized by microorganisms into various lignin-derived aromatics into protocatechuate and catechol. The central intermediates can be eventually converted to value-added compounds via aromatic ring cleavage and TCA cycle (Weng et al. 2021).

Lignin modification could be done to increase its chemical reactivity or solubility for industrial applications. Some of the chemical modifications attempted are nitration, amination, alkylation/dealkylation, carboxylation, and halogenation to promote reactivity or to change the nature of active sites. For functional modification of phenolic and hydroxyl groups, the modifications are esterification, etherification, and reaction with isocyanates and silylation (Bertella and Luterbacher 2020). Sugarcane bagasse lignin was modified by oxidation, hydroxymethylation, and sulfonation into a water-soluble lignosulfonate and the results showed that the aromatic units were maintained and the dispersing efficiency was improved for using as pesticide formulation (Li and Ge 2011). Likewise, in another study, modified sugarcane bagasse lignin (acetylation, epoxidation, and hydroxymethyaltion) possess emulsifier properties with both physical and oxidative stabilities (Brenelli et al. 2021). The epoxy lignin was reported to have antibacterial and antioxidant activities (Kaur et al. 2017).

7.6 Diverse Applications of LDAs across Industries

Lignin-derived chemicals have the potential for high-value products in different fields such as biofuels, cosmetics, materials, pharmacies, pesticides, and textiles. However, the following sections limit the applications relevant to sugarcane-derived lignin.

7.6.1 Cosmetics

Alkaline lignin from sugarcane bagasse with more phenolic content is used as sun protection factor (SPF). It has been reported that under optimal conditions of 7% NaOH, 135 °C, and residence time of 47.92 min, 69.41 ± 0.32 mg gallic acid equivalent g^{-1} extract for total phenolic content, 262.30 ± 2.98 mg Trolox equivalent/g extract for antioxidant activity, and 8.65 ± 0.21 for SPF were obtained. Therefore, sugarcane bagasse lignin extract was proposed as a promising bioactive multi-functional ingredient that can offer antiaging, sun protection, and skin-lightening properties for sun care formulations (Ratanasumarn and Chitprasert 2020). Further, lignin also demonstrated good potential as a natural UV protection ingredient in broad-spectrum (UVA and UVB) sunscreens due to blocking of free radicals (Sadeghifar and Ragauskas 2020).

7.6.2 Nanoparticles for Formulations and Nanomedicine

Sugarcane bagasse-derived micro- or nanosized lignin has been reported as a bioinsecticide (Costa et al. 2017), graphene microcrystal (Tang et al. 2018), nanostructured films, and biosorbent (Peng et al. 2020). Colloidal lignin nanoparticles from sugarcane (alkaline and Organosolv) showed that the preparation process influenced the property of the nanomaterial. The poor yield of lignin nanoparticles needs to be addressed. Despite the diverse applications, the safety demonstration of various lignin grades remains crucial for future applications in food, nanomedicine, and healthcare materials (Sipponen et al. 2019).

7.6.3 Films and Coatings

The hydrophobic nature of lignin finds its application in food packaging in terms of films and coatings (Hua et al. 2019). The ability to absorb UV light, along with antioxidant, and antimicrobial properties is a unique feature for the development of such films and coating materials (Sirvio et al. 2020). Lignin is miscible with other polymers such as PVA and PVP due to migration and intermolecular hydrogen bonding between OH groups. In the study, PVA films blended with lignin up to 3% (w/w) and the PVP films blended with lignin up to 5% (w/w) were reported as suitable for application in the fatty food packaging field (Inna et al. 2018). Further, the antimicrobial properties of the films also enabled to enhance the shelf life of fresh fruits (Jonglertjunya et al. 2014).

7.6.4 Textiles

Lignin caters to the properties of textile materials in its characteristic strength, flexibility, elasticity, heat insulation, durability, water absorbance, and dyeability. Sugarcane bagasse-derived lignin-coated fabric was used as a potential additional fabric in a sanitary mask (Sunthornvarabhas et al. 2017). The experimental results suggested the potential of using lignin in filtration and antimicrobial textile applications. Although lignin has interesting properties to be explored for the textile industry, it requires more research to validate the potential and safety of fabrics.

7.6.5 Resins and Insulation Foams

Raw kraft lignin is used in the manufacture of bio-phenol formaldehyde (BPF) resins and thermal insulation foams by reducing both petroleum-based phenol and formaldehyde by up to 50%. The obtained lignin-based BPF foams exhibited low apparent density (20–80 kg/m^3), strong compressive strength (up to 1.01 MPa), excellent thermal conductivity (0.03–0.048 W/mK), combustion properties (LOI: 32–33%), and satisfactory closed cell structure, comparable to commercial phenol-formaldehyde (PF) foams (Tang et al. 2018). These BPF foams can also act as a flame retardant and insulation materials for construction or retrofitting of buildings. Therefore, future commercialization of these lignin-derived phenolic foams would lead to cleaner production insulation materials using forestry and agricultural residues/waste including sugarcane tops and bagasse.

7.7 Challenges and Opportunities in Lignin Valorization

In the contemporary realm of green chemistry, the transformation of lignocellulosic biomass poses a formidable research challenge, garnering attention both in academic and industrial spheres. This involves the conversion of three primary subcomponents—cellulose, hemicellulose, and lignin—to synthesize chemicals, fuels, and materials. Lignin, with its intricate three-dimensional amorphous structure, originates from the enzymatic dehydrogenative polymerization of three phenylpropanoid monomers: synapyl alcohol, coniferyl alcohol, and p-coumaryl alcohol (Sethupathy et al. 2022). Concerning the chemical valorization of lignin, a key challenge lies in the intricate task of determining the precise composition and structure of the macromolecule due to its inherent complexity and heterogeneity. The variability in lignin composition also stems from factors such as the biomass source, wood type, season, and various growth parameters. The elusive nature of the native lignin structure is further compounded by the necessity of isolating it before analysis, a process contingent on the chosen separation method (Behling et al. 2016). Moreover, the polyphenolic structure of lignin exhibits notable chemical stability, making it resistant to facile modification and necessitating harsh experimental conditions for effective depolymerization. The majority of existing methods for converting lignin into value-added chemicals encounters several experimental limitations. These include (a) the necessity for harsh conditions involving high pressure and/or temperature, often accompanied by extended reaction times, (b) reliance on costly catalysts, particularly those based on noble metals, or the use of toxic catalysts, (c) dependence on non-ecofriendly solvents, and (d) inadequate mass transfer from lignin feed stocks to the catalytic surface, resulting in moderate conversion rates (Xu et al. 2014).

The other challenges include the production of relatively low yields of generated products. For instance, reported yields of total oxidized products from lignin are in the range of 10–11%, with average product yields, including vanillin, typically falling within the 1–5% range. The feasibility of product separation and isolation is complicated by the low quantities of generated products. The recyclability of heterogeneous catalysts involved in the process presents issues related to separation and surface saturation/deactivation of the catalysts due to the generation of alcohols and aromatic compounds (Teleman et al. 2000).

Lignin valorization holds immense potential across multiple sectors, offering opportunities to transform this complex plant polymer into high-value products. One significant avenue is the development of biofuels through lignin depolymerization, providing an eco-friendly alternative to traditional fossil fuels.

Advanced conversion technologies, such as pyrolysis and hydrodeoxygenation, enable the extraction of valuable compounds that can be utilized in biofuel production, contributing to a more sustainable energy landscape. Moreover, lignin-derived biofuels can play a crucial role in mitigating greenhouse gas emissions and reducing dependence on non-renewable resources (Chio et al. 2019).

Agricultural and forestry residues represent the most abundant potential for utilizing lignin in future circular economies. Therefore, it is crucial to recognize the value of agricultural and forestry residues. When utilized in materials and composites, the inclusion of lignin can greatly improve the mechanical properties of products such as adhesives, coatings, and composites. This approach not only broadens the range of raw materials available but also lessens the negative environmental effects of traditional petrochemical-based alternatives. Furthermore, the agricultural and environmental sectors stand to benefit from lignin valorization. Lignin-based products can serve as effective soil amendments, improving soil structure and fertility (Singhvi and Kim 2021; Song et al. 2020).

7.8 Environmental and Economic Dimensions

The use of lignin as a feedstock for producing aromatics has the potential to reduce dependence on fossil resources, which can help to reduce the negative environmental impact caused by traditional petrochemical-based processes. By utilizing a natural and renewable polymer like lignin, we can contribute to the development of more eco-friendly materials and chemicals. Lignin, a natural and renewable polymer, can serve as a green alternative, contributing to the development of more eco-friendly materials and chemicals (Garlapati et al. 2020).

Moreover, the valorization of lignin-derived aromatics aligns with the principles of a circular bioeconomy, where waste products from one industry become valuable feed stocks for another. This circular approach not only minimizes waste but also reduces the overall environmental footprint of industrial processes. In terms of the economic dimension, the commercial viability of lignin-derived aromatics plays a crucial role. The development of cost-effective and scalable processes for the production of these compounds is essential for their widespread adoption in industries. Economic considerations also involve the competitiveness of lignin-derived aromatics compared to their petrochemical counterparts. Factors such as production costs, market demand, and regulatory incentives impact the economic feasibility of incorporating lignin-derived aromatics into various value chains. Lignin-derived aromatics can find applications in various industries, such as polymers, resins, and specialty chemicals, further diversifying the sources of raw materials and promoting sustainable practices. These aromatics are ideally suited for second-generation biorefineries, enabling the extraction of value-added chemicals and hydrocarbons (Wang and Song 2018).

As the second most abundant aromatic feedstock in plant-derived biomass, lignin constitutes approximately 15–40% of the dry mass in terrestrial plants and 40% in terms of energy content (Islam et al. 2021). Globally, an estimated 150 billion tons of lignin are produced annually (Hu et al. 2018). Lignin, particularly derived from the paper and pulp industry as a by-product, contributes 50 million tons per year, with only 2% being converted into lignin-based products, while the remainder is either burned for energy or disposed of in landfills. Currently, the market size of lignin and its products is experiencing annual growth at a rate of 2.28%, projected to increase from 599 million USD in 2014 to 704 million USD by 2022 (Garlapati et al. 2020). Despite lignin's advantages such as greater energy density, a high C: H ratio (above 10:1), and exclusive aromatic monomers that make it suitable for biorefinery applications, its complexity and heterogeneity pose challenges in terms of valorization and practical utilization. It was observed a reduction in ethanol production costs (2.02 USD per gallon) by directing 50% of the generated lignin toward phenolics, which possess significant market value.

To effectively incorporate lignin-derived aromatics into industrial applications, it is crucial to conduct a thorough evaluation of both the environmental and economic aspects. This involves tackling issues surrounding process optimization, market dynamics, and ensuring that the environmental advantages align with economic sustainability. By taking these factors into account, lignin-derived aromatics have the potential to promote a more sustainable and economically sound model in the chemical and materials industries.

7.9 Concluding Insights and Implications

The potential of utilizing lignin as a raw material for the production of aromatics has been widely explored, especially in the context of sugarcane bagasse processing. This strategy is an environmentally friendly and sustainable approach that can enhance the efficient utilization of biomass resources while improving the overall economic viability of the biorefinery. However, the conversion of lignin into aromatics is a complex process, and various methods have been devised to address this challenge, including solvothermal liquefaction, enzymatic degradation, and microbial catalysis.

Solvothermal liquefaction involves the use of a solvent at elevated temperature and pressure to break down lignin into oligomeric aromatic oxides, which can then be further converted into monomeric aromatics through processes like hydrodeoxygenation or hydrogenolysis. Enzymatic degradation, on the other hand, relies on ligninolytic enzymes to convert lignin into aromatics, while microbial catalysis employs lignin-degrading microorganisms.

Despite the significant potential of lignin-derived aromatics for various biorefinery applications, there are persisting difficulties in efficiently converting lignin into valuable products. Researchers are actively exploring novel catalysts, processes, and technologies to improve the yield and selectivity of lignin-derived aromatics. Moreover, efforts are being made to expand the product spectrum and value of lignin-derived aromatics through the design of new synthetic routes and metabolic pathways.

The incorporation of lignin-derived aromatics into the biorefinery process enables the production of a more diverse range of high-value products, encompassing bio-based chemicals, materials, and fuels. This approach aligns with global initiatives aimed at reducing greenhouse gas emissions and decreasing dependence on fossil resources. By utilizing lignin-derived aromatics, the biorefinery industry can contribute to a more sustainable and environmentally friendly future.

References

Abdelaziz OY, Ravi K, Nöbel M, Tuna P, Turner C, Hulteberg CP (2019) Membrane filtration of alkali-depolymerised kraft lignin for biological conversion. Bioresour Technol Rep 7:100250. https://doi.org/10.1016/j.biteb.2019.100250

Ahmad W, Ahmad A, Ostrowski KA, Aslam F, Joyklad P, Zajdel P (2021) Sustainable approach of using sugarcane bagasse ash in cement-based composites: a systematic review. Case Stud Constr Mater 15. https://doi.org/10.1016/j.cscm.2021.e00698

Ahmed PM, Pajot HF, De Figueroa LIC, Gusils CH (2018) Sustainable bioremediation of sugarcane vinasse using autochthonous macrofungi. J Environ Chem Eng 6(4):5177–5185. https://doi.org/10.1016/j.jece.2018.08.007

Arni SAL (2018) Extraction and isolation methods for lignin separation from sugarcane bagasse: a review. Ind Crop Prod 115:330–339). Elsevier. https://doi.org/10.1016/j.indcrop.2018.02.012

Arumugam DP, Uthandi S (2024) Optimization and characterization of laccase (LccH) produced by *Hexagonia hirta* MSF2 in solid-state fermentation using coir pith wastes (CPW). J Environ Manag 356:120625. https://doi.org/10.1016/j.jenvman.2024.120625

Behling R, Valange S, Chatel G (2016) Heterogeneous catalytic oxidation for lignin valorization into valuable chemicals: what results? What limitations? What trends? Green Chem 18(7):1839–1854. https://doi.org/10.1039/c5gc03061g

Benjamin Y, Cheng H, Görgens JF (2013) Evaluation of bagasse from different varieties of sugarcane by dilute acid pretreatment and enzymatic hydrolysis. Ind Crop Prod 51:7–18. https://doi.org/10.1016/j.indcrop.2013.08.067

Bertella S, Luterbacher JS (2020) Lignin functionalization for the production of novel materials. Trends Chem 2(5):440–453. https://doi.org/10.1016/J.TRECHM.2020.03.001

Boerjan W, Ralph J, Baucher M (2003) Lignin biosynthesis. Annu Rev Plant Biol 54:519–546. Annual Reviews. https://doi.org/10.1146/annurev.arplant.54.031902.134938

Brenelli LB, Mandelli F, Mercadante AZ, Rocha GJDM, Rocco SA, Craievich AF, Gonçalves AR, Centeno DDC, de Oliveira Neto M, Squina FM (2016) Acidification treatment of lignin from sugarcane bagasse results in fractions of reduced polydispersity and high free-radical scavenging capacity. Ind Crop Prod 83:94–103. https://doi.org/10.1016/j.indcrop.2015.12.013

Brenelli LB, Mariutti LRB, Villares Portugal R, de Farias MA, Bragagnolo N, Mercadante AZ, Franco TT, Rabelo SC, Squina FM (2021) Modified lignin from sugarcane bagasse as an emulsifier in oil-in-water nanoemulsions. Ind Crop Prod 167:113532. https://doi.org/10.1016/J.INDCROP.2021.113532

Buller LS, Romero CWDS, Lamparelli RAC, Ferreira SF, Bortoleto AP, Mussatto SI, Forster-Carneiro T (2021) A spatially explicit assessment of sugarcane vinasse as a sustainable by-product. Sci Total Environ 765. https://doi.org/10.1016/j.scitotenv.2020.142717

Cheng Y, Yang S, Ji L, Shi H (2020) Oligomeric aromatic oxides (OAO) production from sugarcane bagasse lignin by acid-catalyzed solvothermal liquefaction in methanol. Front Energy Res 8. https://doi.org/10.3389/FENRG.2020.608415/FULL

Chio C, Sain M, Qin W (2019) Lignin utilization: a review of lignin depolymerization from various aspects. Renew Sust Energ Rev 107:232–249. https://doi.org/10.1016/j.rser.2019.03.008

Costa ES, Perlatti B, Da Silva EM, Matos AP, Da Silva MFGF, Fernandes JB, Zuin VG, Da Silva CMP, Forim MR (2017) Use of lignins from sugarcane bagasse for assembling microparticles

loaded with *Azadirachta indica* extracts for use as neem-based organic insecticides. J Braz Chem Soc 28(1):126–135. https://doi.org/10.5935/0103-5053.20160155
da Silva CG, Grelier S, Pichavant F, Frollini E, Castellan A (2013) Adding value to lignins isolated from sugarcane bagasse and Miscanthus. Ind Crop Prod 42(1):87–95. https://doi.org/10.1016/J.INDCROP.2012.04.040
del Río JC, Lino AG, Colodette JL, Lima CF, Gutiérrez A, Martínez ÁT, Lu F, Ralph J, Rencoret J (2015) Differences in the chemical structure of the lignins from sugarcane bagasse and straw. Biomass Bioenergy 81:322–338. https://doi.org/10.1016/j.biombioe.2015.07.006
Dionizio BS, Rabelo CABS, de Jesus HCR, Varesche MBA, de Souza DHF (2022) The deconstruction of the Lignocellulolytic structure of sugarcane bagasse by laccases improves the production of H_2 and organic acids. Appl Biochem Biotechnol 194(7):3145–3166. https://doi.org/10.1007/s12010-022-03905-9
Espinoza-Acosta JL, Torres-Chávez PI, Carvajal-Millán E, Ramírez-Wong B, Bello-Pérez LA, Montaño-Leyva B (2014) Ionic liquids and organic solvents for recovering lignin from lignocellulosic biomass. Bioresources 9(2):3660–3687. https://doi.org/10.15376/biores.9.2.3660-3687
Garlapati VK, Chandel AK, Kumar SPJ, Sharma S, Sevda S, Ingle AP, Pant D (2020) Circular economy aspects of lignin: towards a lignocellulose biorefinery. Renew Sust Energ Rev 130. https://doi.org/10.1016/j.rser.2020.109977
Gil-Chavez J, Gurikov P, Hu X, Meyer R, Reynolds W, Smirnova I (2021) Application of novel and technical lignins in food and pharmaceutical industries: structure-function relationship and current challenges. Biomass Convers Biorefinery 11(6):2387–2403. https://doi.org/10.1007/S13399-019-00458-6
Godshall MA, Legendre BL (2003) SUGAR | sugarcane. In Encyclopedia of Food Sciences and Nutrition (pp. 5645–5651). Elsevier. https://doi.org/10.1016/b0-12-227055-x/01160-3
González G, Ehman N, Felissia FE, Area MC (2023) Strategies towards a green solvent biorefinery: efficient delignification of lignocellulosic biomass residues by gamma-valerolactone/water catalyzed system. Ind Crop Prod 205. https://doi.org/10.1016/j.indcrop.2023.117535
Govindaraj O, Olaganathan GN, Asm R, Uthandi S (2023) An eco-friendly EnZolv pretreatment optimization in banana fiber biomass using response surface methodology (RSM) and its structural characterization. Biomass Convers Biorefinery 1–14. https://doi.org/10.1007/S13399-023-04829-Y/METRICS
Hendriks ATWM, Zeeman G (2009) Pretreatments to enhance the digestibility of lignocellulosic biomass. Bioresour Technol 100(1):10–18. https://doi.org/10.1016/j.biortech.2008.05.027
Hernández C, Farnet Da Silva AM, Ziarelli F, Perraud-Gaime I, Gutiérrez-Rivera B, García-Pérez JA, Alarcón E (2017) Laccase induction by synthetic dyes in *Pycnoporus sanguineus* and their possible use for sugar cane bagasse delignification. Appl Microbiol Biotechnol 101(3):1189–1201. https://doi.org/10.1007/s00253-016-7890-0
Hu J, Zhang Q, Lee DJ (2018) Kraft lignin biorefinery: a perspective. Bioresour Technol 247:1181–1183. https://doi.org/10.1016/j.biortech.2017.08.169
Hua Q, Liu LY, Karaaslan MA, Renneckar S (2019) Aqueous dispersions of esterified lignin particles for hydrophobic coatings. Front Chem 7. https://doi.org/10.3389/fchem.2019.00515
Inna K, Sunthornvarabhas J, Thanapimmetha A, Saisriyoot M, Srinopakun P (2018) Natural antimicrobial lignin in polyvinyl alcohol and polyvinylpyrrolidone film for packaging application. Mater Sci Forum 936 MSF:105–109. https://doi.org/10.4028/www.scientific.net/MSF.936.105
Islam MK, Rehman S, Guan J, Lau CY, Tse HY, Yeung CS, Leu SY (2021) Biphasic pretreatment for energy and carbon efficient conversion of lignocellulose into bioenergy and reactive lignin. Appl Energy 303:117653. https://doi.org/10.1016/j.apenergy.2021.117653
Jonglertjunya W, Juntong T, Pakkang N, Srimarut N, Sakdaronnarong C (2014) Properties of lignin extracted from sugarcane bagasse and its efficacy in maintaining postharvest quality of limes during storage. LWT 57(1):116–125. https://doi.org/10.1016/j.lwt.2013.11.042
Karp SG, Woiciechowski AL, Soccol VT, Soccol CR (2013) Pretreatment strategies for delignification of sugarcane bagasse: a review. Braz Arch Biol Technol 56(4):679–689. https://doi.org/10.1590/S1516-89132013000400019

Katahira R, Mittal A, McKinney K, Ciesielski PN, Donohoe BS, Black SK, Johnson DK, Biddy MJ, Beckham GT (2014) Evaluation of clean fractionation pretreatment for the production of renewable fuels and chemicals from corn stover. ACS Sustain Chem Eng 2(6):1364–1376. https://doi.org/10.1021/SC5001258

Kaur R, Uppal SK, Sharma P (2017) Antioxidant and antibacterial activities of sugarcane bagasse lignin and chemically modified lignins. Sugar Tech 19(6):675–680. https://doi.org/10.1007/s12355-017-0513-y

Klemm D, Philpp B, Heinze T, Heinze U, Wagenknecht W (1998) Comprehensive cellulose chemistry. Volume 1: fundamentals and analytical methods. Wiley-VCH Verlag GmbH

Korasapati R, Babu Basthipati S, Rosangzuala K, Ravinder Reddy P, Gajjala R, Ravindra Babu P, Chandrasekhar C, Malothu R, Reddy Shetty P, Banoth L (2023) Value added products generation from sugarcane bagasse and its impact on economizing biorefinery and sustainability of sugarcane industry. In Sugarcane—its products and sustainability. https://doi.org/10.5772/intechopen.107472

Lan W, Liu CF, Sun RC (2011) Fractionation of bagasse into cellulose, hemicelluloses, and lignin with ionic liquid treatment followed by alkaline extraction. J Agric Food Chem 59(16):8691–8701. https://doi.org/10.1021/JF201508G

Li Z, Ge Y (2011) Extraction of lignin from sugar cane bagasse and its modification into a high performance dispersant for pesticide formulations. J Braz Chem Soc 22(10):1866–1871. https://doi.org/10.1590/S0103-50532011001000006

Li Y, Cai Z, Liao M, Long J, Zhao W, Chen Y, Li X (2017) Catalytic depolymerization of Organosolv sugarcane bagasse lignin in cooperative ionic liquid pairs. Catal Today 298:168–174. https://doi.org/10.1016/j.cattod.2017.04.059

Li J, Liu B, Liu L, Luo Y, Zeng F, Qin C, Liang C, Huang C, Yao S (2023) Pretreatment of poplar with eco-friendly levulinic acid to achieve efficient utilization of biomass. Bioresour Technol 376:128855. https://doi.org/10.1016/J.BIORTECH.2023.128855

Lu Y, Lu Y, Fan X (2020) Structure and characteristics of lignin (pp. 17–75). https://doi.org/10.1007/978-3-030-40663-9_2

Matei JC, Soares M, Bonato ACH, de Freitas MPA, Helm CV, Maroldi WV, Magalhães WLE, Haminiuk CWI, Maciel GM (2020) Enzymatic delignification of sugar cane bagasse and rice husks and its effect in saccharification. Renew Energy 157:987–997. https://doi.org/10.1016/J.RENENE.2020.05.028

Meghana M, Shastri Y (2020) Sustainable valorization of sugar industry waste: status, opportunities, and challenges. Bioresour Technol 303:122929. Elsevier. https://doi.org/10.1016/j.biortech.2020.122929

Moreira Neto J, Costa JM, Bonomi A, Costa AC (2023) A novel kinetic modeling of enzymatic hydrolysis of sugarcane bagasse pretreated by hydrothermal and Organosolv processes. Molecules 28:14. https://doi.org/10.3390/molecules28145617

Muniraj IK, Anbu PV, Parthiban KT, Uthandi S (2023) A new Enzolv process for simultaneous delignification and lignin-derived chemical production from the woody biomass of *Melia dubia*. Biomass Convers Biorefinery 13(16):14557–14571. https://doi.org/10.1007/s13399-022-03084-x

Norgren M, Edlund H (2014) Lignin: recent advances and emerging applications. Curr Opin Colloid Interface Sci 19(5):409–416). Elsevier. https://doi.org/10.1016/j.cocis.2014.08.004

Orduña Ortega J, Mora Vargas JA, Perrone OM, Metzker G, Gomes E, da Silva R, Boscolo M (2020) Soaking and ozonolysis pretreatment of sugarcane straw for the production of fermentable sugars. Ind Crop Prod 145. https://doi.org/10.1016/j.indcrop.2019.111959

Oriez V, Peydecastaing J, Pontalier PY (2020) Lignocellulosic biomass mild alkaline fractionation and resulting extract purification processes: conditions, yields, and purities. Clean Technol 2(1):91–115. https://doi.org/10.3390/cleantechnol2010007

Pandey A, Soccol CR, Nigam P, Soccol VT (2000) Biotechnological potential of agro-industrial residues. I: sugarcane bagasse. Bioresour Technol 74(1):69–80. https://doi.org/10.1016/S0960-8524(99)00142-X

Peng X, Wu Z, Li Z (2020) A bowl-shaped biosorbent derived from sugarcane bagasse lignin for cadmium ion adsorption. Cellulose 27(15):8757–8768. https://doi.org/10.1007/S10570-020-03376-3

Ratanasumarn N, Chitprasert P (2020) Cosmetic potential of lignin extracts from alkaline-treated sugarcane bagasse: optimization of extraction conditions using response surface methodology. Int J Biol Macromol 153:138–145. https://doi.org/10.1016/j.ijbiomac.2020.02.328

Rawal TB, Zahran M, Dhital B, Akbilgic O, Petridis L (2020) The relation between lignin sequence and its 3D structure. Biochim Biophys Acta Gen Subj 1864:5. https://doi.org/10.1016/j.bbagen.2020.129547

Rencoret J, Pereira A, del Río JC, Martínez ÁT, Gutiérrez A (2017) Delignification and saccharification enhancement of sugarcane byproducts by a laccase-based pretreatment. ACS Sustain Chem Eng 5(8):7145–7154. https://doi.org/10.1021/acssuschemeng.7b01332

Ribul M, Lanot A, Tommencioni Pisapia C, Purnell P, McQueen-Mason SJ, Baurley S (2021) Mechanical, chemical, biological: moving towards closed-loop bio-based recycling in a circular economy of sustainable textiles. J Clean Prod 326:129325. https://doi.org/10.1016/j.jclepro.2021.129325

Sadeghifar H, Ragauskas A (2020) Lignin as a UV Light blocker-a review. In Polymers (Vol. 12, Issue 5). https://doi.org/10.3390/POLYM12051134

Saha BC (2003) Hemicellulose bioconversion. J Ind Microbiol Biotechnol 30(5):279–291. https://doi.org/10.1007/S10295-003-0049-X

Saha K, Dasgupta J, Chakraborty S, Antunes FAF, Sikder J, Curcio S, dos Santos JC, Arafat HA, da Silva SS (2017) Optimization of lignin recovery from sugarcane bagasse using ionic liquid aided pretreatment. Cellulose 24(8):3191–3207. https://doi.org/10.1007/S10570-017-1330-X

Sarker TC, Azam SMGG, Bonanomi G (2017) Recent advances in sugarcane industry solid by-products valorization. Waste Biomass Valorization 8(2):241–266). Springer. https://doi.org/10.1007/s12649-016-9665-3

Schmatz AA, Masarin F, Brienzo M (2022) Lignin removal and cellulose digestibility improved by adding antioxidants and surfactants to Organosolv pretreatment of sugarcane bagasse. Bioenergy Res 15(2):1107–1115. https://doi.org/10.1007/S12155-021-10367-0

Sethupathy S, Murillo Morales G, Gao L, Wang H, Yang B, Jiang J, Sun J, Zhu D (2022) Lignin valorization: status, challenges and opportunities. Bioresour Technol 347:126696. https://doi.org/10.1016/j.biortech.2022.126696

Sherpa KC, Ghangrekar MM, Banerjee R (2018) A green and sustainable approach on statistical optimization of laccase mediated delignification of sugarcane tops for enhanced saccharification. J Environ Manag 217:700–709. https://doi.org/10.1016/j.jenvman.2018.04.008

Sindhu R, Kuttiraja M, Elizabeth Preeti V, Vani S, Sukumaran RK, Binod P (2013) A novel surfactant-assisted ultrasound pretreatment of sugarcane tops for improved enzymatic release of sugars. Bioresour Technol 135:67–72. https://doi.org/10.1016/j.biortech.2012.09.050

Sindhu R, Gnansounou E, Binod P, Pandey A (2016) Bioconversion of sugarcane crop residue for value added products—an overview. Renew Energy 98:203–215. https://doi.org/10.1016/j.renene.2016.02.057

Singhvi M, Kim BS (2021) Lignin valorization using biological approach. Biotechnol Appl Biochem 68(3):459–468). John Wiley & Sons, Ltd. https://doi.org/10.1002/bab.1995

Sipponen MH, Lange H, Crestini C, Henn A, Osterberg M (2019) Lignin for Nano- and Microscaled carrier systems: applications, trends, and challenges. ChemSusChem 12(10):2039–2054. https://doi.org/10.1002/cssc.201900480

Sirvio JA, Ismail MY, Zhang K, Tejesvi MV, Ammala A (2020) Transparent lignin-containing wood nanofiber films with UV-blocking, oxygen barrier, and anti-microbial properties. J Mater Chem A 8(16):7935–7946. https://doi.org/10.1039/c9ta13182e

Solomon S (2011) The Indian sugar industry: an overview. Sugar Tech 13(4):255–265. https://doi.org/10.1007/s12355-011-0115-z

Song B, Liang H, Sun R, Peng P, Jiang Y, She D (2020) Hydrogel synthesis based on lignin/sodium alginate and application in agriculture. Int J Biol Macromol 144:219–230. https://doi.org/10.1016/j.ijbiomac.2019.12.082

Subramaniam S, Karunanandham K, Asm R, Uthandi S (2023) Delignification of the cotton stalk and ginning mill waste via EnZolv pretreatment and optimization of process parameters using response surface methodology (RSM). Bioresour Technol 387:129655. https://doi.org/10.1016/j.biortech.2023.129655

Subramaniam S, Karunanandham K, Asm R, Shukla SK, Uthandi S (2024) EnZolv delignification of cotton spinning mill waste and optimization of process parameters using response surface methodology (RSM). Biotechnol Biofuels Bioprod 17(1):1–19. https://doi.org/10.1186/s13068-024-02473-w

Sultan Z, Graca I, Li Y, Lima S, Peeva LG, Kim D, Ebrahim MA, Rinaldi R, Livingston AG (2019) Membrane fractionation of liquors from lignin-first biorefining. ChemSusChem 12(6):1203–1212. https://doi.org/10.1002/cssc.201802747

Suman SK, Malhotra M, Kurmi AK, Narani A, Bhaskar T, Ghosh S, Jain SL (2022) Jute sticks biomass delignification through laccase-mediator system for enhanced saccharification and sustainable release of fermentable sugar. Chemosphere 286. https://doi.org/10.1016/j.chemosphere.2021.131687

Sun RC (2020) Lignin source and structural characterization. ChemSusChem 13(17):4385–4393). NLM (Medline). https://doi.org/10.1002/cssc.202001324

Sun Z, Fridrich B, de Santi A, Elangovan S, Barta K (2018) Bright side of lignin depolymerization: toward new platform chemicals. Chem Rev 118(2):614–678. https://doi.org/10.1021/acs.chemrev.7b00588

Sunthornvarabhas J, Liengprayoon S, Suwonsichon T (2017) Antimicrobial kinetic activities of lignin from sugarcane bagasse for textile product. Ind Crop Prod 109:857–861. https://doi.org/10.1016/j.indcrop.2017.09.059

Suriyachai N, Champreda V, Kraikul N, Techanan W, Laosiripojana N (2018) Fractionation of lignocellulosic biopolymers from sugarcane bagasse using formic acid-catalyzed Organosolv process. 3 Biotech 8(5):1–14. https://doi.org/10.1007/S13205-018-1244-9/METRICS

Tang PD, Du QS, Li DP, Dai J, Li YM, Du FL, Long SY, Xie NZ, Wang QY, Huang RB (2018) Fabrication and characterization of graphene microcrystal prepared from lignin refined from sugarcane bagasse. Nano 8(8):565. https://doi.org/10.3390/nano8080565

Tavares MP, Dutra TR, Morgan T, Ventorim RZ, de Souza Ladeira Ázar RI, Varela EM, Ferreira RC, de Oliveira Mendes TA, de Rezende ST, Guimaraes VM (2022) Multicopper oxidase enzymes from Chrysoporthe cubensis improve the saccharification yield of sugarcane bagasse. Process Biochem 119:68–81. https://doi.org/10.1016/j.procbio.2022.05.012

Teleman A, Lundqvist J, Tjerneld F, Stalbrand H, Dahlman O (2000) Characterization of acetylated 4-O-methylglucuronoxylan isolated from aspen employing ^{1}H and ^{13}C NMR spectroscopy. Carbohydr Res 329(4):807–815. https://doi.org/10.1016/S0008-6215(00)00249-4

Thangavelu K, Desikan R, Taran OP, Uthandi S (2018) Delignification of corncob via combined hydrodynamic cavitation and enzymatic pretreatment: process optimization by response surface methodology. Biotechnol Biofuels 11(1):1–13. https://doi.org/10.1186/s13068-018-1204-y

Thangavelu K, Desikan R, Uthandi S (2020) Delignification of corncob using catalytic hydrodynamic cavitation reactor. Madras Agric J 107(Special):3–6. https://doi.org/10.29321/maj.s.000452

Thangavelu K, Desikan R, Uthandi S (2022) Optimization of combined lime and hydrodynamic cavitation for pretreatment of corncob biomass using response surface methodology for lignin removal. Biomass Convers Biorefinery 13(16):14433–14445. https://doi.org/10.1007/S13399-022-02728-2

Tobimatsu Y, Schuetz M (2019) Lignin polymerization: how do plants manage the chemistry so well? Curr Opin Biotechnol 56:75–81). Elsevier Current Trends. https://doi.org/10.1016/j.copbio.2018.10.001

Vallejos ME, Zambon MD, Area MC, Curvelo AADS (2015) Low liquid-solid ratio fractionation of sugarcane bagasse by hot water autohydrolysis and Organosolv delignification. Ind Crop Prod 65:349–353. https://doi.org/10.1016/J.INDCROP.2014.11.018

Vanholme R, Demedts B, Morreel K, Ralph J, Boerjan W (2010) Lignin biosynthesis and structure. Plant Physiol 153(3):895–905. https://doi.org/10.1104/pp.110.155119

Wahba SMR, Darwish AS, Shehata IH, Abd Elhalem SS (2015) Sugarcane bagasse lignin, and silica gel and magneto-silica as drug vehicles for development of innocuous methotrexate drug against rheumatoid arthritis disease in albino rats. Mater Sci Eng C 48:599–610. https://doi.org/10.1016/J.MSEC.2014.12.054

Wang A, Song H (2018) Maximizing the production of aromatic hydrocarbons from lignin conversion by coupling methane activation. Bioresour Technol 268:505–513. https://doi.org/10.1016/j.biortech.2018.08.026

Watkins D, Nuruddin M, Hosur M, Tcherbi-Narteh A, Jeelani S (2015) Extraction and characterization of lignin from different biomass resources. J Mater Res Technol 4(1):26–32. https://doi.org/10.1016/J.JMRT.2014.10.009

Weng C, Peng X, Han Y (2021) Depolymerization and conversion of lignin to value-added bioproducts by microbial and enzymatic catalysis. Biotechnol Biofuels 14(1):84. https://doi.org/10.1186/s13068-021-01934-w

Xu C, Arancon RAD, Labidi J, Luque R (2014) Lignin depolymerisation strategies: towards valuable chemicals and fuels. Chem Soc Rev 43(22):7485–7500. https://doi.org/10.1039/c4cs00235k

Yang M, Guo X, Liu G, Nan Y, Zhang J, Noyazzesh H, Kuittinen S, Vepsalainen J, Pappinen A (2022) Effect of solvent mixture pretreatment on sugar release from short-rotation coppice *Salix schwerinii* for biobutanol production. Bioresour Technol 344:126262. https://doi.org/10.1016/J.BIORTECH.2021.126262

Zakzeski J, Bruijnincx PCA, Jongerius AL, Weckhuysen BM (2010) The catalytic valorization of lignin for the production of renewable chemicals. ACS Publications 110(6):3552–3599. https://doi.org/10.1021/cr900354u

Zhang H, Wu S (2015) Generation of lignin and enzymatically digestible cellulose from ethanol-based Organosolv pretreatment of sugarcane bagasse. Cellulose 22(4):2409–2418. https://doi.org/10.1007/S10570-015-0678-Z/METRICS

Zhang R, Wang J, Milligan S, Yan Y (2021) Microbial utilization of lignin-derived aromatics: via a synthetic catechol meta -cleavage pathway. Green Chem 23(20):8238–8250. https://doi.org/10.1039/d1gc02347k

Zhao X, Li S, Wu R, Liu D (2017) Organosolv fractionating pre-treatment of lignocellulosic biomass for efficient enzymatic saccharification: chemistry, kinetics, and substrate structures. Biofuels Bioprod Biorefin 11(3):567–590. https://doi.org/10.1002/BBB.1768

Zhou Z, Cheng Y, Zhang W, Jiang J, Lei F (2016) Characterization of lignins from sugarcane bagasse pretreated with green liquor combined with ethanol and hydrogen peroxide. Bioresources 11(2):3191–3203. https://doi.org/10.15376/biores.11.2.3191-3203

Proteomic Approaches for Ligninolytic Enzymes Production for Enhancing Sugarcane Biomass Conversion Efficiency

8

Lakshmi Kasirajan, Keerthana Kamaraj, Sheelamary Sebastiar, Prathima Perumal Thirugnanasambandam, and Elumalai Karpagam

Abstract

Sugarcane is an efficient C4 energy cane. The lignocellulosic biomass is fermented for the production of second-generation biofuel. Lignin, a complex heteropolymer hinders biomass biodegradation in two ways either by acting as a protective cover to polysaccharides against enzymatic depolymerization resulting in biomass recalcitrance or by binding to saccharification enzymes thereby affecting hydrolysis. Hence lignin degradation challenges cost-effective bioethanol production. Delignification enzymes such as lignin peroxidase, manganese peroxidase, and laccase from microbial consortium are currently utilized in bioethanol industries. Microbial enzymes from recombinant microorganisms with optimal physiochemical properties to overcome expensive pre-treatment processes are attempted to increase sugar yields. To harness the maximum breakdown of lignocellulosic biomass, research on engineering novel enzymes through the manipulation of amino acid residues has received more attention to overproduce targeted enzymes for better cocktails. Proteomic approaches are utilized to study the genetics and expression status of microbes at specific conditions for the biodegradation of lignocellulose biomass. Proteomic analysis on the secretome of lignocellulosic biomass-degrading microorganisms and different commercial cocktails on biomass hydrolysis is in progress way in the identification of novel and potential genes to synthesize better formulation of targeted enzyme cocktails for efficient delignification. This chapter summarizes biodegradation of lignin by microorganisms-producing ligninolytic enzymes such as laccases, lignin peroxidase (LiP), manganese peroxidase (MnPs), versatile peroxidase (VP), and dye-

L. Kasirajan (✉) · K. Kamaraj · S. Sebastiar · P. P. Thirugnanasambandam · E. Karpagam
Division of Crop Improvement, ICAR-Sugarcane Breeding Institute, Coimbatore, India
e-mail: K.Lakshmi@icar.gov.in

G. S. Suresha et al. (eds.), *Value Addition and Product Diversification in Sugarcane*, https://doi.org/10.1007/978-981-97-7228-5_8

decolorizing peroxidase (DyP), identification of new lignin hydrolyzing microbes and on proteomic and transcriptome analysis of lignocellulolytic enzymes for enhanced sugarcane bagasse delignification.

Keywords

Enzyme cocktails · Ligninolytic enzymes · Monolignol biosynthesis · Proteomics · Secretomics

8.1 Introduction

Globally, plant-based biofuels existed long before. However, the advent of coal-based fossil fuels at minimal cost made mankind rely on fossil fuels as an energy source. Despite that, the depletion of fossil fuels and escalated crude oil prices have led to the search for green energy resources (Kumar 2020). Biomass from plant-based resources seems to be a substantial alternative source of energy for biofuel production. Bio energy-based biofuel utilization also provides socioeconomic and environmental benefits.

8.1.1 Generations of Biofuels

Based on the type of feedstock, biofuel research and development has been categorized into three different generations. First-generation biofuel (or conventional biofuels) is produced by edible energy crops such as sugar-based crops (sugarcane, sugar beet, and sorghum), starch-based crops (corn, wheat, and barley), or oil-based crops (rapeseed, sunflower, and canola). Through simple biochemical processes that have been developed in the food industries (fermentation), these elements can be modified to become a transportation fuel. However, these crops require intensive agricultural input (fertilizer) and also have raised serious concerns about food supply, food security, and arable land requirements (Phwan et al. 2018). Second-generation biofuels are also called "cellulosic-ethanol" which are produced from a wide range of feedstocks, such as waste vegetable oil, forest residue, industry residue, and sustainable biomass (perennial grasses, and trees).

The crops grown on marginal land areas that are highly erodible or with minimum soil quality may be best used for second-generation biofuel production. Third-generation biofuels are also known as "algae fuel" since they are produced from algae biomass which has a very distinctive growth yield approximately 10 times higher than the second-generation biofuel (Brennan and Owende 2010). The microalgal feedstock is used to produce different biofuels and other valuable co-products (Brennan and Owende 2010; Nigam and Singh 2011). Microalgae-based biofuel has many advantages when compared to higher plants, such as they naturally synthesize large quantities of neutral oils (20–50% dry weight of biomass), need less water, and they can tolerate harsh environmental conditions (Um and Kim 2009; Mata et al. 2010). Interestingly, algae biomass has low hemicellulose content and

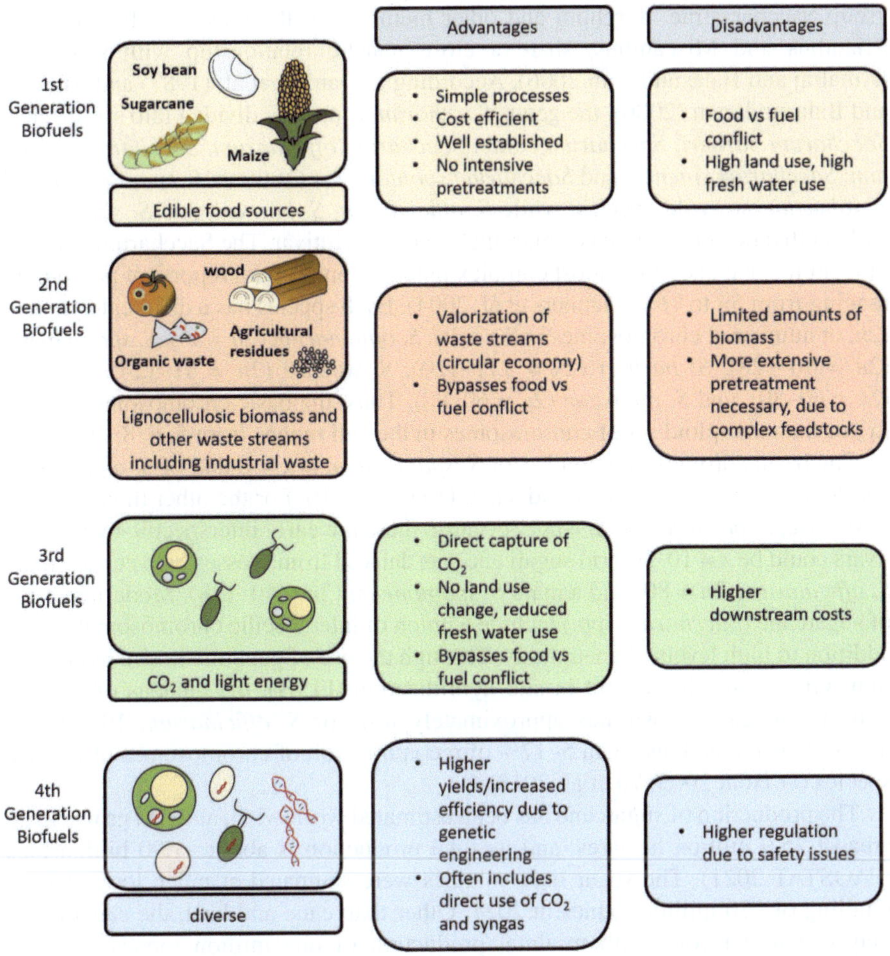

Fig. 8.1 Represents the pictorial diagram of varying biofuel productions along with their feedstocks owning to the advantages and disadvantages respectively. Adapted from Cavelius et al. (2023)

about zero lignin which results in increased hydrolysis and fermentation efficiency (Saqib et al. 2013). A schematic representation of all generations of biofuels along with their advantages and disadvantages is represented below in Fig. 8.1.

8.1.2 Sugarcane Biology and Genetics

According to taxonomy, sugarcane is classified as a member of the genus *Saccharum*, which was created by Carl Linnaeus in 1753. It is a member of the grass family Poaceae, subfamily Panicoideae, tribe Andropogoneae, subtribe Sacharinae, and group Saccharastrae. Sorghum and other members of the grass family, including Erianthus and Miscanthus, share a close genetic relationship with sugarcane (Amalraj and Balasundaram 2006). According to Daniels et al. (1987) and Amalraj and Balasundaram (2006), the genus *Saccharum* is usually divided into six species: *Saccharum barberi*, *Saccharum edule*, *Saccharum officinarum*, *Saccharum robustum*, *Saccharum sinense*, and *Saccharum spontaneum*. Of these, *S. spontaneum* and *S. robustum* are wild species, while *S. officinarum*, *S. barberi*, and *S. sinense* are early cultivars and *S. edule* is a marginal specialty cultivar. The Saccharum genus is thought to have one of the most complex plant genomes, with reports of polyploidy ranging from 5× to 16× (Manners et al. 2004). Each species has a different cytotype (2*n*, or number of chromosomes in the cell): *S. officinarum* ($2n = 80$), *S. spontaneum* ($2n = 40–128$), *S. barberi* ($2n = 111–120$), *S. sinense* ($2n = 81–124$), *S. edule* ($2n = 60–80$), and *S. robustum* ($2n = 60, 80$). Thus, the basic chromosome number "*x*" or the monoploid set of chromosomes in the cell ranges from 5, 6, 8, 10, to 12.

The basic chromosome number of *S. spontaneum* is $x = 8$ and of *S. officinarum* and *S. robustum* is $x = 10$ (Piperidis and D'Hont 2020). For the other three species, *S. sinense*, *S. barberi*, and *S. edule*, because these are early interspecific hybrid cultivars could be $x = 10$. Hybrid sugarcane was derived from crosses between a female *S. officinarum* ($2n = 80$) and a male *S. spontaneum* ($2n = 40–128$). Modern hybrids of sugarcane (*Saccharum* spp.) include a union of interspecific chromosomal sets in addition to high levels of aneuploidy. Through the use of genomic in situ hybridization (GISH) and fluorescent in situ hybridization (FISH), the nucleus of modern hybrid sugarcane constitutes approximately 80% of *S. officinarum*, 10–20% of *S. spontaneum*, and less than 5–17% of recombination of chromosomes of the two species (D'Hont 2005; Yu et al. 2018).

The production of sugarcane has been estimated worldwide, and it is grown in an area of 25.9 million hectares, and its total production is about ~1.90 billion tons (FAOSTAT 2021). The sugar beet volumes were stagnated at much lower levels, totalling of 270 million tonnes in 2021. Other than cane and beet, the sugar crops played a minor role, with minimal production of one million tonnes in 2021. Globally, Brazil was the largest producer of sugar cane in 2021 and putatively with 38 percent of the world's production, followed by India accounted with 22%, China, Pakistan, and Thailand hold to be 4–6% each shown in Fig. 8.2.

8.1.3 Sugarcane for Bioenergy Production

Sugarcane is considered to be one of the most prominent energy efficient crops with a wide adaptability range along with desirable traits of high yield, low input requirement, and better processing potentiality (Verheye 2013). In India, sugarcane is the crucial source for 70% of sugar production. Sugarcane produces nearly 100 t/ha biomass with 25–30% of fibre and yield of 60 t/ha of wet bagasse as against 32 t/ha wet bagasse yield of sugarcane (Kasirajan et al. 2017). The abundant lignocellulosic sugarcane residues is a potential source of second-generation biofuel and will boost the biofuel yields per unit land area compared to current sucrose-based conversion

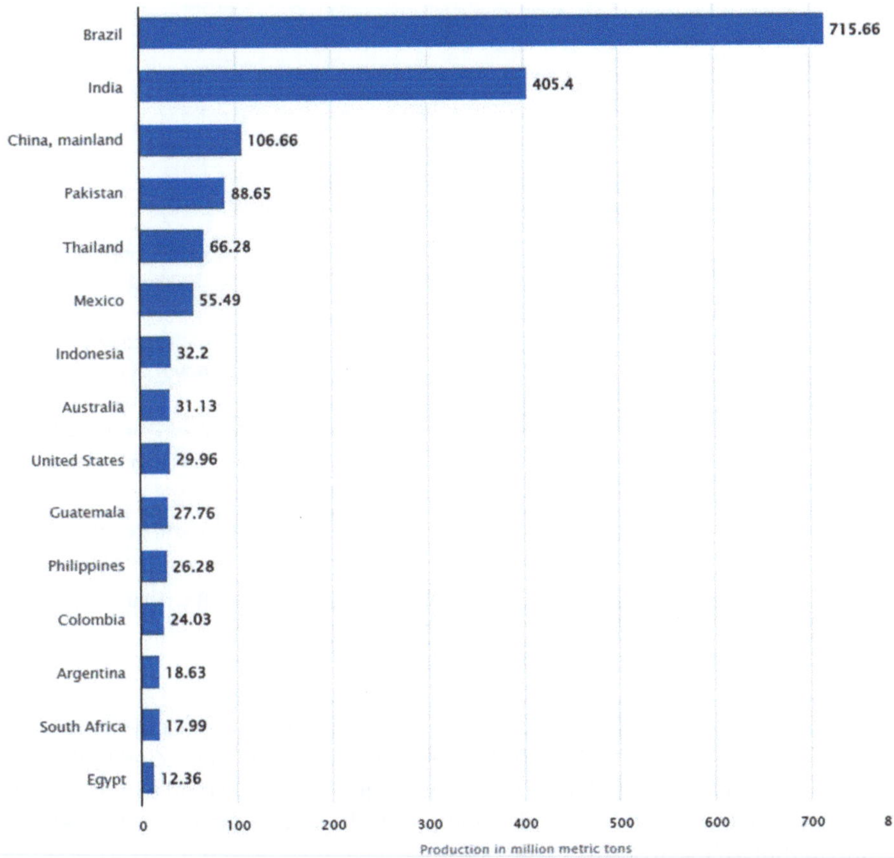

Fig. 8.2 Leading sugar cane producers worldwide in 2021, based on production volume (in million metric tons). Source from: https://www.statista.com/

technologies. The primary limit of sugarcane bagasse usage is its high degree of complexity due to its mixed composition of cellulose 35–50%, hemicellulose 20–35%, and lignin 5–30% (Kasirajan et al. 2017), in which the lignin is the compound most resistant to degradation. To effectively utilize cellulose and hemicellulose from lignocellulosic raw materials, it is essential to release them from lignin bonds. Therefore, removal of the lignin from the complex by pre-treatment is the most important step in bioethanol production. Figure 8.3 shows the schematic diagram that explains the bio-ethanol production from sugarcane.

8.1.3.1 Ethanol Production

Sugar crops is a major feedstock for renewable bioethanol production for use as a transportation fuel. Other feedstocks include starch-rich crops such as corn, wheat, and cassava. Because it is a clean, affordable, and low-carbon biofuel, ethanol from sugar crops has emerged as a leading renewable transportation fuel. Brazil is the

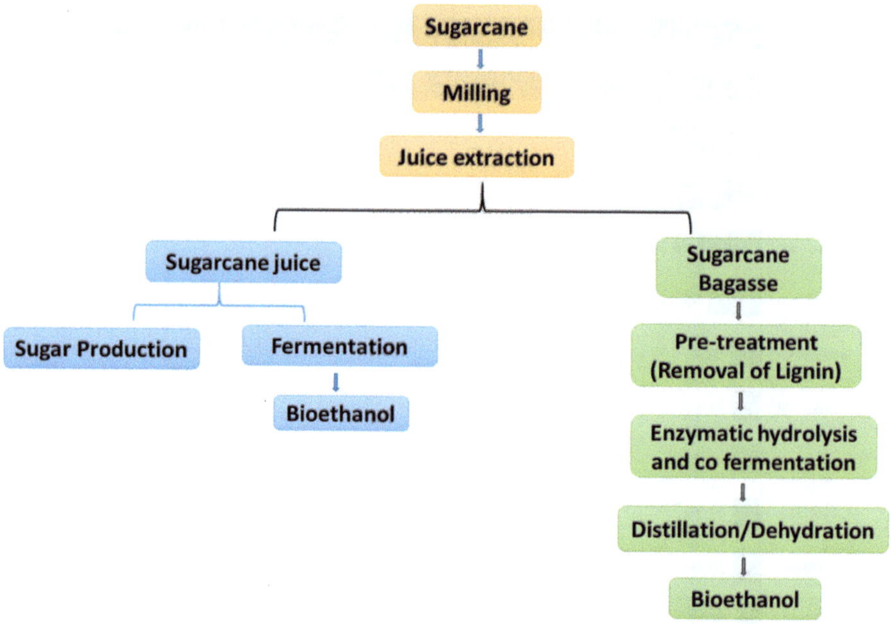

Fig. 8.3 Flow diagram of bioethanol production from the sugarcane

world leader in fuel ethanol production from sugarcane. Global ethanol production in 2021 rose to 104.5 bn litres, from 101.4 bn litres in 2020. Production in the US (+4.1 bn litres) and India (+1.2 bn litres) rose sharply, more than offsetting falls in output in Brazil (−2.8 bn litres), Thailand (−0.15 bn litres), and Colombia (−0.14 bn litres) (https://www.isosugar.org/).

Lignin is an aromatic heteropolymer composed of chemically different monolignol, with large variation in content and composition ratios among species, individuals, and in different tissues of an organism. Three monolignols such as p-coumaryl, coniferyl, and sinapyl alcohols are major proportional constituents of lignin with other different additional components and are cross linked by β-5, α-O-4, β–β, C–H, C=C, C–O, C–C, O–H, and other several types of bonds among different plant sources. This vast variation in composition ratio and strong linkages with cellulose and hemicelluloses makes lignocellulosic biomass, a highly resistant and recalcitrant structure for biodegradation by fungi and few bacteria. Enzymatic hydrolysis with mixture of several enzymes viz., cellulase, hemicellulose, and ligninase from different microorganisms are utilized for conversion of lignocellulosic biomass into fermentable sugars. White-rot fungi produce effective ligninolytic enzymatic system of laccases and peroxidases, such as lignin peroxidase (LiP), manganese peroxidase (MnP), and versatile peroxidase (VP) to degrade lignin and for detoxification after pre-treatments.

Application of combined fungal fermentation, laccase mediator system, and enzymatic delignification had resulted in highest percentage of glucose recovery, and improvements in terms of time and cost are needed to compete with conventional pre-treatment techniques. Laccasses have restricted action with relatively low redox potentials on oxidation of phenolic components, while it has the maximum of 80–90% on non-phenolic compounds. Redox mediators are used as rational approaches for modification of ligninolytic enzymes to improve degradation of non-phenolic compounds. Ligninolytic enzymes oxidize mediators, which carries electrons between the enzyme and substrate and reacts with final substrate to recover lost electrons. This results in an increment of redox potential and thereby improves degradation of recalcitrant compounds, which the enzyme by itself cannot degrade. The most commonly used mediators are 1-hydroxybenzotriazole (HBT), 2,2′-azino-bis (3-ethylbenzothiazoline-6-sulphonic acid) (ABTS), and some natural mediators as syringaldehyde or vanillin. Other ligninolytic enzymes viz., lignin peroxidase (LiP) and manganese peroxidase (MnP) contain protoporphyrin IX as a prosthetic group and a newly identified enzymes called dye-peroxidases (DyP) also participate in the degradation of lignin. The application of secretomics to thermophilic fungal species with the identification of novel thermostable ligninolytic enzymes in future will enhance lignin depolymerization through modification of ligninolytic enzymes than conventional approaches.

A proteomic approach helps to analyse the functional annotation of genes at the protein level, characterization of genes, and enzymes involved in metabolism. The post-transcriptional modification approaches also regulate the gene expression at the RNA level. Conventional methods like ELISA and chromatographic techniques provide the elementary data for gene-expression study. The gel-techniques such as SDS-PAGE (sodium dodecyl sulphate polyacrylamide gel electrophoresis), (2-DGE) two-dimensional gel electrophoresis, and (2D-DIGE) two-dimensional differential gel electrophoresis produce only the relative gene expression data. Currently, the advance of high through-out techniques is feasible, for instances, the methods such as protein array, ICAT (Isotope-coded affinity tag), SILAC (stable isotope labelling with amino acids in cell culture), and iTRAQ (isobaric tag for relative and absolute quantitation) (Aslam et al. 2017) are shown in Fig. 8.4.

For reducing the saccharification recalcitrance of lignocellulosic biomass, selective delignification with cellulose preservation is regarded to be the best lignin degradation strategy. Extensive research efforts from various teams have been dedicated towards the development of a biocatalyst for lignin depolymerisation. A 'microbial sink' has been reported by Salvachúa et al. (2016) from fungal secretomes, which comprises of laccasses and peroxidase activity for partial depolymerization of high-lignin content. as. Most of the research finding are towards the development of microorganisms/ enzymes for cellulosic/hemicellulosic fermentation, while very few or limited efforts are towards the selective delignification. However thermostable enzymatic combination towards depolymerization of sugarcane bagasse of high lignin content is yet to be identified. Kolbusz et al. (2014) compared the transcriptomes and exoproteomes of *Myceliophthora thermophila* a thermophilic fungus. He found a set of 59 genes-encoding CAZymes that were up-regulated in response to both monocot and dicot straws, including nine polysaccharide monooxygenases and GH10. Calderan-Rodrigues et al. (2016) has characterized the cell wall proteome of

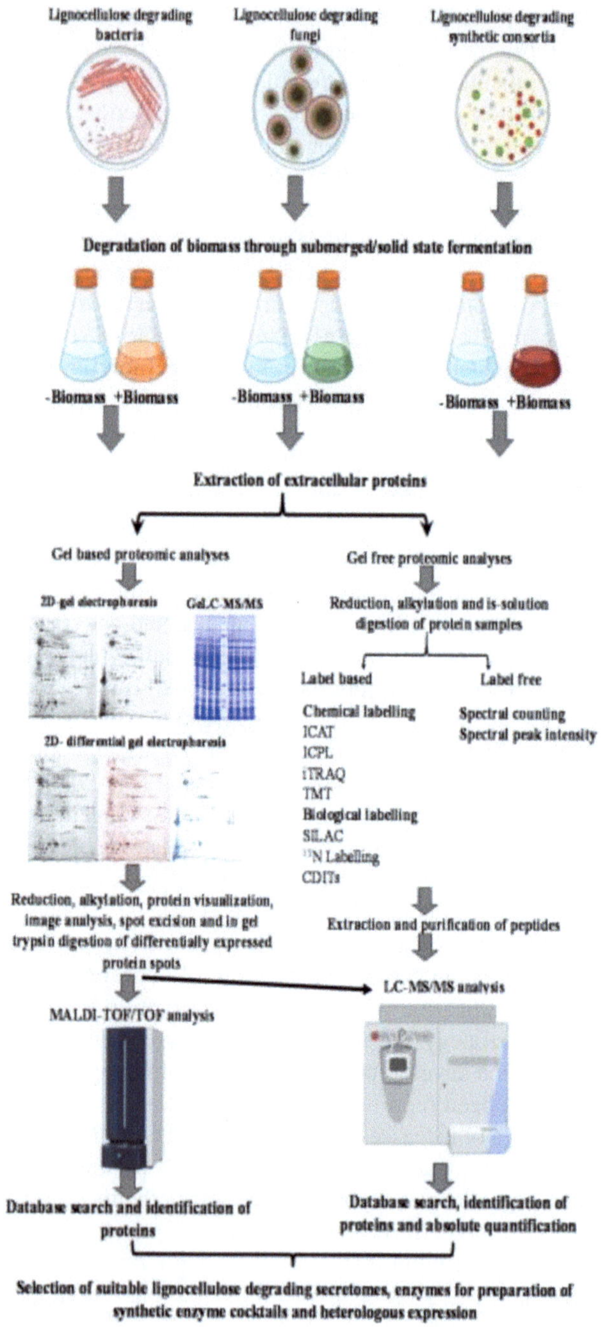

Fig. 8.4 Procedures and tools involved in secretome analysis Adapted from Bianco and Perrotta (2015) and Sethupathy et al. (2021)

young sugarcane culms, to identify proteins involved in cell wall biogenesis. They have identified 10 glycoside hydrolase (GHs) as possible targets to reduce cell walls recalcitrant to deconstruction. Schneider et al. (2016) studied the secretome of the fungi *Penicillium echinulatum*, and a glycoside hydrolase (GH61) endoglucanase (Cel61A) was highly secreted in the secretome derived from fungi grown on sugarcane bagasse. Saykhedkar et al. (2012) performed a time course analysis of the extracellular proteome of *Aspergillus nidulans* on sorghum stover. A total of 294 extracellular proteins were identified with hemicellulases, cellulases, polygalacturonases, chitinases, esterases, and lipases predominating in the secretome.

Dong et al. (2013) studied delignification of sugarcane bagasse with use of three fungi viz., *Phanerochaete chrysosporium* PC2, *Lentinula edode* LE16, and *Pleurotus ostreatus* PO45 and observed *Lentinula edodes* LE16 as best candidate for biodegradation with high manganese peroxidase (MnP) and Polyphenol oxidase (PPO) activities. It is also identified that PPO as a primary component of ligninolytic enzymes in degradation of lignin-cellulosic waste and the function of esterase in detoxification of hemicellulose lignin matrix was also confirmed by the study. Machado and Ferraz (2017) utilized white-rot fungi *Ceriporiopsis subvermispora* for biodegradation of sugarcane bagasse reported 47 recovery percent of glucose. Borin et al. (2015) utilized filamentous fungi *Aspergillus niger and Trichoderma reesei* with two levels of cell wall complexity for pre-treatment of sugarcane biomass. The fungi were grown in culm at natural condition and in pretreated bagasse. The enzyme production due to biomass degradation was monitored by secretome analyses at 6, 12, and 24 h. Polysaccharides such as arabinoxylan and β-glucan were released first and found to be assimilated by both fungi species, and the important enzymes of biomass degradation viz., endoglucanases, xyloglucanases, β-glucosidases, β-xylosidases, cellobiohydrolases, endoxylanases, and α-arabinofuranosidases were identified in secretomes of both fungi grown at abovementioned conditions. Authors inferred that both fungi produced relatively more enzymes when only culm is used as a carbon source.

Borin et al. (2017) conducted RNA-seq comparative transcriptome profiling technique with employing the same above studied fungi (*A. niger* and *T. reesei*) on sugarcane bagasse (carbon source) with particular focus on transcription factors (TFs), CAZymes, sugar transporters, and other related proteins for lignocellulosic detoxification. They found significant regulation in expression of above-mentioned factors mainly at the transcription level and also identified difference in expression of TFs between fungi viz., *A. niger* and *T. reesei*. Sugar transporters, identified specifically induced on biomass degradation by both fungal strains, can be further exploited as new players in lignin delignification. The study also identified several genes-encoding proteins such as pro-oxidants, antioxidants, detoxification enzymes, and some with unknown function in steam exposed bagasse as carbon source that, but their specific roles on lignocellulosic pre-treatment has to be further explored. The above discussed study was the first report on bagasse pretreatment with information on comparison on performance of different fungi along with activity difference at different time intervals using the RNA-seq technology.

To understand the molecular mechanism involved in lignocellulose deconstruction, genomic and transcriptomic analyses of *Irpex lacteus* CD2 were grown with ball-milled corn stover as the carbon source under submerged fermentation (Qin et al. 2018). *I. lacteus* CD2 efficiently decomposed lignin, cellulose, and hemicellulose present in corn stover at the rate of 74.9%, 86.3%, and 83.5%, respectively, within 9 days. Followed by the accumulation of cellulase and hemicellulase manganese peroxidases was rapidly induced. The study showed the importance of free radicals for future designing of novel, robust lignocellulose degrading enzyme cocktails. Su et al. (2018) studied the lignin depletion capability of Myrothecium verrucaria strains and they recorded the level of enzyme activities, such as laccase, lignin peroxidase, and manganese peroxidase were 6.61, 0.78, and 1.31 U g^{-1} dry biomass respectively. The lignin content of corn stover decreased by 42.30% and the conversion rate increased by 123.84% during saccharification. Henceforth, the proteomics imparts highly reliable and direct assess for characterization of genes. Among which sucrose metabolism plays a pivotal target for juice improvement and quality enhancement in sugarcane plants (Ostengo et al. 2021). In this chapter, let us delves deeper into the topic of ligninolytic enzymes used for the degradation of lignin and improvise the sugarcane biomass into a biofuel.

8.2 Ligninolytic Enzymes Fungal, Bacterial, and Other Microbial Sources

The main recalcitrance on the lignocellulosic biomass of sugarcane bagasse is due to the presence of aromatic phenolic lignin complex compounds. Using the biological approach called enzymatic treatment from varying sources such as fungal, bacterial, and other sources mortifies the lignin into simple cyclic compounds (Huang et al. 2013; Mäkinen et al. 2019).

8.2.1 Enzymes from Fungal Sources for Lignin Degradation

Fungi are the most common microbial strains involved in lignin breakdown and degradation especially white-rot fungi from basidiomycetes and other microbes from ascomycetes analogous to the synthesis of ligninolytic enzymes (Singh et al. 2021; Kasirajan et al. 2022). Comparatively, fungi species show a higher rate of delignification and mediate the catalytic conversion (Spence et al. 2021). There are many fungal species such as White-rot fungi, Pleurotus species, Trametes hirsuta, Coniophora puteana, Phanerochaete chrysosporium, Ganoderma tsugae, Penicillium chrysogenum, and Aspergillus niger involved in the ligninolytic enzyme production. Laccase, Manganese peroxidases (MnPs), Lignin peroxidases (LiPs), and versatile peroxidases (VPs) are highly potential enzymes for delignification (Janusz et al. 2020). Apart from that, these enzymes have numerous industrial applications such as complex aromatic textile dye degradation, removal of phenolic chemicals from pollutants, generation of bio-methane, and biomass degradation (Pecková et al. 2020; Zhuo and Fan 2021). Following that, two white-rot fungi namely Pleurotus ostreatus and Trametes hirsuta were explored for their ligninolytic enzymes, the resultant appears to be a two-fold increase in fibrinolysis and exhibited a maximal organic matter composability from 26.5% to 31.2%. The study represents the increased rate of lignin de-polymerization in the tobacco stalks (Yang et al. 2020). Pleurotus species can produce enormous types of peroxidases enzymes to metabolize chemicals like methoxybenzene and benzoic acid (Janusz et al. 2017).

Consecutively, Coniophora puteana is a laccase producer and in the presence of tannic acids it might effectively metabolize the cell walls (Hyde et al. 2019). By combining both the laccases and peroxidases enzymes to work synergistically to depolymerize the complex aromatic lignin into simpler compounds, the outcome could be revolutionized into a high-valued commercial goods (Kameshwar et al. 2018). Similarly, Penicillium chrysogenum, Aspergillus caespitosus, and Aspergillus species were three fungal strains identified for the degradation of lignin in pine wood, sugarcane bagasse, and sycamore (Salem et al. 2019; Kasirajan et al. 2022). According to the study, conducted by Fonseca et al. (2018) manifest a proteomic and genomic research on the Lac I gene extricated from Phlebia brevispora was expressed in Pichia pastoris demonstrated a protein expression and the enzyme with good solvent/salt tolerance. Thereafter, the work done by Jin et al. (2018) explored that the Lac I gene identified from Ganoderma tsugae has ligninolytic activity in knockout tests. In this study, Phanerochaete chrysosporium is a fungal strain that overexpress laccases and peroxidase enzymes during shock wave-induced acoustic cavitation. When the enzyme was treated on different biomass materials like wheat bran and sugarcane bagasse, they represented an efficient lignin metabolizer. The recombinant enzymes showed a 25% increased degradation of lignin and 2.6-fold uplifted peroxidase activity and four-fold laccase activity compared to the control (Coconi Linares et al. 2018).

8.2.2 Enzymes from Bacterial Sources for Lignin Depolymerization

On the contrary, hardly few bacterial strains are identified with ligninolytic enzyme activity in nature. For example, Rhodococcus jostii (RHA1 species) became the wildly investigated bacterial strain for its lignin-metabolizing capabilities and conversion of lignin into value-added products (Spence et al. 2020). A study showed that (Sun et al. 2018) Rhodococcus jostii strain was used for the lignin modification into aromatic dicarboxylic acids and also analysed its metabolic pathway, which has been used for bioplastic production. In yet another research work on Rhodococcus jostii (RAH1), the investigators identified two dye decolorizing peroxidases (DyPs). DypA, B isolated from the periplasmic and non-periplasmic regions from the bacterial strain, respectively. The enzymatic kinetic activity was determined using the Reactive Blue 4 dye showed enhanced activity. This work proposes that the metabolic engineering on RAH1 can elevate the lignin depolymerization enzymes. Pseudomonas spp. can be easily grown on a minimal medium, and it is an excellent model organism for exploring lignin biotransformation. This can be achieved by inclusive detailed research on metabolic engineering (Nikel and de Lorenzo 2018). Likewise, ligninolytic bacteria were discovered and they are elaborated in detail, yet effective screening methods lead to the identification of novel strains with efficient enzyme activity (Gonçalves et al. 2020).

Distinct fungal and bacterial strains were identified to produce the respective enzymes to metabolize lignin. There are a few drawbacks to the use of in vitro enzyme in ligninolytic approaches including the direct contact of microbial species and time consumption during culturing between the substrate and microbe (Li et al. 2018). Moreover, the usage of a single enzyme has been demonstrated in numerous studies to degrade the lignin in lignocellulosic biomass, but it requires further more detailed inquiry and exploitation of molecular mechanisms for multi-enzyme systems (Sharma et al. 2019). Laccases and peroxidases are the predominant enzyme groups that use the oxidative cleavage technique to degrade the lignin. It has been used in in vitro studies as these respective enzymes target more randomly than specific targets like cleaving the links or substrates. The majority of the lignolytic enzymes utilize the non-specific cleavage to break down the aromatic compounds. Apart from that there are few more enzymes like MnP, LiP, DyP, and VP peroxidases with wider applicability (Bilal et al. 2021).

8.2.3 Laccases

Laccase enzyme is a multicopper oxidase enzyme, an eco-friendly one found in a large number of fungal and bacterial species. This enzyme uses an oxidoreductases mechanism as the blue copper phenol-oxidases employ O2 as an electron acceptor to oxidize the phenolic compounds. This enzyme is the "greenest" enzyme as it makes use of O2 and converts it into H_2O as a by-product (Cannatelli and Ragauskas 2017; Kohli et al. 2019). Laccases have minimal oxidative adaptability with least catalysis as the enzyme mediates a number of chemical processes like oxidation of aromatic compounds, ring cleavage and degradation of polymers (Raveendran et al. 2018). In an excess, more than 60 fungal strains with laccases activity were discovered, and the fungal species called brown-rot and white-rot fungi had been stated with maximal and potential laccases producers (Janusz et al. 2020). Apart from that, other laccase-producing species are Trametes versicolor, Pleurotus ostreatus, Dichomitus squalene, Lentinula edodes, Current maxima, and Irpex lacteus (Youn et al. 1995; Asgher et al. 2018; Yesilada et al. 2018). It has been employed in varying processes involving ethanol production, alteration in fibre, bioremediation, biofuel cell production, dye degradation, and effluent treatment (Agrawal et al. 2018).

8.2.4 Lignin Peroxidases (LiP)

Lignin peroxidases (LiP) is the most prevalent ligninolytic enzyme which is heme-containing monomeric protein, relatively non-specific to its substrates by oxidative degradation. It catalyses both the phenolic and non-phenolic compounds in the presence of H_2O_2 (Datta et al. 2017; Khatoon et al. 2019). The enzyme LiP was initially studied in the microorganism called Phanerochaete chrysosporium, and the enzyme was found to be on the extracellular matrix of the strain (Khan et al. 2013). Eventually, the LiP isozymes were identified in various microbial species namely Phlebia radiata, Trametes versicolor, and Phanerochaete sordida (Mäkinen et al. 2018; Kasirajan et al. 2023). Generally, 96% of the species generate manganese peroxidases, whereas the species like Daedaleopsis septentrionalis and Cylindrobasidium secrete two distinct producers of LiP enzyme. The automated screening test for the strains might help to detect the ligninolytic enzymes and further these could be enhanced (Datta et al. 2017). Because of its higher redox potential, it resulted in significantly disrupting the lignin in biomass works alone or in enzyme cocktail with greater selectivity (Pham et al. 2018). Besides, its lignin metabolizing ability, it can also be used to produce bioplastics and numerous value-added by-products (Miyauchi et al. 2018; Asgher et al. 2020; Xie et al. 2020).

8.2.5 Manganese Peroxidases (MnPs)

Manganese Peroxidase (MnP) is a highly flexible enzyme in nature. It is made up of glycosylated protein and the enzyme requires hydrogen peroxide (H_2O_2) to catalyse the reaction (Chio et al. 2019). This enzyme implicates the oxidation of both non-phenolic and phenolic compounds (Bilal et al. 2017). It has numerous industrial and biotechnological applications; henceforth, it has been manufacturing largely all around the world (Bilal et al. 2018; Bilal and Iqbal 2019). MnP is an extracellular enzyme predominantly found in abundant of brown and white-rot fungi along with some bacterial strains. It seems to be a proficient biocatalyst for lignin degradation when couple with enzymes like laccases. In the beginning, the MnP enzyme was discovered in Phanerochaete chrysosporium (Huang et al. 2020). It can act as a substitute to produce a low-cost enzyme with industrial biofuel applications. Aside from that, it has a wide range of applications in the field of bio-bleaching, biopulping, bioremediation, and lignin degradation (Saleem et al. 2018). MnP enzyme production has been announced by various extracellular excreting microbial strains namely *Ceriporiopsis subvermispora, Phanerochaete sordida, Physisporinus rivulosus, Dichomeris squalene,* and *Phlebia radiata* (Chowdhary et al. 2017). According to the research work conducted by Chowdhary et al. (2019), *Phanerochaete chrysosporium* BKM-1767 expressed the MnP production at diverse climatic and nutritional conditions. To improvise the biodegradation efficiency, an explicit investigation is required to invent the inorganic/organic ions tolerance.

8.3 Sugarcane Genetic Transformation for Cell Wall Modification

Sugarcane is an exemplary crop with high biomass content as a biofuel crop to generate bio-ethanol. Especially, the genetically modified sugarcane crop via growth hormones and biomass pathways of sugarcane cell wall components promotes the saccharification efficiency from its soluble sugar and lignocellulosic components (Andrade et al. 2019). The largest germplasm collection of Saccharum species was found in India and America (Tai and Miller 2001; Hemaprabha et al. 2022). Subsequently, this germplasm can be potentially utilized to improve the sugarcane plants with biofuel characteristics by plant breeding. The wild varieties show abundant genetic variations than the domesticated canes (Todd et al. 2014). Because of its genomic complexity and diversity, sugarcane has been explored by using molecular markers for better understanding. DNA-based molecular markers could be a better option for studying sugarcanes' complex hybrids and genetic manipulation of sugarcane for biomass production (Hoang et al. 2015; Manimekalai et al. 2020).

Lately, genetically modified sugarcane plants can generate hydrolytic enzyme, which results in ethanol as they do not necessarily need much chemical treatment to break down the lignocellulosic cell wall polymers. In recent times, researchers have been concentrating on the isolation of genes from bacterial and fungal strains which produce enzymes that have a potential to degrade the cell wall. In that way, enzymes such as endo-glucanases, β-glucosidases, and cello-bio-hydrolases can degrade the cellulose to glucose. Laccase, manganese peroxidase, and lignin peroxidase are very few enzymes that have the ability to degrade lignin in cell wall (Kasirajan et al. 2022; Kasirajan and Kamaraj 2023). The enzyme like endo-glucanases is derived from synthetically designed bacterial gene, and it has been introduced into the sugarcane plants to catalyse the internal part of cellulose polymer (Willis et al. 2016). The genes like Cinnamyl Alcohol Dehydrogenase (CAD), caffeic acid O-methyltransferase (COMT), Phenylalanine Ammonium Lyase (PAL), and Ferulate 5 hydroxylase (F5H) degrade lignin compounds, ultimately improve the biofuel production (Kasirajan et al. 2020). The contemporary exploration of sugarcane gained a name as "energy cane" distinctly focusing on high fibre and biomass varieties to generate bio-ethanol and other by-products (Knoll et al. 2013).

8.4 Investigation on Genes in Monolignol Biosynthesis through Gene Editing Techniques

The gene caffeic acid 3-O-methyltransferase (COMT) plays a pivotal role in the lignin biosynthesis pathway. It modulates the multistep methylation reactions of hydroxylated monomeric lignin, and the enzyme induces the production of sinapaldehyde and sinapyl alcohol (Louie et al. 2010; Kasirajan et al. 2017). Fu et al. (2011) state that the transgenic plants or mutants with COMT suppression reduce the lignin content, which signifies the importance of sinapyl alcohol biosynthesis. Subsequently, the downregulation of COMT lignin biosynthetic gene encourages saccharification efficacies and yields higher lignocellulosic ethanol. Daly et al. (2019) proposed that the suppression of cinnamyl-alcohol dehydrogenase by employing RNA interference (RNAi) might enhance the higher saccharification after acid pre-treatment in barley. Likewise, RNAi downregulates the COMT gene in sugarcane with 67–97%, showing 3.9–13.7% decline level, whereas the syringyl/guaiacyl ratio reduced to 1.27 and 0.79 from its wild-type sugarcane of 4.47. Moderately, the recalcitrance of cane biomass has been limited to 8.4% and resulted in the outturns of directly fermentable glucose by 29% (Jung et al. 2012).

Following those researchers have examined the field performance of downregulated COMT transgenic lines in the United States, where it exhibited a 12% of decline in lignin content and the S subunit curtailed up to 49% in some canes. Additionally, this targeted gene's suppression resulted in 91% of the reduced transcript with a lignin reduction of 6%. The suppression of COMT gene had no adverse effect on biomass yield, brix values, carbohydrate structure, and agronomic characters, with a remarkable outcome of 19–23% saccharification efficiency. As the transgenic lines require only one-third of hydrolysis time and are exploited only with 3–4 fold less enzymes for fermentable sugar than the non-transgenic plants (Jung et al. 2013).

The first report of genome editing in sugarcane seems to be Transcription activator-like effectors (TALEs) a genomic tool utilized to customizable DNA-binding motifs for locus-specific modifications. Especially, TALE nucleases-mediated mutagenesis creates specific modifications in COMT gene which resulted in 74% of transgenic clones being subjected to genome editing. Pyrosequencing showed that 8–99 % of the wild type COMT gene sequence were converted to mutanted COMT gene sequences. This targeted mutation resulted in showed impact in 29–32% of diminished lignin biosynthesis against the wild-type canes (Jung and Altpeter 2016).

Reduced lignin level might enhance the saccharification efficiency with minimal hydrolysis to obtain the fermentable sugars from the lignocellulosic biomass. The study conducted by Jung et al. (2016) preferred the RNAi approach to suppress the 4-coumarate: CoA ligase (4CL) gene and the aftereffect transgenic lines had 16.5% reduced lignin content without affecting the biomass yield. Consequently, the field-grown transgenic lines produced 52–76% of elevated saccharification efficiency than the wild-type sugarcane plants. Yet another report investigated by Bewg et al. (2016) accustomed to RNAi downregulation expression of the three most predominant genes closely associated to lignin biosynthesis in sugarcane plants containing ferulate 5-hydroxylase (F5H), caffeoyl-CoA O-methyltransferase (CCoAOMT), and COMT. The sequent RNAi transgenic lines displayed emphasized sucrose levels and biomass quality in all transgenic lines excluding one that showed reduced content of lignin.

According to the earlier report of Whetten and Sederoff (1995), they have demonstrated that nearly 10 enzymes involved in the lignin biosynthesis, and approximately 28 unigenes were stated to be related to the monolignol biosynthesis (Bottcher et al. 2013). Down-regulating or restricting the crucial genes diminishes the lignin content in the sugarcane cell wall. Additionally, customization of sugarcane cell wall composition might manipulate the key genes entailed in lignin biosynthesis. Mostly, terminal enzymes like cinnamyl alcohol dehydrogenase (CAD) and caffeic acid O-methyltransferase (COMT) are remarkable targets for this kind of above-mentioned study. Because these targeted genes could regulate the lignin level without bothering the growth and development of sugarcane plants (Sticklen 2006; Jung et al. 2012; Hoang et al. 2015).

8.5 Proteomic Analysis Using the Secretome for Lignocellulolytic Enzymes

A study by Adav et al. (2012) proposed that the Phanerochaete chrysosporium has been cultivated in varying lignocellulosic biomass including sugarcane bagasse where it was analysed by iTRAQ technique using LC-MS/MS. The research group has identified novel and diverse hydrolysing and lignocellulolytic enzymes. In total, 329 secreted proteins were quantified, out of which secretome of P. chrysosporium using iTRAQ assisted to quantify nearly 15 of the lignin-degrading proteins such as copper radical oxidases, CDH, glucose oxidase, isoamyl alcohol oxidase, glutathione reductase, glutathione S-transferase, peroxiredoxins, pyranose 2-oxidase, quinone oxidoreductase, iron-containing alcohol dehydrogenase, and cytochrome c degrade lignin through oxidases. The enzyme called copper radical oxidases was crucially upregulated in all the biomass.

The secretome analysis has been done on solid-state fermentation of sugarcane bagasse treated with Trichoderma asperellum S4F8 and Trichoderma reesei Rut C30, and the proteomic data was analysed shows that S4F8 had remarkably higher hemicellulase and β-glucosidase enzyme activities. In this study, nearly 815 proteins had been detected using (LC-MS/MS) Liquid chromatography tandem mass spectrometry analysis of the two fungal secretomes. In-depth analysis explains that Trichoderma asperellum S4F8 produces a complex form of lignocellulolytic cocktail produced—a more complex lignocellulolytic cocktail, with magnified hemicellulose and cellobiose hydrolysis potential, comparatively (Marx et al. 2013). In fact, another study on fungal species called Aspergillus niger and Trichoderma reesei was grown on sugarcane biomass with culm in natura and pretreated bagasse. The biomass degradation was observed critically due to enzyme production in the biomass, which has been detected using secretome analyses. As these biological treatment with microbial species and the secretome support and enhance the saccharification efficiency (Borin et al. 2015).

Similarly, Manavalan et al. (2012) have proposed that the secretome of G. lucidum grown on biomass called sugar cane bagasse apparently exposed the ubiquitous lignin-degrading enzymes such as cellulases, hemicellulases, glycoside hydrolases, proteases, and phosphatases along with hydrolytic activity through proteomic analysis and biochemical assays. In the same way, secretome profiling of G. trabeum and P. ostreatus on sugarcane bagasse displayed 119 and 63 different extracellular proteins, which also affiliated to biomass deconstructing enzymes like oxalate decarboxylase, intracellular AA6 quinone oxidoreductases along with cellulases, hemicellulases and auxiliary proteins such as polysaccharide monooxygenase, laccase and versatile peroxidase (Valadares et al. 2019). The investigation was done by Brenelli et al. (2019), marine Agaricomycete Peniophora sp. CBMAI 1063 produces two laccase enzymes to degrade lignin. Similar to the research work done by Kasirajan et al. (2023), H. volcanii katG (hvo_1778) encoding peroxidase was overexpressed and characterized for its activity and treated on sugarcane bagasse. Table 8.1 explains the lignin-degrading enzymes identified through secretome analysis.

8.6 Secretome and Transcriptome Analysis Used for Lignocellulolytic Enzymes

To recognize the novel and potential genes, the secretome analysis of Aspergillus fumigatus in the presence of sugarcane bagasse showed a maximum activity of cellulases (0.032 U mL^{-1}), endo-1,4-β--xylanase (10.82 U mL^{-1}), and endo-1,3-β glucanases (0.77 U mL^{-1}). They also identified important enzymes for biomass modification, such as endoxylanases, xylo-glucanases, β-xylosidases, LPMOs, α-arabino-furanosidases, cello-biohydrolases, endoglucanases, and β-glucosidases through secretome (de Gouvêa et al. 2018). A study was conducted by Jiménez et al. (2018) to understand the temporal expression dynamics of plant biomass-degrading enzymes through meta-transcriptomics of the synthetic bacterial consortium growing on sugarcane bagasse. They observed a significant peak of extracellular β-xylosidase activity during the initial 48 h and a higher expression of several CAZy families at 12–48 h, suggesting easy access to the main plant polysaccharides.

8.7 Enzyme Cocktail Production for Enhanced Sugarcane Bagasse Delignification

The detailed studies and secretome profiling of lignolytic enzymes from the bacterial/ fungal strains on sugarcane bagasse through varying techniques support the formulation of an enzyme cocktail for enriched biomass degradation (Adsul et al. 2020). The comparative research done by Bussamra et al. (2015) using Celluclast 1.5 L commercially available enzyme cocktail and secretome from *T. reesei* consist of cellobio-hydrolases, endoglucanase, and β-glucosidase showed an efficient destruction of biomass with higher yield of glucose from *T. reesei* secretome.

8.8 Conclusion and Future Prospects

To avoid the energy crisis, biofuel production becomes an inevitable source to replace the fossil fuel demand. Lignocellulosic biomass is a preferable and favourable approach because of its availability, as they are abundant in nature. In a matter of industries, varying methods like physical, chemical, and thermochemical

Table 8.1 Lignin degrading enzymes from respective microbes using secretome analysis. Adapted from Sethupathy et al. (2021), this article is licensed under a Creative Commons Attribution 4.0 International License)

SN	Microbes	Ligninolytic enzymes
1	*Doratomyces stemonitis C8*	4-O-methyl-glucuronoylmethylesterase, GMC oxidoreductases and catalases
2	*A. fumigatus AF293*	Catalase-peroxidase (KatG), cellobiose dehydrogenase (CDH), catalase B (CatB), FAD-oxidase, laccase, and cu–Zn superoxide dismutase
3	*A. fumigatus Z5*	Laccase, Mn-peroxidase and lignin peroxidase
4	*Fusarium verticillioides*	Superoxide dismutase
5	*Ceriporiopsis subvermispora*	Mn-peroxidases, laccase, aryl alcohol oxidase
6	*Ganoderma lucidum*	Mn-peroxidases, laccases, cellobiose dehydrogenase, glutathione reductases and acyl-CoA dehydrogenase-like protein
7	*Penicillium chrysogenum P33*	Oxidoreductases
8	*Trichoderma reesei*	Laccase, peroxidase, Cytosolic Cu/Zn superoxide dismutase, Mn superoxide dismutase, bifunctional catalase-peroxidase, aldehyde dehydrogenase, aldehyde oxidase, glutathione reductase, oxidoreductases, etc
9	*Phanerochaete chrysosporium CIRM-BRFM41*	Mn-peroxidases, lignin peroxidases
10	*Schizophyllum commune SH12*	Glucose oxidases, alcohol oxidases, aryl-alcohol oxidase, glucooligosaccharide oxidase, cellobiose dehydrogenase, 1,4-benzoquinone reductase, and ferric reductase
11	*Penicillium echinulatum 2HH*	Isoamyl alcohol oxidases, FAD-dependent oxidoreductase and 6-hydroxy-d-nicotine oxidase
12	*Lentinula edodes*	Laccase 1, laccase 5, laccase 6, laccase 8, laccase 13, glucooligosaccharide oxidases, glucose oxidases, copper radical oxidase, aryl–alcohol oxidases, Mn-peroxidase 2 and pyranose dehydrogenase
13	*Pleurotus eryngii, Irpex Lacteus and Pleurotus ostreatus*	Laccase, laccase 4, laccase 5, glyoxal oxidases, peroxidase, Mn peroxidase, phenol oxidase and diphenol oxidase-A2, GMC oxidoreductases and glyoxylatedehydrogenase
14	*Phanerochaete chrysosporium*	Glyoxal oxidases, peroxidase, Mn peroxidase, phenol oxidase and diphenol oxidase-A2, GMC oxidoreductases and glyoxylatedehydrogenase
15	*Clonostachys byssicola*	Alcohol oxidase and catalase/peroxidase
16	*Aspergillus fumigatus*	Laccase, isoamyl alcohol oxidases, bifunctional catalase-peroxidase, Cu,Zn super-oxide dismutase, Mn superoxide dismutases, glutathione reductase, oxidoreductases, alcohol dehydrogenase, aryl–alcohol oxidase, cytochrome c peroxidase, and FAD-dependent oxygenase

(continued)

Table 8.1 (continued)

SN	Microbes	Ligninolytic enzymes
17	*Chrysoporthe cubensis*	Laccases, catalase-peroxidase, choline dehydrogenase, alcohol oxidase, pyranose dehydrogenase, isoamyl alcohol oxidases, GMC oxidoreductase, and FAD-binding oxidoreductase
18	*Pleurotus ostreatus*	Laccase, versatile peroxidase, glyoxal oxidase, alcohol oxidases, GMC oxidoreductase, oxidoreductase-FAD binding domain
19	*Peniophora* sp. *CBMAI 1063*	Laccases, glyoxal oxidases
20	*Podospora anserina*	Laccases, cellobiose dehydrogenase, GMC oxidoreductase, glyoxal oxidase, FAD-linked oxidases, and AA7 oxidoreductases
21	*Fusarium solani MYA 4552*	Laccase, Nickel superoxide dismutase, catalase, FAD-oxidoreductase, cellobiose dehydrogenase
22	*Irpex lacteus*	Dyp peroxidases, Mn peroxidases, glyoxal oxidases, and cellobiose dehydrogenases
23	*Phanerochaete chrysosporium*	Lignin peroxidases, GMC oxidoreductases, Mn peroxidases, glyoxal oxidases, cellobiose dehydrogenases, and pyranose 2-oxidase
24	*Pleurotus ostreatus*	Versatile peroxidase, Mn peroxidase, laccases, and glyoxal oxidases
25	*Trichoderma guizhouense NJAU4742*	Catalase-peroxidase 2, isoamyl alcohol oxidases, GMC oxidoreductases, multicopper oxidases, Cu/Zn superoxide dismutases, and oxidoreductase
26	*Trametes versicolor*	Dye-decolorizing peroxidase 1 and 2, laccase 2, 3 and 4, lignin peroxidase1, 2, 6, 9 and 12, Mn peroxidase 1, 2, 3, 4, 5, 6, 9, and 9
27	*Stereum hirsutum*	Laccase 5 and Mn peroxidases
28	*Bacillus ligniniphilus L1*	Multicopper oxidases, catalase-peroxidase, superoxide dismutase and glutathione transferase, etc.
29	*Trametes trogii MT*	Laccase 4, laccase TilA, Mn peroxidase, lignin peroxidase, cellobiose dehydrogenases and pyranose 2-oxidase, phenoloxidase, aryl-alcohol oxidase, etc.
30	*Pleurotus ostreatus*	Laccase 2, laccase 6, laccase 9, laccase 10, versatile peroxidase 1 and Mn peroxidase 1
31	*Trametes hirsuta 072*	Lignin peroxidase 9, versatile peroxidase2, Mn peroxidase 2, 3, 4, 5, 6, and 7
32	*Pseudomonas putida A514*	B subfamily dye-decolorizing peroxidase ($N = 2$)
33	*Phanerochaete chrysosporium*	Glyoxal oxidase, aryl alcohol oxidase, alcohol oxidase (cellobiose dehydrogenase $N = 2$), pyranose oxidase

methods were chosen for biomass utilization. However, the result was not up to the standard, and the biomass was not utilized completely due to the presence of structural complexity and recalcitrant nature. Microbes and their enzymes play a pivotal role in biomass depolymerization, an environmentally eco-friendly and feasible perspective. Contemporarily, proteomic analysis for novel ligninolytic enzymes from the microbial consortia especially fungal and bacterial strains. To interpret the enzyme–substrate kinetics, identification and quantification along with post-translational modifications were explored with the upgraded protein extraction protocols and quantitative approaches like absolute quantitative proteomic approaches such as ITRAQ and TMT. These strategies help the researchers to design specific and tailored enzyme cocktails with a prime combination. The recently available proteomic systems are good enough for microbial secretome studies on different biomass and examine the purity of secreted protein, even though still these proteomic analyse necessitate auxiliary improvement.

In addition to that, they detected ligninolytic key enzymes from numerous microbe's requisite to validate their contribution to biomass deconstruction. The detailed studies on the secretome analysis revealed that ligninolytic enzyme complemented with CAZymes in the fungal and bacterial strains resulted in an effective enzyme cocktail production. Moreover, the development of ligninolytic microbial consortia, manipulation of active enzyme-secreting microbes, and comprehensive screening strategies should be accomplished for better biomass pre-treatment. Nevertheless, the potential candidate enzyme should withstand the wide pH, temperature range, and harsh environmental conditions. The above-mentioned properties can be achieved only by stability and enzyme activity with post-translational modification (PTMs). Most of the secretomes identified by the research group lack PTMs in enzymes. In the future, elaborate experimental studies are necessary to produce the targeted enzyme for heterologous production through suitable hosts. Concurrent production of multiple ligninolytic enzymes by engineering the microbes might reduce the enzyme production cost with a synergistic effect to obtain a complete enzyme cocktail.

References

Adav SS, Ravindran A, Sze SK (2012) Quantitative proteomic analysis of lignocellulolytic enzymes by Phanerochaete chrysosporium on different lignocellulosic biomass. J Proteome 75(5):1493–1504. https://doi.org/10.1016/j.jprot.2011.11.020

Adsul M, Sandhu SK, Singhania RR, Gupta R, Puri SK, Mathur A (2020) Designing a cellulolytic enzyme cocktail for the efficient and economical conversion of lignocellulosic biomass to biofuels. Enzym Microb Technol 133:109442. https://doi.org/10.1016/j.enzmictec.2019.109442

Agrawal K, Chaturvedi V, Verma P (2018) Fungal laccase discovered but yet undiscovered. Bioresour Bioprocess 5(1):1–12. https://doi.org/10.1186/s40643-018-0190-z

Amalraj V, Balasundaram N (2006) On the taxonomy of the members of 'Saccharum complex.'. Genet Resour Crop Evol 53(1):35–41

Andrade LM, Peixoto-Junior RF, Ribeiro RV, Nóbile PM, Brito MS, Marchiori PER, Carlin SD, Martins APB, Goldman MHS, Llerena JPP et al (2019) Biomass accumulation and cell wall structure of rice plants overexpressing a dirigent-jacalin of sugarcane (ShDJ) under varying conditions of water availability. Front Plant Sci 10. https://doi.org/10.3389/fpls.2019.00065

Asgher M, Wahab A, Bilal M, Iqbal HMN (2018) Delignification of Lignocellulose Biomasses by Alginate–Chitosan Immobilized Laccase Produced from Trametes versicolor IBL-04. Waste Biomass Valorization 9(11):2071–2079. https://doi.org/10.1007/s12649-017-9991-0

Asgher M, Nasir I, Khalid N, Qamar SA (2020) Development of biocomposites based on bacterial cellulose reinforced delignified rice husk-PVA plasticized with glycerol. J Polym Res 27(11):1–11. https://doi.org/10.1007/s10965-020-02314-y

Aslam B, Basit M, Nisar MA, Khurshid M, Rasool MH (2017) Proteomics: technologies and their applications. J Chromatogr Sci 55(2):182–196. https://doi.org/10.1093/chromsci/bmw167

Bewg WP, Poovaiah C, Lan W, Ralph J, Coleman HD (2016) RNAi downregulation of three key lignin genes in sugarcane improves glucose release without reduction in sugar production. Biotechnol Biofuels 9(1):1–13. https://doi.org/10.1186/s13068-016-0683-y

Bianco L, Perrotta G (2015) Methodologies and perspectives of proteomics applied to filamentous fungi: from sample preparation to secretome analysis. Int J Mol Sci 16(3):5803–5829. https://doi.org/10.3390/ijms16035803

Bilal M, Iqbal HMN (2019) Lignin peroxidase immobilization on Ca-alginate beads and its dye degradation performance in a packed bed reactor system. Biocatal Agric Biotechnol 20:101205. https://doi.org/10.1016/j.bcab.2019.101205

Bilal M, Asgher M, Iqbal HMN, Hu H, Wang W, Zhang X (2017) Bio-catalytic performance and dye-based industrial pollutants degradation potential of agarose-immobilized MnP using a packed bed reactor system. Int J Biol Macromol 102:582–590. https://doi.org/10.1016/j.ijbiomac.2017.04.065

Bilal M, Rasheed T, Iqbal HMN, Yan Y (2018) Peroxidases-assisted removal of environmentally-related hazardous pollutants with reference to the reaction mechanisms of industrial dyes. Sci Total Environ 644:1–13. https://doi.org/10.1016/j.scitotenv.2018.06.274

Bilal M, Qamar SA, Yadav V, Cheng H, Khan M, Adil SF, Taherzadeh MJ, Iqbal HMN (2021) Exploring the potential of ligninolytic armory for lignin valorization—A way forward for sustainable and cleaner production. J Clean Prod 326:129420. https://doi.org/10.1016/j.jclepro.2021.129420

Borin GP, Sanchez CC, De Souza AP, De Santana ES, De Souza AT, Leme AFP, Squina FM, Buckeridge M, Goldman GH, De Castro Oliveira JV (2015) Comparative secretome analysis of Trichoderma reesei and Aspergillus niger during growth on sugarcane biomass. PLoS One 10(6):e0129275. https://doi.org/10.1371/journal.pone.0129275

Borin GP, Sanchez CC, de Santana ES, Zanini GK, dos Santos RAC, de Oliveira PA, de Souza AT, Dal'Mas RMMTS, Riaño-Pachón DM, Goldman GH et al (2017) Comparative transcriptome analysis reveals different strategies for degradation of steam-exploded sugarcane bagasse by Aspergillus niger and Trichoderma reesei. BMC Genomics 18(1):501. https://doi.org/10.1186/s12864-017-3857-5

Bottcher A, Cesarino I, dos Santos AB, Vicentini R, Sampaio Mayer JL, Vanholme R, Morreel K, Goeminne G, Magalhães Silva Moura JC, Nobile PM et al (2013) Lignification in sugarcane: biochemical characterization, gene discovery, and expression analysis in two genotypes contrasting for lignin content. Plant Physiol 163(4):1539–1557. https://doi.org/10.1104/pp.113.225250

Brenelli LB, Persinoti GF, Paulo J, Cairo LF, Lib MV, Gonçalves TA, Otero IVR, Mainardi PH, Felby C, Sette LD et al (2019) Novel redox-active enzymes for ligninolytic applications revealed from multiomics analyses of Peniophora sp. CBMAI 1063, a laccase hyper-producer strain:1–15. https://doi.org/10.1038/s41598-019-53608-1

Brennan L, Owende P (2010) Biofuels from microalgae-A review of technologies for production, processing, and extractions of biofuels and co-products. Renew Sust Energ Rev 14(2):557–577. https://doi.org/10.1016/j.rser.2009.10.009

Bussamra BC, Freitas S, da Costa AC (2015) Improvement on sugar cane bagasse hydrolysis using enzymatic mixture designed cocktail. Bioresour Technol 187:173–181. https://doi.org/10.1016/j.biortech.2015.03.117

Calderan-Rodrigues MJ, Jamet E, Douché T, Bonassi MBR, Cataldi TR, Fonseca JG, San Clemente H, Pont-Lezica R, Labate CA (2016) Cell wall proteome of sugarcane stems: comparison of a destructive and a non-destructive extraction method showed differences in glycoside hydrolases and peroxidases. BMC Plant Biol 16(1):1–17. https://doi.org/10.1186/s12870-015-0677-0

Cannatelli MD, Ragauskas AJ (2017) Two decades of laccases: advancing sustainability in the chemical industry. Chem Rec 17(1):122–140. https://doi.org/10.1002/tcr.201600033

Cavelius P, Engelhart-Straub S, Mehlmer N, Lercher J, Awad D, Brück T (2023) The potential of biofuels from first to fourth generation. PLoS Biol 21(3):e3002063. https://doi.org/10.1371/JOURNAL.PBIO.3002063

Chio C, Sain M, Qin W (2019) Lignin utilization: A review of lignin depolymerization from various aspects. Renew Sust Energ Rev 107:232–249. https://doi.org/10.1016/j.rser.2019.03.008

Chowdhary P, Yadav A, Kaithwas G, Bharagava RN (2017) Distillery wastewater: A major source of environmental pollution and its biological treatment for environmental safety. Green Technol Environ Sustain:409–435. https://doi.org/10.1007/978-3-319-50654-8_18

Chowdhary P, Shukla G, Raj G, Ferreira LFR, Bharagava RN (2019) Microbial manganese peroxidase: a ligninolytic enzyme and its ample opportunities in research. SN Appl Sci 1(1):1–12. https://doi.org/10.1007/s42452-018-0046-3

Coconi Linares N, Fernández F, Loske AM, Gómez-Lim MA (2018) Enhanced delignification of lignocellulosic biomass by recombinant fungus phanerochaete chrysosporium overexpressing laccases and peroxidases. J Mol Microbiol Biotechnol 28(1):1–3. https://doi.org/10.1159/000485976

D'Hont A (2005) Unraveling the genome structure of polyploids using FISH and GISH; examples of sugarcane and banana. Cytogenet Genome Res 109(1–3):27–33. https://doi.org/10.1159/000082378

Daly P, McClellan C, Maluk M, Oakey H, Lapierre C, Waugh R, Stephens J, Marshall D, Barakate A, Tsuji Y et al (2019) RNAi-suppression of barley caffeic acid O-methyltransferase modifies lignin despite redundancy in the gene family. Plant Biotechnol J 17(3):594–607. https://doi.org/10.1111/pbi.13001

Daniels J, Roach BT, Heinz DJ (1987) Sugarcane improvement through breeding. Taxon Evol:7–84

Datta R, Kelkar A, Baraniya D, Molaei A, Moulick A, Meena RS, Formanek P (2017) Enzymatic degradation of lignin in soil: A review. Sustain For 9(7):1163. https://doi.org/10.3390/su9071163

de Gouvêa PF, Bernardi AV, Gerolamo LE, de Souza SE, Riaño-Pachón DM, Uyemura SA, Dinamarco TM (2018) Transcriptome and secretome analysis of aspergillus fumigatus in the presence of sugarcane bagasse. BMC Genomics 19(1):1–18. https://doi.org/10.1186/s12864-018-4627-8

Dong XQ, Yang JS, Zhu N, Wang ET, Yuan HL (2013) Sugarcane bagasse degradation and characterization of three white-rot fungi. Bioresour Technol 131:443–451. https://doi.org/10.1016/j.biortech.2012.12.182

FAOSTAT (2021) Food and Agriculture Organization of the United Nations, Agricultural production statistics 2000–2021. Agric Prod Stat 2000-2021(60):1–17

Fonseca MI, Molina MA, Winnik DL, Busi MV, Fariña JI, Villalba LL, Zapata PD (2018) Isolation of a laccase-coding gene from the lignin-degrading fungus Phlebia brevispora BAFC 633 and heterologous expression in Pichia pastoris. J Appl Microbiol 124(6):1454–1468. https://doi.org/10.1111/jam.13720

Fu C, Mielenz JR, Xiao X, Ge Y, Hamilton CY, Rodriguez M, Chen F, Foston M, Ragauskas A, Bouton J et al (2011) Genetic manipulation of lignin reduces recalcitrance and improves ethanol production from switchgrass. Proc Natl Acad Sci USA 108(9):3803–3808. https://doi.org/10.1073/pnas.1100310108

Gonçalves CC, Bruce T, Silva CDOG, Fillho EXF, Noronha EF, Carlquist M, Parachin NS (2020) Bioprospecting microbial diversity for lignin valorization: dry and wet screening methods. Front Microbiol 11:1081

Hemaprabha G, Mohanraj K, Jackson PA, Lakshmanan P, Ali GS, Li AM, Huang DL, Ram B (2022) Sugarcane genetic diversity and major germplasm collections. Sugar Tech 24(1):279–297. https://doi.org/10.1007/s12355-021-01084-1

Hoang NV, Furtado A, Botha FC, Simmons BA, Henry RJ (2015) Potential for genetic improvement of sugarcane as a source of biomass for biofuels. Front Bioeng Biotechnol 3(NOV):1–15. https://doi.org/10.3389/fbioe.2015.00182

Huang XF, Santhanam N, Badri DV, Hunter WJ, Manter DK, Decker SR, Vivanco JM, Reardon KF (2013) Isolation and characterization of lignin-degrading bacteria from rainforest soils. Biotechnol Bioeng 110(6):1616–1626. https://doi.org/10.1002/bit.24833

Huang S, Huang D, Wu Q, Hou M, Tang X, Zhou J (2020) Effect of environmental C/N ratio on activities of lignin-degrading enzymes produced by Phanerochaete chrysosporium. Pedosphere 30(2):285–292. https://doi.org/10.1016/S1002-0160(17)60391-6

Hyde KD, Xu J, Rapior S, Jeewon R, Lumyong S, Niego AGT, Abeywickrama PD, Aluthmuhandiram JVS, Brahamanage RS, Brooks S et al (2019) The amazing potential of fungi: 50 ways we can exploit fungi industrially. Fungal Divers 97(1):1–136. https://doi.org/10.1007/s13225-019-00430-9

Janusz G, Pawlik A, Sulej J, Świderska-Burek U, Jarosz-Wilkolazka A, Paszczyński A (2017) Lignin degradation: microorganisms, enzymes involved, genomes analysis and evolution. FEMS Microbiol Rev 41(6):941–962. https://doi.org/10.1093/femsre/fux049

Janusz G, Pawlik A, Świderska-Burek U, Polak J, Sulej J, Jarosz-Wilkołazka A, Paszczyński A (2020) Laccase properties, physiological functions, and evolution. Int J Mol Sci 21:3. https://doi.org/10.3390/ijms21030966

Jiménez DJ, De Mares MC, Salles JF (2018) Temporal expression dynamics of plant biomass-degrading enzymes by a synthetic bacterial consortium growing on sugarcane bagasse. Front Microbiol 9(FEB):299. https://doi.org/10.3389/fmicb.2018.00299

Jin W, Li J, Feng H, You S, Zhang L, Norvienyeku J, Hu K, Sun S, Wang Z (2018) Importance of a laccase gene (Lcc1) in the development of Ganoderma tsugae. Int J Mol Sci 19(2):471. https://doi.org/10.3390/ijms19020471

Jung JH, Altpeter F (2016) TALEN mediated targeted mutagenesis of the caffeic acid O-methyltransferase in highly polyploid sugarcane improves cell wall composition for production of bioethanol. Plant Mol Biol 92(1–2):131–142. https://doi.org/10.1007/s11103-016-0499-y

Jung JH, Fouad WM, Vermerris W, Gallo M, Altpeter F (2012) RNAi suppression of lignin biosynthesis in sugarcane reduces recalcitrance for biofuel production from lignocellulosic biomass. Plant Biotechnol J 10(9):1067–1076. https://doi.org/10.1111/j.1467-7652.2012.00734.x

Jung JH, Vermerris W, Gallo M, Fedenko JR, Erickson JE, Altpeter F (2013) RNA interference suppression of lignin biosynthesis increases fermentable sugar yields for biofuel production from field-grown sugarcane. Plant Biotechnol J 11(6):709–716

Jung JH, Kannan B, Dermawan H, Moxley GW, Altpeter F (2016) Precision breeding for RNAi suppression of a major 4-coumarate:coenzyme A ligase gene improves cell wall saccharification from field grown sugarcane. Plant Mol Biol 92(4–5):505–517. https://doi.org/10.1007/s11103-016-0527-y

Kameshwar AKS, Barber R, Qin W (2018) Comparative modeling and molecular docking analysis of white, brown and soft rot fungal laccases using lignin model compounds for understanding the structural and functional properties of laccases. J Mol Graph Model 79:15–26. https://doi.org/10.1016/j.jmgm.2017.10.019

Kasirajan L, Kamaraj K (2023) Optimization of laccase enzyme from Flammulina velutipes using response surface methodology box-behnken design. Biologia (Bratisl) 78:2273. https://doi.org/10.1007/s11756-023-01378-7

Kasirajan L, Aruchamy K, Thirugnanasambandam PP, Athiappan S (2017) Molecular cloning, characterization, and expression analysis of lignin genes from sugarcane genotypes varying in lignin content. Appl Biochem Biotechnol 181(4):1270–1282. https://doi.org/10.1007/s12010-016-2283-5

Kasirajan L, Thirugnanasambandam PP, Furtado A, Botha FC, Henry RJ (2020). Isolation and characterization of full-length phenylalanine ammonium lyase and cinnamyl alcohol dehydrogenase genes involved in lignin biosynthesis of erianthus arundinaceus:174. doi:https://doi.org/10.3390/proceedings2019036174

Kasirajan L, Kamaraj K, Maupin-Furlow JA, Uthandi S (2022) Isolation, purification, and identification of novel lignin-degrading Aspergillus caespitosus strain S2. Biomass Convers Biorefinery. https://doi.org/10.1007/s13399-022-03510-0

Kasirajan L, Kamaraj K, Maupin-Furlow JA (2023) Molecular cloning and characterization of peroxidase from Haloferax volcanii. Sugar Tech. 25(4):881–892. https://doi.org/10.1007/s12355-023-01253-4

Khan R, Bhawana P, Fulekar MH (2013) Microbial decolorization and degradation of synthetic dyes: A review. Rev Environ Sci Biotechnol 12(1):75–97. https://doi.org/10.1007/s11157-012-9287-6

Khatoon N, Jamal A, Ali MI (2019) Lignin peroxidase isoenzyme: a novel approach to biodegrade the toxic synthetic polymer waste. Environ Technol (United Kingdom) 40(11):1366–1375. https://doi.org/10.1080/09593330.2017.1422550

Knoll JE, Anderson WF, Richard EP, Doran-Peterson J, Baldwin B, Hale AL, Viator RP (2013) Harvest date effects on biomass quality and ethanol yield of new energycane (Saccharum hyb.) genotypes in the Southeast USA. Biomass Bioenergy 56:147–156. https://doi.org/10.1016/j.biombioe.2013.04.018

Kohli K, Prajapati R, Sharma BK (2019) Bio-based chemicals from renewable biomass for integrated biorefineries. Energies 12(2):233. https://doi.org/10.3390/en12020233

Kolbusz MA, Di Falco M, Ishmael N, Marqueteau S, Moisan MC, Baptista CDS, Powlowski J, Tsang A (2014) Transcriptome and exoproteome analysis of utilization of plant-derived biomass by Myceliophthora thermophila. Fungal Genet Biol 72:10–20. https://doi.org/10.1016/j.fgb.2014.05.006

Kumar N (eds) (2020) Biotechnology for biofuels: a sustainable green energy solution. Springer Singapore. https://doi.org/10.1007/978-981-15-3761-5;

Li P, Chen Q, Wang TC, Vermeulen NA, Mehdi BL, Dohnalkova A, Browning ND, Shen D, Anderson R, Gómez-Gualdrón DA et al (2018) Hierarchically engineered mesoporous metal-organic frameworks toward cell-free immobilized enzyme systems. Chem 4(5):1022–1034. https://doi.org/10.1016/j.chempr.2018.03.001

Louie GV, Bowman ME, Tu Y, Mouradov A, Spangenberg G, Noel JP (2010) Structure-function analyses of a caffeic acid O-methyltransferase from perennial ryegrass reveal the molecular basis for substrate preference. Plant Cell 22(12):4114–4127. https://doi.org/10.1105/tpc.110.077578

Machado ADS, Ferraz A (2017) Biological pretreatment of sugarcane bagasse with basidiomycetes producing varied patterns of biodegradation. Bioresour Technol 225:17–22. https://doi.org/10.1016/j.biortech.2016.11.053. https://www.sciencedirect.com/science/article/pii/S0960852416315681

Mäkinen MA, Risulainen N, Mattila H, Lundell TK (2018) Transcription of lignocellulose-decomposition associated genes, enzyme activities and production of ethanol upon bioconversion of waste substrate by Phlebia radiata. Appl Microbiol Biotechnol 102(13):5657–5672. https://doi.org/10.1007/s00253-018-9045-y

Mäkinen M, Kuuskeri J, Laine P, Smolander OP, Kovalchuk A, Zeng Z, Asiegbu FO, Paulin L, Auvinen P, Lundell T (2019) Genome description of Phlebia radiata 79 with comparative genomics analysis on lignocellulose decomposition machinery of phlebioid fungi. BMC Genomics 20(1):1–22. https://doi.org/10.1186/s12864-019-5817-8

Manavalan T, Manavalan A, Thangavelu KP, Heese K (2012) Secretome analysis of Ganoderma lucidum cultivated in sugarcane bagasse. J Proteome 77:298–309. https://doi.org/10.1016/j.jprot.2012.09.004

Manimekalai R, Suresh G, Govinda Kurup H, Athiappan S, Kandalam M (2020) Role of NGS and SNP genotyping methods in sugarcane improvement programs. Crit Rev Biotechnol 40(6):865–880. https://doi.org/10.1080/07388551.2020.1765730

Manners J, McIntyre CL, Casu RE, Cordeiro GM, Jackson M, Aitken KS, Jackson PA, Bonnett GD, Lee LS, Henry RJ (2004) Can genomics revolutionize genetics and breeding in sugarcane. In: Proceedings of the 4th international crop science congress Brisbane Australia. Vol. 26. p. 1–4. http://www.cropscience.org.au/icsc2004/poster/3/1/1793_mannersj.htm

Marx IJ, Van Wyk N, Smit S, Jacobson D, Viljoen-Bloom M, Volschenk H (2013) Comparative secretome analysis of Trichoderma asperellum S4F8 and Trichoderma reesei Rut C30 during solid-state fermentation on sugarcane bagasse. Biotechnol Biofuels 6(1):172. https://doi.org/10.1186/1754-6834-6-172

Mata TM, Martins AA, Caetano NS (2010) Microalgae for biodiesel production and other applications: A review. Renew Sust Energ Rev 14(1):217–232. https://doi.org/10.1016/j.rser.2009.07.020

Miyauchi S, Rancon A, Drula E, Hage H, Chaduli D, Favel A, Grisel S, Henrissat B, Herpoël-Gimbert I, Ruiz-Dueñas FJ et al (2018) Integrative visual omics of the white-rot fungus

Polyporus brumalis exposes the biotechnological potential of its oxidative enzymes for delignifying raw plant biomass. Biotechnol Biofuels 11(1):1–14. https://doi.org/10.1186/s13068-018-1198-5

Nigam PS, Singh A (2011) Production of liquid biofuels from renewable resources. Prog Energy Combust Sci 37(1):52–68. https://doi.org/10.1016/j.pecs.2010.01.003

Nikel PI, de Lorenzo V (2018) Pseudomonas putida as a functional chassis for industrial biocatalysis: from native biochemistry to trans-metabolism. Metab Eng 50:142–155. https://doi.org/10.1016/j.ymben.2018.05.005

Ostengo S, Rueda Calderón MA, Bruno C, Cuenya MI, Balzarini M (2021) A protocol for identifying characteristic sucrose accumulation curves of sugarcane genotypes (Saccharum spp.). Sugar Tech 23(3):519–523. https://doi.org/10.1007/s12355-020-00926-8

Pecková V, Legerská B, Chmelová D, Horník M, Ondrejovič M (2020) Comparison of efficiency for monoazo dye removal by different species of white-rot fungi. Int J Environ Sci Technol 18:21–32. https://doi.org/10.1007/s13762-020-02806-w

Pham LTM, Seo H, Kim KJ, Kim YH (2018) In silico-designed lignin peroxidase from Phanerochaete chrysosporium shows enhanced acid stability for depolymerization of lignin. Biotechnol Biofuels 11(1):1–13. https://doi.org/10.1186/s13068-018-1324-4

Phwan CK, Ong HC, Chen WH, Ling TC, Ng EP, Show PL (2018) Overview: comparison of pretreatment technologies and fermentation processes of bioethanol from microalgae. Energy Convers Manag 173:81–94. https://doi.org/10.1016/j.enconman.2018.07.054. https://www.sciencedirect.com/science/article/pii/S0196890418307891

Piperidis N, D'Hont A (2020) Sugarcane genome architecture decrypted with chromosome-specific oligo probes. Plant J 103(6):2039–2051. https://doi.org/10.1111/tpj.14881

Qin X, Su X, Luo H, Ma R, Yao B, Ma F (2018) Deciphering lignocellulose deconstruction by the white rot fungus Irpex lacteus based on genomic and transcriptomic analyses. Biotechnol Biofuels 11(1):58. https://doi.org/10.1186/s13068-018-1060-9

Raveendran S, Parameswaran B, Ummalyma SB, Abraham A, Mathew AK, Madhavan A, Rebello S, Pandey A (2018) Applications of microbial enzymes in food industry. Food Technol Biotechnol 56(1):16–30. https://doi.org/10.17113/ftb.56.01.18.5491

Saleem R, Khurshid M, Ahmed S (2018) Laccases, manganese peroxidases and xylanases used for the bio-bleaching of paper pulp; an environmental friendly approach. Protein Pept Lett 25(2):180–186. https://doi.org/10.2174/0929866525666180122100133

Salem MZM, Mansour MMA, Elansary HO (2019) Evaluation of the effect of inner and outer bark extracts of sugar maple (Acer saccharum var. saccharum) in combination with citric acid against the growth of three common molds. J Wood Chem Technol 39(2):136–147. https://doi.org/10.1080/02773813.2018.1547763

Salvachúa D, Katahira R, Cleveland NS, Khanna P, Resch MG, Black BA, Purvine SO, Zink EM, Prieto A, Martínez MJ et al (2016) Lignin depolymerization by fungal secretomes and a microbial sink. Green Chem 18(22):6046–6062. https://doi.org/10.1039/c6gc01531j

Saqib A, Tabbssum MR, Rashid U, Ibrahim M, Gill SS, Mehmood MA (2013) Marine macro algae ulva: A potential feed-stock for bioethanol and biogas production. Asian J Agric Biol 1(3):155–163

Saykhedkar S, Ray A, Ayoubi-Canaan P, Hartson SD, Prade R, Mort AJ (2012) A time course analysis of the extracellular proteome of Aspergillus nidulans growing on sorghum stover. Biotechnol Biofuels 5(1):52. https://doi.org/10.1186/1754-6834-5-52

Schneider EJ, Lannin NA, Ada L, Schmidt J (2016) Increasing the amount of usual rehabilitation improves activity after stroke: a systematic review. J Physiother 62(4):182–187. https://doi.org/10.1016/j.jphys.2016.08.006. Epub 2016 Aug 24. PMID: 27637769

Sethupathy S, Morales GM, Li Y, Wang Y, Jiang J, Sun J, Zhu D (2021) Harnessing microbial wealth for lignocellulose biomass valorization through secretomics: a review. Biotechnol Biofuels 14(1):1–31. https://doi.org/10.1186/s13068-021-02006-9

Sharma HK, Xu C, Qin W (2019) Biological pretreatment of lignocellulosic biomass for biofuels and bioproducts: an overview. Waste Biomass Valorization 10(2):235–251. https://doi.org/10.1007/s12649-017-0059-y

Singh AK, Bilal M, Iqbal HMN, Meyer AS, Raj A (2021) Bioremediation of lignin derivatives and phenolics in wastewater with lignin modifying enzymes: status, opportunities and challenges. Sci Total Environ 777:145988. https://doi.org/10.1016/j.scitotenv.2021.145988. https://www.sciencedirect.com/science/article/pii/S004896972101055X

Spence EM, Scott HT, Dumond L, Calvo-Bado L, di Monaco S, Williamson JJ, Persinoti GF, Squina FM, Bugg TDH (2020) The hydroxyquinol degradation pathway in rhodococcus jostii RHA1 and agrobacterium species is an alternative pathway for degradation of protocatechuic acid and lignin fragments. Appl Environ Microbiol 86(19):e01561–e01520. https://doi.org/10.1128/AEM.01561-20

Spence EM, Calvo-Bado L, Mines P, Bugg TDH (2021) Metabolic engineering of Rhodococcus jostii RHA1 for production of pyridine-dicarboxylic acids from lignin. Microb Cell Factories 20(1):1–12. https://doi.org/10.1186/s12934-020-01504-z

Sticklen M (2006) Plant genetic engineering to improve biomass characteristics for biofuels. Curr Opin Biotechnol 17(3):315–319. https://doi.org/10.1016/j.copbio.2006.05.003

Su Y, Yu X, Sun Y, Wang G, Chen H, Chen G (2018) Evaluation of screened lignin-degrading fungi for the biological pretreatment of corn stover. Sci Rep 8(1):5385. https://doi.org/10.1038/s41598-018-23626-6

Sun Z, Bottari G, Afanasenko A, Stuart MCA, Deuss PJ, Fridrich B, Barta K (2018) Complete lignocellulose conversion with integrated catalyst recycling yielding valuable aromatics and fuels. Nat Catal 1(1):82–92. https://doi.org/10.1038/s41929-017-0007-z

Tai PYP, Miller JD (2001) A Core collection for Saccharum spontaneum L. from the world collection of sugarcane. Crop Sci 41:879. https://doi.org/10.2135/cropsci2001.413879x

Todd J, Wang J, Glaz B, Sood S, Ayala-Silva T, Nayak SN, Glynn NC, Gutierrez OA, Kuhn DN, Tahir M et al (2014) Phenotypic characterization of the Miami World Collection of sugarcane (Saccharum spp.) and related grasses for selecting a representative core. Genet Resour Crop Evol 61(8):1581–1596. https://doi.org/10.1007/s10722-014-0132-3

Um BH, Kim YS (2009) Review: A chance for Korea to advance algal-biodiesel technology. J Ind Eng Chem 15(1):1–7. https://doi.org/10.1016/j.jiec.2008.08.002

Valadares F, Gonçalves TA, Damasio A, Milagres AM, Squina FM, Segato F, Ferraz A (2019) The secretome of two representative lignocellulose-decay basidiomycetes growing on sugarcane bagasse solid-state cultures. Enzym Microb Technol 130:109370. https://doi.org/10.1016/j.enzmictec.2019.109370

Verheye W (2013) Growth and production of sugarcane. Encycl Life Support Syst 2(3):1–10

Whetten R, Sederoff R (1995) Lignin biosynthesis. Plant Cell 7(7):1001

Willis JD, Mazarei M, Stewart CN (2016) Transgenic plant-produced hydrolytic enzymes and the potential of insect gut-derived hydrolases for biofuels. Front. Plant Sci 7(may 2016):1–18. https://doi.org/10.3389/fpls.2016.00675

Xie P, Fan L, Huang L, Zhang C (2020) An innovative co-fungal treatment to poplar bark sawdust for delignification and polyphenol enrichment. Ind Crop Prod 157:112896. https://doi.org/10.1016/j.indcrop.2020.112896

Yang L, Yuan H, Yang Y, Wang R, Wang C, Wei X, Chen S, Yu J, Ma X (2020) Enhanced lignin degradation in tobacco stalk composting with inoculation of white-rot fungi Trametes hirsuta and Pleurotus ostreatus. Waste Biomass Valorization 11:3525–3535

Yesilada O, Birhanli E, Geckil H (2018) Bioremediation and decolorization of textile dyes by white rot fungi and laccase enzymes. Mycoremediation Environ Sustain 2:121–153. https://doi.org/10.1007/978-3-319-77386-5_5

Youn HD, Hah YC, Kang SO (1995) Role of laccase in lignin degradation by white-rot fungi. FEMS Microbiol Lett 132(3):183–188. https://doi.org/10.1016/0378-1097(95)00315-V

Yu F, Wang P, Li X, Huang Y, Wang Q, Luo L, Jing Y, Liu X, Deng Z, Wu J et al (2018) Characterization of chromosome composition of sugarcane in nobilization by using genomic in situ hybridization. Mol Cytogenet 11(1):1–8. https://doi.org/10.1186/s13039-018-0387-z

Zhuo R, Fan F (2021) A comprehensive insight into the application of white rot fungi and their lignocellulolytic enzymes in the removal of organic pollutants. Sci Total Environ 778:146132. https://doi.org/10.1016/j.scitotenv.2021.146132

9 Transgenic Sugarcane–Oilcane: An Alternative Feedstock for the Production of Drop-in Fuel and Value-Added Bioproducts

Shraddha Maitra and Vijay Singh

Abstract

The utilization of plants and other agricultural produce can partly offset petroleum dependency for energy requirements and can potentially provide sustainable solutions to global environmental problems. To this end, synthetic biology has shown great potential in developing transgenic bioenergy grasses such as sugarcane, sorghum, miscanthus, and energy cane that hyperaccumulate energy-rich lipid molecules in their vegetative tissues, such as leaves, stems, and roots. These perennial high-biomass transgenic C4 grasses are not targeted to grow on prime agricultural land and can be dedicatedly used to produce biofuels and other value-added bioproducts. Recently, sugarcane has been metabolically engineered to sequestrate carbon from juice towards biosynthesis of lipid molecules in the vegetative tissues. Transgenic sugarcane is referred to as "oilcane." Transgenic sugarcane–oilcane has improved energy density due to an elevated lipid content in the vegetative tissues. Transgenic bioenergy crops cater to both cellulosic sugars and vegetative lipids, hence resulting in higher biofuel yield (biodiesel/renewable diesel and bioethanol) per unit area of cultivable land. In this book chapter, the genetic engineering of sugarcane and its bioprocessing are discussed to illustrate its development and use as an alternative feedstock for the production of biofuels (bio-jet fuel, biodiesel, and bioethanol) and value-added bioproducts.

The original version of the chapter has been revised. A correction to this chapter can be found at https://doi.org/10.1007/978-981-97-7228-5_19

S. Maitra · V. Singh (✉)
Department of Agricultural and Biological Engineering, University of Illinois Urbana-Champaign, Urbana, IL, USA
e-mail: vsingh@illinois.edu

G. S. Suresha et al. (eds.), *Value Addition and Product Diversification in Sugarcane*, https://doi.org/10.1007/978-981-97-7228-5_9

Keywords

Sugarcane · Transgenic oilcane · Bioprocessing · Drop-in fuel · Value-added bioproducts

9.1 Introduction

There has been a shift in renewable energy use from fossil fuels to a legitimate green alternative in the last two decades because of deteriorating climate conditions and the ongoing energy crisis. According to total final energy consumption (TEFC) in 2023, 78.9% of total energy usage was derived from fossil fuels followed by 12.6% of modern renewable energy that includes hydro, ocean, solar, and wind power and biomass, and 8.5% of nuclear energy and traditional biomass in 2021 (REN21 2023). The usage of fossil fuels has declined by approximately 3% in the last decade (from 2011 to 2021) (REN21 2023). However, petroleum-based liquid fuels still play a central role in transportation worldwide. Blending requirements and support from the government by providing financial incentives have boosted biofuel usage in various countries including India. Approximately 80% of the global biofuel expansion is created by the USA, Brazil, Indonesia, and India. The worldwide demand for ethanol increased by 3%, i.e., 3100 million liters during 2021–2022 of which India accounted for more than one-third of the demand (International Energy Agency 2023).

The demand and production of biofuel in India are predicted to increase by 70% over 2022–2027 (International Energy Agency 2023). This is primarily due to Inda's goal of achieving 20% ethanol blending by 2025 and 5% biodiesel blending by 2030 (The Gazette of India 2023). For a long time, India maintained relatively low prices of ethanol compared to other parts of the globe by making ethanol primarily from excess and leftover sugars and molasses. However, in 2022 the average sugar price increased by 50% (International Energy Agency 2023). Therefore, achieving the blending targets for ethanol and biodiesel by 2030 will require an appropriate stock of feedstocks along with various credit programs and financial incentives from the government.

To meet liquid biofuel needs for future transportation sustainably, alternative biomass feedstocks are needed to produce a similar amount of energy as compared to gasoline. Research investigations are still ongoing to add new non-food feedstocks that can be grown in non-prime agricultural land to avert the food–fuel controversy. To this end, an emerging genetic engineering strategy to enhance the biosynthesis and hyperaccumulation of high-energy lipid molecules, primarily in the form of triacylglycerides (TAG), in the vegetative tissues of highly productive C4 perennial grasses, such as sorghum, sugarcane, and energycane has provided a platform to create noble lignocellulosic feedstock that does not compete for human food and agricultural land (Cao et al. 2023; Luo et al. 2022; Parajuli et al. 2020; Vanhercke et al. 2019; Zale et al. 2016). Recent studies by the Department of Energy Center for Advanced Biofuel and Bioproducts Innovation (DOE-CABBI), USA, have reported the development of transgenic sugarcane genotype that has achieved TAG hyperaccumulation up to 4.3% of dry biomass in the stems and 8.0% of dry biomass in the leaves (Parajuli et al. 2020). The transgenic sugarcane is also referred

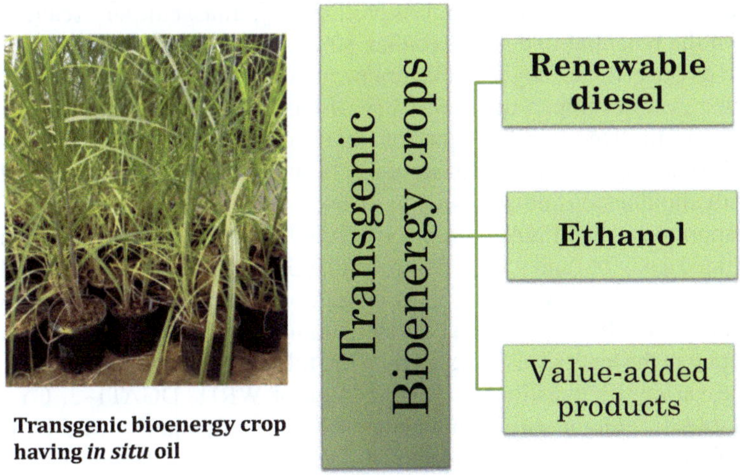

Fig. 9.1 Noble transgenic sugarcane–oilcane caters to vegetative lipids and cellulosic sugars for the production of biofuels and value-added bioproducts

to as "oilcane." The transgenic sugarcane–oilcane is an excellent source for sustainable co-production of cellulosic sugar and vegetative lipids which can be further converted to a wide variety of biochemicals (Fig. 9.1).

Production of biofuels and value-added bioproducts from perennial crops will improve the energy security of the country by reducing the dependency on fossil fuels and increasing carbon sequestration to achieve carbon neutrality in the near future. This chapter is on metabolic engineering performed to produce the transgenic sugarcane–oilcane variety, bioprocessing to co-produce vegetative lipids and cellulosic sugars, and the techno-economic feasibility of the biorefinery design when transgenic sugarcane–oilcane is used as feedstock to produce biofuels and value-added bioproducts.

9.2 Genetic Modification of Sugarcane to Improve Energy Density

Most plants have biosynthesis and storage capacity for TAG molecules but generally do not hyperaccumulate considerable amounts of TAG molecules in their vegetative tissues (Xu and Shanklin 2016). Neutral lipid droplets of TAG molecules have almost twice the energy density as compared to proteins and carbohydrates of equivalent masses. The advancements in genetic engineering provided a platform to develop high-energy alternative feedstocks for biofuel and bioproducts production which can be grown in non-prime agricultural lands. The perennial high-biomass bioenergy crops such as sugarcane, energycane, and sorghum are being metabolically modified to sequestrate carbon from juice and allocate it towards biosynthesis and hyperaccumulation of TAG molecules in vegetative tissues (Cao et al. 2023; Luo et al. 2022; Parajuli et al. 2020; Vanhercke et al. 2019; Zale et al. 2016). The technology has provided a platform to improve the energy density of sugarcane crops. Furthermore, due to high biomass productivity, transgenic sugarcane–oilcane has shown the potential to date to produce 80% of oil yield per hectare of agricultural land as oilseed, i.e., soybean (Maitra et al. 2022b).

The first ever developed transgenic line of metabolically engineered sugarcane included constitutive co-expression of three genes, i.e., WRI1 (Wrinkled1), DGAT1-2 (Diacylglycerol acyltransferase1-2), and CYSOLE1 (Cysteine-oleosin) along with simultaneous co-suppression of two genes, i.e., AGPase (ADP-glucose pyrophosphorylase) and PXAI (peroxisomal ABC transporter1). The transgenic line demonstrated a 43-fold and 95-fold improved TAG accumulation in stem and leaves, respectively, as compared to their corresponding non-transgenic sugarcane variety (Zale et al. 2016). In the following study, Parajuli et al. (2020), reported that the TAG content in the leaves was elevated by >400 fold compared to the non-transgenic sugarcane variety by constitutive co-expression of WRI1, DGAT1–2, CYSOLE1 (cysteine-oleosin) along with simultaneous RNAi-suppression of SDP1 (SUGAR DEPENDENT1) and TGD1 (trigalactosyl diacylglycerol1) (Parajuli et al. 2020). The second genetic modification elevated the total TAG and fatty acid contents to an average of 8% and 13% of dry weight, respectively. An increase in polyunsaturated fatty acids, especially oleic acid (C18:1) was also observed at the expense of palmitic (C16:0) and stearic acid (C18:0) (Parajuli et al. 2020). The functionalities of genes and their expression/suppression for hyperaccumulation of TAG molecules in the vegetative tissues of transgenic sugarcane–oilcane are presented in Table 9.1.

9.3 Biomass Yield, Structural Composition, and Lipid Content of Transgenic Sugarcane

The genetic engineering of sugarcane plants to hyperaccumulate vegetative lipids had a significant metabolic burden, resulting in approximately 50% lower biomass yield, and juice content in transgenic sugarcane–oilcane as compared to non-transgenic lines (Kannan et al. 2022; Parajuli et al. 2020). The extractives analysis of the leaves of transgenic sugarcane–oilcane showed higher extractiveness in the leaves due to hyper-accumulation. The percentages of structural carbohydrates (glucan and xylan) of transgenic sugarcane–oilcane were comparable to those of non-transgenic sugarcane. The total lignin content including acid-soluble and insoluble lignin was similar to its non-transgenic variety. Transgenic sugarcane–oilcane had a higher mineral index in their leaves as compared to bagasse, similar to non-transgenic sugarcane, which was estimated by analyzing ash content (Maitra et al. 2022b).

The percentage of total fatty acids and TAG along with their compositions are critical for the yield and quality of biodiesel/renewable diesel. The comparison of total fatty acid and TAG fatty acid compositions of bagasse, leaves, and juice of transgenic sugarcane–oilcane grown under field conditions and controlled greenhouse conditions showed a distinct variation. Palmitic (C16:0) and linoleic (C18:2) acids were predominant in all the vegetative tissues of transgenic sugarcane–oilcane grown under field conditions followed by linolenic acid (C18:3). Unlike field-grown plants, the leaves of greenhouse-grown plants were rich in linolenic acid (C18:3),

Table 9.1 Genes involved and their functions in the metabolic engineering of sugarcane to develop transgenic sugarcane–oilcane

Genes	Function	Expression/ suppression of gene for TAG accumulation in vegetative tissues	References
WRI1	• Transcriptional factor associated with oil accumulation in maturing seeds • Involved in fatty acid synthesis	Expression	Baud et al. (2007)
DGAT1-2	• Membrane-bound enzymes catalyze diacylglyceride (DAG) esterification to triacylglyceride (TAG)	Expression	Chapman and Ohlrogge (2012)
OLE1	• Oil body proteins help in stabilizing and regulating oil body size • Boost TAG accumulation in vegetative tissues	Expression	Siloto et al. (2006)
CYS-OLE1	• Engineered oleosins containing multiple cysteine residues • Co-expression of DGAT1 and CYS-OLE1 improves TAG accumulation, biomass, and photosynthetic rate	Expression	Winichayakul et al. (2013)
SDP1	• Predominant TAG lipase in vegetative tissues • Limits TAG accumulation	Suppression	Kelly et al. (2013)
TGD1	• ATP binding cassette transporter • Imports phospholipids into plastids that reduce lipid availability for photosynthetic membrane synthesis • Reduces TAG accumulation	Suppression	Xu et al. (2005)
AGPase	• ADP-glucosepyrophosphorylase enzyme mediates carbon flux to fatty acid synthesis • Involved in starch biosynthesis • Reduces TAG accumulation	Suppression	Sanjaya et al. (2011)
PXA1	• Peroxisoml ATP-binding cassette transporter • Part of fatty acid transport cycle and its β-oxidation • TAG mobilization during germination	Suppression	Zolman et al. (2001)

followed by linoleic acid (C18:2) and palmitic acid (C16:0), while bagasse was rich in linoleic acid (C18:2), followed by almost equal amounts of palmitic (C16:0) and oleic acids (C18:1). The fatty acid composition of juice was like that of bagasse in both cases (Maitra et al. 2022b; Parajuli et al. 2020). The transgenic plants grown under the field conditions of Florida and under controlled greenhouse conditions displayed a similar total fatty acid and TAG fatty acid composition. However, the total fatty acid and TAG fatty acid contents and compositions are expected to vary based on the timeline of planting, harvesting, and storage conditions (Kannan et al. 2022; Maitra et al. 2022b; Parajuli et al. 2020). The optimized crop management program and minimal storage periods can potentially prevent the degeneration of lipids and maintain the fatty acid and TAG yields till processing.

9.4 Bioprocessing Transgenic Sugarcane to Recover Cellulosic Sugar and Vegetative Lipids

The transgenic crops require low severity and chemical-free processing to avert the degeneration of vegetative lipids and maximize the recovery of both lipids and cellulosic sugars. This section will discuss the optimized bioprocessing conditions for transgenic sugarcane-oilcane to recover vegetative lipids and cellulosic sugars (Fig. 9.2). Transgenic sugarcane–oilcane plants were harvested at maturity. The leaves and stems were segregated followed by the juicing of stems. The biomass was stored at −20 °C or below until processed further to minimize the degeneration of vegetative lipids. Leaves, bagasse, and juice were processed and analyzed separately to get a deeper insight into the distribution of vegetative lipids among different vegetative tissues. The vegetative lipids in transgenic sugarcane–oilcane are present in lipid–protein complexes at considerably lower concentrations, thus making the lipid extraction process difficult and less efficient as compared to the lipid extraction from oilseeds (Maitra et al. 2021; Parajuli et al. 2020). Therefore, preprocessing and pretreatment are mandatory to maximize the recovery of vegetative lipids and cellulosic sugars.

Chemical-free hydrothermal pretreatment was performed on the bagasse and leaves of transgenic sugarcane–oilcane. The lignocellulosic biomass was pretreated with hot water at 20% solid loading in a fluidized sand bath. Hydrothermal pretreatment was carried out at 180 °C for 10 min for both bagasse and leaves. The pretreatment reaction was quenched in cold water immediately after pretreatment time. After pretreatment, pretreatment liquor was separated from the biomass residue by using a vacuum filter, and pretreated biomass residue was disk milled without neutralization or pH adjustment followed by enzyme hydrolysis to produce fermentable sugars. The pretreatment liquor, pretreated biomass residues, and hydrolysate

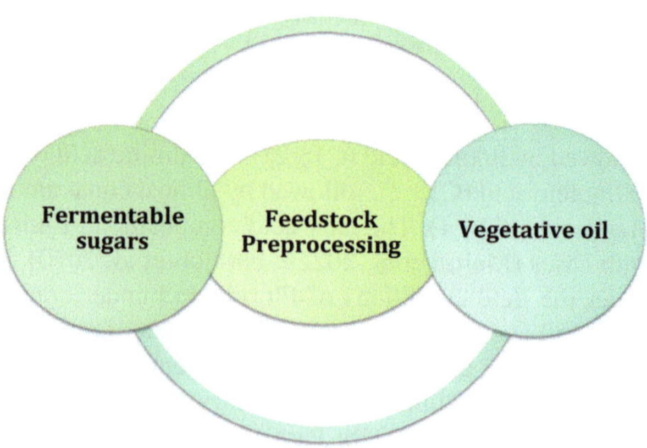

Fig. 9.2 Chemical-free hydrothermal pretreatment prevents the degeneration of vegetative lipids and improves the recovery of both vegetative lipids and cellulosic fermentable sugars

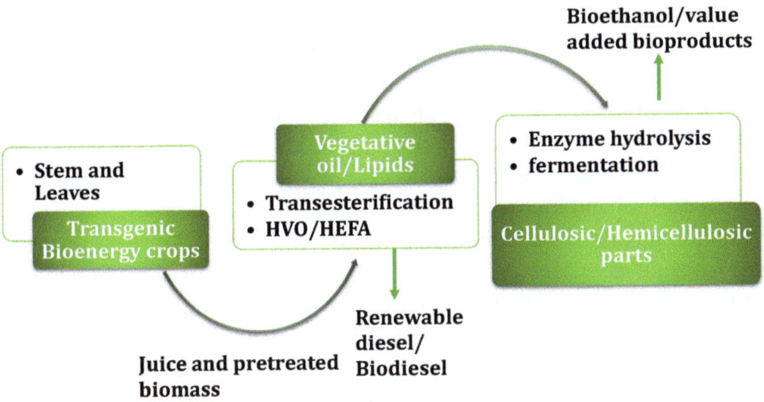

Fig. 9.3 Bioprocessing of transgenic sugarcane–oilcane to co-produce vegetative lipids and cellulosic sugars that can further be processed to produce renewable/bio-diesel, bioethanol, and value-added bioproducts

generated after enzyme hydrolysis were examined for sugars, total lipid, total fatty acid, and TAG compositions (Maitra et al. 2022b). Figure 9.3 presents a schematic of the processing and recovery of products during the processing of transgenic sugarcane–oilcane.

The optimized enzymatic hydrolysis protocols for the recovery of cellulosic sugars from lignocellulosic biomass yielded ≥80% of total sugar (primarily glucose and xylose) upon HMR pretreatment (hydrothermal pretreatment and mechanical refining) (Maitra et al. 2022a). Additionally, chemical-free and low-severity hydrothermal pretreatment prevented the degeneration of vegetative lipids and maximized their recovery. No significant change in the composition of total fatty acids and TAG fatty acids was observed after pretreatment and enzyme hydrolysis. At the first lab-scale attempt, only 55% of the total lipids present in the vegetative tissues were recovered upon processing the whole plant. Hydrothermal pretreatment and saccharification, 25–30% of total lipids were recovered in the pretreatment liquor and hydrolysate. Interestingly, the major fraction of vegetative lipids remained with the biomass residues after each step of bioprocessing. This makes the co-product recovery process convenient and allows the recovery of vegetative lipids at the end of bioprocessing from enriched biomass (Maitra et al. 2021, 2022a, b). Continuous advancements and improvements are being made to the bioprocessing methods to improve lipid recovery efficiency.

9.5 Economic Evaluation of Biofuel from Transgenic Oilcane

Ethanol is the widely used commercial biofuel in internal combustion engines globally (REN21 2020). In India, the main source of bioethanol is sugarcane juice. The other two transportation biofuels are renewable diesel and biodiesel which are produced from oils, fats, and grease. Presently, vegetable oil is the main feedstock for biodiesel and renewable diesel production (U.S. EIA 2023). However, since the feedstocks for the bioethanol, and renewable/bio-diesel are also human and animal food and are limited by arable land availability. Due to this, a deficit between demand and supply is expected, which will likely increase the prices of the commodities, especially food. To avert this situation, the transgenic sugarcane–oilcane has been developed to provide a dedicated supply of vegetative lipids and cellulosic sugars to the biofuel and biochemical industries.

The technological and economical feasibility and competitiveness of transgenic sugarcane–oilcane have been performed for the production of biofuels and bioproducts (Bhagwat et al. 2021; Cortés-Peña et al. 2022; Huang et al. 2016; Li et al. 2021). Huang et al. (2016) reported that the cost of production of biodiesel from transgenic sugarcane–oilcane with varying levels of vegetative lipids (2–20% of dry biomass weight) would be 0.59–0.85 USD/L if ethanol is also coproduced in the biorefinery which is 20–50% less than the production cost of biodiesel from soybean (Huang et al. 2016). Likewise, the cost of production of sustainable aviation fuel (SAF) would vary from 0.73 to 1.79 USD/L if the lipid accumulation in transgenic sugarcane–oilcane varies from 5% to 20% of dry biomass weight (Kumar et al. 2018). The lipid content of transgenic sugarcane–oilcane and its pricing would play a critical role in the economics of the biorefinery (Cortés-Peña et al. 2022). Additionally, the production of platform chemicals such as lactic acid and 3-hydroxy propionic acid (3-HPA) could be economically, environmentally and sustainably produced from the cellulosic sugars obtained from transgenic sugarcane–oilcane in the biorefinery along with renewable and bio-diesel (Bhagwat et al. 2021; Li et al. 2021).

9.6 Future Perspectives

Certain uncertainties in the case of alternative cellulosic feedstock such as the suitability of non-prime agricultural land for the growth of bioenergy crops, advancements in biotechnology and bioprocess engineering, and regulatory laws make the feasibility of cellulosic biorefinery challenging. The optimum parameters for utilizing alternative feedstocks need to be developed simultaneously and in a complementary fashion to achieve an end-to-end solution. To this end, extensive field testing of transgenic sugarcane–oilcane has already been initiated to study the impact of agronomic conditions in favorable warm weather and abbreviated climate conditions (Kannan et al. 2022; Maitra et al. 2022b). Additionally, preprocessing and chemical-free bioprocessing of transgenic bioenergy crops having in-situ vegetative lipids are being optimized at the lab as well as industrially relevant scale for sustainable co-production and to ensure maximum recovery of vegetative lipids and cellulosic sugars for further biochemical conversion into biofuels and value-added bioproducts (*pilot-scale study under review*) (Maitra et al. 2022b).

Although metabolic engineering provides a technology to hyperaccumulate lipids in vegetative tissues to enhance the energy densification of transgenic crops, these engineered crops undergo a physiological futile cycle of TAG degradation after its synthesis phase (Vanhercke et al. 2017). The genetic engineering phase to develop transgenic crops produces a plethora of samples consisting of different tissues (stems, leaves, and roots) from various genotypes, plant maturities, and agronomic conditions. Traditional wet chemistry methods are tedious and time-intensive. Therefore, it is critical to develop a high-throughput phenotyping method for reducing the shelf-life and continue optimizing the processes in real-time. Recently developed Time-Domain NMR spectroscopy method provides a rapid, accurate, convenient and real-time analytical platform for initial screening, characterizing, and quantifying lipids in the vegetative tissues of transgenic crops with minimal sample size at the genetic engineering stage and each bioprocessing step (Maitra et al. 2021, 2022a; Maitra and Singh 2023). Transgenic plants demand regular monitoring of vegetative lipids to decide the right time to harvest. Time Domain NMR spectroscopy can be used to evaluate physiological lipid dynamics in the vegetative tissues of transgenic plants at different developmental stages and help in deciding the ideal growth stage for harvesting transgenic crops (Maitra and Singh 2023).

Sustainable production of biofuels and value-added products from transgenic bioenergy crops can be achieved if scientific and technological advancements and economic considerations begin in every aspect of the proposed advanced cellulosic biorefinery that includes but is not limited to scale-up challenges, supply chain, and logistics, biofuel economy, nonscientific environmental issues (Balan 2014; Kargbo et al. 2021; Lynd et al. 2022). The field trials (Kannan et al. 2022) and industrially relevant bioprocessing of transgenic crops provides proof of concept for the cost-competitive and environmentally sustainable production process of liquid biofuel and value-added bioproduct precursors (Maitra et al. 2024).

9.7 Conclusion

Production of bioenergy and bioproducts from lignocellulosic materials has both environmental and economic benefits in the long run. Next-generation biofuels will help achieve the carbon-neutral goal by sequestrating environmental carbon and decreasing the use of fossil fuels. The development of transgenic sugarcane–oilcane has shown the potential of becoming a viable feedstock for the commercial production of renewable/biodiesel by hyperaccumulating vegetative lipids. The very first lab-scale bioprocessing study of field-grown transgenic sugarcane–oilcane showed its potential to produce 0.35 metric tons/hectare of vegetative lipids which accounts for 80% of the oil yield of soybean. Chemical-free hydrothermal pretreatment at 180 °C for 10 minutes of transgenic sugarcane–oilcane was established to prevent the thermal degeneration of vegetative lipids during pretreatment and improve recovery of vegetative lipids and cellulosic sugars. During bioprocessing, a major fraction of lipids remained in the biomass residues that allowed its recovery at the end of bioprocessing, also called as backend recovery of oil. Further advancements in bioprocessing technologies, genetic engineering platforms, and crop storage management are expected to improve the hyperaccumulation of lipids and their recovery.

Acknowledgements This work was funded by the DOE Center for Advanced Bioenergy and Bioproducts Innovation (U.S. Department of Energy, Office of Science, Biological and Environmental Research Program under Award Number DE-SC0018420). Any opinions, findings, and conclusions or recommendations expressed in this publication are those of the author(s) and do not necessarily reflect the views of the U.S. Department of Energy or the U.S. Department of Agriculture.

References

Balan V (2014) Current challenges in commercially producing biofuels from lignocellulosic biomass. ISRN Biotechnol 2014:1–31

Baud S, Mendoza MS, To A, Harscoët E, Lepiniec L, Dubreucq B (2007) WRINKLED1 specifies the regulatory action of LEAFY COTYLEDON2 towards fatty acid metabolism during seed maturation in Arabidopsis. Plant J 50:825–838. https://doi.org/10.1111/j.1365-313X.2007.03092.x

Bhagwat SS, Li Y, Cortés-Peña YR, Brace EC, Martin TA, Zhao H, Guest JS (2021) Sustainable production of acrylic acid via 3-hydroxypropionic acid from lignocellulosic biomass. ACS Sustain Chem Eng 9:16659–16669. https://doi.org/10.1021/acssuschemeng.1c05441

Cao VD, Kannan B, Luo G, Liu H, Shanklin J, Altpeter F (2023) Triacylglycerol, total fatty acid, and biomass accumulation of metabolically engineered energycane grown under field conditions confirms its potential as feedstock for drop in fuel production. GCB Bioenergy. https://doi.org/10.1111/gcbb.13107

Chapman KD, Ohlrogge JB (2012) Compartmentation of triacylglycerol accumulation in plants. J Biol Chem 287:2288–2294. https://linkinghub.elsevier.com/retrieve/pii/S0021925820531906

Cortés-Peña YR, Kurambhatti C, Eilts K, Singh V, Guest JS (2022) Economic and environmental sustainability of vegetative oil extraction strategies at integrated Oilcane and oil-sorghum biorefineries. ACS Sustain Chem Eng 10:13972–13979

Huang H, Long S, Singh V (2016) Techno economic analysis of biodiesel and ethanol co production from lipid producing sugarcane. Biofuels Bioprod Biorefin 10:299–315. https://doi.org/10.1002/bbb.1640

International Energy Agency (2023) Renewables 2022 analysis and forcast to 2027. https://iea.blob.core.windows.net/assets/ada7af90-e280-46c4-a577-df2e4fb44254/Renewables2022.pdf

Kannan B, Liu H, Shanklin J, Altpeter F (2022) Towards oilcane: preliminary field evaluation of metabolically engineered sugarcane with hyper-accumulation of triacylglycerol in vegetative tissues. Mol Breed 42:64. https://doi.org/10.1007/s11032-022-01333-5

Kargbo H, Harris JS, Phan AN (2021) "Drop-in" fuel production from biomass: critical review on techno-economic feasibility and sustainability. Renew Sust Energ Rev 135:110168. https://linkinghub.elsevier.com/retrieve/pii/S1364032120304597

Kelly AA, Van Erp HV, Quettier AL, Shaw E, Menard G, Kurup S, Eastmond PJ (2013) The SUGAR-DEPENDENT1 lipase limits triacylglycerol accumulation in vegetative tissues of Arabidopsis. Plant Physiol 162:1282–1289

Kumar D, Long SP, Singh V (2018) Biorefinery for combined production of jet fuel and ethanol from lipid-producing sugarcane: a techno-economic evaluation. GCB Bioenergy 10:92–107

Li Y, Bhagwat SS, Cortés-Peña YR, Ki D, Rao CV, Jin Y-S, Guest JS (2021) Sustainable lactic acid production from lignocellulosic biomass. ACS Sustain Chem Eng 9:1341–1351. https://doi.org/10.1021/acssuschemeng.0c08055

Luo G, Cao VD, Kannan B, Liu H, Shanklin J, Altpeter F (2022) Metabolic engineering of energycane to hyperaccumulate lipids in vegetative biomass. BMC Biotechnol 22:24. https://doi.org/10.1186/s12896-022-00753-7

Lynd LR, Beckham GT, Guss AM, Jayakody LN, Karp EM, Maranas C, McCormick RL, Amador-Noguez D, Bomble YJ, Davison BH, Foster C, Himmel ME, Holwerda EK, Laser MS, Ng CY, Olson DG, Román-Leshkov Y, Trinh CT, Tuskan GA, Upadhayay V, Vardon DR, Wang L,

Wyman CE (2022) Toward low-cost biological and hybrid biological/catalytic conversion of cellulosic biomass to fuels. Energy Environ Sci 15:938–990
Maitra S, Singh V (2023) Application of time-domain 1 H NMR for investigating dynamics of vegetative lipids in bioenergy crops at different developmental stages. Ind Biotechnol 19:86–93. https://doi.org/10.1089/ind.2022.0039
Maitra S, Cheng MH, Liu H, Cao VD, Kannan B, Long SP, Shanklin J, Altpeter F, Singh V (2024) Sustainable co-production of plant lipids and cellulosic sugars from transgenic energycane at an industrially relevant scale: A proof of concept for alternative feedstocks. Chem Eng J 1:487:150450
Maitra S, Dien B, Long SP, Singh V (2021) Development and validation of time domain 1 H NMR relaxometry correlation for high throughput phenotyping method for lipid contents of lignocellulosic feedstocks. GCB Bioenergy. https://doi.org/10.1111/gcbb.12841
Maitra S, Long SP, Singh V (2022a) Optimizing chemical-free pretreatment for maximizing oil/lipid recovery from transgenic bioenergy crops and its rapid analysis using time domain-NMR. Front Energy Res 10:1–13. https://doi.org/10.3389/fenrg.2022.840418/full
Maitra S, Viswanathan MB, Park K, Kannan B, Alfanar SC, McCoy SM, Cahoon EB, Altpeter F, Leakey ADB, Singh V (2022b) Bioprocessing, recovery, and mass balance of vegetative lipids from metabolically engineered "Oilcane" demonstrates its potential as an alternative feedstock for drop-in fuel production. ACS Sustain Chem Eng 10:16833–16844. https://doi.org/10.1021/acssuschemeng.2c05327
Parajuli S, Kannan B, Karan R, Sanahuja G, Liu H, Garcia Ruiz E, Kumar D, Singh V, Zhao H, Long S, Shanklin J, Altpeter F (2020) Towards oilcane: engineering hyperaccumulation of triacylglycerol into sugarcane stems. GCB Bioenergy 12:476–490. https://doi.org/10.1111/gcbb.12684
REN21 (2020) Renewables 2020 global status report. Renew. energy policy Netw. 21st Century. https://www.ren21.net/wp-content/uploads/2019/05/GSR_2020_Press_Release_ES.pdf
REN21 (2023) Renewables 2023 global status report:1–75. https://www.ren21.net/wp-content/uploads/2019/05/GSR2023_GlobalOverview_Full_Report_with_endnotes_web.pdf
Sanjaya, Durrett TP, Weise SE, Benning C (2011) Increasing the energy density of vegetative tissues by diverting carbon from starch to oil biosynthesis in transgenic Arabidopsis. Plant Biotechnol J 9:874–883
Siloto RMP, Findlay K, Lopez-Villalobos A, Yeung EC, Nykiforuk CL, Moloney MM (2006) The accumulation of Oleosins determines the size of seed oilbodies in Arabidopsis. Plant Cell 18:1961–1974. https://academic.oup.com/plcell/article/18/8/1961/6115341
The Gazette of India (2023) National biofuel policy—Ministry of Petroleum and Natural Gas
U.S. EIA (2023) Total biofuels operable production capacity
Vanhercke T, Divi UK, El Tahchy A, Liu Q, Mitchell M, Taylor MC, Eastmond PJ, Bryant F, Mechanicos A, Blundell C, Zhi Y, Belide S, Shrestha P, Zhou X-R, Ral J-P, White RG, Green A, Singh SP, Petrie JR (2017) Step changes in leaf oil accumulation via iterative metabolic engineering. Metab Eng 39:237–246. https://linkinghub.elsevier.com/retrieve/pii/S109671761630091X
Vanhercke T, Belide S, Taylor MC, El Tahchy A, Okada S, Rolland V, Liu Q, Mitchell M, Shrestha P, Venables I, Ma L, Blundell C, Mathew A, Ziolkowski L, Niesner N, Hussain D, Dong B, Liu G, Godwin ID, Lee J, Rug M, Zhou X-R, Singh SP, Petrie JR (2019) Up-regulation of lipid biosynthesis increases the oil content in leaves of Sorghum bicolor. Plant Biotechnol J 17:220–232. https://doi.org/10.1111/pbi.12959
Winichayakul S, Scott RW, Roldan M, Hatier J-HB, Livingston S, Cookson R, Curran AC, Roberts NJ (2013) In vivo packaging of triacylglycerols enhances Arabidopsis leaf biomass and energy density. Plant Physiol 162:626–639. https://academic.oup.com/plphys/article/162/2/626/6110760
Xu C, Shanklin J (2016) Triacylglycerol metabolism, function, and accumulation in plant vegetative tissues. Annu Rev Plant Biol 67:179–206. https://www.annualreviews.org/doi/10.1146/annurev-arplant-043015-111641

Xu C, Fan J, Froehlich JE, Awai K, Benning C (2005) Mutation of the TGD1 chloroplast envelope protein affects phosphatidate metabolism in Arabidopsis. Plant Cell 17:3094–3110. https://academic.oup.com/plcell/article/17/11/3094/6114747

Zale J, Jung JH, Kim JY, Pathak B, Karan R, Liu H, Chen X, Wu H, Candreva J, Zhai Z, Shanklin J, Altpeter F (2016) Metabolic engineering of sugarcane to accumulate energy-dense triacylglycerols in vegetative biomass. Plant Biotechnol J 14:661–669

Zolman BK, Silva ID, Bartel B (2001) The Arabidopsis pxa1 mutant is defective in an ATP-binding cassette transporter-like protein required for peroxisomal fatty acid β-oxidation. Plant Physiol 127:1266–1278. https://academic.oup.com/plphys/article/127/3/1266/6110098

Recent Advances in Sugarcane-Mediated Nanocomposites and Its Applications

10

J. Anitha, P. J. Sherin, R. Ranjani, and R. Selvakumar

Abstract

Sugarcane, a highly productive tropical crop known for its C4 photosynthetic efficiency, contributes approximately 70% of global sugar production and holds significant potential for biomass and ethanol-based biofuel production. Countries like Brazil and India lead in sugarcane production, with India experiencing a quadrupled increase in cultivated area and substantial improvements in yield and sugar recovery over the decades. The economic importance of sugarcane in India is profound, contributing significantly to the agricultural GDP and providing raw materials for various by-products, including ethanol, which is increasingly emphasized in energy policies. In addition to its agricultural and economic significance, sugarcane exhibits notable nutritional and medicinal properties. Sugarcane juice is rich in phenolic acids, flavonoids, and glycosides, and is traditionally used in various systems of medicine for its diuretic, aphrodisiac, laxative, and anti-inflammatory properties. Recent advancements in nanotechnology have leveraged sugarcane for the sustainable synthesis of nanoparticles, presenting an environmentally friendly, economical, and scalable approach. Sugarcane-derived nanoparticles, such as silica, silver, copper, titanium, and gold, have demonstrated diverse applications ranging from environmental remediation to biomedical uses. These nanoparticles possess distinctive characteristics, including high surface area-to-volume ratios and quantum mechanical effects, making them suitable for catalysis, drug delivery, and as antimicrobial agents. The synthesis of nanoparticles using sugarcane components offers a sustainable method,

J. Anitha (✉) · P. J. Sherin · R. Ranjani
Department of Biochemistry (PG and Research), Kongunadu Arts and Science College (Autonomous), Coimbatore, India

R. Selvakumar
Department of Science and Humanities, Sri Krishna Polytechnic College, Kovaipudur, Tamil Nadu, India

G. S. Suresha et al. (eds.), *Value Addition and Product Diversification in Sugarcane*, https://doi.org/10.1007/978-981-97-7228-5_10

avoiding toxic chemicals and high energy consumption typically associated with conventional methods. This review highlights the progress and potential of using sugarcane in the sustainable synthesis of nanoparticles, addressing current research gaps and suggesting future directions to harness the utmost potential of this approach. The integration of sugarcane in nanotechnology not only adds value to this important crop but also contributes to the development of sustainable and environmentally friendly technologies.

Keywords

Sugarcane · Nanotechnology · Nanomedicine · Economy · Cane juice

10.1 Introduction

Sugarcane is a highly productive tropical crop with superior growth due to its C4 photosynthetic efficiency. Further, it accounts for approximately 70% of global sugar production and has significant potential for both biomass and bioethanol production. In 2018, the worldwide production of sugarcane surpassed 1.9 billion metric tons (MMT), with Brazil and India as the top producers (FAO 2020, 2021). The genus Saccharum separated from maize approximately 15 to 20 million years ago and from sorghum as recently as 6–9 million years ago (Paterson et al. 2004). For more than 2000 years, cane sugar has historically been the favored sweetener. However, due to the rising sugar and bioenergy raw materials requirements, farmers have been exploiting the sugarcane gene pool to its maximum potential to satisfy the demands of a rapidly multiplying population.

In contemporary agriculture, both conventional breeding methods and transgenic technologies play pivotal roles in enhancing sugarcane plants to achieve higher yields and increased resilience against various environmental challenges. The extensive breeding efforts by Dutch colonizers in the Dutch East Indies have notably expanded the genetic diversity of cultivated sugarcane, resulting in varieties that offer improved productivity and robustness. Sugarcane holds significant global importance, cultivated across approximately 20 million hectares in more than 90 countries, particularly in semi-tropical and tropical regions. Belonging to the Poaceae family, which includes economically vital crops like wheat, maize, rice, and sorghum, sugarcane contributes essential dietary components such as carbohydrates and other crucial nutrients.

10.2 Cultivation

Sugarcane, a significant crop grown in tropical and subtropical regions, is cultivated on nearly 27 million hectares across over 120 countries worldwide. It accounts for over 75% of the global sugar production (Budeguer et al. 2021). The territory and profitability of sugarcane significantly varies from one country to another. Brazil has the largest area for sugarcane cultivation (5.34 million hectares) while Australia has the highest yield (85.1 tonnes per hectare). Brazil, India, China, and Pakistan are the largest producers of sugarcane, collectively contributing to more than half of the world's production (Qureshi and Afghan 2005). Brazil, specifically, leads globally in sugarcane production, cultivating approximately 4.5 million hectares across its northeastern and southeastern regions (Galvao et al. 2005). In Brazil, sugarcane serves as a raw material for sugar production, as well as for anhydrous alcohol (used as a gasoline additive) and hydrous alcohol (a natural, renewable fuel for vehicles), fulfilling both domestic and international markets (De Oliveira 2002). To improve the productivity of the sugarcane industry in Brazil, ongoing efforts focus on developing new varieties with higher yields. Ideally, these varieties should be adaptable to environmental variations, such as climate, soil types, and harvesting seasons (Badaloo et al. 1999; Bissessur et al. 2000). They should also exhibit resistance to pests, diseases, and drought, while maintaining a high concentration of sucrose in their tissues (Croft et al. 2000).

10.3 Economic Significance in India

Sugarcane holds a significant position as a cash crop in India, predominantly cultivated in tropical and subtropical regions. It serves as the principal raw material for sugar production in the Indian subcontinent. Nearly five million farmers are engaged in its cultivation, making India the second-largest producer of sugarcane. Sugarcane occupies 2.57% of the total cultivated area and contributes 1.1% to the national GDP. Its share in the agricultural GDP has risen from 5% in 1990–91 to 10% in 2010–11. Sugar industries, typically situated in rural areas, play a vital role in the local economy by employing approximately 4% of the population. Besides sugar, these industries produce various by-products such as jaggery, ethanol, electricity, paper, and high-value biomolecules, all contributing to economic growth. Bagasse, vinasse, and molasses—key by-products of sugarcane processing—are widely utilized in rural India. Bagasse is primarily used for heat and power generation in boilers and cogeneration units during milling and distillation. Molasses, known for their high energy content, are commonly used as a food ingredient and raw material for ethanol production (Dotaniya et al. 2016; Santos et al. 2020).

Vinasse is mainly employed in irrigation to improve soil fertility (Steffi Formann et al. 2020). The area dedicated to sugarcane agriculture in India has expanded significantly, increasing from 1.17 million hectares in 1930–31 to 5.01 million hectares by 2013–14, leading to the lowest sugar prices globally (Vision 2050, SBI, ICAR 2015). The sugar productivity has risen from 31 tonnes per hectare to 68 tonnes per hectare. As a result of this enhanced productivity, sugarcane production has risen from 3.7 million metric tonnes to 35.5 million metric tonnes, and sugar production has increased from 0.12 million metric tonnes to 27.9 million metric tonnes. Additionally, sugar recovery has improved to 10.27% from 9.05%. The operational sugar factories have risen significantly from 29 to over 527. The growth in sugarcane and sugar production is credited to two main factors: a fourfold expansion in cultivated areas and productivity improvements exceeding 100%. These accomplishments were facilitated by the introduction of new, well-suited varieties and advancements in crop production and protection technologies developed by the ICAR-SBI since 1912 (Vision 2050, SBI, ICAR 2015; Solomon 2016).

Currently, India combines 5% of ethanol with petrol and aims to increase it to 10% by 2022. However, this is much lower compared to Brazil's 27%. Therefore, there is a need to shift the Indian energy sector's policy toward higher ethanol utilization. The production of ethanol from molasses is insufficient to meet the demand, necessitating the redirection of surplus sugar production towards ethanol production directly from sugarcane juice. Additionally, there is a need to explore the establishment of energy plantations for ethanol and biomass production. Due to limitations in expanding sugarcane cultivation areas, it is necessary to establish new plantations in marginal and underutilized lands using related varieties such as Erianthus, energy canes, and high-biomass sugarcane clones. This strategy aims to meet the increasing demands for sugar, ethanol, and power. To achieve this objective, sugarcane production needs to increase to 55 million metric tons with a sugar recovery rate of 11%. According to forecasts, sugarcane production has been rising by five million tons annually, accompanied by improved sugar recovery rates. However, concerns persist about stagnant productivity, high cultivation costs, natural resource depletion, and climate change impacts. The primary challenge for the country's sugarcane research and development institutions is to enhance productivity despite these constraints, to reach the projected production targets by 2050. Addressing these challenges can be achieved through the development of enhanced sugarcane hybrids and genetically modified varieties with desirable agricultural traits.

10.4 The Nutritional Content and Therapeutic Qualities of Sugarcane

In several sugarcane-producing nations like India, sugarcane juice serves as a favored non-alcoholic beverage, especially during hot seasons. The juice comprises approximately 75–85% water, 10–21% sucrose (a non-reducing sugar), 10–15% fiber, 0.3–3% reducing sugars (such as glucose and fructose), and various other organic and inorganic substances (Fig. 10.1) (Swaminathan 1995). Sugarcane juice and its unprocessed derivatives contain high levels of phenolic acids, flavonoids, and other glycosides. Cane juice is the input for producing purified white sugar and unrefined products like Jaggery and Khandsari. Sugarcane juice is recognized for its therapeutic properties in multiple traditional medicinal systems like Ayurveda, Unani, and Siddha (Basu and Chaudhuri 1991; Kaushik and Dhiman 1999; Sammbamurty 2006; Warrier 2010).

Ayurveda recommends cane sugar juice as a diuretic and a nutritional drink (Karthikeyan and Samipillai 2010). Drinking juice regularly is believed to promote clear and fast urinary flow, which can improve kidney function (Van Wyk and Wink 2005). For enhanced results, sugarcane juice or jaggery is mixed with dry ginger is consumed with lime or ginger to alleviate hiccups (Duke and Wain 1981). It is also utilized as an aphrodisiac (stimulating sexual desire) (Chopra et al. 1956), laxative

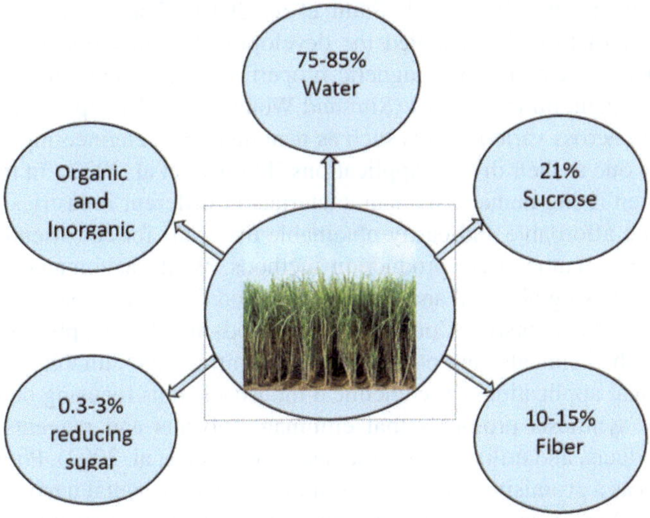

Fig. 10.1 Shows the phytoconstituents of sugarcane

(promoting bowel movements), coolant (beneficial for fatigue, thirst and burning sensations), demulcent (soothing irritation or inflammation) (Kaushik and Dhiman 1999), purifies the blood, antiseptic, and tonic (Chopra et al. 1956; Khare 2008). It is also recommended for leprosy and intestinal troubles (Basu and Chaudhuri 1991; Van Wyk and Wink 2005). It is highly recommended against tridosh (Pitta, Kapha and Vata) in Ayurveda (Basu and Chaudhuri 1991).

This remedy has laxative, diuretic, fattening, aphrodisiac, and blood-purifying properties, and is beneficial for the lungs according to Unani medicine (Anis and Iqbal 1986; Cáceres et al. 1987; Vedavathy et al. 1991; Khare 2008; Koh et al. 2009; Karthikeyan and Samipillai 2010). Drinking a substantial amount of sugarcane juice is thought to benefit the liver, making it strongly recommended for quickly relieving jaundice (Khare 2008; Karthikeyan and Samipillai 2010) Sugarcane enhances the natural immune response against viral, bacterial, and protozoan infections in the host (Lo et al. 2005). It is suggested for individuals experiencing low blood pressure. A stem decoction is given to children with diarrhea. It is an essential component of the medicines used to treat skin and mucous membrane ulcers, particularly in Cambodia. (Basu and Chaudhuri 1991; Kaushik and Dhiman 1999; Van Wyk and Wink 2005).

10.5 Nanotechnology

Nanotechnology encompasses the scientific and engineering discipline focused on creating, analysing, and utilizing materials and devices whose smallest functional unit measures on the nanometer scale—equivalent to one billionth of a meter—in at least one dimension (Silva 2004; Saini et al. 2010). The ability to manipulate atomic-level structures has enabled the development of nanomaterials exhibiting unique optical, electrical, and magnetic properties at the nanoscale, governed by principles of quantum mechanics (Sim and Wong 2021). Nanoparticles have seen increased use across various fields such as nanomedicine, engineering, electronics, and industry due to their diverse applications (Iwuozor et al. 2022). In recent years, there has been rising demand for nanomaterials in different industries, prompting efforts to find affordable and easily obtainable materials for eco-friendly, reliable, and low-energy nanoparticle production methods. While nanomaterials are typically produced using physical and chemical methods, these approaches often prove costly and energy-intensive. Conventional methods involve capping agents, toxic solvents, harsh chemicals, and other additives, limiting their suitability in biomedical and clinical applications. The chemical industry is thus focusing on developing eco-friendly synthesis protocols that eliminate solvents and reagents, minimize waste byproducts, and utilize renewable resources (Lee et al. 2002). Photosynthesis has emerged as a promising method for synthesizing noble metal nanoparticles (Ag, Pd, Pt, and Au) using entire plants, gaining significant attention over the past decade (Akhtar et al. 2013; Kharissova et al. 2013; Mittal et al. 2013). This approach is valued for its applications in catalysis, drug delivery, biomedical sensors, and other fields.

Nanotechnology is valued for its cost-effectiveness, safety, cleanliness, and potential for significant financial benefits, blending technology and science to focus on practical applications (El Naschie 2006). Sugarcane was the world's largest crop by production in 2019 (El Chami et al. 2020). This review examines recent developments in the environmentally friendly synthesis of nanoparticles using sugarcane. It highlights key discoveries, identifies areas needing further research, and suggests future directions. Utilizing sugarcane for nanoparticle synthesis is a promising and innovative approach. Various parts of sugarcane have been utilized to produce nanoparticles of diverse materials including iron, copper, silica, silver, carbon quantum dots, gold, and zinc oxide. Sugarcane serves as a precursor, stabilizing agent, and reducing agent during the synthesis process. This method is environmentally friendly, cost-effective, and operates at room temperature without generating toxic by-products. It is also straightforward, quick, safe, reproducible, economical, and compatible with biological systems. Hence, there is a call for more extensive research to develop nanoparticles with broad-ranging applications.

10.5.1 Silica Nanoparticles

Successful research into synthesis and surface science has resulted in advancements in ceramic nanoparticles such as silica, alumina, titania, zirconia, silicon nitride, and silicon carbide, enhancing their properties (Rahman and Padavettan 2012). Silicon, constituting about 28% of the Earth's crust, is the second most abundant element (Ghorbani et al. 2013). Silica nanoparticles have been widely applied in various fields including sensors (Bonyadi and Ghanbari 2021), as fillers in polymers

(Li et al. 2020), in drug delivery systems (Porrang et al. 2021), and dye-sensitized solar cells (Fang et al. 2012), based on previous research. Nanoparticles of SiO2 were synthesized using sugarcane bagasse ash as a silica source via a sol-gel process, achieving over 96% purity by adjusting key synthesis parameters (Falk et al. 2019). In a 2018 study by Rovani et al., silica nanoparticles were produced and employed as an adsorbent to eliminate acridine orange 8 dye, demonstrating their efficacy. In another study, these same silica nanoparticles proved effective in removing methylene blue (Rovani et al. 2019).

Scientific research conducted by D. S. Arthur and his team in the year 2023, discovered that exposure to sugarcane ash and its by-products, specifically sugarcane ash-derived nanoparticles, can lead to mitochondrial dysfunction and affect the metabolic activity of human Proximal Convoluted Tubule (PCT) cells. On the other hand, biogenic silica nanoparticles derived from sugarcane bagasse were found to be biocompatible, as they reduced the percentage of fibroblasts in the G1 phase. This suggests that they could have potential functions in the area of biomedicine (Athinarayanan et al. 2017).

10.5.2 Silver Nanoparticles

Silver nanoparticles were produced using ultrasonic irradiation in the presence of cellulose derived from sugarcane bagasse. These silver nanoparticles made from cellulose exhibited considerable promise in removing heavy metals from water during remediation processes. (Yong et al. 2020). Furthermore, nanocomposites composed of silver nanoparticles derived from sugarcane bagasse have been employed as effective adsorbents for sequestering the Bisphenol A micropollutant from water bodies (Mpatani et al. 2022).

In another application, a nanocomposite was developed using a sugarcane-based liquid (poly vinyl alcohol) ionic polymer along with silver nanoparticles. This nanocomposite found utility in fuel cells, wearable sensors, actuators, and supercapacitors (Panwar et al. 2022). Ag nanoparticles synthesized by bacteria, fungi, and sugarcane husk demonstrated anti-fungal activity against sugarcane phytopathogens *Colletotricum falcatum* and *Fusarium moniliforme* (Amna et al. 2021). Sugarcane leaf extract reduced silver nitrate to silver (Srikhao et al. 2021).

10.5.3 Copper Nanoparticles

Copper is a crucial micronutrient essential for numerous enzymatic processes in both plants and animals (White and Brown 2010; Izydorczyk et al. 2020). Copper nanoparticles find diverse applications, including plant disease control, electronics, textile coatings, and antimicrobial uses (Bhagat et al. 2021; Archana et al. 2021). Researchers have recently managed to synthesize biogenic nanoparticles by using extracts from different plant seeds. (Khalil et al. 2021; Khan et al. 2021; Naseer et al. 2020). Research by Mary et al. 2019 has documented the synthesis of copper nanoparticles using sugarcane juice, which is a sustainable method. Copper nanoparticles were studied against *Escherichia coli*, *Pseudomonas aeruginosa*, *Staphylococcus aureus*, and *Bacillus subtilis*. The results showed that the inhibition zone against these bacteria was 5, 8, and 9 mm, respectively, at a concentration of 10 ml of sugarcane juice stabilized copper nanoparticles. The FTIR spectrum of sugarcane juice with reduced copper nanoparticles revealed the presence of molecular-weight polysaccharides and protein moieties. The peaks observed at 3263 (high molecular weight), 2933, 1633, 1048, 1411, 1346, and 1267 cm^{-1} correspond to various molecules and functional groups, such as OH, C–H stretching, C=O, O–C=O, C–H stretching of the aromatic ring, phenolic OH, and C–C stretching (Camargo et al. 2012). Small peaks at 103 and 887–827 cm^{-1} indicate β-glycosidic linkage between sugars and correspond to protein and polysaccharide bands (Guilherme et al. 2015). It is a well-established fact that sugarcane juice contains glucose and fructose as its major components, along with some minor organic acids (Bhattacharjee and Ahmaruzzaman 2016). Research has demonstrated that sugar molecules and organic acids can serve as modifiers on surfaces, acting as stabilizers and capping agents. This capability assists in regulating the size, shape, and structure of the material (Sharma et al. 2015). Copper oxide nanoparticles have been effectively employed for sensing and targeting applications in both in vivo and in vitro settings. However, several significant challenges must be addressed before these applications can be widely commercialized (Verma and Kumar 2019).

10.5.4 Titanium Nanoparticles

TiO_2 is a non-toxic, light-stable, chemically and biologically inert material. It is commonly used as a photocatalyst to eliminate organic compounds such as colorants, antibiotics, pharmaceuticals, etc. (Gusain et al. 2019; Haider et al. 2019). $Pt/TiO_2/Ti_2Fe_2O_7$ nanocomposites demonstrated a degradation efficiency of over 95% for toluene, a performance four times greater than that of Pt/C and Pt/TiO_2 (Mohamed and Ali 2023). Another study utilized titanium oxide-coated carbon aerogel (TiO_2/CA) synthesized from sugarcane bagasse for the degradation of methylene blue dye, achieving an impressive efficiency of 93.86% at pH 9 (Lin et al. 2023). Singh et al. (2024) reported the use of various natural substances present in sugarcane juice, such as proteins, enzymes, and phytochemicals, as reducing and capping agents in the synthesis of TiO2 nanoparticles. These nanoparticles act as stabilizers, preventing excessive growth and aggregation during their formation. Sugarcane juice contains hydroxide ions that can be adsorbed onto the surface of carbon fibers or TiO_2 nanoparticles. These hydroxide ions are then oxidized by positively charged holes, leading to the reduction of Titanium nanoparticles (Wu et al. 2010). Meanwhile, oxygen can react with electrons in the conduction band to produce superoxide, which can then react with protons to generate hydroperoxyl radicals (Wang et al. 2006).

10.5.5 Gold Nanocomposite

Gold nanoparticles (Au NPs) are noble metal nanoparticles that have garnered considerable interest for their role as catalysts in environmental engineering. Their small size at the nanoscale provides them with a substantial surface area relative to their volume, leading to enhanced catalytic and antimicrobial properties (Liu et al. 2018). Commonly used chemical procedures for synthesizing Au (0) typically incorporate reagents that are either toxic to humans or environment for reducing and stabilizing purposes. These reagents may include hexadecyltrimethylammonium bromide, formaldehyde, N-dimethylformamide (DMF), and hydrazine hydrate, etc. (Narayanan et al. 2015; Tian et al. 2020). Hence, it is important to create a fast and sustainable method for synthesizing gold nanoparticles. Gold nanoparticles (Au-NPs) were synthesized using sugarcane bagasse extract, a waste product of the sugarcane industry, through a biogenic green synthesis process. The bagasse is stirred in aqueous medium for 20 min to obtain the extract. The study examined how varying pH levels of the medium influence the formation of gold nanoparticles during synthesis (Naeem et al. 2023). A sugar that can donate electrons to oxidants is called a reducing sugar, thanks to its carbonyl group. In sugarcane, the glycosidic bond prevents the ring opening of its constituent monosaccharides, glucose, and fructose. This characteristic makes it challenging for sugarcane to function as a reducing sugar in the synthesis of metal nanoparticles without a process that involves breaking open the ring structure of either glucose or fructose's hemiacetal groups. (Filippo et al. 2010). Zhang et al. (2021) employed a three-dimensional sugarcane membrane coated with gold nanoparticles for the removal of methylene blue, showcasing superior catalytic efficiency. Additionally, biogenic gold nanoparticles derived from sugarcane bagasse have demonstrated effectiveness in industrial applications, particularly in the removal of nitroarenes and organic dyes.

10.5.6 Quantum Dots

The carbon material is a wide-ranging classification that encompasses functional materials like graphene quantum dots (GQDs) (Zhang et al. 2018), carbon nanotubes (Khoerunnisa et al. 2019), graphene (Yoo et al. 2019), activated carbon (Ren et al. 2017), and activated carbon fibers (Buchmeiser et al. 2019). Graphene Quantum Dots (GQDs) were initially identified and have subsequently demonstrated numerous advantages compared to conventional semiconductor quantum dots. These include excellent photo-stability, high solubility in water, chemical stability, biocompatibility, low toxicity, and facile modification. Biomass material, which contains a high level of carbon and heteroatoms, has become an important raw material for GQDs. However, existing research often overlooks the structure, composition, and formation mechanism of biomass quantum dots, thereby hindering progress in understanding the preparation of GQDs (Ramanan et al. 2018). Carbon dots were produced from bagasse, and the incorporation of an increased number of these carbon dots into a polyester composite was found to enhance surface conductivity, as reported by Saravanan et al. (2023). Another research project concentrated on environmentally friendly production of highly luminescent carbon quantum dots (CQDs) with an impressive quantum yield of 17.98%. Sugarcane bagasse pulp was used as the starting material for these CQDs, and their synthesis was achieved via a hydrothermal method. Subsequently, the CQDs were evaluated for antibacterial activity against aquatic Gram-positive (*Benthesicymus cereus* and *Staphylococcus aureus*) and Gram-negative (*Pseudomonas aeruginosa*, *Vibrio cholerae*, and *Escherichia coli*) microbial organisms (Pandiyan et al. 2020). Moreover, natural carbon quantum dots (NCQDs) derived from bagasse have been utilized as an efficient fluorescent probe for multicolor live cell imaging, as highlighted by Kasinathan et al. (2022).

10.6 Conclusion

Numerous researchers focussed on the various aspects of sugarcane in maintaining the natural world and well-being through the formation of metal nanoparticles from sugarcane and its parts by non-toxic green methods which find its application in dye degradation, live cell imaging, textile industry, etc. Further, the metabolites of sugarcane could be utilized to generate highly efficient drugs for mankind.

References

Akhtar MS, Panwar J, Yun YS (2013) Biogenic synthesis of metallic nanoparticles by plant extracts. ACS Sustain Chem Eng 1(6):591–602

Amna MT, Khan UN, Amin B, Javed MT, Mehmood S, Farooq MA, Sultan T, Munis MF, Chaudhary HJ (2021) Characterization of bio-fabricated silver nanoparticles for distinct antifungal activity against sugarcane phytopathogens. Microsc Res Tech 84(7):1522–1530

Anis M, Iqbal M (1986) Antipyretic utility of some Indian plants in traditional medicine. Fitoterapia 57(1):52–55

Archana KM, Rajagopal R, Krishnaswamy VG, Aishwarya S (2021) Application of green synthesized copper iodide particles on cotton fabric-protective face mask material against COVID-19 pandemic. J Mater Res Technol 15:2102–2116

Athinarayanan J, Periasamy VS, Alhazmi M, Alshatwi AA (2017) Synthesis and biocompatibility assessment of sugarcane bagasse-derived biogenic silica nanoparticles for biomedical applications. J Biomed Mater Res B Appl Biomater 105(2):340–349

Badaloo GH, Domaingue R, Ramdoyal K (1999) A critical review of parental choice and cross prediction techniques in the MSIRI sugarcane breeding programme. InProceedings third annual meeting of agricultural scientists. Food and Agricultural Research Council Reduit, Mauritius pp. 47–54

Basu AN, Chaudhuri AN (1991) Preliminary studies on the antiinflammatory and analgesic activities of Calotropis procera root extract. J Ethnopharmacol 31(3):319–324

Bhagat M, Anand R, Sharma P, Rajput P, Sharma N, Singh K (2021) Multifunctional copper nanoparticles: synthesis and applications. ECS J Solid State Sci Technol 10(6):063011

Bhattacharjee A, Ahmaruzzaman M (2016) CuO nanostructures: facile synthesis and applications for enhanced photodegradation of organic compounds and reduction of p-nitrophenol from aqueous phase. RSC Adv 6(47):41348–41363

Bissessur D, Tilney-Bassett RA, Chong LL, Domaingue R, Julien MH (2000) Family× environment and genotype× environment interactions for sugarcane across two contrasting marginal environments in Mauritius. Exp Agric 36(1):101–114

Bonyadi S, Ghanbari KH (2021) Development of highly sensitive and selective sensor based on molecular imprinted polydopamine-coated silica nanoparticles for electrochemical determination of sunset yellow. Microchem J 167:106322

Buchmeiser MR, Muks E, Schowner R, Frank E, Hageroth U, Henzler S, Spörl J, Ota A, Beyer R, Müller A (2019) Structure evolution in all-aromatic, poly (p-phenylene-vinylene)-derived carbon fibers. Carbon 144:659–665

Budeguer F, Enrique R, Perera MF, Racedo J, Castagnaro AP, Noguera AS, Welin B (2021) Genetic transformation of sugarcane, current status and future prospects. Front Plant Sci 12:768609. https://doi.org/10.3389/fpls.2021.768609

Cáceres A, Girón LM, Alvarado SR, Torres MF (1987) Screening of antimicrobial activity of plants popularly used in Guatemala for the treatment of dermatomucosal diseases. J Ethnopharmacol 20:223–237

Camargo FA, Innocentini-Mei LH, Lemes AP, Moraes SG, Durán N (2012) Processing and characterization of composites of poly (3-hydroxybutyrate-co-hydroxyvalerate) and lignin from sugar cane bagasse. J Compos Mater 46(4):417–425

Chopra RN, Nayar SI, Chopra IC (1956) Glossary of Indian medicinal plants. Publication and information directorate, C.S.I.R, New Delhi

Croft B, Magarey R, Whittle P (2000) Disease management. Manual of cane growing. Bureau of Sugar Experiment Stations, Brisbane, pp 263–289

De Oliveira JP (2002) The policymaking process for creating competitive assets for the use of biomass energy: the Brazilian alcohol programme. Renew Sust Energ Rev 6(1–2):129–140

Dotaniya ML, Datta SC, Biswas DR, Dotaniya CK, Meena BL, Rajendiran S, Regar KL, Lata M (2016) Use of sugarcane industrial by-products for improving sugarcane productivity and soil health. Int J Recycl Org Waste Agric 5:185–194

Duke JA, Wain KK. (1981) Medicinal plants of the world. Computer index with more than 85000 entries. 3 vols.

El Chami D, Daccache A, El Moujabber M (2020) What are the impacts of sugarcane production on ecosystem services and human Well-being? A review. Ann Agric Sci 65(2):188–199

El Naschie MS (2006) Nanotechnology for the developing world. Chaos, Solitons Fractals 30(4):769–773

Falk G, Shinhe GP, Teixeira LB, Moraes EG, de Oliveira AN (2019) Synthesis of silica nanoparticles from sugarcane bagasse ash and nano-silicon via magnesiothermic reactions. Ceram Int 45(17):21618–21624

Fang Y, Zhang J, Zhou X, Lin Y, Fang S (2012) "Soggy sand" electrolyte based on COOH-functionalized silica nanoparticles for dye-sensitized solar cells. Electrochem Commun 16(1):10–13

FAO (2020) World Food and Agriculture—Statistical Yearbook 2020, pp 58

FAO (2021) World Food and Agriculture—Statistical Yearbook 2021. pp 60

Filippo E, Serra A, Buccolieri A, Manno D (2010) Green synthesis of silver nanoparticles with sucrose and maltose: morphological and structural characterization. J Non-Cryst Solids 356(6–8):344–350. https://doi.org/10.1016/j.jnoncrysol.2009.11.021

Formann S, Hahn A, Janke L, Stinner W, Sträuber H, Logroño W, Nikolausz M (2020) Beyond sugar and ethanol production: value generation opportunities through sugarcane residues. Front Energy Res 8:579577. https://doi.org/10.3389/fenrg.2020.579577

Galvao LS, Formaggio AR, Tisot DA (2005) Discrimination of sugarcane varieties in Southeastern Brazil with EO-1 Hyperion data. Remote Sens Environ 94(4):523–534

Ghorbani F, Younesi H, Mehraban Z, Çelik MS, Ghoreyshi AA, Anbia M (2013) Preparation and characterization of highly pure silica from sedge as agricultural waste and its utilization in the synthesis of mesoporous silica MCM-41. J Taiwan Inst Chem Eng 44(5):821–828

Guilherme AA, Dantas PV, Santos ES, Fernandes FA, Macedo GR (2015) Evaluation of composition, characterization and enzymatic hydrolysis of pretreated sugar cane bagasse. Braz J Chem Eng 32:23–33

Gusain R, Gupta K, Joshi P, Khatri OP (2019) Adsorptive removal and photocatalytic degradation of organic pollutants using metal oxides and their composites: a comprehensive review. Adv Colloid Interf Sci 272:102009

Haider AJ, Jameel ZN, Al-Hussaini IH (2019) Review on: titanium dioxide applications. Energy Procedia 157:17–29

Iwuozor KO, Ogunfowora LA, Oyekunle IP (2022) Review on sugarcane-mediated nanoparticle synthesis: a green approach. Sugar Tech 4:1186–1197

Izydorczyk G, Sienkiewicz-Cholewa U, Baśladyńska S, Kocek D, Mironiuk M, Chojnacka K (2020) New environmentally friendly bio-based micronutrient fertilizer by biosorption: from laboratory studies to the field. Sci Total Environ 710:136061

Karthikeyan J, Samipillai SS (2010) Sugarcane in therapeutics. J Herb Med Toxicol 4(1):9–14

Kasinathan K, Samayanan S, Marimuthu K, Yim JH (2022) Green synthesis of multicolour fluorescence carbon quantum dots from sugarcane waste: investigation of mercury (II) ion sensing, and bio-imaging applications. Appl Surf Sci 601:154266

Kaushik P, Dhiman AK (1999) Medicinal plants and raw drugs of India. Doon phototype printers Dehradun 560–564

Khalil AT, Iqbal J, Shah A, Haque MZ, Khan I, Ayaz M, Ahmad I, Tasneem S, Shah H (2021) The bio–nano interface as an emerging trend in assembling multi-functional metal nanoparticles, SPR_Nanoscience : Volume-7 (1–24):Royal Society of Chemistry

Khan M, Khan AU, Moon IS, Felimban R, Alserihi R, Alsanie WF, Alam M (2021) Synthesis of biogenic silver nanoparticles from the seed coat waste of pistachio (Pistacia vera) and their effect on the growth of eggplant. Nanotechnol Rev 10(1):1789–1800

Khare CP (2008) Indian medicinal plants: an illustrated dictionary. Springer Science & Business Media, p 739

Kharissova OV, Dias HR, Kharisov BI, Pérez BO, Pérez VM (2013) The greener synthesis of nanoparticles. Trends Biotechnol 31(4):240–248

Khoerunnisa F, Futamura R, Mukai S, Konishi T, Fujikawa T, Kaneko K (2019) X-ray absorption anomaly of well-characterized multiwall carbon nanotubes. Carbon 145:209–217

Koh HL, Chua TK, Tan CH (2009) Guide to medicinal plants, a: An illustrated scientific and medicinal approach. World Sci 13

Lee SW, Mao C, Flynn CE, Belcher AM (2002) Ordering of quantum dots using genetically engineered viruses. Science 296(5569):892–895

Li X, Yang L, Shao D, Luo K, Liu L, Wu Z, Luo Z, Wang X (2020) Preparation and application of poly (ethylene oxide)-based all solid-state electrolyte with a walnut-like SiO2 as nano-fillers. J Appl Polym Sci 137(24):48810

Lin TH, Thang TQ, An H, Hai ND, Duy PH, Duyen DT, Oanh DT, Van Tuyen N, Nam HM, Phong MT, Hieu NH (2023) Synthesis of TiO2-doped carbon aerogel from sugarcane bagasse for high efficiency of photodegradation of methylene blue in the water. Vietnam J Chem 61(2):227–237

Liu W, Wang L, Wang J, Du J, Jing C (2018) New insights into microbial-mediated synthesis of Au@ biolayer nanoparticles. Environ Sci Nano 5(7):1757–1763

Lo DY, Chen TH, Chien MS, Koge K, Hosono A, Kaminogawa S, Lee WC (2005) Effects of sugar cane extract on the modulation of immunity in pigs. J Vet Med Sci 67(6):591–597

Mary AA, Ansari AT, Subramanian R (2019) Sugarcane juice mediated synthesis of copper oxide nanoparticles, characterization and their antibacterial activity. J King Saud Univ Sci 31(4):1103–1114

Mittal AK, Chisti Y, Banerjee UC (2013) Synthesis of metallic nanoparticles using plant extracts. Biotechnol Adv 31(2):346–356

Mohamed EF, Ali H (2023) New Pt/TiO2/Ti2Fe2O7 nanocomposite using sugarcane bagasse agro-waste for photodegradation of toluene gas pollutant under sunlight. Mater Sci Eng B 295:116583

Mpatani FM, Kombo UM, Kheir KR, Nassor NK, Joshua D, Mussa SS, Mohamed SA, Mbarak SA, Hamad AS, Ahmada SA (2022) Impregnation of Silver Nanoparticles onto Polymers Based on Sugarcane Bagasse for the Remediation of Endocrine Disruptor–Bisphenol A from Water. Adsorption Science Technology, Hindawi Limited

Naeem M, Irfan A, Begum R, Farooqi ZH (2023) One step biogenic sugarcane bagasse mediated synthesis of gold nanoparticles and their catalytic applications in removing environmental pollutants. Z Phys Chem 237(6):675–688

Narayanan KB, Park HH, Han SS (2015) Synthesis and characterization of biomatrixed-gold nanoparticles by the mushroom Flammulina velutipes and its heterogeneous catalytic potential. Chemosphere 141:169–175

Naseer A, Ali A, Ali S, Mahmood A, Kusuma HS, Nazir A, Yaseen M, Khan MI, Ghaffar A, Abbas M, Iqbal M (2020) Biogenic and eco-benign synthesis of platinum nanoparticles (Pt NPs) using plants aqueous extracts and biological derivatives: environmental, biological and catalytic applications. J Mater Res Technol 9(4):9093–9107

Pandiyan S, Arumugam L, Srirengan SP, Pitchan R, Sevugan P, Kannan K, Pitchan G, Hegde TA, Gandhirajan V (2020) Biocompatible carbon quantum dots derived from sugarcane industrial wastes for effective nonlinear optical behavior and antimicrobial activity applications. ACS Omega 5(47):30363–30372

Panwar V, Anoop G, Sharma M, Gaur SS, Park S (2022) Sugarcane liquid-generated silver nanoparticles connected ionic polymer nanocomposite for enhanced electrical and wearable sensing signals. Mater Today Chem 26:101195

Paterson AH, Bowers JE, Chapman B (2004) Ancient polyploidization predating divergence of the cereals, and its consequences for comparative genomics. Proc Natl Acad Sci 101(26):9903–9908

Porrang S, Rahemi N, Davaran S, Mahdavi M, Hassanzadeh B, Gholipour AM (2021) Direct surface modification of mesoporous silica nanoparticles by DBD plasma as a green approach to prepare dual-responsive drug delivery system. J Taiwan Inst Chem Eng 123:47–58

Qureshi MA, Afghan S (2005) Sugarcane cultivation in Pakistan. Sugar Book Pub. Pakistan Society of Sugar Technologist

Rahman IA, Padavettan V (2012) Synthesis of silica nanoparticles by sol-gel: size-dependent properties, surface modification, and applications in silica-polymer nanocomposites—a review. J Nanomater 2012:8

Ramanan V, Siddaiah B, Raji K, Ramamurthy P (2018) Green synthesis of multifunctionalized, nitrogen-doped, highly fluorescent carbon dots from waste expanded polystyrene and its application in the fluorimetric detection of Au3+ ions in aqueous media. ACS Sustain Chem Eng 6(2):1627–1638

Ren X, Li H, Chen J, Wei L, Modak A, Yang H, Yang Q (2017) N-doped porous carbons with exceptionally high CO2 selectivity for CO2 capture. Carbon 114:473–481

Rovani S, Santos JJ, Corio P, Fungaro DA (2018) Highly pure silica nanoparticles with high adsorption capacity obtained from sugarcane waste ash. ACS Omega 3(3):2618–2627

Rovani S, Santos JJ, Corio P, Fungaro DA (2019) An alternative and simple method for the preparation of bare silica nanoparticles using sugarcane waste ash, an abundant and despised residue in the Brazilian industry. J Braz Chem Soc 30:1524–1533

Saini R, Saini S, Sharma S (2010) Nanotechnology: the future medicine. J Cutan Aesthet Surg 3(1):32–33

Sammbamurty AV (2006) Dictionary of medicinal plants. CBS Publishers & Distributors, pp 263–264

Santos F, Eichler P, de Queiroz JH, Gomes F (2020) Production of second-generation ethanol from sugarcane. In: Sugarcane biorefinery, technology and perspectives. Academic Press, pp 195–228. https://doi.org/10.1016/b978-0-12-814236-3.00011-1

Saravanan K, Jayabalakrishnan D, Bhaskar K, Madhu S (2023) Thermally reduced sugarcane bagasse carbon quantum dots and in-plane flax fiber unsaturated polyester composites: surface conductivity and mechanical properties. Biomass Convers Biorefinery 12:1

Sharma JK, Akhtar MS, Ameen S, Srivastava P, Singh G (2015) Green synthesis of CuO nanoparticles with leaf extract of Calotropis gigantea and its dye-sensitized solar cells applications. J Alloys Compd 632:321–325
Silva GA (2004) Introduction to nanotechnology and its applications to medicine. Surg Neurol 61(3):216–220
Sim S, Wong NK (2021) Nanotechnology and its use in imaging and drug delivery. Biomed Rep 14(5):1–9
Singh P, Sharma S, Srivastava P (2024) Natural Saccharum officinarum (sugarcane) juice assisted TiO2 nanoparticle synthesis for high performance dye-sensitized solar cell. Opt Mater 147:114685
Solomon S (2016) Sugarcane production and development of sugar industry in India. Sugar Tech 18(6):588–602
Srikhao N, Kasemsiri P, Lorwanishpaisarn N, Okhawilai M (2021) Green synthesis of silver nanoparticles using sugarcane leaves extract for colorimetric detection of ammonia and hydrogen peroxide. Res Chem Intermed 47:1269–1283
Swaminathan MS (1995) Food science, chemistry and experimental foods. Bangalore Print. & Publishing Company
Tian F, Zhou J, Fu R, Cui Y, Zhao Q, Jiao B, He Y (2020) Multicolor colorimetric detection of ochratoxin A via structure-switching aptamer and enzyme-induced metallization of gold nanorods. Food Chem 320:126607
Van Wyk BE, Wink M (2005) Medicinal plants of the world. S Afr J Bot 71(1):127–131
Vedavathy S, Rao KN, Rajaiah M, Nagaraju N (1991) Folklore information from Rayalaseema region, Andhra Pradesh for family planning and birth control. Int J Pharmacogn 29(2):113–116
Verma N, Kumar N (2019) Synthesis and biomedical applications of copper oxide nanoparticles: an expanding horizon. ACS Biomater Sci Eng 5(3):1170–1188
Vision 2050, SBI, ICAR (2015). https://icar.org.in/files/ICAR-SBIVision2050-FinalDir-SBI.pdf
Wang G, Wu F, Zhang X, Luo M, Deng N (2006) Enhanced TiO2 photocatalytic degradation of bisphenol E by β-cyclodextrin in suspended solutions. Journal of hazardous materials 133(1–3):85–91
Warrier PK (2010) Indian medicinal plants: a compendium of 500 species. Orient Blackswan
White PJ, Brown P (2010) Plant nutrition for sustainable development and global health. Ann Bot 105(7):1073–1080
Wu RJ, Chen CC, Lu CS, Hsu PY, Chen MH (2010) Phorate degradation by TiO2 photocatalysis: parameter and reaction pathway investigations. Desalination 250(3):869–875
Yong BC, Abdul Raman AA, Buthiyappan A, Zainal Abidin MI (2020) Synthesis and characterization of sugarcane bagasse cellulose-capped silver nanoparticle using ultrasonic irradiation for the adsorption of heavy metal. Asia Pac J Chem Eng 15(3):e2433
Yoo S, Jeong SY, Lee JW, Park JH, Kim DW, Jeong HJ, Han JT, Lee GW, Jeong SY (2019) Heavily nitrogen doped chemically exfoliated graphene by flash heating. Carbon 144:675–683
Zhang Z, Lei Y, Zhao L, Jiang Z, Ouyang Z (2018) Graphene quantum dots decorated Al-doped ZnS for improved photoelectric performance. Materials 11(8):1452
Zhang Q, Li M, Luo B, Luo Y, Jiang H, Chen C, Wang S, Min D (2021) In situ growth gold nanoparticles in three-dimensional sugarcane membrane for flow catalytical and antibacterial application. J Hazard Mater 402:123445

Transgenic Sugarcane for Biopharming: Challenges and Prospects

11

Gauri Nerkar and C. Appunu

Abstract

Sugarcane is a potential candidate for biopharming to generate high-value bioproducts because of its high biomass content, rapid growth rate, highly efficient carbon fixation pathway, and well-developed storage tissue system in the form of culms. Transgenic approaches have significantly contributed to generating new varieties by biolistic and *Agrobacterium*-mediated transformation methods. However, important considerations in breeding programs and transformation technologies and biosafety issues should be considered to explore the potential of sugarcane as a biofactory. In this chapter, we underline the most recent biotechnological advances in sugarcane towards the production of pharmaceutical proteins, biopolymers, and high-value carbohydrates, and also discuss the challenges and prospects of transgenic sugarcane for biopharming. The chapter concludes that the progress in transgenic technology towards sugarcane biopharming demonstrates that sugarcane has tremendous potential to contribute in the area of value addition and product diversification in sugarcane which may play a crucial role in the growing bio-economy.

Keywords

Sugarcane · Biopharming · Sub-cellular targeting · Recombinant proteins · Biofactory

G. Nerkar · C. Appunu (✉)
ICAR-Sugarcane Breeding Institute, Coimbatore, Tamil Nadu, India

G. S. Suresha et al. (eds.), *Value Addition and Product Diversification in Sugarcane*, https://doi.org/10.1007/978-981-97-7228-5_11

11.1 Introduction

Sugarcane contributes to 80% sucrose and 40% of biofuel production worldwide. Sugarcane belongs to the genus *Saccharum* L., tribe Andropogoneae belonging to the grass family, Poaceae. Sugarcane has a complex taxonomy as it encompasses plants from the five genera forming a closely related interbreeding group: "*Saccharum* complex." The *Saccharum* complex comprises *Saccharum, Erianthus* Michx section *Ripidium* Henrard, *Miscanthus* Andersson section *Diandra* Keng, *Narenga* Borrand *Sclerostachya* A. Camus (Daniels and Roach 1987). Sugarcane has a polyploid–aneuploid type of genome (unbalanced number of chromosomes) posing a challenge for molecular biologists and taxonomists (Daniels and Roach 1987; Sreenivasan et al. 1987). The genus *Saccharum* comprises of six species: *S. officinarum* L., classified as the noble cane; *S. spontaneum* L., *S. robustum* E.W. Brandes & Jeswiet ex Grassl, and *S. edule* Hassk., classified as wild species; and *S. sinense* Roxb., *S. barberi* Jeswiet, classified as ancient hybrids (Daniels and Roach 1987; D'Hont and Layssac 1998). The commercially grown sugarcane cultivars are the interspecific hybrids between *Saccharum officinarum* and *Saccharum spontaneum.* Sugarcane is an allopolyploid with a large number of chromosomes and a (100–130) and a highly polyploid–anueploid genome (x = 10–13) with a genome size of ~10 Gb. This presents a challenge for breeders as their statistical genetic approaches are based on models established for diploid organisms (Gomez-Merino et al. 2014).

Improving sugarcane yield and accelerating selection for desirable traits are crucial in addressing the increasing need for sugar and biofuel. Conventional breeding and improved agronomic practises have played a significant role in improving the sugarcane yield (Ram et al. 2022). However, sugarcane improvement using these approaches is limited due to its complex genome structure, poor fertility and long-term breeding programmes (Dal-Bianco et al. 2012). Unlike the conventional breeding programmes, introduction of superior allelesinto elite sugarcane cultivars by genetic engineering does not require screening of thousands of segregating progenies. The first report on sugarcane genetic transformation came in 1990s (Bower and Birch 1992) after the particle delivery system was introduced by Klein et al. (1987). Since then, genetic transformation technology has been used to introduce a number of genes of commercial value into sugarcane such as those conferring, resistance to biotic and abiotic stresses, herbicide tolerance, increased sucrose content, and improvement in sugar quality (Altpeter and Karan 2018; Nerkar et al. 2018).

Sugarcane is a C4 monocot and has large biomass, rapid growth rate, efficient carbon fixation pathway, and a well-developed above-ground storage tissue system in the form of parenchyma cells; making it an ideal candidate for biopharming applications (Palaniswamy et al. 2016; Appunu et al. 2017). Sugarcane is predominantly a vegetatively propagated crop providing a higher transgene containment which makes it a "safe" platform for producing recombinant proteins. A rapid in vitro regeneration protocol and an efficient transformation protocol make sugarcane an ideal crop for biopharming among other crops (Arencibia et al. 1998; Lakshmanan et al. 2005; Arvinth et al. 2010; Nerkar et al. 2018). Nonetheless, the complex polyploid–aneuploid genome of sugarcane has impeded crop improvement programs through breeding approaches. It therefore becomes crucial to review the possibilities of utilizing sugarcane as a platform for producing high-value molecules.

This chapter gives an update on the current status of transgenic sugarcane for biopharming especially for the production of bioplastics, alternative sugars, recombinant proteins, and cellulolytic enzymes.

11.2 Sugarcane for the Production of Liquid Crystal Polymers

The earliest attempt to use sugarcane for the expression of value-added products was the production of transgenic sugarcane accumulating high levels of pHBA (McQualter et al. 2004). Liquid crystal polymers (LCPs) are thermotropic polyesters and are potentially advantageous over conventional resins being electrically resistant, flame retardant, and the ability to withstand higher temperatures and can therefore be applied in electrical and optical connectors, integrated circuit boards, vehicle ignition components, and mobile phone components (McQualter et al. 2004). pHBA (p-hydroxy benzoic acid) is the main monomer in LCPs and a precursor for the chemical synthesis of parabens, mostly used as a preservative in the food and cosmetic industry. The industrial production of pHBA is relatively expensive. McQualter et al. (2004) suggested the use of tissue-specific promoters to enhance the production of recombinant protein in the stalks rather than the use of conventional constitutional promoters thus providing valuable insights into approaches that need to be taken into account for the sugarcane pHBA production process that is commercially viable.

11.3 Sugarcane for the Production of Bioplastics

PHAs are biodegradable polymers made by bacteria. There have been considerable efforts for the production of polyhydroxybutrate (PHB) in sugarcane (Brumbley et al. 2004, 2007; Petrasovits et al. 2012). Till the first report of PHB production in sugarcane, the three gene pathways, PhaA, PhaB, and PhaC had been successfully engineered into several plant species. These reports indicated either low levels of PHB accumulation or adverse effects of the products from the PHB biosynthetic pathway on the transgenic plants. According to Brumbley et al. (2004), lines that were targeted to the chloroplast accumulated PHB at levels of up to 1.88% of dry weight, whereas lines that were targeted to the cytosol accumulated PHB at a level of 0.01%. These lines were products of the *Ralstonia eutropha* PHB biosynthetic pathway. However, no PHB accumulation was seen in the mitochondria-targeted lines. In the glasshouse trial of six separate lines of PHB-producing sugarcane, PHB levels remained unchanged during the duration of the trial. Although, the PHB accumulation had no detrimental effects on the important agronomic characteristics like height, weight, and sugar levels; as a result of switching from the carbon pathway to the phenylpropanoid pathway, lignin composition changes, a decrease in leaf chlorogenic acid, and an increase in phenylalanine ammonia-lyase were noted (Brumbley et al. 2007).

Out of the total number of transgenic sugarcane lines generated only 20% lines produced PHB at quantifiable levels by high-performance liquid chromatography (HPLC) and only 4% of the lines showed levels over 1% dry weight in mature leaf tips. Thus, a relatively low frequency of "high producing lines" was achieved. Purnell et al. (2007) aimed to the characterization of the spatio-temporal pattern of PHB accumulation in transgenic sugarcane lines (whole-plants) for the identification of factors limiting the production of PHB in sugarcane, and determination of the adverse effects of PHB production on agronomic performance of sugarcane. They observed a vertical PHB concentration gradient throughout the plant with the polymer concentration being lowest in the youngest leaves and increasing with leaf age. Additionally, a horizontal gradient was observed along the length of a leaf, with the PHB concentration increasing from the youngest part of the leaf (the base) to the oldest (the tip). Despite significant variations in absolute PHB concentration, the transgenic lines exhibited a consistent spatiotemporal pattern of relative PHB accumulation. The maximum PHB concentration observed was 1.77% of leaf dry weight, with no deleterious effect on the agronomic traits of the plants viz., the plant height, total aerial biomass, and culm-internode sugar content. However, the maximum total-plant PHB yield was only 0.79% (11.9 g PHB in 1.51 kg dry weight).

Plant peroxisomes contain the substrate molecules and necessary reducing power for PHB biosynthesis but peroxisomal PHB production remained unexplored in soil-grown transgenic plants until Tilbrook et al. (2011) for the first time highlighted the use of peroxisomal metabolism for PHA biosynthesis for reaching commercial production levels of PHAs in crop plants. Transgenic sugarcane (*Saccharum* sp.) lines were generated with the three-enzyme *Ralstonia eutropha* PHA biosynthetic pathway targeted to peroxisomes and it was observed that PHB accumulated throughout most leaf cell types in both peroxisomes and vacuoles in sugarcane at levels up to 1.6–1.8% dry weight with no obvious deleterious ramification on plant growth (Tilbrook et al. 2011).

Petrasovits et al. (2012) used a multi-promoter [the maize ubiquitin promoter (Ubi1), rice ubiquitin 2 promoter (rubi2), and the maize chlorophyll A / B-binding protein promoter (cab-m5)] multi-gene approach for enhanced PHB production in sugarcane to levels equivalent to or more than those attained in switchgrass. This work utilized the cab-m5 promoter for the first time in sugarcane to drive the transgene expression. Total leaf protein analysis from plants grown in glasshouses revealed that expression from the cab-m5 promoter produced significantly higher levels of *PhaA* and *PhaB* than expression from the ubiquitin promoters in maize and rice. However, the immunological detection of *PhaC* was not reported in this study (Petrasovits et al. 2012). Petrasovits et al. (2012) also proposed that in sugarcane, mesophyll cells appear to be attractive targets for PHB production as they occupy approximately 2.4 times more leaf volume compared to bundle sheath cells.

11.4 Sugarcane for the Production of Alternative Sugars

As sucrose (a-D-glucopyranosyl-1, 2-D-fructofuranose), the primary storage product of sugarcane, is a valuable source for feedstocks and sustainable production of value-added biomaterials, sugarcane is a desirable candidate for metabolic engineering (Birch 2007). The most common sugar in the world, sucrose is a non-reducing glucose-fructose disaccharide that is essential to plants as a transport carbohydrate. It may have originated much earlier as a sign of food safety and nutritional value and has been consumed by humans for tens of thousands of years (van der Poel et al. 1998). Certain microorganisms transform sucrose into structural isomers, such as trehalose (a-D-glucopyranosyl-1,1-D-fructose) and isomaltulose (palatinose, a-D-glucopyranosyl-1,6-D-fructofuranose), which may sequester the sugar in a way that gives them an advantage over rival species (Watzlawick and Mattes 2009). These sucrose isomers are naturally occurring sugars that combine the sweet taste profile of sucrose with potential health benefits for consumers, such as slower digestion into monosaccharides and carcinogenicity (Lina et al. 2002). They also have greater acid stability than sucrose.

11.4.1 Trehalulose

The high solubility of Trehalulose makes it useful for making foods like jams and for industrial applications that call for extremely concentrated sugar solutions. Due to the costly microbial or enzymatic conversion from the more common, plant-derived sucrose, the use of sucrose isomers is restricted (Schiweck et al. 1991). Through a multi-step mechanism, sucrose isomerase (SI) enzymes can convert dissolved sucrose to isomers without the need for cofactors. The conversion process is remarkably complete due to the lower free energy of the isomers. More effective isomer production from sucrose in genetically modified plants is now possible by cloning of the genes encoding the microbial Sis (Sucrose Isomerases). Transgenic sugarcane plants expressing the vacuole-targeted trehalose synthase, modified from the gene in "*Pseudomonas mesoacidophila* MX-45" and driven by the maize ubiquitin (Ubi-1) promoter, were produced by Hamerli and Birch (2011). Trehalose and isomaltulose, two isomers of sucrose, were produced by the transgenic line. As internodes matured, the concentration of trehalose in the juice grew and reached around 600 mM. In the most mature internodes, the sucrose was nearly completely converted. Under both glasshouse and field environments, the transgenic sugarcane plants continued to grow vigorously, and the production of trehalose in certain lines remained stable over several vegetative generations.

According to O'Neill et al. (2012), trehalose metabolism was engineered in sugarcane through the introduction of two transgenes: an RNAi transgene specific for trehalase, which revokes trehalose catabolism, and trehalose-6-phosphate synthase-phosphatase (TPSP), which increases trehalose biosynthesis. Trehalase expression was disrupted in the RNAi-expressing lines, but trehalase activity did not drop. There was increased trehalase activity in the TPSP lines. Co-integration of the TPSP and RNAi transgenes did not occur. They proposed that trehalase activity contributes to the reduction of trehalose metabolism's lethal effects on embryos. In transgenic sugarcane plants where sucrose metabolism is engineered, a strong correlation between sucrose and trehalose content may be used to understand trehalose metabolism and determine whether RNAi transgene can increase tolerance to abiotic stress. Research on the expression profiling of *ScTPS* genes in sugarcane and genome-wide analysis of the Trehalose-6-phosphate synthase (TPS) gene family (Hu et al. 2020) have demonstrated the role of *ScTPS* genes in sugarcane adaptation to environmental stimuli. These genes may also be used as a molecular marker in screening sugarcane germplasm for increased stress resistance.

11.4.2 Sorbitol

Making the most of the plant's capacity to produce sucrose can be achieved by combining the synthesis of substitute products with the metabolic intermediates of sucrose metabolism to create an efficient *in planta* sugarcane-based production system. According to Chong et al. (2007), *Saccharum* spp.'s intermediates in the metabolism of sucrose can provide striking yields of alternative products. For the purpose of sorbitol synthesis in sugarcane, they reported transgenic sugarcane plants expressing the *Malus domestica* sorbitol-6-phosphate dehydrogenase gene (*MdS6PDH*). In the leaf lamina, the most abundant transgenic line had mean levels of sorbitol of 120 mg/g dry weight (or 61% of the soluble sugars) and 10 mg/g dry weight in the stalk pith. Enzymes involved in sucrose synthesis and cleavage were found in higher concentrations in the leaves of plants accumulating sorbitol, but this had no effect on sucrose accumulation within the culm. Compared to control lines, sugarcane lines that accumulated sorbitol exhibited 30–40% lower aerial biomass and were 10–30% shorter. The leaves displayed a recognizable pattern of early senescence, and the amount of sorbitol that accumulated in relation to the severity of necrosis in the leaves was proportionate. Plants that were co-expressed the *Zymomonas mobilis* glucokinase (ZmGLK) gene with *MdS6PDH* to produce more glucose-6-phosphate were smaller, suggesting that the reduced growth was not due to a glucose-6-phosphate deficiency. Thus, sugarcane growth and metabolism were impacted by sorbitol hyperaccumulation. All the same, the result did not kill the plant. Abnormal leaf physiology may result from disturbances in the allocation and transport of carbohydrates (Sheveleva et al. 1998). As a result, GC-MS metabolic profiling was used to analyze the changes in leaf carbohydrates. This method systematically measures the deviation from metabolic homeostasis that can happen in plants that are exposed to abiotic stress, infection, and genetic modification (Roessner et al. 2001), as well as metabolic changes that are associated with sugarcane culm development (Glassop et al. 2007). Chong et al. (2010) examined the polar metabolites in leaf extracts derived from sorbitolcane which unveiled the presence of a novel sorbitol glucoside, which they structurally characterized and found to be a sorbitol–glucose conjugate.

11.4.3 Isomaltulose

One structural isomer of sucrose called isomaltulose (IM) has some potential benefits, such as a lower glycaemic index due to its delayed digestibility and decreased cariogenicity. An expensive industrial conversion from sucrose using immobilized SI enzymes limits its production. Under greenhouse conditions, transgenic sugarcane lines with vacuole-targeted IM synthase (*ims* gene) demonstrated high accumulation of IM (Wu and Birch 2007). According to Basnayake et al. (2012), transgenic sugarcane plants expressing a vacuole-targeted isomaltulose synthase were evaluated in the field over a three-year period in seven recipient genotypes (elite cultivars). The results showed that the concentration of isomaltulose was directly correlated with internode maturity and made up as much as 217 mm (33% of total sugars) in whole-cane juice. Although there was no overall drop in the amount of sugars, the transgenic lines displayed a similar decrease in sucrose concentration. The development of stools, sett germination, and stalk height and diameter measurements of vigor did not reveal any negative effects from IM accumulation. These transgenic lines exhibited low levels of sucrose isomerase activity. With the help of appropriate developmental promoters and gene sequence alterations, higher-level production can be accomplished in order to produce consistent patterns of SI activity and improved IM profiles along stalks. All through the season of commercial harvesting.

Mudge et al. (2013) showed that large IM yields over several generations in field-grown plants are possible by developmentally regulated expression of a silencing-resistant gene for vacuolar SI activity. Whole-cane juice from plants that were 13 months old showed high yields of IM (up to 483 mm, or 81% of total sugars). IM levels were consistent between stalks and stools within a transgenic line and over several vegetative field generations of tested high-isomer lines, using sugarcane promoters to drive expression primarily in the sugarcane stem (ScLSG). Up to the tested level of about 500 mM in flanking stem internodes, IM accumulation had no effect on the germination and early growth of plants from setts. Thus, using the sugarcane stem promoters appears to be a promising strategy for achieving higher IM levels as well as for other applications in sugarcane molecular improvement.

11.5 Sugarcane for the Production of Recombinant Proteins

11.5.1 Human Cytokine Granulocyte-Macrophage Colony-Stimulating Factor GM-CSF19

The human body produces GM-CSF, a cytokine, in small amounts from a variety of cell types (Metcalf 1991). Purified protein is given to patients receiving bone marrow transplants to lower the risk of infection by speeding up the neutrophil response. It may find use in the treatment of neutropenia and aplastic anemia (Shin et al. 2003). Granulocyte-macrophage colony-stimulating factor (GM-CSF), a human cytokine, is produced by over 300 transgenic sugarcane plants representing approximately 200 independent lines. Wang et al. (2005) provided the first report on the field production of GM-CSF in sugarcane by analyzing the plants for recombinant protein accumulation and activity levels. The researchers employed distinct expression constructs that comprised distinct promoters, namely maize polyubiquitin 1 (Ubi-1) or sugarcane polyubiquitin 9 (ubi9); additionally, they examined the presence or absence of a C-terminal HDEL tag for ER retention and a 6X Histidine tag for metal ion affinity purification. The amount of GM-CSF protein accumulated varied from 0.02% of total soluble protein to undetectable. Promoters did not differ significantly, but higher accumulation levels were associated with the ER retention tag. Human bone marrow cells (TF-1), which require GM-CSF for cell division, were used to test the transgenic sugarcane extracts. It was shown that these cells proliferated well when growth media was supplemented with transgenic sugarcane extracts. The activity levels of the pure, commercially manufactured GM-CSF and the protein derived from sugarcane were nearly equal. GM-CSF accumulation levels was steady during the course of a 14-month field study.

11.5.2 Canecystatins (Cysteine Protease Inhibitors) CaneCP-1, CaneCP-2, and CaneCP-3

Plants naturally produce phytocystatins, which are cysteine proteinase inhibitors that control protein turnover, programmed cell death, and pathogen defense mechanisms. The first cystatin identified in sugarcane was canecystatin (Soares-Costa et al. 2002; Oliva et al. 2004). The purification and heterologous expression of two novel sugarcane cystatins were reported by Gianotti et al. (2006). Using the pET28 expression vector, they subcloned and expressed two ORFs encoding cystatins (CaneCPI-2 and CaneCPI-3). Through affinity chromatography, large concentrations of pure recombinant proteins (about 20 mg/L) were produced in a single purification step. Immunodetection of endogenous proteins in plant tissues was made feasible by the expression of polyclonal antibodies against the recombinant Canecystatins. In a fluorometric assay, the proteins inhibited papain with K(i) values of 0.25 μM for CaneCPI-3 and 0.2 μM for CaneCPI-2. These results stimulate further activity and structural investigations of these proteins and advance our knowledge of the function of sugarcane cystatins.

11.5.3 Bovine Lysozyme (BvLz)

An effective antimicrobial enzyme used in the food, cosmetics, and agricultural industries is called bovine lysozyme (BvLz). Its effectiveness against Gram-negative bacteria and fungi at concentrations as low as 25 ppm and its six-fold higher chitinase activity compared to chicken lysozyme makes it preferred over other lysozymes. Damaj et al. (2020) reported a novel approach to achieve enhanced expression of recombinant proteins through the use of a combinatorial stacked promoter system. They demonstrated co-expression of BvLz under the control of multiple constitutive and culm-regulated promoters on separate expression vectors and combinatorial plant transformation. Compared to single promoter BvLz lines, where BvLz accumulation was 0.07% of TSP (0.56 mg/kg), stacked multiple promoter BvLz lines showed BvLz accumulation of 1.4% of total soluble protein (TSP), and 10.0 mg BvLz/kg culm mass. Through event stacking, BvLz accumulation was further increased to 11.5% of TSP (82.5 mg/kg) by repurposing the stacked promoter—BvLz lines with extra BvLz expression vectors. Interestingly, transgenic lines with combinatorial promoter stacking exhibited greater transgene expression stability, which is a significant step toward proving that sugarcane can be used as a biofactory to produce high-value proteins and bioproducts.

11.6 Vacuolar Targeting in Sugarcane for Expression of Recombinant Proteins

The majority of previous attempts (Wang et al. 2005) produced extremely low yields (0.02% of total soluble protein) of recombinant proteins in sugarcane as opposed to tobacco (2%) and rice (1.3%) (Jackson et al. 2007). Compared to subcellular targeting to other compartments like the cytoplasm, endoplasmic reticulum, or apoplasm, targeting the recombinant proteins to vacuolar compartments seems to be a more effective tactic (Appunu et al. 2017). In sugarcane stem parenchyma cells, in particular, the size of the vacuoles can greatly increase the protein yield (Appunu et al. 2017). Palaniswamy et al. (2016) described a process to separate the recombinant protein from sugarcane juice as well as the vacuolar targeting of GUS and aprotinin in sugarcane. Green fluorescent protein (*GFP*) gene was fused with a putative vacuolar targeting sequence (78 bp) from the N-terminal domain of unknown function (DUF) in Triticum aestivum 6-SFT (sucrose: fructan 6-fructosyl transferase) and used under strong constitutive promoter (Port ubi882) for efficient transgene expression. This work demonstrated a streamlined procedure for the recovery of the r-proteins from the juice and opened the possibility of localizing recombinant proteins to the lytic vacuole using a novel vacuolar targeting determinant (Palaniswamy et al. 2016). The expression of the three recombinant cellulolytic enzymes in the mature leaves of transgenic sugar cane was first reported by Harrison et al. (2011). The constitutive maize phosphoenolpyruvate carboxylase (*Zm-PepC*) promoter was responsible for driving the transgenes. The accumulation of cellulase enzymes in various subcellular compartments, including vacuoles, the endoplasmic reticulum, and chloroplasts, was compared in this study. The use of cellulolytic enzyme expression in plants to enhance the economics of lignocellulosic ethanol production from sugar cane biomass was a major contribution of this work.

11.7 Conclusion and Prospects

Sugarcane is the most productive cultivated field crop due to it efficient carbon fixation pathway, large biomass, and storage in the form of parenchymatous tissue. However, the technical constraints need to be addressed it is a challenging area in research and development for the potential development of alternative sugarcane industries worldwide. The complex polyploid genome has limited the exploitation of sugarcane variability thereby hampering sugarcane improvement by conventional breeding approaches. Sugarcane improvement through transgenic approaches has made considerable progress over the last three decades. However, the polyploidy genome of sugarcane remained a major bottleneck in developing stable transgenic lines. Fortunately, the huge number of target copies per cell is advantageous for genome editing as it may increase the probability of homologous recombination thereby providing a unique opportunity for crop improvement using gene targeting. The development of transgenic approaches to turn sugarcane into a biofactory, which are covered in this chapter, shows how new discoveries in knowledge, technology, and resources can help sugarcane expand its uses beyond food production and bioenergy feedstock to include using it as a biofactory to express high-value proteins. Addressing the biosafety concerns should be an important consideration in reaching our goal of development of a novel sugarcane biofactory in the future.

References

Altpeter F, Karan R (2018) Genetic improvement of sugarcane by transgenic, intragenic and genome editing technologies. In: Rott P (ed) Achieving sustainable cultivation of sugarcane, vol 2. Burleigh Dodds Science Publishing, Cambridge, pp 133–154

Appunu C, Ram B, Subramonian N (2017) Sugarcane: an efficient platform for molecular farming. In: Mohan C (ed) Sugarcane biotechnology: challenges and prospects. Springer, Cham. https://doi.org/10.1007/978-3-319-58946-6_7

Arencibia AD, Carmona ER, Téllez P, Chan MT, Yu SM, Trujillo LE, Oramas P (1998) An efficient protocol for sugarcane (*Saccharum spp. L.*) transformation mediated by *agrobacterium tumefaciens*. Transgenic Res 7(3):213–222

Arvinth S et al (2010) Genetic transformation and pyramiding of aprotinin-expressing sugarcane with cry1Ab for shoot borer (*Chilo infuscatellus*) resistance. Plant Cell Rep 29(4):383–395. https://doi.org/10.1007/s00299-010-0829-5

Basnayake SW, Morgan TC, Wu L, Birch RG (2012) Field performance of transgenic sugarcaneexpressing isomaltulose synthase. Plant Biotechnol J 10:217–225. https://doi.org/10.1111/j.1467-7652.2011.00655.x

Birch RG (2007) Metabolic engineering in sugarcane: assisting the transition to a bio-based economy. In: Verpoorte RA, Alfermann W, Johnson TS (eds) Applications of plant metabolic engineering. Springer, Berlin, pp 249–281

Bower R, Birch RG (1992) Transgenic sugarcane plants via microprojectile bombardment. Plant J 2:409–416

Brumbley SM, Purnell MP, Petrasovits LA, O'Sheal MG, Nielsen LK (2004) Development of sugarcane as a biofactory for biopolymers. In: Proceedings of plant and animal genomes XII conference, San Diego, CA. pp 10–14 Jan 2004 http://www.intl-pag.org/12/abstracts/W27_PAG12_117.html

Brumbley SM, Purnell MP, Petrasovits LA, Nielsen LK, Twine PH (2007) Developing the sugarcane biofactory for high-value biomaterials. Int Sugar J 109:5–15

Chong BF, Bonnett GD, Glassop D, O'Shea MG, Brumbley SM (2007) Growth and metabolism in sugarcane are altered by the creation of a new hexose-phosphate sink. Plant Biotechnol J 5:240–253

Chong BF, Abeydeera WPP, Glassop D, Bonnett GD, O'Shea MG, Brumbley SM (2010) Coordinated synthesis of gentiobiitol and sorbitol, evidence of sorbitol glycosylation in transgenic sugarcane. Phytochemistry 71:736–741

D'Hont A, Layssac M (1998) Analysis of cultivars genome structure by molecular cytogenetics and the study of introgression mechanisms. France: Centre de coope ´ration internationale en recherche agronomique pour le de'veloppement, Annual Crops Department (CIRAD-CA), 9–10

Dal-Bianco M, Carneiro SM, Hotta CT, Chapola RG, Hoffmann HP, Garcia AA, Souza GM (2012) Sugarcane improvement: how far can we go? Curr Opin Biotechnol 23:265–270. https://doi.org/10.1016/j.copbio.2011.09.002

Damaj MB, Jifon JL, Woodard SL et al (2020) Unprecedented enhancement of recombinant protein production in sugarcane culms using a combinatorial promoter stacking system. Sci Rep 10:13713. https://doi.org/10.1038/s41598-020-70530-z

Daniels J, Roach BT (1987) Taxonomy and evolution. In: Heinz DJ (ed) Sugarcane improvement through breeding. Elsevier, Amsterdam, pp 7–84

Gianotti A, Rios WM, Soares-Costa A, Nogaroto V, Carmona AK, Oliva MLV, Henrique-Silva F (2006) Recombinant expression, purification, and functional analysis of two novel cystatins from sugarcane (*Saccharum officinarum*). Protein Expr Purif 47(2):483–489. https://doi.org/10.1016/j.pep.2005.10.026

Glassop D, Roessner U, Bacic A, Bonnett GD (2007) Changes in the sugarcane metabolome with stem development. Are they related to sucrose accumulation? Plant Cell Physiol 48:573–584

Gomez-Merino FC, Trejo-Tellez LI, Senties-Herrera HE (2014) Sugarcane as a novel biofactory:potentialities and challenges. In: Guevara-Gonzalez R, Torres-Pacheco I (eds) Biosystemsengineering: biofactories for food production in the century XXI. Springer, Cham, pp 129–149

Hamerli D, Birch RG (2011) Transgenic expression of trehalulose synthase results in high concentration of the sucrose isomer trehalulose in mature stems of field-grown sugarcane. Plant Biotech J 9:32–37. https://doi.org/10.1111/j.1467-7652.2010.00528.x

Harrison MD, Geijskes J, Coleman HD, Shand K, Kinkema M, Palupe A, Hassall R, Sainz M, Lloyd R, Miles S, Dale JL (2011) Accumulation of recombinant cellobiohydrolase and endoglucanase in the leaves of mature transgenic sugar cane. Plant Biotechnol J 9(8):884–896. https://doi.org/10.1111/j.1467-7652.2011.00597.x

Hu X, Zhuan-Di W, Luo Z-Y, Burner DM, Pan Y-B, Cai-Wen W (2020) Genome-wide analysis of the trehalose-6-phosphate synthase (TPS) gene family and expression profiling of ScTPS genes in sugarcane. Agronomy 10(7):969. https://doi.org/10.3390/agronomy10070969

Jackson MA, Rae AL, Casu RE, Grof CPL, Bonnet GD, Maclean DJ (2007) A bioinformatics approach to the identification of a conserved domain in of a sugarcane legumain that directs GFP to the lytic vacuole. Funct Plant Biol 34:633–644

Klein TM, Wolf ED, Wu R, Sanford JC (1987) High-velocity microprojectiles for delivering nucleic acids into living cells. Nature 327:70–73

Lakshmanan P, Geijskes RJ, Aitken KS, Grof CLP, Bonnett GD, Smith GR (2005) Sugarcane biotechnology: the challenges and opportunities. In Vitro Cell Dev Biol Plant 41(4):345–363

Lina BAR, Jonker D, Kozianowski G (2002) Isomaltulose (Palatinose): a review of biological and toxicological studies. Food Chem Toxicol 40:1375–1381

McQualter RB, Chong BF, Meyer K et al (2004) Initial evaluation of sugarcane as a production platform for p-hydroxybenzoic acid. Plant Biotechnol J 3(1):29–41. https://doi.org/10.1111/j.1467-7652.2004.00095.x

Metcalf D (1991) Control of granulocytes and macrophages: molecular, cellular, and clinical aspects. Science 254(5031):529–533

Mudge SR, Basnayake SW, Moyle RL, Osabe K, Graham MW, Morgan TE, Birch RG (2013) Mature-stem expression of a silencing-resistant sucrose isomerase gene drives isomaltuloseaccumulation to high levels in sugarcane. Plant Biotechnol J 11:502–509. https://doi.org/10.1111/pbi.12038

Nerkar G, Thorat A, Sheelavantmath S, Babu KH, Devarumath R (2018) Genetic transformation of sugarcane and field performance of transgenic sugarcane. In: Biotechnologies of crop improvement, vol 2. Springer Nature, Switzerland AG. https://doi.org/10.1007/978-3-319-90650-8_9

O'Neill BP, Purnell MP, Nielsen LK, Brumbley SM (2012) RNAi-mediated abrogation of trehalase expression does not affect trehalase activity in sugarcane. Springer Plus 1:74. https://doi.org/10.1186/2193-1801-1-74

Oliva MLV, Carmona AK, Andrade SS, Cotrin SS, SoaresCosta A, Henique-Silva F (2004) Inhibitory selectivity of canecystatin: a recombinant cysteine peptidase inhibitor from sugarcane. Biochem Biophys Res Commun 320:1082–1086

Palaniswamy H, Syamaladevi DP, Mohan C, Philip A, Petchiyappan A, Narayanan S (2016) Vacuolar targeting of r-proteins in sugarcane leads to higher levels of purifiable commercially equivalent recombinant proteins in cane juice. Plant Biotechnol J 14:791–807

Petrasovits LA, Zhao L, McQualter RB, Snell KD, Somleva MN, Patterson NA, Nielsen LK, Brumbley SM (2012) Enhanced polyhydroxybutyrate production in transgenic sugarcane. Plant Biotechnol J 10:569–578. https://doi.org/10.1111/j.1467-7652.2012.00686.x

Purnell MP, Petrasovits LA, Nielsen LK, Brumbley SM (2007) Spatio-temporal characterization of polyhydroxybutyrate accumulation in sugarcane. Plant Biotechnol J 5:173–184. https://doi.org/10.1111/j.1467-7652.2006.00230.x

Ram B, Hemaprabha G, Singh BD et al (2022) History and current status of sugarcane breeding, germplasm development and molecular biology in India. Sugar Tech 24:4–29. https://doi.org/10.1007/s12355-021-01015-0

Roessner U, Luedemann A, Brust D, Fiehn O, Linke T, Willmitzer L, Fernie AR (2001) Metabolic profiling allows comprehensive phenotyping of genetically or environmentally modified plant systems. Plant Cell 13:11–29

Schiweck H, Munir M, Rapp KM, Schneider B, Vogel M (1991) New developments in the use of sucrose as an industrial bulk chemical. In: Lichtenthaler FW (ed) Carbohydrates as organic raw materials. Wiley-VCH, Weinheim, pp 57–94

Sheveleva EV, Marquez S, Chmara W, Zegeer A, Jensen RG, Bohnert HJ (1998) Sorbitol-6-phosphate dehydrogenase expression in transgenic tobacco. High amounts of sorbitol lead to necrotic lesions. Plant Physiol 117:831–839

Shin YJ, Hong SY, Kwon TH, Jang YS, Yang MS (2003) High level of expression of recombinant human granulocyte-macrophage colony stimulating factor in transgenic rice cell suspension culture. Biotechnol Bioeng 82(7):778–783

Soares-Costa A, Beltramini LM, Thiemann F, Henrique-Silva A (2002) Sugarcane cystatin: recombinant expression, puriWcation, and antifungal activity. Biochem Biophys Res Commun 296:1194–1199

Sreenivasan TV, Ahloowalia BS, Heinz DJ (1987) Cytogenetics. Chapter 5. In: Heinz DJ (ed) Sugarcane improvement through breeding. Elsevier, Amsterdam, pp 211–253

Tilbrook K, Gebbie L, Schenk PM, Poirier Y, Brumbley SM (2011) Peroxisomal polyhydroxyalkanoate biosynthesis is a promising strategy for bioplastic production in high biomass crops. Plant Biotechnol J 9:958–969. https://doi.org/10.1111/j.1467-7652.2011.00600.x

van der Poel P, Schiweck H, Schwartz T (1998) Sugar technology: beet and cane sugar manufacture. Verlag Dr. Albert Bartens KG, Berlin

Wang ML, Goldstein C, Su W, Moore PH, Albert HH (2005) Production of biologically active GM-CSF in sugarcane: a secure biofactory. Transgenic Res 14:167–178. https://doi.org/10.1007/s11248-004-5415-6

Watzlawick H, Mattes R (2009) Gene cloning, protein characterization, and alteration of product selectivity for the trehalulose hydrolase and trehalulose synthase from "*Pseudomonas mesoacidophila*" MX-45. Appl Environ Microbiol 75:7026–7036

Wu L, Birch RG (2007) Doubled sugar content in sugarcane plants modified to produce a sucrose isomer. Plant Biotechnol J 5:109–117

Sugarcane Bagasse: Transforming Waste into Renewable and Bio-Compostable Materials for a Sustainable Future

12

Amaresh, Aswini Nunavath, Vinayaka, Shweta Kumari, G. Avinash, G. J. Abhishek, V. G. Dhanya, R. T. Maruthi, Gopalareddy Krishnappa, Huskur Kumaraswamy Mahadeva Swamy, Giriyapura Shivalingamurthy Suresha, and K. Elayaraja

Abstract

Sugarcane bagasse (SCB), the fibrous residue left after juice extraction from sugarcane stalks, has emerged as a promising solution to address the dual challenges of waste management and sustainable material sourcing. This chapter explores the potential of SCB as a renewable and bio-compostable material. With the global sugar industry generating millions of tons of bagasse annually, its utilization offers an opportunity to mitigate waste disposal issues while promoting eco-friendly alternatives to conventional materials. The composition of SCB,

Amaresh · Vinayaka · S. Kumari (✉) · R. T. Maruthi · K. Elayaraja
ICAR-Sugarcane Breeding Institute, Coimbatore, India
e-mail: shweta.kumari1@icar.gov.in

A. Nunavath
ICAR-National Institute of Biotic Stress Management, Raipur, India

G. Avinash
ICAR-Indian Agricultural Statistics Research Institute, New Delhi, India

G. J. Abhishek
ICAR-National Bureau of Plant Genetic Resources, New Delhi, India

V. G. Dhanya
ICAR-Sugarcane Breeding Institute, Regional Centre, Kannur, India

G. Krishnappa · H. K. M. Swamy
Genetics and Plant Breeding, ICAR-Sugarcane Breeding Institute, Coimbatore, Tamil Nadu, India

G. S. Suresha
Biochemistry (Plant Science), ICAR-Sugarcane Breeding Institute, Coimbatore, Tamil Nadu, India

G. S. Suresha et al. (eds.), *Value Addition and Product Diversification in Sugarcane*, https://doi.org/10.1007/978-981-97-7228-5_12

rich in cellulose, hemicellulose, and lignin, provides a versatile platform for the development of various bioproducts. Through innovative processing techniques such as enzymatic hydrolysis, pyrolysis, and fermentation, bagasse can be transformed into bioplastics, biofuels, and biochemicals, thus reducing reliance on fossil fuels and mitigating carbon emissions. Moreover, the bio-compostable nature of SCB-derived materials presents a sustainable end-of-life solution, offering the potential for composting and organic recycling. This characteristic aligns with circular economy principles, facilitating the transition towards a more resource-efficient and environmentally conscious society. Challenges associated with the widespread adoption of SCB-based materials include technological optimization, economic viability, and infrastructure development for processing and distribution. However, ongoing research and development efforts are addressing these obstacles, driving innovation and scalability in the utilization of bagasse as a renewable resource. In a nutshell, SCB holds immense promise as a source of renewable and bio-compostable material, offering a sustainable alternative to traditional materials while contributing to waste reduction and environmental conservation efforts on a global scale.

Keywords

Sugarcane bagasse · Biofuel · Bioenergy · Renewable source · Sustainability

12.1 Introduction

Sugarcane (*Saccharum* spp.) is one of the most cultivated crops worldwide, primarily grown for sugar production. With the extraction of sugar from sugarcane, vast quantities of fibrous residue, known as sugarcane bagasse (SCB) are generated as a byproduct. An estimated 700 million tonnes of SCB are produced per year worldwide, with Brazil and India being the two biggest producers (Karp et al. 2022). Bagasse from sugarcane is also produced in large quantities in other nations that produce sugarcane, including China, Thailand, and Pakistan. Because sugarcane is widely grown throughout the world, there is an abundance of SCB. Furthermore, the sugar sector is continuously growing, which ensures a steady supply of bagasse from sugarcane. Traditionally, SCB has been considered a waste material and often burned as a means of disposal, contributing to air pollution and greenhouse gas emissions. However, in recent years, there has been a paradigm shift towards recognizing bagasse as a valuable resource with untapped potential. The significance of SCB lies in its renewable nature and abundance. Being one of the largest producers of agricultural residues in the world, it is a promising sustainable alternative to non-renewable resources (Loh et al. 2013). It is a rough-surfaced, fibrous material that is usually light brown in color. It has a high moisture content, which can vary from 45% to 70% based on how it is dried and stored. Moreover, its composition, primarily are lignocellulosic substances, such as cellulose, hemicellulose, and lignin, which make up roughly 40–50%, 20–30%, and 20–25% of the dry weight, respectively (Iwuozor et al. 2021), renders it suitable for various applications ranging from biofuel production to bio-composites and bio-based packaging. Its additional trace elements consist of proteins, extractives, and ash. When it comes to using bagasse in microbial cultures, it has many advantages over other crop residues such as wheat and rice straw, which have ash concentrations of 11.0% and 17.5%, respectively. The age and variety of the sugarcane, the time of year it is harvested, and the processing techniques used all affect the chemical makeup of SCB. Bagasse exhibits enhanced mechanical qualities, including tensile strength, flexural strength, flexural modulus, hardness, and impact strength, when suitable modifications and production processes are used. Bagasse's high yields and capacity for annual regeneration make it a valuable solar energy reservoir when compared to other agricultural wastes. The manufacturing of ethanol, or alcohol, uses more than 9% of bagasse (Pandey et al. 2000). Not only is ethanol a suitable substitute for fossil fuels but also has a low ecological impact. Aside from this, ethanol is a very adaptable chemical raw material that may be used to create a wide range of compounds (Pandey et al. 2000). Because it is recyclable, biodegradable, and reusable, it also meets environmental standards (Loh et al. 2013). By harnessing the potential of SCB, it is possible to reduce reliance on fossil fuels, mitigate environmental impacts, and promote a circular economy. In continuation, the subsequent sections highlight its important properties, as well as its eco-friendly advantages.

12.2 Sugarcane Bagasse: A Renewable Source

12.2.1 Textile Industry

Utilizing SCB in the textile industry is a burgeoning field that capitalizes on the fibrous nature of SCB and its potential as a sustainable alternative to traditional textile materials. This section will delve into the various aspects of sugarcane-based utilization in the textile industry, including fiber extraction, processing methods, applications, and sustainability implications (Costa et al. 2013; Sirlene et al. 2013; Sakthivel et al. 2021).

12.2.2 Fiber Extraction

Sugarcane fibers can be extracted from bagasse, the fibrous residue left behind after sugarcane is crushed to extract juice. There are several methods for extracting fibers from bagasse, including mechanical, chemical, and enzymatic processes. Mechanical extraction involves separating the fibers from the lignocellulosic matrix through mechanical means such as grinding, milling, or pressing. This method is relatively simple and cost-effective but may result in shorter fibers with reduced strength. Chemical extraction involves treating bagasse with chemicals such as alkalis or acids to dissolve lignin and hemicellulose, leaving behind cellulose fibers. This method produces longer and stronger fibers but may require extensive processing and pose environmental concerns due to chemical usage and waste disposal.

Enzymatic extraction utilizes enzymes to selectively degrade lignin and hemicellulose, leaving behind cellulose fibers. This method offers the advantage of being more environmentally friendly and producing high-quality fibers but may be slower and costlier than other methods.

12.2.3 Processing Methods

Once extracted, sugarcane fibers can undergo various processing methods to prepare them for use in textile applications. These methods include cleaning, refining, spinning, and weaving or knitting. Cleaning involves removing impurities and debris from the fibers through washing and screening processes to ensure purity and uniformity. Refining involves further refining the fibers to enhance their quality and properties, such as length, strength, and fineness. This may involve processes such as carding, combing, or blending with other fibers. Spinning transforms the refined fibers into yarns or threads through processes, such as ring spinning, open-end spinning, or rotor spinning, depending on the desired characteristics of the final product. Weaving or knitting involves interlacing the spun yarns or threads to create fabrics or textiles with desired patterns, textures, and properties.

12.2.3.1 Applications in the Textile Industry

12.2.3.1.1 Sugarcane Fibers Advantages for Textile Applications

- *Softness and comfort*: Sugarcane fibers are known for their softness and comfort, making them suitable for clothing and apparel applications.
- *Breathability*: Sugarcane fibers have good breathability, allowing air to circulate and moisture to evaporate, which enhances comfort and performance in clothing.
- *Durability*: Sugarcane fibers are durable and strong, making them suitable for use in various textile products, including garments, upholstery, and technical textiles.
- *Biodegradability*: Sugarcane fibers are biodegradable and compostable, offering environmental benefits compared to synthetic fibers derived from petroleum-based sources.

12.2.3.1.2 Textile Applications

- *Apparel*: Sugarcane fibers can be spun into yarns and woven or knitted into fabrics for clothing, including shirts, pants, dresses, and outerwear.
- *Home Textiles*: Sugarcane fibers can be used in home textile products, such as bed linens, towels, curtains, and upholstery fabrics.
- *Technical Textiles*: Sugarcane fibers can be incorporated into technical textiles for applications such as automotive interiors, filtration, geotextiles, and medical textiles.

12.2.4 Sustainability Implications

Utilizing sugarcane fibers in the textile industry offers several sustainability benefits:

- *Renewability*: Sugarcane is a renewable resource that can be harvested annually, making sugarcane fibers a sustainable alternative to nonrenewable textile materials.
- *Biodegradability*: Sugarcane fibers are biodegradable and compostable, reducing the environmental impact of textile waste compared to synthetic fibers.
- *Carbon Sequestration*: Sugarcane cultivation can sequester carbon dioxide from the atmosphere, contributing to climate change mitigation.
- *Resource Efficiency*: Sugarcane fibers can be produced using byproducts from the sugar industry, reducing waste and maximizing resource efficiency.

12.2.5 Paper Industry

Utilizing sugarcane in the paper industry is an innovative approach that leverages the fibrous characteristics of SCB, the residue left over after juice extraction from sugarcane, as a sustainable alternative to traditional wood-based pulp. This section will delve into the various aspects of sugarcane-based utilization in the paper industry, including pulp production, papermaking processes, applications, and sustainability implications (Kasiviswanathan 1998; Rainey and Covey 2016; Sayed et al. 2022; Azmi and Othman 2023).

12.2.5.1 Pulp Production from SCB

SCB serves as an excellent source of pulp for papermaking due to its high cellulose content and fibrous structure. SCB is first subjected to pulping processes to separate the cellulose fibers from lignin and other non-cellulosic components. Pulping methods commonly used for bagasse include mechanical pulping, chemical pulping, and semi-chemical pulping. Mechanical pulping involves mechanically grinding or refining the bagasse fibers to separate them from the lignocellulosic matrix. This method is relatively simple and cost-effective but may result in shorter fibers with reduced strength. Chemical pulping methods, such as kraft pulping or sulfite pulping, involve treating the bagasse with chemicals to dissolve lignin and separate the cellulose fibers. Chemical pulping produces longer and stronger fibers compared to mechanical pulping but may require extensive processing and pose environmental concerns due to chemical usage and waste disposal. Semi-chemical pulping combines mechanical and chemical processes to remove lignin and other impurities from bagasse fibers. This method offers a balance between fiber quality and process efficiency, making it suitable for certain paper grades.

12.2.5.2 Papermaking Processes

Once pulped, SCB fibers can undergo papermaking processes to produce a wide range of paper and paperboard products. These processes typically include:

- *Fiber Refining*: The pulped bagasse fibers may undergo refining processes to enhance their quality and properties, such as length, strength, and flexibility. Refining can involve mechanical or chemical treatments to modify fiber characteristics.
- *Sheet Formation*: The refined bagasse fibers are then mixed with water to form a slurry, which is deposited onto a wire mesh screen to form a continuous sheet. The sheet may undergo further processing, such as pressing and drying, to remove excess water and consolidate the fibers into a solid paper product.
- *Finishing*: Once dried, the paper product may undergo additional finishing processes, such as calendaring or coating, to enhance its surface properties, smoothness, and printability.

12.2.5.3 Applications in the Paper Industry

SCB-based paper products offer several advantages for various applications, including:

- *Printing and Writing Papers*: Sugarcane-based papers are suitable for printing and writing applications, including office papers, notebooks, and stationery products. These papers offer good printability, smoothness, and opacity, making them ideal for high-quality printing and writing.
- *Packaging Papers*: Sugarcane-based papers can be used in packaging applications, such as corrugated boxes, cartons, and bags. These papers offer strength, durability, and moisture resistance, making them suitable for protecting and transporting goods.
- *Specialty Papers*: Sugarcane-based papers can be tailored to specific requirements for specialty applications, such as tissue papers, specialty packaging, and industrial papers. These papers can be engineered to meet specific performance criteria, such as strength, absorbency, and chemical resistance.

12.2.5.4 Sustainability Implications

Utilizing SCB in the paper industry offers several sustainability benefits:

- *Resource Efficiency*: SCB is a byproduct of the sugar industry, making it readily available and abundant. By using bagasse as a raw material for paper production, it helps reduce waste and maximize resource efficiency.
- *Carbon Sequestration*: Sugarcane cultivation can sequester carbon dioxide from the atmosphere, contributing to climate change mitigation. Using bagasse-based paper products further enhances carbon sequestration by locking carbon in paper fibers.
- *Renewability*: Sugarcane is a renewable resource that can be harvested annually, making sugarcane-based paper products a sustainable alternative to paper made from wood pulp.
- *Biodegradability*: Sugarcane-based paper products are biodegradable and compostable, offering environmental benefits compared to synthetic paper products derived from petroleum-based sources.

12.2.6 Bio-Compostable Material

Utilizing SCB in the production of bio-compostable cups, cutleries, boxes, and furniture is an innovative approach that leverages the fibrous nature of sugarcane waste to create sustainable alternatives to traditional plastic and wood-based products. This section will explore in detail the various aspects of SCB utilization in these applications, including material properties, manufacturing processes, product examples, and sustainability implications (Narancic and O'Connor 2017; Kusuma et al. 2023).

12.2.6.1 Material Properties of SCB

These natural components of SCB give unique properties that make it suitable for various applications, such as biodegradability, compostability, strength and durability, and heat resistance.

12.2.6.2 Manufacturing Processes

The production of bio-compostable cups, cutleries, boxes, and furniture from SCB typically involves the following processes:

- Pulping: SCB is pulped to separate the cellulose fibers from the lignin and other non-cellulosic components. This can be done using mechanical, chemical, or enzymatic methods to break down the bagasse into a pulp.
- Molding or Forming: The bagasse pulp is then molded or formed into the desired shapes using molds or presses. This can be done using heat and pressure to create cups, cutleries, boxes, or furniture components.
- Drying: The molded products are dried to remove excess moisture and strengthen the structure. This can be done using air drying or heat drying methods.
- Finishing: The finished products may undergo additional finishing processes, such as polishing or coating, to enhance their appearance and durability.

12.2.6.3 Applications of SCB as Bio-Compostable Products

SCB-based products find applications in various sectors, including:

- *Bio-Compostable Cups*: Bio-compostable cups made from SCB are suitable for hot and cold beverage applications, including coffee, tea, and soft drinks. These cups are heat resistant, sturdy, and environmentally friendly, making them an ideal alternative to traditional plastic cups.
- *Bio-Compostable Cutleries*: Bio-compostable cutleries made from SCB are suitable for use with hot and cold foods. These cutleries are sturdy, heat resistant, and biodegradable, making them an eco-friendly alternative to plastic cutleries.
- *Bio-Compostable Boxes*: Bio-compostable boxes made from SCB are suitable for packaging food items such as takeaway meals, snacks, and bakery products. These boxes are sturdy, leak-proof, and biodegradable, making them an environmentally friendly choice for food packaging.

- *Bio-Compostable Furniture*: Bio-compostable furniture made from SCB is suitable for indoor and outdoor use. These furniture pieces are sturdy, durable, and weather-resistant, making them an eco-friendly alternative to traditional wood-based furniture.

12.2.6.4 Sustainability Implications

Utilizing SCB in the production of bio-compostable cups, cutleries, boxes, and furniture offers several sustainability benefits:

- *Reduction of Waste*: SCB-based products help reduce waste by utilizing a byproduct of the sugar industry that would otherwise be discarded or burned.
- *Carbon Sequestration*: Sugarcane cultivation helps sequester carbon dioxide from the atmosphere, contributing to climate change mitigation. Using SCB-based products further enhances carbon sequestration by locking carbon in the products.
- *Biodegradability and Compostability*: SCB-based products are biodegradable and compostable, offering environmental benefits compared to traditional plastic and wood-based products that may persist in the environment for hundreds of years.

12.2.7 Compostable Manure

Utilizing SCB in the production of compostable manure is an innovative approach that capitalizes on the organic nature of bagasse and its potential to enrich soil fertility. This section will explore in detail the various aspects of SCB utilization in compostable manure, including its composition, benefits for soil health, composting process, applications, and sustainability implications (Saranraj and Stella 2014; Nakhla et al. 2017; Diallo et al. 2017; Ajala et al. 2021; Alabi et al. 2023).

12.2.7.1 Composition and Properties of SCB

The composition of bagasse makes it an ideal candidate for composting and soil enrichment due to its properties. SCB has a high carbon content, which provides a source of energy for microorganisms involved in the composting process. Bagasse has a porous structure, which allows air and moisture to penetrate the compost pile, promoting aerobic decomposition. Bagasse contains various nutrients, such as nitrogen, phosphorus, and potassium, which are essential for plant growth and soil fertility. Bagasse is biodegradable and can be broken down by microorganisms in the soil, releasing nutrients and organic matter that benefit plant growth.

12.2.7.2 Benefits for Soil Health

Utilizing SCB in compostable manure offers several benefits for soil health and fertility:

- *Nutrient Enrichment*: Bagasse contains organic matter and nutrients that can improve soil fertility and provide essential elements for plant growth.
- *Microbial Activity*: Compostable manure made from bagasse encourages microbial activity in the soil, promoting the decomposition of organic matter and the release of nutrients.
- *Soil Structure*: Bagasse-derived compost improves soil structure by enhancing water retention, aeration, and drainage, leading to healthier and more productive soils.
- *pH Balance*: Bagasse compost helps maintain the pH balance of the soil, creating optimal conditions for plant growth and nutrient uptake.

12.2.7.3 Composting Process

The composting process involves the decomposition of organic materials, including SCB, under controlled conditions to produce compostable manure. The composting process typically consists of the following stages:

- *Collection and Shredding*: SCB is collected and shredded into smaller pieces to facilitate decomposition and increase surface area.
- *Mixing*: Bagasse is mixed with other organic materials such as food waste, yard trimmings, and manure to create a balanced compost pile with the proper carbon-to-nitrogen ratio.
- *Aeration and Moisture Management*: The compost pile is regularly turned or aerated to provide oxygen to aerobic microorganisms and ensure even decomposition. Moisture levels are monitored and adjusted to maintain a damp but not saturated environment.
- *Temperature Monitoring*: The compost pile undergoes thermophilic (heat-generating) phases, during which temperatures rise to kill pathogens and weed seeds. Temperature monitoring ensures that composting conditions are optimal for microbial activity.
- *Maturation*: After several weeks to months, depending on environmental conditions and composting methods, the compost matures into a stable, dark, crumbly material known as compostable manure, ready for use in soil enrichment.

12.2.7.4 Applications of Compostable Manure

Compostable manure made from SCB has various applications in agriculture, landscaping, and gardening:

- *Soil Amendment*: Compostable manure can be incorporated into soils to improve fertility, structure, and moisture retention, promoting healthy plant growth and crop yields.
- *Mulching*: Bagasse compost can be used as a mulch to suppress weeds, conserve soil moisture, and regulate soil temperature, enhancing plant health and reducing water usage.

- *Fertilizer*: Compostable manure serves as a natural fertilizer, providing essential nutrients and organic matter to plants without the risk of chemical runoff or soil contamination.

12.2.7.5 Sustainability Implications

Utilizing SCB in compostable manure offers several sustainability benefits:

- *Waste Reduction*: Bagasse compost utilizes a byproduct of the sugar industry that would otherwise be discarded or burned, reducing waste and environmental pollution.
- *Soil Health*: Bagasse compost improves soil health and fertility, reducing the need for synthetic fertilizers and pesticides while promoting sustainable agriculture practices.
- *Carbon Sequestration*: Composting bagasse sequesters carbon dioxide from the atmosphere, mitigating climate change by reducing greenhouse gas emissions, and enhancing soil carbon storage.

With growing concerns about soil degradation, water pollution, and climate change, bagasse-based composting represents a promising strategy for sustainable agriculture and land management.

12.2.8 Production of Biofuels and Bioproducts Using SCB

SCB can be utilized for producing various biofuels and bioproducts through hydrolysis after pre-treatment, enabling microbial activity for fermentation. Enzyme-catalyzed hydrolysis is crucial for breaking down complex sugars in SCB into simpler forms like glucose, making them accessible for fermentation by microorganisms. Commercial cellulases, such as Spezym, Novozym, and viscozyme, as well as cultured enzymes like Cellulomonas flavigena and Trichoderma reesei, are employed for efficient hydrolysis of SCB. The optimal conditions for enzymatic hydrolysis of SCB include a temperature range of 40–50 °C, acidic medium (pH 4.8–6.0), moderate agitation speed, and around 72 hours of hydrolysis to yield high concentrations of glucose and other hydrolysates. Acid hydrolysis using inorganic acids like sulphuric acid or phosphoric acid is another method to obtain glucose syrup from SCB. Reportedly acid hydrolysis was to be preferred over enzymatic hydrolysis, as it is a faster process and highly efficient.

12.2.8.1 Biofuels

SCB serves as a valuable feedstock for producing biofuels like bio-ethanol, bio-methane, bio-hydrogen, and bio-butanol, contributing to sustainable energy production. Studies have reported SCB as a feedstock for bio-methane production, with acid pre-treatment or enzymatic pre-treatment to achieve optimal bio-methane yields. The production of bio-methane from SCB presents a sustainable pathway for energy generation, aligning with the global shift towards renewable energy sources and environmental sustainability (Ajala et al. 2021; Ling et al. 2021). Bio-ethanol, a biofuel derived from the fermentation of reducing sugars obtained from SCB, offers renewability, cost-effectiveness, and environmental friendliness as advantages over fossil fuels. Studies suggested that various microorganisms like *Saccharomyces cerevisiae*, *Pichia stipites*, and *Escherichia coli* have been utilized for fermenting SCB hydrolysate to produce bio-ethanol, with yields ranging from 0.2 to 0.3 g/g under specific conditions. Different microorganisms and process conditions have been explored for bio-ethanol production from SCB, with maximum bio-ethanol yields achieved under specific temperature, pH, and incubation time conditions. Bio-hydrogen, another biofuel derived from SCB, presents advantages in terms of high energy yield and non-release of greenhouse gases post-combustion, highlighting its potential as a clean energy source. Bio-butanol, a promising biofuel, can also be produced from SCB, contributing to the diversification of biofuel options and promoting sustainable energy solutions (Ajala et al. 2021).

12.2.8.2 Bioproducts

The bioprocessing of SCB is a popular method for producing various biochemical substances, utilizing microorganisms like fungi and bacteria to digest the substrate and generate valuable metabolic products. SCB can be utilized to produce Poly-3-hydroxybutyrate through the action of *Burkholderia cepacia* and *Burkholderia sacchari* (Silva et al. 2004), showcasing its potential for biopolymer production. Xylan, a valuable component, can be obtained from SCB through alkaline and enzymatic approaches, with alkaline methods yielding higher xylan content and enzymatic methods producing xylan with lower lignin and glucan contamination (Sporck et al. 2017). Lipids can be produced from SCB substrate by cultivating *Yarrowia lypolytica*, with studies reporting a maximum lipid yield of 6.68 g/L under specific conditions (Tsigie et al. 2011). Enzymes such as cellulase, hemicellulase, lipase, and others can be produced using SCB as a substrate, highlighting its potential as a source for enzyme production in various bioprocessing applications. The diverse range of bioproducts derived from SCB emphasizes its versatility and potential in generating value-added biochemicals, biopolymers, and enzymes, contributing to the sustainable utilization of this abundant biomass resource (Ajala et al. 2021).

12.2.8.3 Source of Biofuel

Utilizing SCB as fuel is a sustainable practice that maximizes the energy potential of this abundant agricultural residue while reducing greenhouse gas emissions and dependence on fossil fuels. This section will delve into the various aspects of SCB utilization as fuel, including its composition, energy content, conversion technologies, applications, and sustainability implications (Khanal 2008; Chandel et al. 2012; Hoang et al. 2015; Huang et al. 2020).

12.2.8.3.1 Composition and Energy Content of SCB

The composition of bagasse gives it significant energy content, with a calorific value typically ranging from 17 to 20 MJ/kg (megajoules per kilogram). Cellulose is the primary component of bagasse and serves as a source of energy when burned.

Hemicellulose contributes to the energy content of bagasse and can be broken down into sugars during combustion. Lignin provides structural support to plant cell walls and contributes to the calorific value of bagasse when burned.

12.2.8.3.2 Conversion Technologies

There are several technologies available for converting SCB into useful forms of energy

- *Direct Combustion*: The most common method of utilizing bagasse as fuel is through direct combustion in boilers to produce steam for electricity generation or process heat for industrial applications. Bagasse is burned in a controlled manner, with the heat released used to produce steam that drives turbines or heats water for various industrial processes.
- *Cogeneration*: Cogeneration, also known as combined heat and power (CHP), is a highly efficient process that utilizes bagasse as both fuel for electricity generation and as a source of heat for industrial processes. Cogeneration systems can achieve high overall efficiencies by maximizing the use of bagasse energy content.
- *Anaerobic Digestion*: Bagasse can be converted into biogas through anaerobic digestion, a biological process in which microorganisms break down organic matter in the absence of oxygen. Biogas produced from bagasse can be used directly as fuel for heating or electricity generation or upgraded to biomethane for injection into natural gas pipelines or used as vehicle fuel.

12.2.8.3.3 Applications of Bagasse as Fuel

SCB fuel has various applications in energy production and industrial processes:

- *Electricity Generation*: Bagasse is commonly used as fuel in sugar mills to generate steam for electricity generation through steam turbines. Excess electricity generated can be exported to the grid or used to meet onsite energy demands.
- *Process Heat*: Bagasse can provide process heat for various industrial applications, including drying, evaporation, distillation, and sterilization in industries such as food processing, textile, and chemical manufacturing.
- *Biofuel Production*: Bagasse-derived biogas or biomethane can be used as a renewable transportation fuel or blended with natural gas for use in heating or electricity generation.

12.2.8.3.4 Sustainability Implications

Utilizing SCB as fuel offers several sustainability benefits:

- *Renewability*: Bagasse is a renewable resource that is continuously generated as a byproduct of sugarcane processing, making it a sustainable alternative to finite fossil fuels.

- *Greenhouse Gas Reduction*: Burning bagasse for energy reduces greenhouse gas emissions by avoiding the combustion of fossil fuels and preventing methane emissions from bagasse decomposition.
- *Waste Reduction*: Utilizing bagasse as fuel reduces waste disposal and pollution associated with landfilling or open burning, contributing to environmental conservation.
- *Energy Independence*: Bagasse-based energy production reduces reliance on imported fossil fuels and enhances energy security by utilizing locally available renewable resources.

Utilizing SCB as fuel offers a sustainable solution to energy production and industrial heat generation needs. With its high energy content and renewable nature, bagasse represents a valuable resource for reducing greenhouse gas emissions, waste disposal, and dependence on fossil fuels. By investing in efficient conversion technologies and promoting bagasse utilization, industries can contribute to a more sustainable energy future while supporting rural economies and agricultural sustainability.

12.2.9 Animal Feed

Utilizing SCB as animal feed is a practice that leverages the fibrous residue left over after sugarcane processing to provide nutrition to livestock. This section will delve into the various aspects of SCB utilization as animal feed, including its composition, nutritional value, processing methods, applications, and sustainability implications (Selman-Housein et al. 2000; Anandan and Sampath 2012; Harrison 2016; de Almeida et al. 2018; Fangueiro et al. 2023).

12.2.9.1 Composition and Nutritional Value of SCB

SCB is primarily composed of cellulose, hemicellulose, lignin, and other structural components of the sugarcane plant. While bagasse is not as nutritionally dense as other feed ingredients like grains or forages, it still offers some nutritional value to animals. The nutritional composition of SCB can vary depending on factors such as the variety of sugarcane, harvesting practices, and processing methods. Bagasse is rich in dietary fiber, including cellulose and hemicellulose, which can aid in digestion and promote gastrointestinal health in animals. Lignin is a structural component of plant cell walls that provides rigidity and support. While lignin is not digestible by most animals, it can still contribute to the overall fiber content of bagasse. SCB contains only modest levels of protein, which may vary depending on the presence of leaves and other plant materials in the bagasse. Bagasse is low in fat, which makes it a suitable option for animals that require low-fat diets.

12.2.9.2 Processing Methods

While SCB can be fed to animals in its raw form, processing methods may be employed to improve its palatability, digestibility, and nutritional value:

- *Chopping or Grinding*: Bagasse can be chopped or ground into smaller particles to enhance its digestibility and reduce the risk of digestive issues in animals.
- *Chemical Treatments*: Some chemical treatments, such as alkaline or acid hydrolysis, may be applied to bagasse to break down its fiber components and improve digestibility. However, these treatments may alter the nutritional composition of bagasse and should be used with caution.
- *Enzyme Supplementation*: Enzymes can be added to bagasse to break down complex carbohydrates and improve nutrient availability to animals.

12.2.9.3 Applications as Animal Feed

SCB can be utilized as animal feed in various ways, including:

- *Ruminant Diets*: Bagasse is often used as a roughage source in ruminant diets, where its high fiber content can support rumen function and promote microbial fermentation. However, due to its low protein content, bagasse is typically supplemented with other protein-rich feeds in ruminant diets.
- *Bedding Material*: Bagasse can also be used as bedding material for livestock, providing a comfortable and absorbent surface for animals to rest on. After use, the soiled bedding can be composted and recycled as fertilizer.

12.2.9.4 Sustainability Implications

Utilizing SCB as animal feed offers several sustainability benefits:

- *Waste Reduction*: By utilizing bagasse as animal feed, the sugar industry can reduce waste and minimize environmental pollution associated with bagasse disposal.
- *Resource Efficiency*: Bagasse can serve as a feed ingredient, reducing the need for other feed sources and conserving valuable agricultural resources.
- *Circular Economy*: Bagasse can be recycled as animal bedding or composted after use, contributing to the circular economy by returning nutrients to the soil and reducing the need for synthetic fertilizers.

With proper processing and supplementation, bagasse can be incorporated into animal diets to improve digestibility and meet nutritional requirements. By leveraging bagasse as a feed resource, the agricultural industry can reduce waste, conserve resources, and promote sustainable farming practices.

12.2.10 Biocoagulants

SCB-based biocoagulants have several potential applications, particularly in industries and processes where water treatment and purification are essential. Here are some potential applications:

12.2.10.1 Water Treatment

SCB-based biocoagulants can be used in water treatment plants for removing suspended solids, organic matter, and other contaminants from wastewater. They can help in the purification of water for reuse in industrial processes or safe discharge into the environment (Maurya and Daverey 2018; Iwuozor et al. 2022a).

12.2.10.2 Drinking Water Purification

Biocoagulants derived from SCB can be utilized in the treatment of drinking water to remove turbidity, pathogens, and other impurities. This can improve the quality of drinking water in communities where access to clean water is a challenge (Sillanpaa et al. 2018).

12.2.10.3 Textile Industry

In textile manufacturing processes, biocoagulants can be employed for treating wastewater generated from dyeing and finishing operations. They aid in the removal of dyes, chemicals, and other pollutants, thus reducing the environmental impact of textile production (Iwuozor et al. 2022b; Ewuzie et al. 2022).

12.2.10.4 Pharmaceutical Industry

The use of bio-coagulants based on SCB has advantages for the pharmaceutical sector as well. Suspended particles, organic matter, and additional pollutants can be eliminated from the wastewater produced while making pharmaceuticals (Alazaiza et al. 2022).

12.2.10.5 Paper and Pulp Industry

SCB-based biocoagulants can assist in the treatment of wastewater from paper and pulp mills, where large quantities of organic matter and lignin are present (Grötzner et al. 2018). They help in clarifying the wastewater before it is discharged, minimizing pollution to water bodies.

12.2.10.6 Mining Industry

Biocoagulants can be used in the mining industry for treating wastewater contaminated with heavy metals, suspended solids, and other pollutants. They can aid in the removal of these contaminants, allowing for safer discharge or reuse of water in mining operations (Ighalo et al. 2022; Adeniyi et al. 2022).

12.2.10.7 Agricultural Runoff Management

In agricultural areas where runoff water contains excess nutrients and pesticides, biocoagulants can be applied to treat the water before it enters natural water bodies. This helps in reducing nutrient pollution and mitigating the impact of agricultural activities on aquatic ecosystems (Turunen et al. 2019).

12.2.10.8 Aquaculture

Biocoagulants can be utilized in aquaculture operations to improve water quality by removing suspended solids, algae, and organic matter (Ogunfowora et al. 2021;

Aibinu et al. 2022; Emenike et al. 2022). This creates a healthier environment for aquatic organisms and enhances overall productivity in fish and shrimp farming.

12.2.10.9 Landfill Leachate Treatment

SCB-based biocoagulants can also be used in the treatment of leachate generated from landfills. They aid in the removal of pollutants such as organic matter, heavy metals, and ammonia, reducing the environmental risks associated with landfill operations.

12.2.10.10 Food and Beverage Sector

Bio-coagulants based on SCB have additional advantages for the food and beverage sector. The wastewater produced during the manufacture of food and beverages can benefit from the use of bio-coagulants to assist remove organic debris, suspended particles, and other pollutants (Iwuozor and Gold 2018). Overall, the versatility and eco-friendly nature of SCB-based biocoagulants make them promising for various water treatment and purification applications across different industries.

12.3 Conclusions and Future Prospects

In conclusion, SCB stands out as a promising source of renewable and bio-compostable material. Through innovative processes, it can be transformed into a range of products, from bioplastics to biofuel, offering a sustainable alternative to traditional petroleum-based materials. Its abundance, renewability, and biodegradability make it an attractive option for addressing environmental concerns associated with plastic pollution and nonrenewable resource depletion. Moreover, utilizing SCB in composting not only reduces waste but also enriches soil health, contributing to sustainable agricultural practices.

Looking ahead, the utilization of SCB holds vast potential for further innovation and application. Continued research and development efforts could lead to advancements in processing techniques, enhancing the efficiency and scalability of bagasse-based products. Collaboration between industries, academia, and policymakers is essential to foster a supportive ecosystem for the widespread adoption of bagasse-derived materials. Additionally, raising awareness among consumers about the benefits of bio-compostable products and incentivizing their use could drive market demand and accelerate the transition toward a more sustainable economy. With concerted efforts, SCB can play a significant role in shaping a greener, more circular future.

References

Adeniyi AG, Emenike EC, Iwuozor KO, Okoro HK, Ige OO (2022) Acid mine drainage: the footprint of the Nigeria mining industry. Chem Afr 5(6):1907–1920

Aibinu AM, Folorunso TA, Saka AA, Ogunfowora LA, Iwuozor KO, Ighalo JO (2022) Green synthesis of CuO nano composite from watermelon (*Citrulluslanatus*) rind for the treatment of aquaculture effluent. Reg Stud Mar Sci 52:102308

Ajala EO, Ighalo JO, Ajala MA, Adeniyi AG, Ayanshola AM (2021) Sugarcane bagasse: a biomass sufficiently applied for improving global energy, environment and economic sustainability. Bioresour Bioprocess 8:1–25

Alabi D, Coopoosamy R, Naidoo K (2023) Composted bagasse: an impact on agricultural crop production. J Agric Sci Food Res 13(2):1000483

Alazaiza MY, Albahnasawi A, Ali GA, Bashir MJ, Nassani DE, Maskari TAI, Amr SSA, Abujazar MSS (2022) Application of natural coagulants for pharmaceutical removal from water and wastewater: a review. Water 14(2):140

Anandan S, Sampath KT (2012) Scope for utilizing sugar cane bagasse as livestock feed–an Asian perspective. Biofuel Co-Products as Livestock Feed 291

Azmi AA, Othman SA (2023) Future of sugarcane bagasse paper: a review. In: AIP conference proceedings, Vol. 2544, no. 1. AIP Publishing

Chandel AK, da Silva SS, Carvalho W, Singh OV (2012) Sugarcane bagasse and leaves: foreseeable biomass of biofuel and bio-products. J Chem Technol Biotechnol 87(1):11–20

Costa SM, Mazzola PG, Silva JC, Pahl R, Pessoa A Jr, Costa SA (2013) Use of sugar cane straw as a source of cellulose for textile fiber production. Ind Crop Prod 42:189–194

de Almeida GAP, de Andrade FM, de Lima SJ, Chagas JCC, Véras ASC, de Barros LJA, de Almeida GLP (2018) Sugarcane bagasse as exclusive roughage for dairy cows in smallholder livestock system. Asian Australas J Anim Sci 31(3):379

Diallo ND, MBengue M, NGuer M, Kâ M, Tine E, Mbaye CT (2017) Composting of sugar cane bagasse by Bacillus strains. Afr J Biotechnol 16(3):113–123

Emenike EC, Iwuozor KO, Anidiobi SU (2022) Heavy metal pollution in aquaculture: sources, impacts and mitigation techniques. Biol Trace Elem Res 200:4476–4492. https://doi.org/10.1007/s12011-021-03037-x

Ewuzie U, Saliu OD, Dulta K, Ogunniyi S, Bajeh AO, Iwuozor KO, Ighalo JO (2022) A review on treatment technologies for printing and dyeing wastewater (PDW). J Water Process Eng 50:103273

Fangueiro JF, de Carvalho NM, Antunes F, Mota IF, Pintado ME, Madureira AR, Costa PS (2023) Lignin from sugarcane bagasse as a prebiotic additive for poultry feed. Int J Biol Macromol 239:124262

Grötzner M, Melchiors E, Schroeder LH, dos Santos AR, Moscon KG, de Andrade MA, Martinelli SHS, Xavier CR (2018) Pulp and paper mill effluent treated by combining coagulation-flocculation-sedimentation and Fenton processes. Water Air Soil Pollut 229:1–7

Harrison MD (2016) Sugarcane-derived animal feed. Sugarcane-based Biofuels and Bioproducts 281–310

Hoang NV, Furtado A, Botha FC, Simmons BA, Henry RJ (2015) Potential for genetic improvement of sugarcane as a source of biomass for biofuels. Front Bioeng Biotechnol 3:182

Huang J, Khan MT, Perecin D, Coelho ST, Zhang M (2020) Sugarcane for bioethanol production: potential of bagasse in Chinese perspective. Renew Sust Energ Rev 133:110296

Ighalo JO, Kurniawan SB, Iwuozor KO, Aniagor CO, Ajala OJ, Oba SN, Iwuchukwu FU, Ahmadi S, Igwegbe CA (2022) A review of treatment technologies for the mitigation of the toxic environmental effects of acid mine drainage (AMD). Process Saf Environ Prot 157:37–58

Iwuozor KO, Gold EE (2018) Physico-chemical parameters of industrial effluents from a brewery industry in Imo state, Nigeria. Adv J Chem Sect A 1(2):66–78

Iwuozor KO, Ighalo JO, Emenike EC, Ogunfowora LA, Igwegbe CA (2021) Adsorption of methyl orange: a review on adsorbent performance. Curr Res Green Sustain Chem 4:100179

Iwuozor KO, Ogunfowora LA, Oyekunle IP (2022a) Review on sugarcane-mediated nanoparticle synthesis: a green approach. Sugar Tech 24(4):1186–1197

Iwuozor KO, Oyekunle IP, Emenike EC, Okoye-Anigbogu SM, Ibitogbe EM, Elemile O, Ighalo JO, Adeniyi AG (2022b) An overview of equilibrium, kinetic and thermodynamic studies for the sequestration of Maxilon dyes. Clean Mater 6:100148

Karp SG, Burgos WJ, Vandenberghe LP, Diestra KV, Torres LA, Woiciechowski AL, Letti LA, Pereira GV, Thomaz-Soccol V, Rodrigues C (2022) Sugarcane: a promising source of green carbon in the circular bioeconomy. Sugar Tech 24(4):1230–1245

Kasiviswanathan KS (1998) Utilisation of bagasse for paper making-an overview. IPPTA 10(3):9–14

Khanal SK (2008) Anaerobic biotechnology for bioenergy production. Principles and Application Willey and Blackwell, pp 161–186

Kusuma HS, Permatasari D, Umar WK, Sharma SK (2023) Sugarcane bagasse as an environmentally friendly composite material to face the sustainable development era. Biomass Convers Biorefinery:1–14

Ling R, Wu W, Yuan Y, Wei W, Jin Y (2021) Investigation of choline chloride-formic acid pretreatment and Tween 80 to enhance sugarcane bagasse enzymatic hydrolysis. Bioresour Technol 326:124748

Loh YR, Sujan D, Rahman ME, Das CA (2013) Sugarcane bagasse—the future composite material: a literature review. Resour Conserv Recycl 75:14–22

Maurya S, Daverey A (2018) Evaluation of plant-based natural coagulants for municipal wastewater treatment. 3 Biotech 8(1):77

Nakhla DA, Mahmoud Y, El Haggar S (2017) Production of compost and organic fertilizer from sugarcane residues. Nations Natural Resources and Environment, FAO, Rome

Narancic T, O'Connor KE (2017) Microbial biotechnology addressing the plastic waste disaster. Microb Biotechnol 10(5):1232

Ogunfowora LA, Iwuozor KO, Ighalo JO, Igwegbe CA (2021) Trends in the treatment of aquaculture effluents using nanotechnology. Clean Mater 2:100024

Pandey A, Soccol CR, Nigam P, Soccol VT (2000) Biotechnological potential of agro-industrial residues. I: sugarcane bagasse. Bioresour Technol 74(1):69–80

Rainey TJ, Covey G (2016) Pulp and paper production from sugarcane bagasse. Sugarcane-Based Biofuels and Bioproducts 259–280

Sakthivel JC, Brindha S, Gowthamraj G, Sabna J (2021) Studies on the properties of sugarcane bagasse fibre for sustainable textile applications. In: Proceedings of the first international conference on combinatorial and optimization, ICCAP 2021, December 7–8 2021, Chennai, India

Saranraj P, Stella D (2014) Composting of sugar mill wastes: a review. World Appl Sci J 31(12):2029–2044

Sayed EAM, Hamad KH, Eid Y, Alaa A, Hassan A, Mohsen A, Safwat A, Atta K, Benbella M, Abdel Tawab S (2022) Production of paper using chemical pulping process of sugarcane bagasse. Int J Ind Sustain Dev 3(1):57–65

Selman-Housein G, Lopez MA, Ramos O, Carmona ER, Arencibia AD, Menéndez E, Miranda F (2000) Towards the improvement of sugarcane bagasse as raw material for the production of paper pulp and animal feed. In: Developments in plant genetics and breeding, Vol. 5. Elsevier, pp 189–193

Sillanpaa M, Ncibi MC, Matilainen A, Vepsalainen M (2018) Removal of natural organic matter in drinking water treatment by coagulation: a comprehensive review. Chemosphere 190:54–71

Silva LF, Taciro MK, Michelin Ramos ME, Carter JM, Pradella JGC, Gomez JGC (2004) Poly-3-hydroxybutyrate (P3HB) production by bacteria from xylose, glucose and sugarcane bagasse hydrolysate. J Ind Microbiol Biotechnol 31(6):245–254

Sirlene M, Costa S, Pahl R, Mazzola P (2013) Textile fiber produced from sugarcane bagasse cellulose: an agro-industrial residue. Int J Text Fashion Technol 3:15–28

Sporck D, Reinoso FA, Rencoret J, Gutiérrez A, Del Rio JC, Ferraz A, Milagres AM (2017) Xylan extraction from pretreated sugarcane bagasse using alkaline and enzymatic approaches. Biotechnol Biofuels 10(1):296

Tsigie YA, Wang CY, Truong CT, Ju YH (2011) Lipid production from Yarrowialipolytica Po1g grown in sugarcane bagasse hydrolysate. Bioresour Technol 102(19):9216–9222

Turunen J, Karppinen A, Ihme R (2019) Effectiveness of biopolymer coagulants in agricultural wastewater treatment at two contrasting levels of pollution. SN Appl Sci 1:210. https://doi.org/10.1007/s42452-019-0225-x

Economics of Sugarcane Value Chain and Policy Capsules on Value-Added Products

13

Murali Palanichamy, Hari Kuppusamy, Giriyapura Shivalingamurthy Suresha, and Prakash P

Abstract

Sugarcane is cultivated as a multi-utility crop for food, fodder, fuel, energy and fibre. It is a renewable resource that can be sustainably grown for sugar and value-added products. The sugarcane value chain comprises price, farmers, producers, traders and end consumers. The study has reported different outputs of sugarcane production in India with a focus on the optimum value creation. The value chain includes sugar, Gur, Khandsari and Jaggery, molasses, bagasse, ethanol, sugarcane tops/trash and others. Sugarcane value chains are much more complex than other crops. This analysis exposes inefficient activities and production expertise which hampers value chain activities. Moreover, the vicious cycle of sugar prices contributes to uneven income dispersal. Additionally, the existing policies do not adequately support sugarcane farmers for high-value chain approach. Therefore, the study has recommended policies for enhancing value chain activities and improving efficiency in sugarcane production through implementing recent cultivation methods, employing artificial intelligence, increasing sugarcane production capacity through precision farming, rejuvenating off-farm activities for value chain production and increasing market share by different

M. Palanichamy (✉)
Agricultural Economics, ICAR-Sugarcane Breeding Institute, Coimbatore, Tamil Nadu, India

H. Kuppusamy
Agricultural Microbiology, ICAR-Sugarcane Breeding Institute, Coimbatore, Tamil Nadu, India

G. S. Suresha
Biochemistry (Plant Science), ICAR-Sugarcane Breeding Institute, Coimbatore, Tamil Nadu, India

Prakash P
Agricultural Economics, ICAR-Central Tuber Crops Research Institute, Thiruvananthapuram, Kerala, India

G. S. Suresha et al. (eds.), *Value Addition and Product Diversification in Sugarcane*, https://doi.org/10.1007/978-981-97-7228-5_13

value chain approach is necessary in the coming decade in order to sustain sugarcane value chain for benefits of the end users and producers.

Keywords

Sugarcane · Value chain · White sugar · Jaggery · Ethanol

13.1 Introduction

Sugarcane is cultivated as a multi utility crop for food, fodder, fuel, energy and fibre. Sugarcane, as a C_4 plant, is one of the most efficient field crops in terms of using solar radiation. Sugarcane may produce more than 100 tonnes of produce per hectare which is more than double the yield of most other food and commercial crops. It is the most prominent sugar crop in many parts of the world and is used to produce a wide range of value-chain products. It is a renewable resource that can be sustainably grown and processed to meet out the sugar and sugar-based derivatives. Though sugarcane is used as a source of sugar, it is the most demanded raw material production of higher value bio-fuels because it can produce biofuel through molasses and juice and a combustible fuel from the fibrous part (bagasse) of the sugarcane crop.

The Brazilian sugarcane sector produces 40% of the world's sugarcane (ProTerra Foundation 2020; UNICA n.d.). Sugarcane ranks second most commercial crop after cotton in India which shares 20% of the global cane production. It contributes 0.62% of India's GDP per year which is significant in terms of the gross cropped area (2.58%) and shares 3.37% of the Agricultural GDP in India. The growth of the sugarcane sector was about 42% and the sugar sector has recorded 96% growth which is significantly higher than the overall agriculture sector in India since 2000. Nearly 50–55 lakh sugarcane farmers and their dependents are doing sugarcane farming. Besides, the sugar industry also offers direct/indirect employment for about 4% of the rural population in India.

Sugar is an indispensable food for mass consumption and an economical source of energy, providing around 10% of the calorie intake of daily consumption. Apart from sugar, sugarcane also supports the energy sector through bio-ethanol and cogeneration of power. The contribution of sugarcane as food, fodder and fuel is significant in the domestic and international markets. The market information on sugar, ethanol and cogeneration is well documented internationally necessitates and needs to focus on sugarcane and its derivatives.

The value and benefits of sugarcane production are extensive and varied. India is numero uno in sugar consumption in the world and its domestic requirement is about 27.5–28.0 million tonnes. The demand increases by 0.5–1.0 million tonnes on a yearly basis. The value chain of sugarcane comprises employment creation, forex revenue, quasi–Foreign Direct Investment (FDI), import substitution of gasoline fuels, revenue earnings through the service sector, cogeneration of power, minimization of pollution level, carbon trade etc. The value chain also contributes to social upliftment and environmental protection. The multiple benefits of sugarcane value chain are a lead contributor to the economic growth of the leading sugar producers in the world.

Value chain refers to estimation/analysing consumer demand in the first stage and based on requirement, the production/service from conception to end use in a value chain, which begins from input supply to delivery of end products to consumers. In this process, value is being added to the product or service at each step. Value chain approach to production means production of each product as per demand/consumer choice at multiple levels of the value chain. Sugarcane is a multifaceted crop which produces wide ranges of products. In sugarcane, value chains are much more complex than other crops.

The value chain probably has many more links which not only include a vertical line but also outward activities that influence activities of the value chain. For the sake of simplicity, these external parts of the value chain are depicted in the Fig. 13.1 in our value chain, downstream activities related to the outputs from the sugar mills and sugar complex were analysed for sugarcane production system.

13.2 Centrifugal (White) Sugar

White sugar is a pure form of sucrose which provide cheapest source of carbohydrate, which fulfils about 10% of the total calorific needs of human being in the world. It is particularly suited as food for workman doing heavy work and all engaged in active exercise. Sugar is not a luxury but an indispensable food. Sugar has a great effect in retarding fatigue. No wonder that the constant activity of childhood craves interminable supplies of sweets.

White sugar is normally made from the sugar mills. Raw sugar is further refined into white sugar as per the requirement of the end users. In the sugar mill, the sugar is centrifuged to remove the remaining liquid from the sugar crystals. Sugar crystals are ground into fine, regular and large crystals, depending on it's ultimate use. The white sugar is a main output sugar mill. Production of white sugar depends on sugarcane production, demand and domestic consumption. The white sugar production is mainly depended on sugar recovery percentage of cane. On average, every tonne of sugarcane produces 100 kg of white sugar with an average sugar recovery is about 10%. Currently, India is producing about 33–35 million tonnes of white sugar which is valued at about 1 lakh crores. Globally, about 180 million tonnes of white sugar is produced from cane and sugar beet, the production is growing gradually due to population and income growth in the developing countries.

The demand for sugarcane and sugar beet is increasing mainly due to the value chain approach. The gross cropped area of both sugar crops has increased than the previous decade, though less than in the 1990s and 2000s in the globe. Of late the sugar crops were also used for the production of first-generation biofuel. It is expected that the diversion of sugar crops towards bio-ethanol production may pose a challenge to sugar production which ultimately helps to increase the monetary values of white sugar.

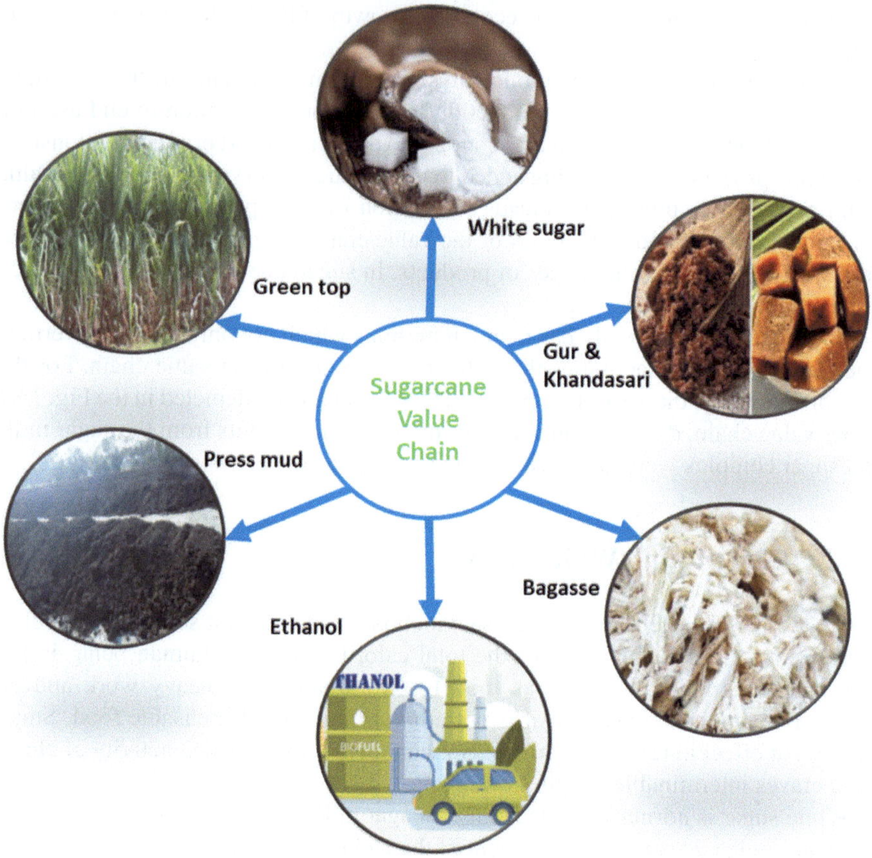

Fig. 13.1 Value chain of sugarcane-based production system

The major share of sugar production is likely to come from developing countries. Brazil and India continue to remain in first and second position in sugar production in the world. These two countries will expect to produce about 21–25% and 18–20% of the world's total sugar throughput by 2030, respectively. The global per-capita consumption is predicted to increase in the current decade due to income growth, changing consumer preference and urbanisation. Major growth of sugar consumption is predicted in Asiatic nations and shares more than 50% of the world consumption by 2030. It is majorly shared by sugar-rich confectionery products and soft drinks. Similarly, consumption growth is expected in Africa mainly driven by population and income growth, however, in real terms, the rate of consumption is projected to be much lower than Asia.

13.3 Gur, Khandsari and Jaggery Production

Jaggery is a raw sugar that is produced in the traditional cottage industries without any chemical processing. India produces about 70% of the total jaggery production in the world. Jaggery is widely known as the 'remedial sugar' and is comparable with honey in terms of health benefits. It is used as a sweetener in Ayurvedic Medicine before Christ's birth. Nowadays Bakery and confectionery products are prepared by using jaggery. It offers a variety of delicious sweets and new vistas and ingredients for baked products.

Gur is low low-calorie sweetener compared to white sugar. It is known to be nutritionally superior to centrifugal/refined sugar. Jaggery, Gur and Khandsari sugars comprise about 70–80% sucrose, less than 10% glucose and fructose, 5% minerals with less than 5% moistness. Jaggery, Gur and Khandsari deliver instantaneous energy to living things. It is widely used in cattle feed formulation, distilleries, organic pesticides and plant growth promotors. The usage of Gur is also seen in leather and tobacco processing. Traditionally, it is used in cement industries and coal mines. It is used as supplement to workers in order to protect them from dust allergies.

Jaggery is considered to be a wholesome sugar since it retains more minerals, vitamins and salts than table sugar. It has a significant quantum of calcium, iron, potassium and magnesium. In Ayurveda medicine, it is used as a blood purifier, treatment of rheumatism, bile disorders and lung infection. Recent scientific research has confirmed the benefits of Jaggery use to human beings. India is a leader in Gur, Jaggery and Khandsari production and is known for its quality, which has wide acceptability in the export destinations and fetches increasing demand in the value chain.

Jaggery consumption is fluctuating due to changing consumption patterns, standard of living and globalisation. Of late, health consciousness has revived Gur, Khandsari and jaggery consumption. Gur, Khandsari and Jaggery is produced at major sugarcane growing regions through cottage industries. India is predominant in exports of Jaggery to the global market due to its share in global production. India traded 761,640.19 MT of Jaggery and Confectionery products to the global market which valued about Rs. 4330.07 Crores/537.01 USD Millions in 2022–2023. It is exported to more than 75 countries and major export destinations are Sudan, Indonesia, Tanzania, UAE and Kenya.

13.4 Bagasse

Bagasse is a by-product of sugarcane processing and it is available in abundance in the sugar mills. Bagasse, also known as sugarcane pulp, is a by-product of sugarcane production. It's the fibrous waste from the sugarcane stalks. As a by-product, it has a multitude of uses. It doesn't require additional land and energy to grow and doesn't increase deforestation. Traditionally, sugar mills process bagasse for valorisation. Bio-based chemical manufacturing is superior to biomass disposal by incineration and landfill. The conversion of bagasse for bio-based products is more profitable with the minimum environmental impact on Earth.

Bagasse from sugar mills is either utilised in cogeneration facility or it will be sold out for different byproducts industries that uses bagasse as raw material. But, in the Gur, Khandasari and jaggery production, it is mostly used as fuel material for Gur production. Though it is a byproduct, it has significant economic value in the sugarcane production system. The bagasse production from organised sector was computed and the value chain was depicted in Fig. 13.2.

The data depict the annual quantity and value of bagasse derived from total sugarcane, with a specific focus on the contribution from the Co 0238 variety from 2013–2014 to 2021–2022 (Fig. 13.2). Over the years, there is a discernible trend in both bagasse quantity and value, with fluctuations in each parameter. Notably, the quantity of bagasse from total sugarcane witnessed a decline from 2013–2014 to 2016–2017 (7154 to 5822 thousand tonnes), followed by a steady increase until 2021–2022 (10,692 thousand tonnes). Conversely, the value of bagasse shows a consistent upward trajectory, indicating a positive economic impact, starting from Rs. 1431 crores in 2013–2014 to Rs. 2673 crores in 2021–2022. Despite fluctuations in begasse production, it remains a substantial contributor in the sugarcane value chain, reflecting its consistent economic importance within the context of the sugarcane industries.

13.5 Cogeneration/Power

The concept of synchronized generation of electricity and thermal energy is called cogeneration. Bagasse is a valuable byproduct of sugar production through cane. It produces two forms of energy from a single fuel source. One of the forms of energy is heat which is used for boilers and steam generation and the other is electrical or mechanical energy which is commonly used for sugar as well as electrification

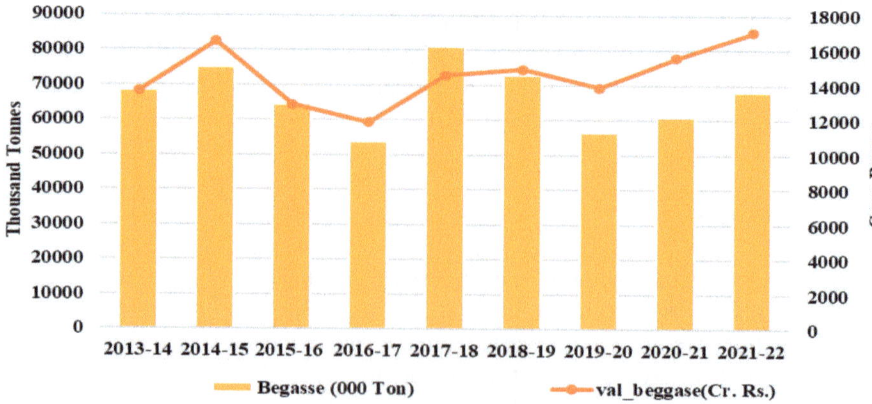

Fig. 13.2 Total quantity and value of bagasse between 2013–2014 and 2021–2022 in India

production. It is quite common for sugar mills for producing electricity by using bagasse as a raw material. After its consumption, the surplus electricity is supplied for sale to the grid.

Over the years (2013–2022), electricity production (Cogen) from sugarcane has exhibited a significant upward trend, starting at 472 MWh in 2013–2014 and reaching 4100 MWh in 2021–2022. The corresponding revenue from electricity generation has also seen a consistent rise, starting from Rs.456 Crores in 2013–2014 and peaking at Rs 8713 Crores in 2021–2022 (Fig. 13.2). The total production fluctuates during 2016–2017 due to a fall in sugarcane production. But remains a noteworthy component, illustrating its impact on the overall cogeneration landscape of cogeneration power in India.

13.6 Green Tops/Trash

Green leaves with tops are by-products of sugarcane since the grand growth phase (Devendra 1985). The harvested sugarcane biomass includes stalks, tops, dead and dying leaves, stubble and roots (Ortiz-Rubio et al. 2007). Sugarcane green tops and trash comprises about 15–25% of the total biomass. It consists of green leaves, bundle sheath and variable amounts of immature cane (Naseeven 1988). Generally, the cutting point is at the highest fully-formed node (Suttie 2000). Sugarcane tops contain phenols, amino acids and soluble polysaccharides. Sugarcane tops from sugar production are harvested at maturity, which coincides with the dry season greatly helps to provide green matters to livestock with significant value addition to the sugarcane by products.

Sugarcane tops are used fresh, dried or ensiled for feeding livestock. Their nutritive value is highly variable and depends on pre-harvesting methods, stalk cutting point, plant maturity and the number of dry leaves (Mckenzie & Griffiths (2007); Ortiz-Rubio et al. 2007). Sugarcane tops are low cost and abundant materials when other green fodder is not available (Suttie 2000; Rangnekar 1988). However, as they are not available year-round, ensiling them increases availability (Naseeven 1988). Sugarcane tops ferment readily and can be used for ethanol or lactic acid production. Yields depend on cultivars, management practices and growing conditions and are about 18% of the aerial biomass. On average, 50% of the total sugarcane tops are used as fodder for livestock. A crop can provide 3–5 tonnes/ha of dry matter (Suttie 2000). In India, on average, sugarcane is produced about 425–450 million tonnes per year and 25–30 million tonnes of green trash and tops are produced which is valued about Rs. 1250–Rs. 1500 Crores annually in India.

13.7 Press Mud and Distillery Spent Wash

Press mud is the residual byproduct of sugarcane processing and it is also called filter cake or press cake. The Press mud is collection of impurities in the clarification process of juice during white sugar production. The recovery of filter cake varies from 3 to 5 kg (wet basis) per tonnes of cane. Seventy-five percent of the world press mud production comes from Brazil, India and China (FAO 2011). The wet press mud is processed to reduce moisture content to about 10% and pelletizing is done making it suitable for storage (Chen and Chou 1993). Initially, press mud is an issue in terms of disposal, later it is converted to bio-based fertilizers by the sugar mills. Otherwise, it is burnt in brick kilns, resulting in the loss and wastage of valuable nutrients, which ultimately degrades the environment. Bio composting of press mud helps to improve its nutrient contents and is used as biofertilizer for crops (Nasir 2006).

The distillery spent wash (DSW) is a liquid waste from the sugarcane industry. It comprises both organic and inorganic substances. The most prominent use of press mud is for bio-composting. Unfortunately, India has 100% import of potassium to the tune of 2.5–3 million Mt. We have about 5.0 million hectares of area under cane plantation. The average potassium requirement of plants is about 90–100 kg/ha which translates to about 0.53 million Mt, i.e. about 20% of the import could be saved by effective use of spent wash as a fertilizer via bio-compost or any other route of recycling spent wash to agriculture fields.

Press mud could be converted to pellets for combustion after drying by sugar mills in the near future due to restrictions on composting and the viability of the Compressed Bio Gas (CBG) plant. Press mud like high ash fuels will pose its own challenges in disposal but due to the disproportionate increase in coal prices, use of press mud for fuel gained a lot of attraction in the recent past. In the future, press mud has greater scope to increase its value addition in the sugarcane value chain.

13.8 Ethanol Value Chain

Ethanol (CH_3CH_2OH) is a clear, colourless liquid is commonly called as ethyl alcohol. Ethanol can be produced in either starch or sugar-based feedstocks, such as maize, sugarcane or other cellulosic feedstocks (such as wood chips or crop residues). Ethanol has a marginally higher value of octane number than gasoline facilitates blending with petrol. Minimum octane number is basic requirement for petrol to prevent engine knocking and safeguard moveability. At times lesser octane petrol is blended with ethanol to attain the standard octane number. In principle, Ethanol has less energy content than petrol per unit and efficiency varies with the percentage of ethanol in the blending. The fuel economy is dependent on the percentage of ethanol content in the fuel and whether an engine is adjusted to run on ethanol-blended petrol. Vehicles manufactured in recent years are calibrated for ethanol-blended petrol.

Ethanol is an ideal liquid chemical with multiple uses. It is used as called rectified spirit and is used as the intoxicating ingredient in alcoholic beverages with 95% purity. Basically, anhydrous ethanol is used for blending with petrol with more than 99.5% purity. Both products are industrially manufactured by using molasses. The sugar mills crush matured sugarcane which has a minimum total fermentable sugars (TFS) content of 14%. The TFS component consists of sucrose along with glucose and fructose. Most of this component gets crystallised into sugar and the residual part is called molasses.

Three major types molasses (A, B, C) are produced during sugar production. The final 'C' molasses is the most un-crystallised and non-recoverable about constitutes 4.0–4.5% of the cane. Thus, mills can produce 110 kg of sugar (at 11.0% recovery) and 45 kg of molasses (18 kg TFS) that yield 11–12 L of ethanol per tonnes of sugarcane.

New biofuel policy allows direct sugarcane juice to ethanol. By using juice/concentrated syrup, about 65–70 L of fuel ethanol from 1 tonne of cane and 0 kg of sugar. But there is a complementary way of producing sugar and fuel ethanol through 'B Heavy' molasses (Fig. 13.3).

By this route, sugar mill is producing 90–95 kg of white sugar along with doubling the fuel ethanol production (20–22 L/tonne) in comparison with 'C molasses' route. It does not require environmental clearance to produce ethanol through B-heavy molasses because of low pollution load among the three routes of ethanol production. Production of ethanol through B-heavy molasses is technically feasible and economically viable for sugar mills and sugar complexes in India.

13.9 Conclusion and Policy Implications

The study has reported different outputs of the sugarcane production in India with a focus on value chain management. Sugarcane is most widely used to produce a variety of products, including sugar, molasses, bagasse, ethanol and others. It can be sustainably grown and processed without damaging environment. It is significant

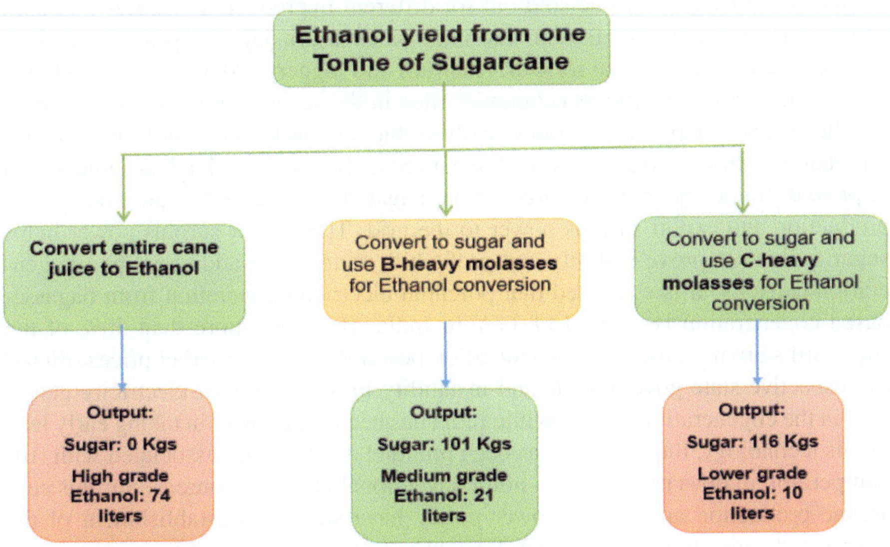

Fig. 13.3 Ethanol output from different feedstock of sugarcane

contributor of employment for many people in the developing countries, and has a well-established supply chain and distribution network in India.

Sustainable sugarcane varietal improvement by ICAR-Sugarcane Breeding Institute, Coimbatore has leveraged a strong technological base for this impressive growth, which has benefitted the farmers sugar mills and end users. Positive demand growth for sweeteners, cogeneration of power and ambitious fuel-ethanol blending program have opened new vistas for the sugarcane-based value chain.

Global sugar production is increasing with consumption demand, the production has touched 135 million tonnes from sugarcane with major shares come from Brazil followed by India. Sugar production is about 330 lakh tonnes in sugar year (SS) 2022–2023 with the diversion of 40 lakh tonnes of sugar equivalent (cane juice) towards fuel ethanol production. This increasing diversion towards high realisation ethanol is likely to support 8–12% revenue growth in the current value chain. It is expected that by 2025, the sugar mill may divert 60 lakh tonnes of excess sugar towards fuel ethanol and enhances the value chain. It would help to improve the demand and supply imbalance in the global market, leading to stable market and prices for white sugar in future periods.

The jaggery consumption is fluctuating due to changing consumption patterns, standard of living as well as urbanisation. Of late, health consciousness has revived the consumption of the Gur, Khandasari and jaggery consumption in India. Jaggery production is taking place in major sugarcane production states through cottage industries for domestic and global markets.

Bagasse is a by-product of sugarcane processing and it is available in abundance in the sugar mills. Almost all sugar mills are traditionally using bagasse as a fuel for cogeneration production, bagasse is byproduct of sugar mills and Gur, Khandasari and jaggery production units. Bagasse from sugar mills is either utilised in cogeneration facilities or it will be sold out for different byproducts industries that uses bagasse as raw material. But, in the Gur, Khandasari, jaggery production, it is mostly used as fuel material for Gur, Khansari and Jaggery production. Though it is a byproduct, it has significant economic value in the sugarcane production system.

The value chain provides economically viable options for sugar mills to diversify into bagasse-based cogeneration of electricity. Technological advancement and improved efficiency in power production, sugar mills to could double the power production and export surplus power to the grid. The sale of surplus power helps sugar mills to improve viability and profitability and create additional power generation capacity. It is estimated that potential electricity generation from bagasse-based cogeneration is about 13.8 GW in India. However, more than 50% of the sugar mills are not producing cogeneration power due to low market prices offered by respective state governments and instability in the pricing of electricity generated in the cogeneration plants. While India has been the pioneer in taking early lead in this technology, the progress has been retarded due to issues associated with stiff competition in power tariffs from non-conventional energy resources. Surplus sugarcane production and a new power policy have supported establishment of the aggregated capacity of 10,632 MW for power generation through biomass project. Currently, sugar mills in India totally generate around 8000 MW of power, out of which 6500 MW is supplied to the power grid excluding its own consumption. The private sector in India is also looking at creating advancements in technology through various means including the increased efficiency of biomass energy conversion, second-generation raw materials and improved fuel processing technologies. Although India has the potential to produce ecologically sustainable bioenergy to meet fast-growing energy needs, progress need to be fastened for achieving the value chain targets.

Sugarcane tops are first by-products of sugarcane. Sugarcane tops are used fresh, dried or ensiled as animal feeds. Their nutritive value with harvesting methods, cane maturity and the number of dry leaves. In India, on average, sugarcane is produced about 425–450 million tonnes per year and 25–30 million tonnes of green trash and tops are produced. Of late, green fodder availability is scarce, sugarcane tops are used as green fodder in major cane-growing states in India. On average, 50% of the total sugarcane tops is used as fodder for livestock. The growth of the livestock sector and fodder availability creates increasing demand for sugarcane-based fodders in the coming years. Considering increasing cane area helps to partial fulfilment of green fodder requirement of the livestock in the cane growing countries in the world.

Press mud is the residue of the filtration of sugarcane processing in the sugar mills. The Press mud is a collection of impurities in the clarification process of juice during white sugar production. The yield of the filter cake is varying from 3 to 5 kg (wet basis) per tonnes of cane. Initially, press mud is an issue in terms of disposal, later it is converted to bio-based fertilizers by the sugar mills. Otherwise, it is burnt in brick kilns, resulting in the loss and wastage of valuable nutrients, which ultimately degrades the environment. Bio composting of press mud helps to improve its fertilizer value, treatment with microorganisms and mixing with distillery effluents.

There is a change in bio-ethanol production from different types of molasses and sugarcane juice. While the production of fuel ethanol from cane juice was nil due to statutory control and new biofuel policy paved way for direct juice-to-ethanol program. Though it remained negligible during 2018–2019, it has substantially increased to more than 20% of the total fuel ethanol production in 2022–2023. Substantial contribution has come from B-heavy molasses due to its production process. Both ethanol and white sugar could be produced, it has contributed a major share in the fuel ethanol production since 2020 in India. The new biofuel policy has helped to accelerate the bio alcohol production particularly 2019–2020 and 2020–2021, witnessed a substantial surge in total ethanol production from sugarcane, reaching 157.09 and 302.30 Crore litres, respectively. The values associated with ethanol production also exhibit a rising trend, with an ascent in 2022–2023 at Rs. 27,695 Crores. This upward trajectory indicates the growing significance of ethanol as a valuable product derived from different sources within the sugarcane processing industry.

India has increased fuel ethanol production for achieving the ambitious goal of E20 (Ethanol 20% blend) by 2025 which was originally planned to be implemented by 2030. However, there are limitations on feedstock availability to reach the target levels by 2025. The new biofuel policy primely depends on molasses and sugarcane juice as the primary feedstocks, other feedstocks such as broken rice and other coarse grains will support to accelerate domestic production. Sugar mills are investing and developing ethanol production units from sugarcane juice through government aided schemes. During 2022–2023, about 74% of fuel ethanol was produced by sugarcane, and this share might increase or decrease depend on feedstock availability. However, given accelerating petrol demand, the blending rate may reach 20% in in 2025–26. Ethanol production is expected to be 13 billion litres in 2032. Achieving 20% EBP in 2025 needs moderation and target revision.

This value chain assessment is based mainly on significant contribution of sugarcane in the value chain model. This study has assessed contribution of various value chain products in relation sugarcane production system. To conclude, value chain analysis is an essential process that helps a sugarcane production system on value-creating activities and eliminate wasteful ones to reinforce value proposition as well as bring more benefits to the production frontier. It helps to manipulate demand-driven and forecast-push to make their value chain more efficient and economical. It leads to the increase in profit and enhance marketability of the products.

References

Chen JCP, Chou CC (1993) Cane sugar handbook: a manual for cane sugar manufacturers and their chemists. John Wiley and Sons, New York. 1090 p

Devendra C (1985) Non-conventional feed resources in Asia and the Pacific, 2nd edn. FAO Regional Office for Asia and the Pacific, Bangkok

FAO (2011) FAOSTAT. Food and Agriculture Organization of the United Nations, Rome

Mckenzie J, Griffiths C (2007) Cane tops as cattle fodder. New South Wales Department of Primary Industries, Primefacts no 314

Naseeven R (1988) Sugarcane tops as animal feed. In: Sansoucy R, Aarts G, Preston TR (eds) Sugarcane as feed. FAO Anim Health Prod Pap 72, pp 106–122

Nasir NM (2006) Better management practices for cotton and sugarcane. WWF-Pakistan, Lahore

Ortiz-Rubio MA, Orskov ER, Milne J, Galina HMA (2007) Effect of different sources of nitrogen on in situ degradability and feed intake of Zebu cattle fed sugarcane tops (Saccharum officinarum). Anim Feed Sci Technol 139(3–4):143–158

ProTerra Foundation (2020) Sustainable sugarcane. https://www.proterrafoundation.org/wpcontent/uploads/2022/02/PROTERRA-SUGARCANE.pdf

Rangnekar DV (1988) Availability and intensive utilization of sugar cane by-products. In: Nonconventional feed resources and fibrous agricultural residues: strategies for expanded utilization. International Development Research Center, Indian Council of Agricultural Research, New Delhi, pp 76–93

Suttie JM (2000) Hay and straw conservation for small-scale farming and pastoral conditions. In: FAO plant production and protection series 29. FAO, Rome

UNICA (n.d.) Sustainable energy and food from Brazil to the world: sugar. https://unica.com.br/en/sugarcane-sector/sugar/

Sugarcane-Based Traditional Sweeteners and Health Benefits

14

K. Chandran, M. Nisha, R. Gopi, B. Mahendran, S. C. Keerthi, C. Dilsha, Giriyapura Shivalingamurthy Suresha, Hari Kuppusamy, and Hemaprabha Govindakurup

Abstract

Sugarcane, an ancient crop, has been cultivated for its sweet nectar-like juice since ancient times, as evidenced by mentions in Indian literature dating back to 1400–1000 BC. The plant, *Saccharum officinarum*, is mentioned in Ayurvedic texts for its medicinal properties. Primarily grown for sugar production, sugarcane is also valued for its non-refined sugars like jaggery and Khandsari. Sugarcane cultivation spans tropical to subtropical regions and has gained attention as a bioenergy crop and for its traditional sweeteners, known for their health benefits. Chewing sugarcane, a practice prevalent in the Pacific Islands, is known for its oral health benefits and nutritional value. Ornamental varieties of sugarcane, known for their attractive colors and stripes, have historical significance and are gaining popularity as ornamental plants globally. Sugarcane juice, a popular nonalcoholic energy drink, is rich in carbohydrates, proteins, vitamins, and minerals. Its nutrient composition makes it a healthy beverage, particularly popular in tropical countries. Preservation of sugarcane juice poses challenges due to its susceptibility to spoilage. Various preservation techniques, including pasteurization and the addition of chemical preservatives, are employed to main-

K. Chandran (✉) · M. Nisha · R. Gopi · B. Mahendran · S. C. Keerthi · C. Dilsha
ICAR-Sugarcane Breeding Institute Research Centre, Kannur, Kerala, India

G. S. Suresha
Biochemistry (Plant Science), ICAR-Sugarcane Breeding Institute, Coimbatore, Tamil Nadu, India

H. Kuppusamy
Agricultural Microbiology, ICAR-Sugarcane Breeding Institute, Coimbatore, Tamil Nadu, India

H. Govindakurup
Genetics and Plant Breeding, ICAR-Sugarcane Breeding Institute, Coimbatore, Tamil Nadu, India

G. S. Suresha et al. (eds.), *Value Addition and Product Diversification in Sugarcane*, https://doi.org/10.1007/978-981-97-7228-5_14

tain its freshness and taste. Jaggery, a traditional sweetener, is widely consumed in many countries and is known for its health benefits. The production of jaggery, however, has seen a decline in recent years despite its nutritional value and economic significance. Other value-added products derived from sugarcane, such as liquid jaggery, granular/powder jaggery, sugarcane wine, cookies, cakes, and ice cream, offer diverse market potential. These products, fortified with other nutritious ingredients, cater to the demand for healthier sweetening options. However, challenges in preserving the freshness and taste of sugarcane juice remain, highlighting the need for further research and technological advancements in processing and preservation techniques.

Sugarcane and its products hold significant medicinal importance in traditional medicine across various cultures. Historically, they have been used in countries like China, the Philippines, La Reunion Guiana, and Brazil, as well as in the Ayurveda and Unani systems of medicine in India. The Latin species name "*officinarum*," meaning "of dispensaries or pharmacy," reflects their widespread traditional medicinal use. Sugarcane stems and roots have been used since ancient times in Ayurveda and Unani systems, with raw sugarcane being recommended for maintaining a healthy body. Various parts of the sugarcane plant, either alone or in combination with other ingredients, have been used to treat ailments such as joint pain and skin problems.

In traditional medicine practices, sugarcane juice is used for a variety of purposes, including as a nutritional drink, and diuretic, and to improve kidney function. It is also used to treat hiccups, improve bowel movements, relieve thirst and fatigue, reduce burning sensations, and as a demulcent for irritation and inflammation. Additionally, sugarcane juice is known for its blood-purifying, antiseptic, and tonic properties. Jaggery, a product derived from sugarcane, is valued for its medicinal properties, including its ability to prevent anemia, reduce the toxicity of arsenic, stimulate the digestive system, improve bowel movements, cleanse the liver, and boost immunity. Jaggery is also rich in antioxidants and minerals, making it effective against free-radical damage and menstrual problems, and beneficial for respiratory health, and weight loss.

Technology for the production of a variety of sweeteners and other valued added products from sugarcane juice and its products are available. To sustain sugarcane agriculture, product diversification is inevitable as the crop is vulnerable to a boom-bust cycle coupled with biotic and abiotic stresses. Hence, product diversification and value addition for its therapeutic and nutritional value need to be exploited on a large scale.

Keywords

Sugarcane traditional sweeteners · Jaggery · Health benefits · *Saccharum officinarum* · Value addition · Sugarcane wine · Medicinal value

14.1 Introduction

Sugarcane is an ancient crop and was initially consumed for sweet nectar-like juice. The earliest mention of sugarcane cultivation is found in Indian literature of the period between 1400 and 1000 BC (Barber 1931). Sugarcane and its products are mentioned in "Vedas" and various literature of the ancient period. The seven Oceans depicted in the Puraṇas represent a fantasy of good things in life, of which the second river Iksu rasoda represents the river of sugarcane juice (Stella 2012). The plant sugarcane (*Saccharum officinarum*) is mentioned in the ayurvedic treaties of Charaka and Susruta in many places. Sugarcane is cultivated mainly for sugar and non-refined sugars like Jaggery, Khandsari, etc. *S. officinarum* also known as "noble cane" has its center of origin and rich diversity in Pacific islands (Fig. 14.1). The cultivation of sugarcane extends from the tropical to subtropical regions of the world. In recent years, the crop has gained more attention as a bioenergy crop for ethanol production and for co-generation of power. Traditional sweeteners are also in increasing demand due to the health benefits associated with sugarcane juice and its products. The various forms of sweeteners that are nutrition-wise superior to white sugar could be manufactured locally and can be a better option for the farmers to generate more income through sugarcane-based entrepreneurship by utilizing the available resources (Singh 2021; Nisha et al. 2023a).

Fig. 14.1 *S. officinarum* collections from center of origin in the field gene bank

14.2 Chewing Cane

Saccharum officinarum was the first domesticated species and was widely used for chewing by the indigenous people of the Pacific Islands. This species is also known as noble canes on account of its attractive rind color and stripe (Fig. 14.2), softness, and juice content (Chandran et al. 2018, 2021). This species is also characterized by thick juicy canes, high sugar content, and less fiber content. However, the cultivation of these soft rind canes is now restricted to isolated pockets in countries like India, Indonesia, and New Guinea. Now commercial hybrid varieties that can withstand adverse climatic conditions, pests, and diseases and higher cane yield and better sugar content replaced the traditional ones. Apart from the nutritional quality, chewing sugarcane regularly increases the strength of the teeth and is widely recommended for maintaining oral health. It is more relevant in the present day's food habits that prefer fast foods which are excessively soft and sticky and increases dental problems. The chewing of sugarcane provides exercise to the facial muscles, cleans the teeth, and keeps them in a healthy state (Karthikeyan and Samipillai 2010). In traditional medicines, the juice obtained while chewing is considered cooler and preferred over the juice obtained through crushers for better digestibility and better healing properties. Rakkiyappan (2003) studied the chewing quality of some *S. officinarum* clones and some improved *S. officinarum* clones, and identified the genotypes Waxy Red, PiO 90-196, PiO 90-99, PiO 90-224, and PiO 88-76 promising in terms of chewing quality.

Fig. 14.2 *Saccharum officinarum* clone-chewing cane

14.3 Ornamental Chewing Canes

Saccharum officinarum "the noble cane" originated in Papua New Guinea and was grown by the natives of this region as an ornamental plant and as chewing cane in home gardens. These cultivated garden canes are known for their ornamental value besides the soft rind, juicy, sweet, and thick canes. During 1920s the Dutch Scientists coined the term "Noble Cane" to describe this species with bizarre morphological and agronomical characteristics that can "ennoble" wild inferior forms of *Saccharum* by hybridization. The attractive rind color, stripes, and purple-colored leaves make *S. officinarum* a good ornamental plant coupled with remarkable food value. Hackel (1889) divided *S. officinarum* into three groups "genium" with yellow–green sugarcane, "violaceum" with purple-colored leaves, and "litteratum" with yellow-green rind with red stripe.

Saccharum officinarum clone "Pele Smoke" ornamental sugarcane along with purple sugarcane, Burgunda sugarcane, Black magic, Akoki, and Hawaiian Sugar plants were marketed as ornamental plants to a limited extent. Less popularity of sugarcane as an ornamental plant may be due to the lack of modern varieties with attractive rind stripes and/or pigmentation and the unavailability of the noble canes under cultivation. However, literature shows that the primitive tribes in New Guinea were attracted to a bright and unbelievable array of colors of these *S. officinarum* plants (Fig. 14.3) that might have resulted in brilliant attractive colored and striped varieties which are characteristics of Papuan garden canes (Stevenson 1965). Across the globe, growing of chewing sugarcane as an ornamental plant is gaining importance and clones like Florida Red, and Java Green (Green canes with red stripes, and ribbon canes) are being raised for beautiful landscapes. The ornamental chewing canes can be easily grown in households and even in a pot which makes them good ornamental with edible value.

14.4 Raw Juice

Sugarcane juice is a nonalcoholic energy drink popular in India and many other sugarcane-growing countries, especially during the summer season. It is a popular drink because of its characteristic flavor, refreshing ability, and sweet taste, and provides instant energy (Fig. 14.4). The compounds isolated through chromatography showed 4-hydroxy-3-methoxy benzaldehyde (vanillin), 4-nonanolide, 4-vinylphenol, and 2,6-dimethoxy phenol (Tokitomo et al. 1984) and were found to contribute to the characteristic aroma of the sugarcane juice. The juice contains 10–21% sucrose, 0.10–1.5% fiber, 0.3–3% reducing sugars, 75–85% water, and other organic and inorganic compounds (Swaminathan 1995).

Fig. 14.3 Attractive variation for rind color in *S. officinarum* clones

14.5 Nutrient Composition Juice

Sugarcane juice is rich in carbohydrates and proteins and contains minerals like potassium, zinc, phosphorous, calcium, iron, vitamin A, vitamin B, vitamin C and vitamin B complex which make it a very healthy drink. The nutrient analysis of sugarcane juice based on the analytical, calculated and/or derived values are reported by Longvah et al. (2017).

14.5.1 Vitamins

Water Soluble Vitamins These are vitamins that cannot be stored in the body on absorption and hence regular intake is required for better growth. Vitamins C and B vitamins (thiamine B1, riboflavin B2, niacin B3, pantothenic acid B5, pyridoxine B6, biotin B7, folate/folic acid B9, and cobalamin B12) are falling in this group (Longvah et al. 2017). Vitamin B complex and vitamin C are found in many foods,

Fig. 14.4 Bottled sugarcane juice

especially vegetables and fruits, as well as dairy, meat, legumes, peas, liver, eggs, and fortified grains and cereals.

Fat Soluble Vitamins Vitamins A, D, E, and K are easily stored in fats on absorption, among these, Ergocalciferol (0.47 ± 0.06 μg) and alpha Tocopherols (0.04 ± 0.01 mg) were estimated from sugarcane juice (Longvah et al. 2017).

14.5.2 Carotenoids

Carotenoids are tetra-terpene pigments, that exhibit yellow, orange, red, and purple colors and are usually present in algae, some groups of photosynthetic bacteria, fungi, animals, and higher plants. Beta carotene (7.87 ± 0.37 μg/100 g) and Lutein, (8.36 ± 2.40 μg/100 g), are reported to be present in sugarcane (Longvah et al. 2017).

14.5.3 Minerals and Trace Elements

Sugarcane is very rich in minerals and large variations were observed between genotypes for the presence of copper, magnesium, zinc, and phosphorus in juice.

Thangavelu and Rao (1997) studied 30 sugarcane genetic stocks for mineral composition and found significant differences among varieties and stages of growth and also their interaction. Potassium in juice is found to decrease from 173 mg at 6 months to 133 mg at 12 months of age. The Juice potassium content reported a significant negative correlation with brix, sucrose, purity, CCS %, and a positive association with reducing sugars, calcium, magnesium, sulfate, chloride, sodium, ash, electrical conductivity, and colloids of juice. Studies of phosphate content at various growth stages revealed that phosphate in juice increased from 70 mg at 6 months to 210 mg at 12 months. Phosphate in the juice was associated significantly and positively with brix, sucrose, purity, CCS %, starch, and color of juice (Thangavelu and Rao 2000). Sugarcane juice also contains a high concentration of magnesium which is correlated to the yield of ethanol. The Magnesium content in juice (Thangavelu and Rao 2001) and sodium content (Thagavelu et al. 2003) also vary significantly between genotypes.

14.5.4 Starch and Sugars

Starch is found in sugarcane as a storage polysaccharide. The concentration of starch and sugars varies widely depending on the geographic location, variety, developmental stage, and growth conditions. Sugarcane starch is characterized by its structure, granule size, gelatinization temperature, filterability, and susceptibility to glucoamylase, pullulanase, and α-amylase enzymes. The available starch and sugars are CHO 0.85 g/100 g, fructose 0.55 g/100 g, glucose 2.27 g/100 g, and sucrose 10.03 g/100 g (Longvah et al. 2017).

14.5.5 Fats and Oil

Sugarcane oil contains fatty acids, heptane, insoluble acids, sterols, resinous hydrocarbons, alcohols, and glycerine. The major fatty acids composition is linoleic (36.1%), palmitic (25.0%) oleic (10.2%), linolenic (6.9%), and arachidic (7.6%) acids (Whyte and Hengeveld 1950). Longvah et al. (2017) reported fatty acid constituents per 100 g of sugarcane juice viz., Palmitic 119 ± 4.4 mg, Stearic 20.80 ± 2.38 mg, Oleic 29.70 ± 2.99 mg, Linoleic 145 ± 12.5 mg, alpha-linoleic 28.34 ± 4.27 mg, total saturated 140 ± 3.1 mg, total mono un-saturated 29.70 ± 2.99 mg, and total poly-unsaturated 173 ± 16.1 mg.

14.5.6 Amino Acids

Twenty-three amino acids have been reported in cane juice (Parish 1965). The following amino acids were estimated from sugarcane juice (Longvah et al. 2017).

14.6 Juice Preservation

The juice is extracted by crushing the cane through roller crushers. The addition of lemon and ginger during crushing provides better flavor and taste to the juice, which can be consumed as such or by cooling. Sugarcane juice is also used in blending with fruit juices, Kinnow (10–15%), Aonla (5%), Lemon (2–2.5%), and Ginger (2.5–3%). Fresh juice marketed by street vendors always lacks hygienic conditions, which creates unhealthy situations for consumers. Packed and stored juice is available in the market but fails to maintain its freshness, aroma, taste, and refreshing ability. The processing method for packing sugarcane juice involves pasteurization (82 °C) and the addition of chemical preservative agents such as potassium meta bi-sulfate, sorbic acid, sodium benzoate, etc. Storing at below ambient temperature (4 °C) and carbonation improves the taste and shelf life for a few hours (Sahla 2023). Agarkar and Aggarwal (2019) found that slight decrease in total soluble solids (1.6%), pH (3.63%), ascorbic acid (22.5%), total sugars (2.38%), and viscosity (1.71%) after 3 months of storage while the reducing sugars were increased by 7.9%. In the meantime, the total polyphenol and total flavonoids were decreased by 20% and antioxidant activity by 25%. Pasteurized juice carbonated at 80 psi level and the addition of sodium benzoate at 60 ppm was reported to have maintained a good taste. The refrigerated (4 °C) and carbonated juice is better on account of refreshing taste compared to that stored under ambient conditions (Agarkar et al. 2019). Among the treatments for juice preservation, the addition of acidulants, preservatives, herb extracts, and stabilizers and the application of filtration, pasteurization, sterilization, irradiation, and high-pressure processing etc., are in use. Despite the availability of several methods of treatment and packing, none of the methods can maintain fresh organoleptic properties of juice (Kohli et al. 2019).

Preservation of sugarcane juice without losing, aroma and taste is a real challenge. There are several technologies on the run and many more are in the process of development. However, the available technologies either require high establishment costs or do not completely fulfil the requirements to retain the freshness, aroma, taste, and quality. Oxidation of the polyphenols is the major factor contributing to the quality of juice under storage. Eissa et al. (2010) reported that the spoilage of fresh sugarcane juice occurred due to the increase in the activity of PPO enzymes. During the maturity of sugarcane, the polyphenol oxidase activity was reported to be reduced. Qudsieh et al. (2002) studied the effects of the maturation of sugarcane juice on chlorophyll, tannin, polyphenol oxidase, and color change and found that chlorophyll, tannin, and color decreased with maturity. Hence, juice from fully mature canes may have better storability. The storage temperature also adversely affects the sugarcane juice and, in a study, to find a suitable temperature for the storage of sugarcane juice, Krishnakumar and Chellamuthu (2013) demonstrated that the juice stored in a closed container at a low temperature (10 °C) maintained quality for 10 days, at 5 °C for 4 days and at 30 °C spoiled within a day.

The different non-thermal and thermal techniques and their combinations were studied for maintaining the aroma and nutrients and can reduce the oxidative enzyme activity (Jadhav et al. 2021), however nutritive value is affected (Petruzzi et al. 2017). The techniques based on pasteurization (Kohli et al. 2019; Chew et al. 2018), ozonation (Rodrigues et al. 2018), ohmic heating (Saxena et al. 2016), gamma irradiation (Kapoor et al. 2017), thermosonication (Kulmann de Medeiros et al. 2021), pulsed electric field (Kayalvizhi et al. 2016), high hydrostatic pressure (Sreedevi and Rao 2018), and microfluidization (Tarafdar et al. 2021) are various methods available for juice preservation.

Non-thermal techniques have been proven effective and showed better results than novel thermal treatments as they were found to reduce microbial load and retain nutrition. Among non-thermal treatments, high hydrostatic pressure is the most efficient technique for the preservation of sugarcane juice while ohmic heating is preferable among thermal techniques due to less nutritional loss. Thermal treatments are effective methods for reducing deterioration and spoilage rate in sugarcane juice but they cause undesirable changes like reduction of nutrients, loss of natural flavor and aroma, and change in color (Petruzzi et al. 2017).

Non-thermal techniques also have limitations viz., ultrasound treatment affects physicochemical properties such as bitter flavor, color change, and destruction of the sugarcane juice (Majid et al. 2015). Irradiation faces problems in regulation and hurdles in consumer perception of radioactivity (Galati et al. 2019). The pulse electric field method involves a high initial cost and does not impact enzymes and microbial spores (Abbas et al. 2017; Chacha et al. 2021). Ultraviolet ray treatment is effective if the sample turbidity is high (Guerrero-Beltrán and Barbosa-Cánovas 2004). High hydrostatic pressure technique requires expensive equipment which limits its applications in the food industry (Nabi et al. 2021). In ohmic heating, monitoring, and control of the process are difficult (Sakr and Liu 2014). Membrane processing has drawbacks of foul-smelling and low yield of juice products (Cassano et al. 2021). Ozone processing has the toxicity of ozone gas at a higher level and instability of the process (Ari et al. 2020).

14.7 Jaggery

Jaggery is widely consumed in Asian countries, the Caribbean Islands, and African and Latin American countries. The name "jaggery" is derived from the Portuguese word *jagara*, which means "sugar." India is the second-largest producer of sugarcane in the world cultivating the crop in an area of five million hectares. India is the largest producer and consumer of jaggery and it contributes to more than 70% of the world's jaggery production (Rao et al. 2007). "Gur" is the Indian word for jaggery and referred to as a sweetener in Ayurvedic medicines for more than 3000 years.

Statistics show a negative trend in jaggery production since the 1970s (Fig. 14.5). Twenty percent of the sugarcane produced was used for jaggery production. However, the trend is down in the recent past and reached up to 13% in 2019. The market for organic jaggery is likely to go up as a functional food. The jaggery supplements in the diet, help in developing immunity and preventing diseases and also act as nutraceutical, functional, or fortified food (Rao et al. 2021; Rao and Singh

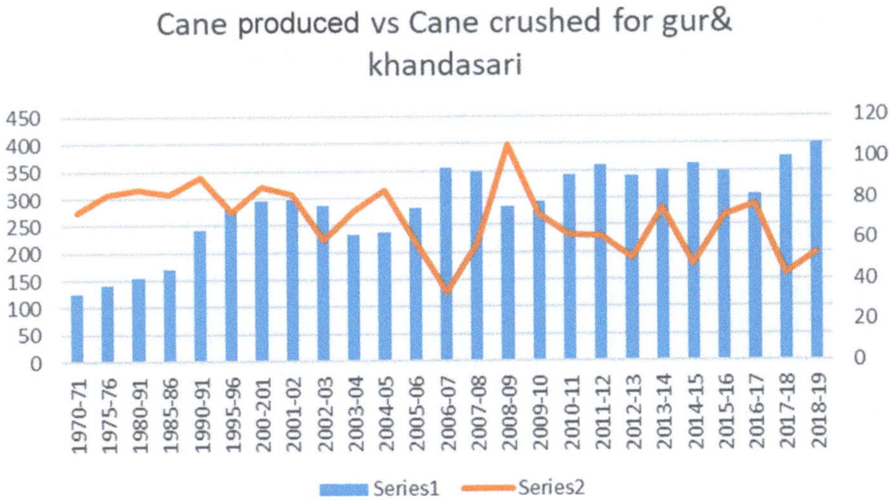

Fig. 14.5 Sugarcane production and cane crushed for Gur and Khandasari

2022). Similarly, these products will provide additional opportunities for increasing the income of sugarcane farmers.

The consumption of total sugar in India almost remains static, however, the per capita consumption of raw sugar is reduced (Table 14.1). The reason may be due to the non-availability of hygienic and good-quality jaggery. It is the healthiest form of sugar as it contains various minerals and vitamins (Madan et al. 2004).

Jaggery production stands out as a major agro-processing industry, offering employment to approximately 2.3 million people in the cooperative sector of rural areas (Nerkar 2004). Jaggery serves as a vital raw material for various cottage industries, being used in the production of sweets, such as reori, gazak, chikki, patti, and ramdana. It also finds application in animal feed mixtures. Jaggery comes in three forms: solid, liquid, and granular. To produce the solid form, sugarcane juice is clarified with herbal clarificant to yield light-colored jaggery by eliminating impurities and coloring compounds from the juice. The juice is then boiled in a triple pan furnace and concentrated to achieve the desired size and shape upon molding (Singh et al. 2011).

High-quality jaggery typically contains more than 70% sucrose, less than 10% glucose and fructose, less than 5% minerals, and less than 3% moisture (Pattnayak and Misra 2004). Liquid jaggery or semi-liquid syrup is obtained during the concentration of purified sugarcane juice for solid jaggery production. It contains 40–60% sucrose, 15–25% invert sugar, 30–36% water, 8.5–10 mg/100 mg iron, 0.3% calcium, 0.5 mg/100 mg phosphorus, and 0.1 mg/100 mg protein (Singh 2008; Singh et al. 2013). The preparation of granular jaggery is similar to solid jaggery up to the concentration of the juice. The concentrated slurry is agitated with a wooden scraper to form granules, which are then cooled, sieved to obtain granules smaller than 3 mm, and dried to reduce the moisture content to less than 2%. The composition of

Table 14.1 Consumption of sugar and Khandsari

Year	Per capita consumption of sugar (kg/year)	Per capita consumption of Gur and Khandsari (kg/year)	Total (kg/year)
2001–2002	16.1	8.0	24.1
2002–2003	17.3	5.4	22.7
2003–2004	16.0	6.6	22.6
2004–2005	16.9	7.5	24.4
2005–2006	16.7	5.2	21.9
2006–2007	17.7	3.0	20.7
2007–2008	19.2	4.5	23.7
2008–2009	19.9	9.3	29.2
2009–2010	18.2	6.2	24.4
2010–2011	17.5	5.1	22.6
2011–2012	18.3	4.7	23.0
2012–2013	18.9	3.7	22.6
2013–2014	20.1	5.7	25.8
2014–2015	20.5	3.3	23.8
2015–2016	19.6	3.1	22.7
2016–2017	18.8	4.1	22.9
2017–2018	19.3	4.1	23.4
2018–2019	19.1	4.1	23.2

100 g of granular jaggery includes 80–90 g sucrose, 5–9 g reducing sugar, 0.4 g protein, 0.1 g fat, 9 mg calcium, 4 mg phosphorus, and 12 mg iron (Singh et al. 1978).

The jaggery obtained from concentrated juice is considered wholesome food due to its mineral and vitamin content, in addition to sucrose, which serves as an energy source. It consists of long chains of sucrose that are digested slowly, providing sustained energy and posing no harm to the body. Jaggery is known to generate heat and provide instant energy to the human body. During its preparation, jaggery accumulates a significant amount of iron from the pans, making it beneficial for health, especially for anemic patients.

Traditionally jaggery is prepared under unhygienic conditions and found often adulterated with various chemicals which hamper the jaggery quality under storage. Jaggery making is an easy industry though some level of drudgery is involved. Everywhere the urge to make jaggery more hygienic is the call. The technology for liquid jaggery production reduces the drudgery of processing and requires less expertise (Malavika and Chandran 2021). The powder jaggery processing has enhanced shelf life and is more consumer-friendly. These products also have very high demand among the health-conscious urban population (Ribisha and Chandran 2021). This sector also has increased job opportunities among rural people in production, packaging, and transport and also provides mid-term income for the farmers.

Fig. 14.6 Freeze preservation of sugarcane juice with organic additives

14.8 Sugarcane Juice Based Other Value-Added Products

14.8.1 Freeze Preservation of Sugarcane Juice with Organic Additives

Raw sugarcane juice is carbohydrate-rich, but low in acid and is vulnerable to the growth of molds, yeast, spoilage bacteria, and pathogenic bacteria. Freshly crushed juice loses its flavor, ferments very quickly, and cannot be preserved even for a few hours in ambient conditions. Its color changes to greyish green and then to brownish green and loses its quality and taste very quickly. A study by Chandran et al. (2023) showed that the quick freezing of sugarcane juice and the addition of various organic additives can improve the shelf life under frozen conditions. The addition of lemon juice and ginger at the time of crushing and natural additives, such as bird eye chillies extract, mango leaf extract, sarsaparilla extract, and beetroot extract had the maximum palatability (Fig. 14.6). Among the genotypes studied the soft rind canes (*S. officinarum*) 51 NG 14, NC 83, and commercial hybrids Co 86032, Co 62175 was reported to maintain better freshness and taste under frozen preservation. The frozen juice block of 1 kg can be stored under ambient conditions for 3–4 h without losing the freshness. This method of processing has a very high prospect in the sugarcane juice industry for marketing as a better health drink that can be consumed more hygienically.

14.8.2 Sugarcane Wine

The sugarcane wine industry is at the infant stage only. Sugarcane juice wine prepared independently lacks the aroma, color, and low alcohol content <10%. Juhiya and Chandran (2023) reported that blending of various fruits (grapes, pomegranate, banana, jack fruit, Mango, and pineapple) with heat-processed sugarcane juice

Fig. 14.7 Sugarcane wine with grape and banana pulp

resulted in good quality wine (Fig. 14.7). The sugarcane juice-based wine with fruit blends had better nutrient composition (minerals and vitamins), better flavor, appealing color, and better acceptability.

14.8.3 Granular/Powder Jaggery with Organic Clarification

Granular jaggery has a better shelf life of more than 2 years. In the powder jaggery-making process, the striking temperature is critical, which ranges from 120 to 130 °C. The concentrate is allowed to cool by continuous stirring and poured onto a wooden pan/platform and allowed for crystallization (Babu and Solomon 1995). Soon after solidification, a hot mass is stirred using a wooden scrapper to make powder. The powder is sieved using 1–3 mm sieves and dried to reduce the moisture content to 1%. In a study conducted by Verma et al. (2019), powder jaggery was found to be more crystalline and stable, compared to solid jaggery at all the tested storage conditions. Rao et al. (2007), Singh et al. (2011), Anwar (2009), and Ribisha and Chandran (2021) studied different methods of powder jaggery preparation and reported that the lower moisture content than solid jaggery is beneficial for increased shelf life (Fig. 14.8). Verma and Narain (1990) reported that at high humidity solids absorb moisture that makes a change in physicochemical properties hence the packaging plays a crucial role in maintaining the quality of jaggery. Improper packaging material and storage conditions can increase reducing sugars (RS) and moisture, which can result in undesirable changes such as liquefaction, microbial contamination, color and taste changes (Singh et al. 2012). All the studies on powder jaggery reveal that powder jaggery is better in terms of its crystallinity and hence the

Fig. 14.8 Powder jaggery

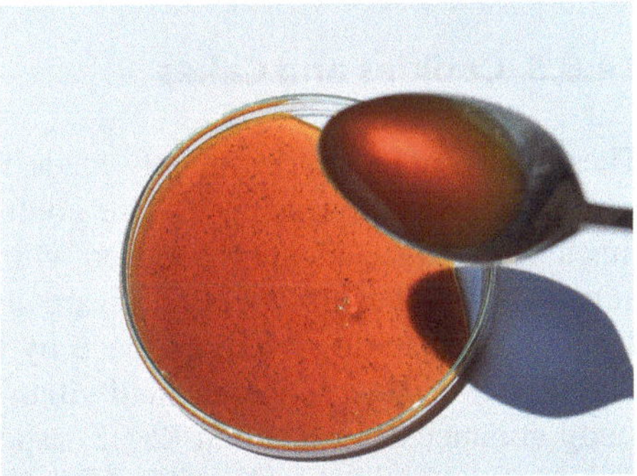

Fig. 14.9 Liquid jaggery

absorption characteristics and differs drastically from the powder made by crushing or milling the solid blocks.

14.8.4 Liquid Jaggery Without Chemical Additives

The liquid jaggery is made by boiling the juice, concentrating liquid to a suitable striking temperature (105–110 °C), cooling to room temperature, and then packing in sterilized bottles in a syrup form. The reduction of moisture content below 12% can improve the shelf life to 8 months (Malavika and Chandran 2021). To reduce the crystallization and enhance the color, the required quantity of lemon juice/citric acid is added to sugarcane juice while boiling (Fig. 14.9). To improve the shelf life of liquid jaggery preservatives like benzoic acid or potassium metabisulphite (at 125 ppm) is added. The liquid jaggery contains a high proportion of inverted sugar compared to other forms of jaggery and it imparts a honey taste. Lowering the pH and mixing juice of different varieties also reduce the crystallization problem. Different flavors and ingredients like ginger and cardamom can be added to liquid jaggery for better palatability. Further fortification of liquid jaggery with other ingredients (millets, peanuts, cashew nuts, etc) will help in developing different food products of economic importance. The final product is known in different names like Chikki, Patti, Reori, Gajak based on the ingredients added to it. To enhance the medicinal values of jaggery, ingredients like amla, ginger, turmeric, caraway, etc., can be incorporated. Instead of white sugar liquid jaggery is used for making chocolate that fetches a good price. The liquid jaggery quality mainly depends upon the quality and composition of the juice, the type of clarification used, and the striking temperature (Patil et al. 2005). High moisture content in jaggery promotes the growth of microbes, resulting in undesirable color, smell, and taste. Shahi (1999) and a study reported that proper packing and transportation of jaggery for marketing is very crucial since jaggery tends to have quality deterioration and physical changes based on the storage conditions. Patil et al. (1994) reported that hygroscopicity of jaggery affects the keeping quality adversely and the percentage of reducing sugars negatively correlates to keeping quality and vice versa.

14.8.5 Cookies and Cakes

These are confectioneries mainly made using refined wheat flour ("Maida") with white sugar as a sweetener. These confectionery items are regular snacks in tea times. Refined sugars are associated with many lifestyle diseases and hence, more nutritious and healthy sweeteners are in demand in the confectionery industry. Jaggery is a natural sweetener made by the concentration of sugarcane juice and rich in antioxidants, minerals, and vitamins that can boost the immune system. A study conducted by Hasnath (2022) reported sugarcane raw jaggery-based cakes and cookies fortified with edible fiber from rice bran and flour of millets can be prepared at ease and will have high nutrient value (Fig. 14.10). Since cereals-based baked products are part of our daily diet the nutria cookies and cakes will be an excellent vehicle for a balanced diet, especially when raw sugar and ingredients like millets and rice bran are used.

14.8.6 Ice Cream

Ice cream is an energy-rich confectionery item and white sugar is additive. White sugar's role in lifestyle diseases is well known and an alternative healthy sweetener either in the form of sugarcane juice or different forms of jaggery has been reported by Nisha et al. (2023b). Sugarcane juice is used as a sweetener in ice cream. The ice cream made using sugarcane juice has good flavor. Sugarcane juice is processed to enhance the sugar level to the ice cream standard, and the powder jaggery is to be added (Fig. 14.11). Another method is that of completely replacing the white sugar with powder jaggery. It can be a suitable replacement for white sugar, but it may lack the aroma of sugarcane juice.

Fig. 14.10 Jaggery-based cookies fortified with rice bran

Fig. 14.11 Ice cream based on sugarcane juice and jaggery

14.9 Medicinal Value of Sugarcane and Its Products

The traditional medicinal uses of sugarcane and its products have been deeply rooted in various cultures and have a long history of being used for their therapeutic properties. In many countries, including China, the Philippines, La Reunion Guiana, and Brazil, sugarcane and its derivatives are considered key ingredients in traditional medicine (Kirtikar and Basu 1991). The Latin species name "*officinarum*," which denotes "of dispensaries or pharmacy," indicates its widespread use in traditional medicine (Zakay-Rones et al. 1995; Miraj 2016; Duarte-Almeida et al. 2006). In Ayurveda and Unani systems of medicine in India, sugarcane stems and roots have been used since ancient times. Chewing raw sugarcane is recommended for maintaining a sound and healthy body. Sugarcane is often used either as a single drug or in combination with other plant parts or products to cure various diseases (Nadkarni 1976; Kirtikar and Basu 1991; Kaushik and Dhiman 1999; Warrier et al. 2010; Karthikeyan and Samipillai 2010). For instance, the roots of sugarcane are used in the treatment of joint pain, and a special oil prepared from fresh roots, dried ginger, and sesame oil is used for skin problems of the feet (Karthikeyan and Samipillai 2010).

In Malaysia, decoction from sugarcane roots is used for whooping cough, while in India, the dew that collects on sugarcane leaves is used for various eye problems. Sugarcane-based Ayurvedic treatments are well-documented in the folklore medicines of several countries, highlighting their significance in traditional medicine. Sugarcane juice fermented to vinegar and mixed with water is a digestive drink after meals. Paste made of equal parts of sugar and yellow soap is used against boils. Sugarcane vinegar mixed with water in a ratio of 1:2 is given for the treatment of extreme urticaria. Sugarcane vinegar is also applied as a lotion for headaches. A decoction of sugarcane root with "Gokhru" is recommended to remove urinary bladder calculi (Nadkarni 1976; Kaushik and Dhiman 1999).

The medicinal properties of sugarcane and its products can be attributed to the presence of phytochemicals like polyphenols (Chen et al. 2008; Singh et al. 2015). The total poly phenols range from 280 to 320 mg/100 g (Sahu and Saxena 1994). These compounds are mainly responsible for the antioxidant and cytoprotective functions observed in sugarcane (Duarte-Almeida et al. 2006). Sugarcane juice is also rich in phytosterols, terpenoids, flavonoids, and glycosides, all of which contribute to its therapeutic properties (Deshmane and Dev 1971; Mabry et al. 1984; McGhie 1993; Vila et al. 2008; Duarte-Almeida et al. 2007; Colombo et al. 2009; Singh et al. 2015).

The light yellowish-green color of the juice is very appealing but changes its quality and color very fast on oxidation of the phenolic compounds present in the juice (Yusof et al. 2000) even under refrigeration. The color of the juice is primarily due to the presence of plant pigments, like, chlorophylls, anthocyanins, and xanthophylls. Amino acids, colloids, intractable acidity, calcium, sulfur, phosphorus, chloride, total nitrogen, and polyphenols in mature juice influenced juice color (Thangavelu and Rao 1987). The occurrence of starch, proteins, gums, colloidal silica-based substances, and waxes make the juice turbid and provide an opaque texture (Perry 1988). The unrefined fresh sugarcane juice contains of high amounts phenolic compounds, flavonoids, and different glycosides. Plant phenolics are considered strong natural antioxidants that play a crucial role in different biological functions and have pharmacological importance such as liver protection, anti-inflammatory, anti-cancer, antimicrobial, anti-allergic, and antiviral properties (Kumar and Goel 2019). An evaluation of the phenolic profile of fresh sugarcane juice conducted by Naira et al. (2021) revealed the presence of 32 phenolic compounds and 17 were quantified which includes flavones (38–49 mg/L) followed by dillignols (22–29 mg/L), and phenolic acid derivatives (17–30 mg/L). Vila et al. (2008) quantified 5 flavonoids in sugarcane juice and found a total polyphenol content of 241 mg/L. Payet et al. (2006) reported the presence of the following phonolic acids in sugarcane juice viz., *p*-coumaric acid (54% of the total phenolics), ferulic acid (28%), *p*-hydroxybenzoic acid (6.3%), caffeic acid (3.3%), vanillin (3.3%), protocatechuic acid (2.5%), and syringic acid (2.0%). Eggleston (2018) reported that commercial sugarcane juices have a total phenolic content between 418 and 567 mg gallic acid equivalents/L which is similar to the range found in the fruits viz., apple, grapefruit, orange, and pineapple.

Sugarcane juice is used for various treatments in Ayurveda, Unanni, and Sidha (Nadkarni 1976; Kirtikar and Basu 1991; Kaushik and Dhiman 1999; Warrier et al. 1997). In the Unani system of medicine, sugarcane juice is used blood purifier, aphrodisiac, laxative, diuretic, fattening, and proper function of the lungs, and consuming a large quantity of sugarcane juice is a remedy for quick relief for jaundice (Anis and Iqbal 1986; Khare 2007). In Ayurvedic medicine, juice is treated as a nutritional drink as well as recommended for the major ailments Vata, Pitta, and Kapha. It is used as a diuretic, to improve kidney function (Nadkarni 1976; Ross 2005). Sugarcane juice along with dried ginger is used to treat hiccup (Duke and Wain 1981; Agarwal 1986) and is used as an aphrodisiac to improve bowel movements (laxative), relieve thirst, and fatigue, and reduce burning sensations, as a demulcent for relieving irritation and inflammation (Chopra et al. 1956; Kaushik and Dhiman 1999). Sugarcane juice is known for purifying blood, antiseptic, and a tonic (Chopra et al. 1956; Khare 2007). Traditional healers suggest sugarcane juice for treating intestinal disorders and leprosy (Nadkarni 1976; Ross 2005).

Sugarcane juice helps to build natural immunity against bacterial, viral, and protozoan infections (Lo et al. 2005). Sugarcane juice is used for treating low blood pressure and a decoction is given as a remedy for diarrhea in children (Nadkarni 1976; Ross 2005). In Cambodians use sugarcane juice to treat ulcers on skin and mucous membranes (Nadkarni 1976; Kaushik and Dhiman 1999; Ross 2005). Jaggery, a product derived from sugarcane, is also highly valued for its medicinal properties. It is rich in iron, which helps prevent anemia, and is used to reduce the toxicity of arsenic. Jaggery is also recommended for the treatment of hepatic and renal toxicity. Its high invert sugar content makes it easily assimilated by the human digestive system, making it suitable for patients suffering from jaundice and anemia.

Jaggery stimulates the digestive system, improves bowel movements, and thereby helps to relieve constipation. It acts as a diuretic, a natural body cleanser, and cleanses the liver by removing harmful toxins from the body. Jaggery consumption is highly recommended for people working in heavily polluted areas such as factories and coal mines. It fights against colds and coughs and can purify the blood. Jaggery is rich in antioxidants and minerals which help to avoid free-radical damage and also boost resistance. Due richness in many essential nutrients, jaggery is an effective natural treatment for menstrual problems.

Richness in iron and folate content ensures the maintenance of normal levels of red blood cells hence preventing anemia. The high magnesium content in Jaggery helps to boost intestinal strength, maintains normal body temperature, and keeps the stomach cool. The presence of high levels of potassium and sodium in jaggery plays a key role in maintaining acid levels in the body and helps to maintain blood pressure. It can prevent many respiratory problems such as asthma, bronchitis, etc. It is a rich source of potassium that helps in the balance of electrolytes that act as an aid for weight loss as well as building muscles and boosting metabolism. In conclusion, sugarcane and its products have been an integral part of traditional medicine systems across various cultures. Their medicinal value, attributed to phytochemicals and other compounds, highlights the importance of these natural resources in promoting health and well-being.

14.10 Conclusions

Sugarcane is cultivated across the world primarily for table sugar. In recent years the crop has emerged as an energy crop due exclusive use for bioethanol production. Though sugarcane is a feedstock primarily for large industries, its potential for the cottage industry is well-known. The production of various forms of raw sugars viz. jaggery Khandsari that are superior to white sugar in nutritional status could help to improve the income of sugarcane farmers and give employment opportunities to the young generation. Sugarcane-based agripreneurship can be a good handy start-up with less input cost. These value-added products can be handled in the rural sector by the cottage and small industries or even at the farmers' level. The bottleneck in the juice industry is the lack of economical packing and storage methods without losing the aroma, taste, and quality of the juice. The high cost of cultivation of sugarcane is another hindrance to sustaining sugarcane production. To sustain sugarcane agriculture intercropping and product diversification are inevitable. In comparison to other cash crops, the net return from sugarcane is low as it is vulnerable to a boom-bust cycle coupled with biotic and abiotic stresses. Hence, product diversification and value addition for its therapeutic and nutritional value need to be exploited on a large scale.

References

Abbas SQ, Ishaq A, Ur Rahman U, Aslam S, Shukat R (2017) Pulsed electric field technology in food preservation: a review. J Nutr Health Food Eng 6(5):168–172

Agarkar BS, Aggarwal P (2019) Development of non-carbonated sugarcane juice beverages blended with fruit juices. J Pharmacogn Phytochem 8(4):2907–2915

Agarkar BS, Aggarwal P, Yaqoob M (2019) Preparation of carbonated sugarcane juice beverages blended with fruit juices. Agric Res J 56(1):124

Agarwal VS (1986) Economic plants of India. Kailash prakashan, Calcutta, pp 414

Anis M, Iqbal M (1986) Antipyretic utility of some Indian plants in traditional medicine. Filoterpia 57:52–55

Anwar SI, Singh RD, Singh J (2009) Value addition of jaggery (gur) through natural source of vitamin C. India J Sugar Technol 24(1 & 2):48–51

Ari B, Budak NH, Seydim AC, Guzel-Seydim Z (2020) Effects of ozonation on apple juice quality. Int J Fruit Sci 20:S1578

Babu B, Solomon S (1995) Nutritive sweeteners from sugarcane crop, development of jaggery and khandsari and syrup industry in India. In: Singh GB, Solomon S (eds) Sugarcane: agroindustrial alternatives. IBH Publishing, Oxford, pp 289–312

Barber CA (1931) History and development of sugarcane as a world crop. J Sugarcane 1:1–5

Cassano A, Conidi C, Drioli E (2021) Integrated membrane operations in fruit juice processing. In: Cassano A, Drioli E (eds) Membrane systems in the food production-wellness ingredients and juice processing. De Gruyter, Berlin, pp 139–174

Chacha JS, Zhang CE, Ofoedu RA, Suleiman JM, Dotto JM, Roobab U, Agunbiade AO, Duguma HT, Mkojera BT, Hossaini SM, Rasaq WA, Shorstkii I, Okpala COR, Korzeniowska M, Guiné RPF (2021) Revisiting non-thermal food processing and preservation methods—action mechanisms, pros and cons: a technological update (2016–2021). Food Secur 10(6):1430

Chandran K, Nisha M, Gireesan PP (2018) Ornamental chewing canes. J Sugarcane Res 7(2):169–173

Chandran K, Nisha M, Gopi R, Mahendran B (2021) Digital database: Saccharum officinarum. Extension publication no. 298. Sugarcane Breeding Institute, Coimbatore

Chandran K, Keerthi C, Dhanya M, Nisha M, Mahendran B, Gopi R (2023) Freeze preservation of sugarcane juice with natural additives. In: Gomathi R, Malathi P, Alarmelu S, Appunu C (eds) Proceedings of the national seminar on research imperatives or sustaining sugarcane, sugar and ethanol production (SugarSem-2023). ICAR-Sugarcane Breeding Institute, Coimbatore, p 152

Chen ZY, Jiao R, Ma KY (2008) Cholesterol lowering nutraceuticals and functional foods. J Agric Food Chem 56:8761–8773

Chew SK, Noor MN, Maizura M, Tan TC, Rusul G (2018) Effect of pasteurization treatment and calamansi (Fortunella japonica) juice on the physicochemical, microbiological, and sensory characteristics of black stem sugarcane juice. Int Food Res J 25(3):1007–1015

Chopra RN, Nayar SI, Chopra IC (1956) Glossary of Indian medicinal plants. Publication and Information Directorate, C.S.I.R., New Delhi

Colombo R, Yariwake JH, Queroz EF, Ndjoko K, Hostettmann K (2009) Online identification of minor flavones from sugarcane juice by LC/UV/MS and post-column derivatization. J Braz Chem Soc 20:1574–1579

Deshmane SS, Dev S (1971) Higher isoprenoids-II: triterpenoids and steroids of Saccharum officinarum Linn. Tetrahedron 27:1109–1118

Duarte-Almeida JM, Novoa AV, Linares AF, Lajolo FM, Genovese MI (2006) Antioxidant activity of phenolic compounds from sugar cane (*Saccharum officinarum* L.) juice. Plant Foods Hum Nutr 61:187–192

Duarte-Almeida JM, Negri G, Salatino A, de Carvalho JE, Lajolo FM (2007) Antiproliferative and antioxidant activities of a tricin acylated glycoside from sugarcane (*Saccharum officinarum*) juice. Phytochemistry 68:1165–1171

Duke JA, Wain KK (1981) Medicinal plants of the world. Computer index with more than 85000 entries, 3 volumes. Springer, Berlin

Eggleston G (2018) Positive aspects of cane sugar and sugar cane derived products in food and nutrition. J Agric Food Chem 66:4007–4012

Eissa HA, Shehata AN, Ramadan MT, Ali HS (2010) Preservation of sugarcane juice by canning effect of thermal and chemical pre-treatments on the enzymatic browning of sugarcane juice. J Am Sci 6(9):883–888

Galati A, Moavero P, Crescimanno M (2019) Consumer awareness and acceptance of irradiated foods: the case of Italian consumers. Br Food J 121:1398–1412

Guerrero-Beltrán JA, Barbosa-Cánovas GV (2004) Advantages and limitations on processing foods by UV light. Food Sci Technol Int 10:137–147

Hackel E (1889) Andropogaoneae. In: de Candolle A, de Candolle C (Eds) Monograph of phanerogams, vol 6

Hasnath PK (2022) Standardization of method for sugarcane juicebased cakes and cookies. MSc thesis submitted to Department of Food Science and Technology, School of Health Sciences, University of Calicut

Jadhav HB, Annapure US, Deshmukh RR (2021) Non-thermal technologies for food processing. Front Nutr 8:657090

Juhiya CB, Chandran K (2023) Optimization of wine making from sugarcane juice blended with grape and banana fruits. In: Gomathi R, Malathi P, Alarmelu S, Appunu C (eds) Proceedings of the national seminar on research imperatives for sustaining sugarcane, sugar and ethanol production (SugarSem-2023), 25 August 2023. ICAR-Sugarcane Breeding Institute, Coimbatore, p 148

Kapoor K, Garg N, Diwan RK, Varshney L, Tyagi AK (2017) Study the effect of gamma radiation pretreatment of sugarcane bagasse on its physcio-chemical morphological and structural properties. Radiat Phys Chem 141:190–195

Karthikeyan J, Samipillai SS (2010) Sugarcane in therapeutics. J Herb Med Toxicol 4(1):9–14

Kaushik P, Dhiman AK (1999) Medicinal plants and raw drugs of India. Doon Phototype Printers, Dehradun, p 635

Kayalvizhi V, Pushpa AJS, Sangeetha G, Antony U (2016) Effect of pulsed electric field (PEF) treatment on sugarcane juice. J Food Sci Technol 53:1371–1379

Khare CP (2007) Indian medicinal plants: an illustrated dictionary. Springer-Verlag, New York, p 739

Kirtikar KR, Basu BD (1991) In: Blatter E, Caius JF, Mhaskar KS (eds) Indian medicinal plants. Lalit Mohan Basu, Allahabad. ID 645288579

Kohli G, Jain G, Bisht A, Upadhyay A, Kumar A, Dabir S (2019) Effect of non-thermal hurdles in shelf life enhancement of sugarcane juice thermosonication for peroxidase inactivation in sugarcane juice. LWT Food Sci Technol 112:10823. https://doi.org/10.1016/j.lwt.2019.05.131

Krishnakumar TD, Chellamuthu T (2013) Effect of delayed extraction and storage on quality of sugarcane juice. Afr J Agric Res 8:930–935

Kulmann de Medeiros J, Sarkis JR, Jaeschke DP, Mercali GD (2021) Thermosonication for peroxidase inactivation in sugarcane juice. LWT Food Sci Technol 140:110730. https://doi.org/10.1080/10408398.2022.2161464

Kumar N, Goel N (2019) Phenolic acids: natural versatile molecules with promising therapeutic applications. Biotechnol Rep 24:e00370. https://doi.org/10.1016/j.btre.2019

Lo DY, Chen TH, Chien MS, Koge K, Hosono A, Kaminogawa S (2005) Effects of sugarcane extract on modulation of immunity in pigs. J Vet Med Sci 67(6):591–597

Longvah T, Ananthan R, Bhaskarachary K, Venkaia K (2017) Indian food composition tables 2017. National Institute of Nutrition, Indian Council of Medical Research, Hyderabad

Mabry TM, Liu YL, Pearce J, Dellamonica G, Chopin J, Markham KR (1984) New flavonoids from sugarcane (*Saccharum*). J Nat Prod 47:127–130

Madan HK, Jaiswal UK, Kumar JS, Khanna SK (2004) Improvement in gur (jaggery) making plant for rural areas. J Rural Technol 1:194–196

Majid I, Nayik GA, Nanda V (2015) Ultrasonication and food technology: a review. Cogent Food Agric 1(1):1071022. https://doi.org/10.1080/23311932.2015.1071022

Malavika SR, Chandran K (2021) Liquid jaggery processing: a comparative study between species of Saccharum having different sucrose level. In: Palaniswamy C, Hemaprabha G, Viswanathan R, Karuppaiyan et al (eds) Proceedings of international conference on sugarcane research: sugarcane for sugar and beyond, June 19–21, 2021, vol 793. ICAR Sugarcane Breeding Institute, Coimbatore, p 743

McGhie TK (1993) Analysis of sugar cane flavonoids by capillary zone electrophoresis. J Chromatogr 634:107–112

Miraj S (2016) Pharmacological effects of *Saccharum officinarum*. L Der Pharmacia Lettre 8(13):223–225

Nabi BG, Mukhtar K, Arshad RN, Radicetti E, Tedeschi P, Shahbaz MU, Walayat N, Nawaz A, Inam-Ur-Raheem M, Aadil RM (2021) High-pressure processing for sustainable food supply. Sustainability 13(24):13908. https://doi.org/10.3390/su132413908

Nadkarni KM (1976) Indian materia medica, vol 1. Popular Prakashan, Bombay, p 1142

Naira PR, Bethania B, Jucelio Kulmann de Medeiros, Ligia Damasceno Ferreira Marczak a, Giovana Domeneghini Mercali (2021) Phenolic profile of sugarcane juice: Effects of harvest season and processing by ohmic heating and ultrasound Food Chem 347(4):129058 https://doi.org/10.1016/j.foodchem.2021.129058

Nerkar YS (2004) Present scenario and thrust areas for increasing sugarcane productivity in India. Financial agriculture

Nisha M, Chandran K, Gopi R, Mahendran B, Gireesan PP (2023a) Status of sugarcane cultivation in Kerala vis-a-vis product diversification option. J Sugarcane Res 11:130–136. https://doi.org/10.37580/JSR.2021.2.11.130-136

Nisha M, Maria W, Chandran K, Gopi R, Mahendran B (2023b) Standardization of the method of ice cream preparation from sugarcane juice syrup. In: Gomathi R, Malathi P, Alarmelu S et al (eds) Proceedings of the national seminar on research imperatives for sustaining sugarcane, sugar and ethanol production (SugarSem-2023), 25 August 2023. ICAR-Sugarcane Breeding Institute, Coimbatore, p 149

Parish DH (1965) The amino-acids of sugar cane. I.—the amino-acids of cane-juice and the effect of nitrogenous fertilisation on the levels of these substances. J Sci Food Agric 16(5):240–242. https://doi.org/10.1002/jsfa.2740160502

Patil JP, Wandre S, More N, Jadhav HD, Hasabnis A (1994) Influence of different varieties and harvesting stages of sugarcane on quality of jaggery. Cooper Sugar 25:377–380

Patil JE, Shinde US, Nevkar GS, Singh J (2005) Clarification efficiency of synthetic and herbal clarificants in quality jaggery production. Sugar Tech 7:77–81

Pattnayak PK, Misra MK (2004) Energetic and economics of traditional gur preparation: a case study in Ganjam District of Orissa, India. Biomass Bioenergy 26:79–88

Payet A, Shum Cheong Sing A, Smadja J (2006) Comparison of the concentrations of phenolic constituents in cane sugar manufacturing products with their antioxidant activities. J Agric Food Chem 54(19):7270–7276

Perry RH (1988) Chemical engineering handbook. McGraw-Hill Chemical Engineering Services, New York, pp 333–334

Petruzzi L, Campaniello D, Speranza B, Corbo MR, Sinigaglia M, Bevilacqua A (2017) Thermal treatments for fruit and vegetable juices and beverages: a literature overview. Compr Rev Food Sci Food Saf 16:668–691

Qudsieh YM, Yusof S, Osman A, Abdul Rahman R (2002) Effect of maturity on chlorophyll, tannin, color, and polyphenol oxidase (PPO) activity of sugarcane juice (Saccharum officinarum Var. Yellow Cane). J Agric Food Chem 50:1615–1618. https://doi.org/10.1021/JF010959L

Rakkiyappan P, Chiranjivi Rao K, Jayamani V (2003) Evaluation of Saccharum officinarum clones for chewing. Sugar Tech 5(3):155–160

Rao GP, Singh P (2022) Value addition and fortification in non-centrifugal sugar (Jaggery): a potential source of functional and nutraceutical foods. Sugar Tech 24(2):387–396. https://doi.org/10.1007/s12355-021-01020-3

Rao JPVK, Das M, Das SK (2007) Jaggery: a traditional Indian sweetener. Indian J Tradit Knowl 6(1):95–102

Rao GP, Singh P, Chauhan A (2021) Significance of value addition and fortification in jaggery: a new health initiative. RAASA J Sci Soc 3(1):15–21

Ribisha S, Chandran K (2021) Comparison of organic clarificants in "powder jaggery" processing from sugarcane. In: Palaniswamy C, Hemaprabha G, Viswanathan R, Karuppaiyan et al (eds) Proceedings of international conference on sugarcane research: sugarcane for sugar and beyond (pp. 793, 721–722), June 19–21, 2021. ICAR Sugarcane Breeding Institute, Coimbatore

Rodrigues R, Sperandio LCC, Andrade CMG (2018) Investigation of color and turbidity in the clarification of sugarcane juice by ozone. J Food Process Eng 41:e12661. https://doi.org/10.1111/JFPE.12661

Ross IA (2005) Medicinal plants of the world. Humana Press, Totowa

Sahla N (2023) Standardization of methods for fresh sugarcane juice preservation. MSc thesis, Department of Food Technology, SAFI Institute of Advanced Study, University of Calicut

Sahu AP, Saxena AK (1994) Enhanced translocation of particles from lungs by jaggery. Environ Health Perspect 102:211–214

Sakr M, Liu SA (2014) Comprehensive review on applications of ohmic heating (OH). Renew Sust Energ Rev 39:262–269. https://doi.org/10.1016/J.RSER.2014.07.061

Saxena J, Makroo HA, Srivastava B (2016) Optimization of time-electric field combination for PPO inactivation in sugarcane juice by ohmic heating and its shelf life assessment. LWT Food Sci Technol 71:329–338. https://doi.org/10.1016/J.LWT.2016.04.015

Shahi HN (1999) Sugarcane: diversification in order. In: The Hindu Survey of Indian Agriculture, vol 10. Kasturi & Sons Limited, Chennai, pp 17–103

Singh J (2008) Nutritive and eco-friendly Jaggery. In Singh J and Singh RD (eds) Processing, handling and storage of sugarcane jaggery. IISR Lucknow, India. p 3-10.

Singh P (2021) Sugarcane and sugar diversification: opportunities for small-scale entrepreneurship. In: Sugar and sugar derivatives: changing consumer preferences. Springer, Singapore. https://doi.org/10.1007/978-981-15-6663-9_15

Singh M, Bhardwaj KM, Aggarwal ML (1978) Storage of jaggery. Coop Sug 3:14–17

Singh RK, Jha A, Singh CK, Singh K (2012) Optimization of process and physico-chemical properties of Ready-to-Serve (RTS) beverage of cane juice with curd. Sugar Tech 14(4):405–411

Singh J, Singh RD, Anwar SI, Solomon S (2011) Alternative sweeteners production from sugarcane in India: lump sugar (Jaggery). Sugar Tech 13(4):366–371

Singh J, Solomon S, Kumar D (2013) Manufacturing jaggery, a product of sugarcane, as health food. Agrotechnology S11:007. https://doi.org/10.4172/2168-9881.S11-007

Singh A, Lal UR, Mukhtar HM, Singh PS, Shas G, Dhawan R (2015) Phyto chemical profile of sugarcane and its potential health aspects. Pharmacognosy 9(17):45–54

Sreedevi P, Rao PS (2018) Microbial destruction kinetics of high-pressure-processed sugarcane juice (*Saccharum officinarum*). J Food Process Eng 41:e12850. https://doi.org/10.1111/JFPE.12850

Stella S (2012) The seven oceans in the Purāṇas and elsewhere. Acta Orientalia 73:151–172

Stevenson GC (1965) Genetics and breeding of sugarcane, Tropical science series. Longmans, Green and Co Ltd, London, p 278

Swaminathan V (1995) Food science, chemistry and experimental foods. Bangalore Printing and Publishing Co. Ltd, Bangalore

Tarafdar A, Kaur BP, Pareek S (2021) Effect of microfluidization on deteriorative enzymes, sugars, chlorophyll, and color of sugarcane juice. Food Bioprocess Technol 14:1375–1385

Thangavelu S, Rao KC (1987) Effect of chemical components on the colour of sugarcane juice. In: Proceedings of joint convention of Sugarcane Technologists Association of India, vol 50, pp 103–108

Thangavelu S, Rao KC (1997) Potassium content in juice of certain sugarcane genetic stocks and its relationship with other traits. Indian Sugar 46:793–796

Thangavelu S, Rao CK (2000) Phosphate content in juice of Saccharum cultivars and its relationship with other characters. Bharatiya Sugar 25:51–56

Thangavelu S, Rao CK (2001) Magnesium content in juice of Saccharum cultivars and its relationship with other parameters at different stages of maturity. Indian Sugar 51:621–625

Thangavelu S, Rao CK, Rakkiyappan P (2003) Sodium content in juice of sugarcane clones and its relationship with other traits at different stages of maturity. Sugar Tech 5(1&2):85–88

Tokitomo Y, Kobayashi A, Yamanishi T (1984) Aroma components of fresh sugar cane juice. Agric Biol Chem 48(11):2869–2870

Verma VK, Narain M (1990) Moisture absorption isotherms of jaggery. J Stored Prod Res 26(2):61–66

Verma P, Shah GN, Sanjay M, Narain M (2019) Why jaggery powder is more stable than solid jaggery blocks. LWT Food Sci Technol 110:299–306

Vila FC, Colombo R, de Lira TO, Yariwake JH (2008) HPLC microfractionation of flavones and antioxidant (radical scavenging) activity of Saccharum officinarum L. J Braz Chem Soc 19:903–908

Warrier PK, Nambiaar VPK, Ramankutti C (2010) Indian medicinal plants a compendium of 500 species -Vaidyaratnam PS warriers Arya Vaidya sala universities press India private limited vol 5:26-44

Warrier PK, Nambiar VPK, Ramankutty C (1997) Indian medicinal plants: a compendium of 500 medicinal plants. Arya Vaidya Sala, Oriental Longman, Chennai

Whyte DE, Hengeveld B (1950) Chemical examination of sugar cane oil. J Am Oil Chem Soc 27:57–60

Yusof S, Shaian LS, Osman A (2000) Changes in quality of sugar-cane juice upon delayed extraction and storage. Food Chem 68(4):395–401

Zakay-Rones Z, Varsano N, Zlotnik M, Manor O, Regev L, Schlesinger M (1995) Inhibition of several strains of influenza virus in vitro and reduction of symptoms by an elderberry extract (Sambucus nigra L.) during an outbreak. J Altern Complement Med 1(4):361–369

A Recent Update on Metabolic Engineering in Plants, Its Futuristic Applications with Special Inference to Sugarcane Value Addition

15

P. M. Namratha, Prathima P. Thirugnanasambandam, A. Swaliha, and Lakshmi Kasirajan

Abstract

Metabolic engineering is an emerging technology that manipulates the metabolic pathways of different organisms to improve the production of the metabolites of our interest. Although many studies have been reported for metabolic engineering in plants, modifying the plant metabolic pathways is difficult due to their complexity and compartmentalization. A maximum number of metabolites must be profiled using different metabolic profiling techniques to better understand the entire metabolites and their pathways. Primary and secondary metabolites of importance are engineered in plants according to the need. Mainly secondary metabolites are targeted as they are not directly involved in the growth and development of plants. Primary plant metabolic engineering can be performed only after a thorough study of the metabolic pathways involved and their network and flux analysis. The metabolic engineering of primary metabolites resulting in stunted growth and development of plants has been reported in many studies. The plant metabolic pathways are highly interconnected and compartmentalized which needs to be carefully untangled and studied in detail for each step of metabolic alteration. Moreover, growing plants and study of metabolic engineering at a particular development stage are tedious and time-consuming. Different strategies, methods, tools, challenges, and applications of plant metabolic engineering are discussed in this review along with a few examples of experiments conducted in manipulating primary metabolic pathways. Sugarcane is a cash crop belonging to the Poaceae family cultivated worldwide for different products, mainly sucrose. It is a highly polyploid plant species with a complex genetic makeup, making it a tough crop for genetic and metabolic manipulation. Few metabolic

P. M. Namratha · P. P. Thirugnanasambandam (✉) · A. Swaliha · L. Kasirajan
ICAR-Sugarcane Breeding Institute, Coimbatore, Tamil Nadu, India

G. S. Suresha et al. (eds.), *Value Addition and Product Diversification in Sugarcane*, https://doi.org/10.1007/978-981-97-7228-5_15

engineering experiments are conducted in sugarcane with encouraging results as a result of metabolic alteration.

Keywords

Plant metabolic engineering · Metabolomics · Primary and secondary metabolites · Metabolic profiling · Primary metabolic engineering · Sugarcane engineering

15.1 Introduction

Metabolic engineering is defined as the process of manipulating the metabolic pathways of organisms to increase the production of certain metabolites by using different techniques in molecular biology. It slightly differs from genetic engineering which involves the modification of genes or enzymes and does not alter any pathways or networks related to enzymes or metabolites (Sangwan et al. 2018). Metabolic engineering was first defined by Bailey in 1991 as "the improvement of cellular activities by manipulating enzymatic, regulatory, and transport functions of the cell with the use of recombinant DNA technology" (Bailey 1991). Many recombinant DNA techniques in molecular biology along with bioinformatics tools are being used alone or in combination for metabolic engineering. The production of metabolic intermediates is mainly focused on this method of engineering. Many metabolites are being produced by different organisms, which are unknown to humans until now. In recent times, different strategies have come up for engineering the metabolic pathways and their related networks depending on the product requirement and its application. The pathway or metabolic engineering can be performed through different methods (a) directly by manipulating the genes encoding the enzymes catalyzing the reactions in the pathways, or (b) indirectly by altering the regulatory pathways affecting gene expression and enzyme activity. After finalizing the method, an appropriate target must be selected for the reaction or engineered to get the desirable results. Knowledge of the target protein or metabolite and its pathways and networks in detail is the utmost requirement to achieve a successful technique. A major part of metabolic engineering focuses on analyzing and predicting the consequences of genetic modifications and alterations in metabolic pathways.

15.2 Plant Metabolic Engineering

Approximately, 300,000 higher plant species have been documented, and around 200,000 unique entities or metabolites identified in plants (Wu and Chappell 2008). The metabolism of plants is of great importance as they are a source of enormous bioactive compounds which have pharmacological and biotechnological importance (Baghalian et al. 2014). The natural entities or metabolites produced in plants can be divided into primary and secondary metabolites. Primary metabolites are essential for the survival of the plant (proteins, lipids, carbohydrates, etc.) and have a major role in its growth and development. Secondary metabolites are generally not essential for the growth and developmental process of plants but their absence results in a long-term impairment of the organism's survival ability and potential. Despite the lack of knowledge of intricate metabolic pathways and their interdependencies, several successful attempts were made to engineer plants. The substrate-product relationship in plant metabolic pathways was first predicted in 1975 using radiotracer studies (DellaPenna 2001). Erythromycin A was produced in *Saccharopolyspora erythraea* by overexpressing *eryK* and *eryG* enzymes with a copy number ratio of 3:2 (Chen et al. 2008).

Nicotianamine, an antihypertensive drug for cardiovascular disease was engineered to be produced in *Saccharomyces cerevisiae* by the heterologous expression of the *Arabidopsis* AtNAS2 gene (Wada et al. 2006). The phenylpropanoid metabolism was altered to improve the phenylalanine content which is a precursor of polyketide biosynthesis, cell wall lignan, and anthocyanin (Wu and Chappell 2008). The co-expression of anthocyanidin reductase and Myb transcription factor for the overproduction of proanthocyanidin in *Arabidopsis thaliana* (Xie et al. 2006). Sugarcane is a grass of the *Saccharum* genus with modern cultivars derived from a hybridization process involving *Saccharum officinarum* (or "noble cane") and *Saccharum spontaneum* (wild cane). The commercial sugarcane varieties cultivated are highly complex heterozygous and polyploid in nature. Metabolic pathways are not fully elucidated and very few metabolites are identified in the crop (Kole and Hall 2008). Sucrose, an osmotically active storage metabolite in sugarcane intersects the primary metabolic pathway within the crop and increases its metabolic complexity (Birch 2007). The introduction of tailored Sucrose isomerase gene (SI) to vacuoles in sugarcane lines increased the accumulation of isomaltulose without a reduction in sucrose levels (Wu and Birch 2007).

15.3 Targets and Strategies of Plant Metabolic Engineering

The major goal or target for metabolic engineering in plants is to produce a plant with improved quality. Enhancement of quality can be achieved by improving the yield, refining color and taste, providing tolerance to biotic and abiotic stress conditions, etc. The target of metabolic engineering can be an end product, an intermediate, or a novel product from another organism, which is unknown in plants. The methods or strategies for achieving these are (1) Engineering of a particular pathway for channeling the flux. Itaconic acid was produced in increased amounts using the RNAi strategy in *S. cerevisiae*. The engineering of the target enzyme ade3 of the itaconic acid pathway directed the flux which resulted in improved production of the acid (Crook et al. 2014). (2) Targeting multiple steps in the same pathway with different enzymes which are regulated by the same transcription factors. Multiple knockdowns of succinate semi-aldehyde dehydrogenase (*sad*), succinyl-CoA synthetase (*sucD, sucC*), and succinate dehydrogenase (*sdhA, sdhB)* genes using the CRISPRi technique, increased the 4-hydroxybutyrate fraction in PHAs in *E. coli.*

Enzymes in the rate-limiting pathways can be modified using different methods (Lv et al. 2015). Listed here are some of the strategies not limited to plants as there are very few attempts in plants.

(a) Enhancing the activity of the rate-limiting enzyme: The β alanine synthesis was increased in *E. coli* by replacing the rate-limiting enzyme pyruvate-dependent aspartate decarboxylase panD with heterologous aspartate decarboxylase from *Corynebacterium glutamicum* CgpanD (Xu et al. 2021). Overexpression of the Apoplastic peroxidase gene (CrPrx) in *Catharanthus roseus* increased the production of indole serpentine by fivefold (Jaggi et al. 2011).
(b) Overexpression of enzymes by increasing precursor supply: The production of Lacto-*N*-neotetraose in *Bacillus subtilis* was increased by constructing a synthetic pathway of LNnT and by co-expressing the precursors like lactose permease (LacY) β-1,3-*N*-acetylglucosaminyltransferase (LgtA) and β-1,4-galactosyltransferase (LgtB) (Dong et al. 2019). Constitutive overexpression of 3-hydroxy-3-methylglutaryl-CoA reductase (HMGR) in tobacco and enhancing the precursor supply for isoprenoid biosynthesis will increase the sterol levels by threefold (Aharoni et al. 2005).
(c) Overexpression of transcription factors: Overexpressing a NAM, ATAF, and CUC (NAC) transcription factor enhances drought resistance and salt tolerance in rice (Hu et al. 2006).The overproduction of proanthocyanidin was achieved by co-expression of anthocyanidin reductase and Myb transcription factor in *Arabidopsis thaliana* (Xie et al. 2006) Vacuole-directed modified trehalulose synthase gene (*Pseudomonas mesoacidophila* MX-45) and transcription factor ubiquitin (Ubi-1) from maize increased the levels of trehalulose which is a low-calorie sweetener in Sugarcane (Hamerli and Birch 2011).
(d) Downregulation or knocking out of genes encoding enzymes in competitive pathways: β-amyrin was produced in *Saccharomyces cerevisiae* by simultaneously controlling the precursor supply and knocking out the key enzymes in the competitive pathways like alcohol dehydrogenase (ADH1, ADH4, ADH5, ADH6), citrate synthase (CIT2), malate synthase (MLS2), and lanosterol synthase, ERG synthase (ERG7) (Ni et al. 2019). The downregulation of the regulator of the telomere elongation helicase 1 (*RTEL1*) gene using RNAi technology in Barley led to a significant increase of cross-overs at distal chromosomal regions without a change in their total number (Abdellah et al. 2021). The amount of sucrose in Sugarcane can be increased by diverting the carbon flux in to sucrose synthetic pathway. The competitive pathway of sucrose synthesis which uses the precursor UDP glucose is cell wall synthesis. The downregulation of the key rate-limiting enzyme UDP glucose dehydrogenase in the cell wall synthesis will divert more glucose into the sucrose synthetic pathway and hence increase the accumulation of sucrose (Pi Bekker and Kossmann 2007).
(e) Multigene engineering methods: (a) use of restriction enzyme-based assembly:
    - A simple and flexible strategy was developed, unique nucleotide sequence-guided nicking endonuclease (UNiE)-mediated DNA assembly (UNiEDA), for efficient cloning of long DNAs and multigene stacking. UNiEDA was

introduced into a modified Cre/loxP recombination-mediated TransGene Stacking II (TGSII) system to generate an improved multigene stacking system. The biosynthesis of betanin in *Nicotiana benthamiana* leaves and transgenic *Oryza sativa* using multigene stacking constructs based on TGSII-UNiE was engineered and validated (Zhao et al. 2022).

- CRE site-specific recombination for multigene stacking: Cre-lox system was used to stack a set of two constitutive genes, green fluorescence protein (GFP) and Glucuronidase (GUS), a heat-induced gene dehydration-responsive element-binding (DREB1a1), and a cold-induced gene Red Fluorescent Protein (pporRFP1) cold-induced gene in rice for improving multiple traits in rice like tolerance to heat and cold climate (Pathak et al. 2022).
- Gateway combination method: A recent study reported an *Agrobacterium*-based GAANTRY system (gene assembly in *Agrobacterium* by nucleic acid transfer using recombinase technology) for assembling multigene into the T-DNA of an *Agrobacterium rhizogenes* pRi virulence plasmid by in vivo MultiRound Gateway BP recombination herbicide and pest tolerance traits (Collier et al. 2018).
- Overlapping DNA sequence assembly: Using the modified Gibson method of gene assembly, a copy of the pristinamycin II gene was produced, and synthesis of pristinamycin was improved by *Streptomyces pristinaespiralis* (Li et al. 2015). An in-fusion-based gene stacking strategy (IGS) for multigene stacking has been developed to assemble six mogrosides synthase genes from monk fruit in *Nicotiana benthamiana* and *Arabidopsis thaliana* for producing mogroside, a sweetener (Liao et al. 2022).
- Combining the above methods for maximizing the formation of the target product. An example of this combined approach is multivitamin corn (M37W), an inbred maize variety that accumulates three different vitamins (ascorbate, β carotene, and folate). Four transgenes from three different metabolic pathways of different organisms using different strategies were used for the development of this maize variety. The maize PSY1 and *Pantoea ananatis* CRTI genes for carotenoid biosynthesis, the rice DHAR gene for ascorbate recycling, and the *E. coli* folE gene (encoding GTPCHI) for folate synthesis were metabolically engineered into the M37W maize variety. The best-performing transgenic line produced 59.3 μg/g DW β-carotene (169-fold increase), 106.9 μg/g DW ascorbate (6.1-fold increase), and 1.9 μg/g DW folate (twofold increase) [28–30]. A single metabolic engineering experiment in Sugarcane which combined both expression (Wrinkled1 and Diacyl glycerol transferase1-2) and silencing (Oleosin1 and ABC transporter1) of four different genes for increasing the level of triacylglycerol (TAG) was performed by Zale et al. in 2016.

15.4 Tools for Plant Metabolic Engineering

After applying the strategies for modifying the metabolic pathways for the desired benefit, it is important to use different tools for monitoring the changes occurring inside the plant cell using different techniques. Different tools used for plant metabolic engineering are mentioned in Fig. 15.1 (Wu and Chappell 2008) The major tools used are network analysis, metabolic flux analysis, metabolic control analysis, and kinetic modeling. Many databases give information regarding metabolic network pathways like PLANTCYC, KEGG, METACYC, PMN (plant metabolic network), etc. Few other databases like the Golm database and KEGG give us data about the various metabolites present in plant systems. Other plant-specific databases are available which can be used as sources for different information-related datasets required for metabolic engineering experiments. The genetic and molecular biology-related data for the higher model plant *Arabidopsis thaliana* is maintained in a database "The Arabidopsis Information Resource (TAIR)." The Arabidopsis genome database (AtGDB) database of the plant genome database (PlantGDB) includes the BLAST-able datasets of *Arabidopsis thaliana* genomic DNA, splice-aligned transcripts and proteins, gene models, and other aligned sequences. MaizeGDB, is a database, specifically for Maize containing genomic data and different tools for analyzing the data presented on the website. The Sugarcane genome hub is a database specific for Sugarcane developed by CIRAD (French Agricultural Research and Cooperation Organization).

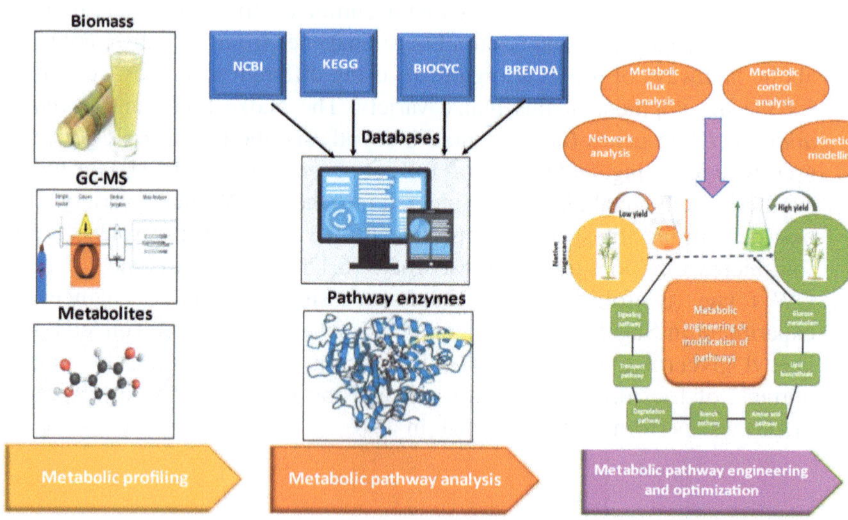

Fig. 15.1 Different tools for plant metabolic engineering

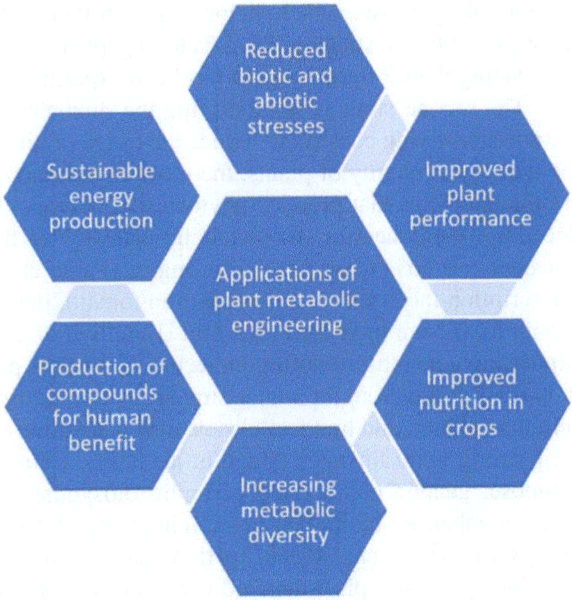

Fig. 15.2 Applications of plant metabolic engineering

15.5 Applications of Plant Metabolic Engineering

Much progress has been achieved in the field of plant metabolic engineering. Crop nutrition improvement through biofortification techniques, production of plants that satisfy their nitrogen requirement, improving photosynthetic efficiency, engineering the lignin content of the crop for improving the biofuel production efficiency of the crop example as in sugarcane oil production (Parajuli et al. 2020) increased resistance to pests (Sanghera and Kumar 2018; Srikanth et al. 2011) production of plastics and vaccines from plants, etc. are a few examples of successful plant metabolic engineering experiments. Plant metabolites (primary and secondary) can be produced on a large scale by different techniques of metabolic engineering. High-value carbohydrates, sugar substitutes, biopolymers, medicines, alcohol, waxes, pigments, etc. are examples of some compounds that are produced in plants. The major applications of plant metabolic engineering are represented in Fig. 15.2 (Yang et al. 2007).

15.6 Engineering of Secondary Metabolic Pathways

Even though many metabolic engineering experiments are conducted, most of them are performed in the secondary metabolic pathways as it does not affect the growth and development of plants. Vitamin E, an important nutrient for humans, is also essential for plants. Whole pathway regulation using transcription factors, and multigene assembly techniques was used to alter the metabolic pathways for improved tocopherol synthesis which helps plants tolerate abiotic stress in *Arabidopsis* (Kinney 2006). A strategy for developing plants that are nitrogen producers by introducing the *nitrogenase* gene has been experimented with in *Arabidopsis thaliana*. The second strategy is modifying the rhizome symbiosis between plants and the nitrogen-fixing bacteria to reduce the usage of nitrogen fertilizer. For improving the nutritional quality of plants, the most chosen targets for metabolic engineering are the pathways that produce phytoalexins, flavonoids, etc. Golden rice was produced by re-engineering the metabolic pathway for β carotene production using a *phytoene synthase (psy) gene* from maize (Paine et al. 2005). By expressing two transcription factors from snapdragon in tomato, the levels of the flavonoid anthocyanin have been increased threefold (Butelli et al. 2008). A different approach is expressing a health-promoting molecule in the host plant. To make plants efficient in biofuel production, the major strategy is to modify the biomass composition of cellulose, lignin, and lignocelluloses, which are quite difficult to break down due to their strong bond linkages (Bugg et al. 2011). Using fungi to degrade lignin and cellulose, genetic engineering of lignin biosynthetic pathways are the strategies used for enhanced biofuel production in plants (Zakzeski et al. 2010). Biopolymers like para hydroxybenzoic acid (pHBA) have been produced in sugarcane lines by introducing a chloroplast-targeted version of *E. coli* chorismate pyruvate-lyase and a 4-hydroxycinnamoylCoA hydratase/lyase from *Pseudomonas fluorescens* (Brumbley et al. 2014).

15.7 Engineering of Primary Metabolic Pathways

The modification of the primary metabolic pathways may affect the growth and development of the plant (Aharoni and Galili 2011; Stitt 2013). Due to this reason, primary metabolic pathway engineering is still in the initial stages of experimentation and the success rate is less compared to secondary metabolic pathway engineering studies. The primary metabolites are important for plant growth and they immediately get converted into different forms (intermediates). They are highly interconnected and compartmentalized. The strategies should be designed such that the engineering does not interfere with the flow of metabolites in the complementary pathways unless that is also a part of the strategy.

Common strategies used in the engineering of primary metabolic pathways include enhancement, suppression, or sequestration of metabolites in plants. These strategies can be used singly or in combinations for more promising results. Strategies depend upon the metabolite that is targeted, if it is the end product, intermediate, or more than one metabolite, etc., and whether it is produced in a linear, branching, or cyclic pathway, etc. (Farré et al. 2014). CRISPR/Cas mediated gene editing, computational modeling of metabolic networks with kinetic proteomic and fluxomic data, metabolomic network flux prediction using stochiometric genome-scale metabolic models (GEMs), and flux balance analysis are the tools used for metabolic engineering techniques. Iteration of methods is done in different combinations and replicas using the testbed method. For example, for improving the photosynthetic efficiency of plants, rewiring of the central metabolic pathways must be performed. The re-programming of the chloroplast or other plastids needs to be done which is practically impossible due to obligate phototrophy, extended time to create stable non-chimeric transplastomic lines, and the technical challenges associated with the regeneration of whole plants. For this purpose, a unicellular green alga, *Chlamydomonas reinhardtii* is a facultative heterotroph that allows for extensive modification of chloroplast function, including non-photosynthetic designs is used as a testbed for the experiment (Jackson et al. 2021). However, there are problems in corroborating these results with a much more complex plant system. Metabolic phenotyping using metabolic profiling or isotope labeling is an aid for metabolic engineering studies as it will help to identify the metabolites in different metabolic pathways in plants (Sweetlove et al. 2017). A few studies have been done by different scientists to modify the primary metabolic pathways. Enhancing the production capacity of natural products like lipids, vitamins, amino acids, carbohydrates, etc. uses genetic engineering and metabolic engineering methods in combination.

15.8 Engineering of Carbohydrate Metabolism

Increasing the amount of starch was experimented with in many plants like potatoes, wheat, rice, etc. either by overexpressing sugar transporters or by introducing genes from other plants like maize (Liu et al. 2003; Smidansky et al. 2002). Metabolic engineering experiments have been conducted in the cellulose biosynthetic pathway for producing many polymers and fructans (Zhong et al. 2003; Ruan et al. 2003). In *Arabidopsis thaliana*, the knockout mutant of caffeoyl shikimate esterase (lignin biosynthetic gene), gave a fourfold increase in the efficiency of saccharification compared to the wild type. The carbon fixation systems in plants are modified to improve photosynthetic efficiency. The strategies followed are actively concentrating CO_2 or reducing the oxygenase activity of Rubisco to develop rice with the C_4 pathway (Hibberd et al. 2008). For increasing the sucrose content in sugarcane culm, the enzyme invertase was targeted. Decreasing the levels of invertase will reduce the hydrolysis of sucrose to glucose and fructose. Knockdown and reducing the invertase activity were done in sugarcane (Rossouw et al. 2007; Zhu et al. 1997), sugarbeet (Godt and Roitsch 2006), and melons (Yu et al. 2008). In addition to the invertase enzyme, sucrose synthase (Gutiorrez-Miceli et al. 2002), sucrose phosphate synthase (Tian et al. 2010), and sucrose phosphate phosphatase (Winter et al. 2000) combinations were targeted for sucrose accumulation. Metabolic flux alterations like reducing feedback inhibitions and increasing the substrate and constant product removal have been tried by different engineers. Sugarcane transgenic lines which produce isomaltulose have improved the sugar yield (Wu and Birch 2007). Sucrose isomerase enzymes were cloned into sugarcane lines for better conversion of sucrose to its isomers which might reduce feedback inhibition and increase the sugar content. The hexose content and sugar alcohols increment which are also having a sweet taste are also conducted in several experiments. Sucrose oligomers like fructan and other simple sugar derivatives like trehalose, maltose, raffinose, etc. have been introduced by gene alteration or metabolically modified for improved sugar content in various sucrose-producing plants. Sucrose accumulation was increased in sugarcane by reducing the activity of soluble acid invertase (SAI) and neutral invertase (NI) which converts sucrose into glucose and fructose (Uys et al. 2007; Zhu et al. 1997).

15.9 Engineering of Amino Acid Metabolism

The engineering of amino acid pathways is usually performed to increase the number of essential amino acids in food and feed crops. RNAi technique has been used to modify the *opaque 2* genes in maize for enhancing the amount of lysine in maize (Segal et al. 2003). The expression of a bacterial feed-back sensitive *dihydrodipicolinate synthase* (DHPS) transgene or knocking out of lysine catabolic pathway in *Arabidopsis* seeds, resulted in 12- or fivefold gains in lysine, respectively (Zhu and Galili 2003). Rice and tomato with increased levels of polyamines have been engineered by introducing the *S-adenosylmethionine decarboxylase* gene from other plants and organisms (Thu-Hang et al. 2002; Mehta et al. 2002). Both the knocking down of the amino acid catabolic pathway and transgene introduction for enhancing the abundance of amino acids can be done together for getting better results (Zhu and Galili 2003). Expressing Dof1 induced the up-regulation of genes encoding enzymes for carbon skeleton production, a marked increase in amino acid contents, and a reduction of the glucose level in transgenic *Arabidopsis* (Yanagisawa et al. 2004).

15.10 Engineering of Fatty Acid/Lipid Metabolism

Industries producing detergents, fuels, paints, etc. are interested in work focusing on the metabolic engineering of lipids or fatty acids. Production of desaturated fatty acids was increased by introducing desaturase enzymes in different ratios (Drexler et al. 2003; Liu et al. 2001). Soybean genes have been introduced into rice to increase the linolenic acid content (Anai et al. 2003). Triacylglycerol accumulation was increased in sugarcane by co-expression of wrinkled 1, diacylglycerol acyl transferase 1,2, oleosin 1, and ABC transporter1 gene (Zale et al. 2016). The nutritionally undesirable fatty acid "erucic acid" was channelized to make useful fatty acids by incorporating *FatB thioesterase* from *Diploknema butyracea* into *Brassica juncea* and by silencing the fatty acid elongase enzyme at the genetic level by incorporation of hairpin RNA (Sinha et al. 2007). Co-expression of fatty acid hydroxylase 12 and type-2 acyl-coenzyme A: diacylglycerol acyltransferase (RcDGAT2) of *Ricinus communis* in *Arabidopsis* cell lines increases the ricinoleate levels in seed oil from 17% to around 30% (Burgal et al. 2008). In transgenic sugarcane, the constitutive co-expression of WRINKLED1 (*WRI1*), diacylglycerol acyltransferase1-2 (DGAT1-2), oleosin1 (OLE1), and simultaneous co-suppression of ADP-glucose pyrophosphorylase (AGPase), and a subunit of the peroxisomal ABC transporter1 (PXA1) elevated triacylglycerides (TAG) accumulation in leaves or stems by 95- or 43-fold to 1.9% or 0.9% of dry weight (Zale et al. 2016). The expression of 3-ketoacyl-ACP synthases (KASIII) from *Spinacia oleracea* in *Brassica napus*, increased palmitic acid proportion through decreased total fatty acids by 5–10% (Thelen and Ohlrogge 2002). The down-regulation of FAD2 desaturase and *FatB* hydrolase in *Glycine max* has increased the oleic acid levels by 85% (Lardizabal et al. 2001). Elimination of fatty acid-modifying enzymes like FAD and FAE in *Arabidopsis* fae1 fad2 double mutant produced greater than 65% oleyl-oleate. Oleyl-oleate is a molecular wax ester species whose content is around 5% in the wild-type *Arabidopsis* (Heilmann et al. 2012).

15.11 Constraints of Plant Metabolic Engineering

The metabolic networks of plants are highly complex and very extensive compared to other organisms (Kruger and Ratcliffe 2012). Plant metabolic pathways prediction is full of uncertainties mainly due to the presence of many metabolites. Despite the importance of plant metabolites, only a small portion of these complicated molecules are explored (Aharoni and Galili 2011; Tatsis and O'Connor 2016). The distinct compartmentalization of plants with large vacuoles and active plastids is very challenging to plant biologists, which is described in Fig. 15.3 (Heinig et al. 2013). Due to the compartmentalization of different biosynthetic pathways, the transport of intermediate metabolites becomes a problem. In addition, the availability of different types of transport and the selectivity of molecules across the organelle membranes also add to the complexity (Roytrakul and Verpoorte 2007). The other major constraints of metabolic engineering are lack of knowledge of metabolic pathways, public acceptance and safety issues, regeneration of genetically altered plants, and viability of the altered metabolism in plants. The hidden problems that might occur during the process of metabolic engineering are (a) interaction among precursors if using more than one (b) chemical or spatial deactivation of the enzymes involved in the reaction (c) silent metabolism. The problems of multiple compartments and the complexity of the components can be solved by techniques like serial sampling or by expressing the transgenes under developmental or experimentally regulated promoters (Carrari et al. 2003). Genome editing and metabolic engineering are very difficult in sugarcane due to its polyploid nature, the large size of the genome, complex compartmentalization, low transformation efficiency, and lack of high-throughput tools and transgene silencing. Very few metabolites are profiled in the crop and the knowledge of metabolic pathways is partial (Meena et al. 2022).

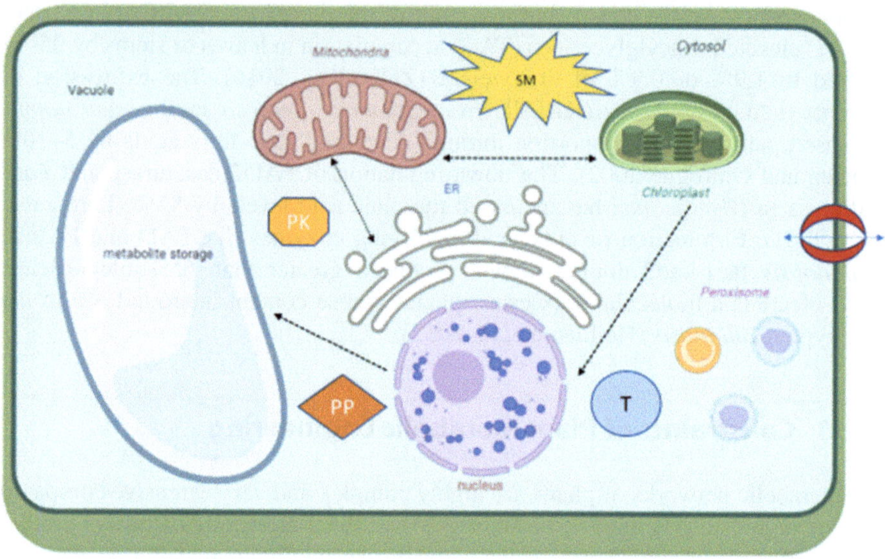

Fig. 15.3 Simplified diagram showing compartmentalization in the plant cell. The major metabolic pathways like glycolysis, gluconeogenesis, TCA cycle, amino acid, fatty acid biosynthesis, α and β oxidation of fatty acids, etc. occur in multiple compartments. Membrane transporters, signaling molecules (SM), transporter proteins (T), protein kinases (PK), protein phosphatases (PP), etc. are needed for the smooth transport of metabolites between compartments inside the plant cell

15.12 Future Prospects

Metabolic engineering in plants is an emerging field with a lot of scope and applications. The complexity of metabolic pathways and compartmentalization pose the major constraints of this technique. Different strategies and tools of metabolic engineering have to be used in combinations according to the requirements of the experiment. More metabolic profiling studies and metabolome data have to be generated to aid metabolic engineering research. Primary metabolic pathway engineering is more difficult to work with as its alteration may lead to plant growth and development changes. With structured planning strategies and the support of modern techniques introduced in metabolic engineering, better results can be achieved for developing plants with different applications.

References

Abdellah YA, Li T, Chen X, Cheng Y, Sun S, Wang Y, Jiang C, Zang H, Li C (2021) Role of psychrotrophic fungal strains in accelerating and enhancing the maturity of pig manure composting under low-temperature conditions. Bioresour Technol 320:124402. https://doi.org/10.1016/j.biortech.2020.124402

Aharoni A, Galili G (2011) Metabolic engineering of the plant primary-secondary metabolism interface. Curr Opin Biotechnol 22(2):239–244. https://doi.org/10.1016/j.copbio.2010.11.004

Aharoni A, Jongsma MA, Bouwmeester HJ (2005) Volatile science? Metabolic engineering of terpenoids in plants. Trends Plant Sci 10(12):594–602. https://doi.org/10.1016/j.tplants.2005.10.005

Anai T, Koga M, Tanaka H, Kinoshita T, Rahman SM, Takagi Y (2003) Improvement of rice (Oryza sativa L.) seed oil quality through introduction of a soybean microsomal omega-3 fatty acid desaturase gene. Plant Cell Rep 21:988–992. https://doi.org/10.1007/s00299-003-0609-6

Baghalian K, Hajirezaei MR, Schreiber F (2014) Plant metabolic modeling: achieving new insight into metabolism and metabolic engineering. Plant Cell 26(10):3847–3866. https://doi.org/10.1105/tpc.114.130328

Bailey JE (1991) Toward a science of metabolic engineering. Science (New York, N.Y.) 252(5013):1668–1675. https://doi.org/10.1126/science.2047876

Birch RG (2007) Metabolic engineering in sugarcane: assisting the transition to a bio-based economy. In: Verpoorte R, Alfermann AW, Johnson TS (eds) Applications of plant metabolic engineering. Springer, Heidelberg

Brumbley SM, Purnell MP, Petrasovits L, Nielsen LK (2014) Developing the sugarcane biofactory for high-value biomaterials. https://www.researchgate.net/publication/43481440

Bugg TDH, Ahmad M, Hardiman EM, Rahmanpour R (2011) Pathways for degradation of lignin in bacteria and fungi. Nat Prod Rep 28(12):1883–1896. https://doi.org/10.1039/c1np00042j

Burgal J, Shockey J, Lu C, Dyer J, Larson T, Graham I, Browse J (2008) Metabolic engineering of hydroxy fatty acid production in plants: RcDGAT2 drives dramatic increases in ricinoleate levels in seed oil. Plant Biotechnol J 6(8):819–831. https://doi.org/10.1111/j.1467-7652.2008.00361.x

Butelli E, Titta L, Giorgio M, Mock HP, Matros A, Peterek S, Schijlen EG, Hall WM, Bovy RD, Luo AG, Martin C (2008) Enrichment of tomato fruit with health-promoting anthocyanins by expression of select transcription factors. Nat Biotechnol 26(11):1301–1308. https://doi.org/10.1038/nbt.1506

Carrari F, Urbanczyk-Wochniak E, Willmitzer L, Fernie AR (2003) Engineering central metabolism in crop species: learning the system. Metab Eng 5(3):191–200. https://doi.org/10.1016/S1096-7176(03)00028-4

Chen Y, Deng W, Wu J, Qian J, Chu J, Zhuang Y et al (2008) Genetic modulation of the overexpression of tailoring genes eryK and eryG leading to the improvement of erythromycin A purity and production in Saccharopolyspora erythraea fermentation. Appl Environ Microbiol 74(6):1820–1828. https://doi.org/10.1128/AEM.02770-07

Collier R, Thomson JG, Thilmony R (2018) A versatile and robust Agrobacterium-based gene stacking system generates high-quality transgenic Arabidopsis plants. Plant J 95:573. https://doi.org/10.1111/tpj.13992

Crook NC, Schmitz AC, Alper HS (2014) Optimization of a yeast RNA interference system for controlling gene expression and enabling rapid metabolic engineering. ACS Synth Biol 3(5):307–313. https://doi.org/10.1021/sb4001432

DellaPenna D (2001) Plant metabolic engineering. Plant Physiol 125(1):160–163. www.plantphysiol.org

Dong X, Li N, Liu Z, Lv X, Li J, Du G et al (2019) Modular pathway engineering of key precursor supply pathways for lacto-N-neotetraose production in Bacillus subtilis. Biotechnol Biofuels 12(1):212. https://doi.org/10.1186/s13068-019-1551-3

Drexler H, Spiekermann P, Meyer A, Domergue F, Zank T, Sperling P et al (2003) Metabolic engineering of fatty acids for breeding of new oilseed crops: strategies, problems and first results. J Plant Physiol 160:779. http://www.urbanfischer.de/journals/jpp

Farré G, Blancquaert D, Capell T, van der Straeten D, Christou P, Zhu C (2014) Engineering complex metabolic pathways in plants. Annu Rev Plant Biol 65:187–223. https://doi.org/10.1146/annurev-arplant-050213-035825

Godt D, Roitsch T (2006) The developmental and organ specific expression of sucrose cleaving enzymes in sugar beet suggests a transition between apoplasmic and symplasmic phloem unloading in the tap roots. Plant Physiol Biochem 44(11–12):656–665. https://doi.org/10.1016/j.plaphy.2006.09.019

Gutiorrez-Miceli FA, Rodriguez-Mendiola MA, Ochoa-Alejo N, Mendez-Salas R, Dendooven L, Arias-Castro C (2002) Relationship between sucrose accumulation and activities of sucrose-phosphatase, sucrose synthase, neutral invertase and soluble acid invertase in micropropagated sugarcane plants. Acta Physiol Plant 24(4):441–446

Hamerli D, Birch RG (2011) Transgenic expression of trehalulose synthase results in high concentrations of the sucrose isomer trehalulose in mature stems of field-grown sugarcane. Plant Biotechnol J 9(1):32–37. https://doi.org/10.1111/j.1467-7652.2010.00528.x

Heilmann M, Iven T, Ahmann K, Hornung E, Stymne S, Feussner I (2012) Production of wax esters in plant seed oils by oleosomal cotargeting of biosynthetic enzymes. J Lipid Res 53(10):2153–2161. https://doi.org/10.1194/jlr.M029512

Heinig U, Gutensohn M, Dudareva N, Aharoni A (2013) The challenges of cellular compartmentalization in plant metabolic engineering. Curr Opin Biotechnol 24(2):239–246. https://doi.org/10.1016/j.copbio.2012.11.006

Hibberd JM, Sheehy JE, Langdale JA (2008) Using C4 photosynthesis to increase the yield of rice-rationale and feasibility. Curr Opin Plant Biol 11(2):228–231. https://doi.org/10.1016/j.pbi.2007.11.002

Hu H, Dai M, Yao J, Xiao B, Li X, Zhang Q et al (2006) Overexpressing a NAM, ATAF, and CUC (NAC) transcription factor enhances drought resistance and salt tolerance in rice. PNAS 103(35):12987–12992. https://doi.org/10.1073/pnas.0604882103

Jackson HO, Taunt HN, Mordaka PM, Smith AG, Purton S (2021) The algal chloroplast as a testbed for synthetic biology designs aimed at radically rewiring plant metabolism. Front Plant Sci 12:708370. https://doi.org/10.3389/fpls.2021.708370

Jaggi M, Kumar S, Sinha AK (2011) Overexpression of an apoplastic peroxidase gene CrPrx in transgenic hairy root lines of Catharanthus roseus. Appl Microbiol Biotechnol 90(3):1005–1016. https://doi.org/10.1007/s00253-011-3131-8

Kinney AJ (2006) Metabolic engineering in plants for human health and nutrition. Curr Opin Biotechnol 17(2):130–138. https://doi.org/10.1016/j.copbio.2006.02.006

Kole C, Hall TC (2008) Compendium of transgenic crop plants. Wiley-Blackwell, Hoboken

Kruger NJ, Ratcliffe RG (2012) Pathways and fluxes: exploring the plant metabolic network. Journal of experimental botany 63(6):2243–2246. https://doi.org/10.1093/jxb/ers073

Lardizabal KD, Mai JT, Wagner NW, Wyrick A, Voelker T, Hawkins DJ (2001) DGAT2 is a new diacylglycerol acyltransferase gene family. J Biol Chem 276(42):38862–38869. https://doi.org/10.1074/jbc.M106168200

Li L, Zhao Y, Ruan L, Yang S, Ge M, Jiang W, Lu Y (2015) A stepwise increase in pristinamycin II biosynthesis by Streptomyces pristinaespiralis through combinatorial metabolic engineering. Metab Eng 29:12–25. https://doi.org/10.1016/j.ymben.2015.02.001

Liao J, Liu T, Xie L, Mo C, Huang X, Cui S, Jia X, Lan F, Luo Z, Ma X (2022) Plant metabolic engineering by multigene stacking: synthesis of diverse mogrosides. Int J Mol Sci 23(18):10422. https://doi.org/10.3390/ijms231810422

Liu J-W, Demichele S, Bergana M, Bobik E, Hastilow C, Chuang L-T, Mukerji P, Huang Y-S (2001) Characterization of oil exhibiting high γ-linolenic acid from a genetically transformed canola strain. J Am Oil Chem Soc 78(5):489–493

Liu Q, Wang Z, Chen X, Cai X, Tang S, Yu H, Zhang J, Hong M, Gu M (2003) Stable inheritance of the antisense waxy gene in transgenic rice with reduced amylose level and improved quality. Transgenic Res 12:71

Lv L, Ren YL, Chen JC, Wu Q, Chen GQ (2015) Application of CRISPRi for prokaryotic metabolic engineering involving multiple genes, a case study: controllable P(3HB-co-4HB) biosynthesis. Metab Eng 29:160–168. https://doi.org/10.1016/j.ymben.2015.03.013

Meena MR, Appunu C, Arun Kumar R, Manimekalai R, Vasantha S, Krishnappa G, Kumar R, Pandey SK, Hemaprabha G (2022) Recent advances in sugarcane genomics, physiology, and phenomics for superior agronomic traits. Front Genet 13:854936. https://doi.org/10.3389/fgene.2022.854936

Mehta RA, Cassol T, Li N, Ali N, Handa AK, Mattoo AK (2002) Engineered polyamine accumulation in tomato enhances phytonutrient content, juice quality, and vine life. Nat Biotechnol 20(6):613–618. http://biotech.nature.com

Ni J, Zhang G, Qin L, Li J, Li C (2019) Simultaneously down-regulation of multiplex branch pathways using CRISPRi and fermentation optimization for enhancing β-amyrin production in Saccharomyces cerevisiae. Synth Syst Biotechnol 4(2):79–85. https://doi.org/10.1016/j.synbio.2019.02.002

Paine JA, Shipton CA, Chaggar S, Howells RM, Kennedy MJ, Vernon G, Wright SY, Hinchliffe E, Adams JL, Silverstone AL, Drake R (2005) Improving the nutritional value of Golden Rice through increased pro-vitamin A content. Nat Biotechnol 23(4):482–487. https://doi.org/10.1038/nbt1082

Parajuli S, Kannan B, Karan R, Sanahuja G, Liu H, Garcia-Ruiz E, Kumar D, Singh V, Zhao H, Long S, Shanklin J, Altpeter F (2020) Towards oilcane: engineering hyperaccumulation of triacylglycerol into sugarcane stems. GCB Bioenergy 12(7):476–490. https://doi.org/10.1111/gcbb.12684

Pathak B, Nandy S, Srivastava V (2022) Multigene transformation through Cre-lox mediated site-specific integration in rice. Methods Mol Biol 2408:293–302. https://doi.org/10.1007/978-1-0716-1875-2_19

Pi Bekker J, Kossmann J (2007) Genetic manipulation of the cell wall composition of sugarcane. University of Stellenbosch, Stellenbosch

Rossouw D, Bosch S, Kossmann J, Botha FC, Groenewald JH (2007) Downregulation of neutral invertase activity in sugarcane cell suspension cultures leads to a reduction in respiration and growth and an increase in sucrose accumulation. Funct Plant Biol 34(6):490–498. https://doi.org/10.1071/FP06214

Roytrakul S, Verpoorte R (2007) Role of vacuolar transporter proteins in plant secondary metabolism: Catharanthus roseus cell culture. Phytochem Rev 6(2–3):383–396. https://doi.org/10.1007/s11101-006-9022-4

Ruan YL, Llewellyn DJ, Furbank RT (2003) Suppression of sucrose synthase gene expression represses cotton fiber cell initiation, elongation, and seed development. Plant Cell 15(4):952–964. https://doi.org/10.1105/tpc.010108

Sanghera GS, Kumar A (2018) A review on mechanisms, screening and engineering for pest resistance in sugarcane (Saccharum spp). J Pharmacogn Phytochem 4:187–194

Sangwan NS, Jadaun JS, Tripathi S, Mishra B, Narnoliya LK, Sangwan RS (2018) Plant metabolic engineering. In: Omics technologies and bio-engineering: volume 2: towards improving quality of life. Elsevier, Amsterdam, pp 143–175. https://doi.org/10.1016/B978-0-12-815870-8.00009-7

Segal G, Song R, Messing J (2003) A new opaque variant of maize by a single dominant RNA-interference-inducing transgene. Genetics 165:387

Sinha S, Jha JK, Maiti MK, Basu A, Mukhopadhyay UK, Sen SK (2007) Metabolic engineering of fatty acid biosynthesis in Indian mustard (Brassica juncea) improves nutritional quality of seed oil. Plant Biotechnol Rep 1(4):185–197. https://doi.org/10.1007/s11816-007-0032-5

Smidansky, E. D., Clancy, M., Meyer, F. D., Lanning, S. P., Blake, N. K., Talbert, L. E., & Giroux, M. J. (2002). Enhanced ADP-glucose pyrophosphorylase activity in wheat endosperm increases seed yield Proc Natl Acad Sci 99(3):1724-1729). www.pnas.orgcgi 10.1073pnas.022635299

Srikanth J, Subramonian N, Premachandran MN (2011) Advances in transgenic research for insect resistance in sugarcane. Trop Plant Biol 4(1):52–61. https://doi.org/10.1007/s12042-011-9077-2

Stitt M (2013) Progress in understanding and engineering primary plant metabolism. Curr Opin Biotechnol 24(2):229–238. https://doi.org/10.1016/j.copbio.2012.11.002

Sweetlove LJ, Nielsen J, Fernie AR (2017) Engineering central metabolism—a grand challenge for plant biologists. Plant J 90(4):749–763. https://doi.org/10.1111/tpj.13464

Tatsis EC, O'Connor SE (2016) New developments in engineering plant metabolic pathways. Curr Opin Biotechnol 42:126–132. https://doi.org/10.1016/j.copbio.2016.04.012

Thelen JJ, Ohlrogge JJ (2002) Metabolic engineering of fatty acid biosynthesis in plants. Metab Eng 4(1):12–21. https://doi.org/10.1006/mben.2001.0204

Thu-Hang P, Bassie L, Safwat GM, Trung-Nghia P, Christou P, Capell T (2002) Expression of a heterologous S-adenosylmethionine decarboxylase cDNA in plants demonstrates that changes in S-adenosyl-L-methionine decarboxylase activity determine levels of the higher polyamines spermidine and spermine. Plant Physiol 129(4):1744–1754. https://doi.org/10.1104/pp.010966

Tian H, Ma L, Zhao C, Hao H, Gong B, Yu X, Wang X (2010) Antisense repression of sucrose phosphate synthase in transgenic muskmelon alters plant growth and fruit development. Biochem Biophys Res Commun 393(3):365–370. https://doi.org/10.1016/j.bbrc.2010.01.124

Uys L, Botha FC, Hofmeyr JHS, Rohwer JM (2007) Kinetic model of sucrose accumulation in maturing sugarcane culm tissue. Phytochemistry 68(16–18):2375–2392. https://doi.org/10.1016/j.phytochem.2007.04.023

Wada Y, Kobayashi T, Takahashi M, Nakanishi H, Mori S, Nishizawa NK (2006) Metabolic engineering of Saccharomyces cerevisiae producing nicotianamine: potential for industrial biosynthesis of a novel antihypertensive substrate. Biosci Biotechnol Biochem 70(6):1408–1415. https://doi.org/10.1271/bbb.50660

Winter H, Huber SC, Sharkey T (2000) Regulation of sucrose metabolism in higher plants: localization and regulation of activity of key enzymes. Crit Rev Biochem Mol Biol 35(4):253–289

Wu L, Birch RG (2007) Doubled sugar content in sugarcane plants modified to produce a sucrose isomer. Plant Biotechnol J 5(1):109–117. https://doi.org/10.1111/j.1467-7652.2006.00224.x

Wu S, Chappell J (2008) Metabolic engineering of natural products in plants; tools of the trade and challenges for the future. Curr Opin Biotechnol 19(2):145–152. https://doi.org/10.1016/j.copbio.2008.02.007

Xie DY, Sharma SB, Wright E, Wang ZY, Dixon RA (2006) Metabolic engineering of proanthocyanidins through co-expression of anthocyanidin reductase and the PAP1 MYB transcription factor. Plant J 45(6):895–907. https://doi.org/10.1111/j.1365-313X.2006.02655.x

Xu J, Zhu Y, Zhou Z (2021) Systematic engineering of the rate-limiting step of β-alanine biosynthesis in Escherichia coli. Electron J Biotechnol 51:88–94. https://doi.org/10.1016/j.ejbt.2021.03.002

Yanagisawa, S., Akiyama, A., Kisaka, H., Uchimiya, H., & Miwa, T. (2004). Metabolic engineering with Dof1 transcription factor in plants: improved nitrogen assimilation and growth under low-nitrogen conditions. Biol Sci 101 (20) 7833-7838 www.pnas.orgcgi 10.1073pnas.0402267101

Yang S-T, Liu X, Zhang Y (2007) Chapter 4: Metabolic engineering-applications, methods, and challenges. In: Bioprocessing for value-added products from renewable resources. Elsevier, Amsterdam

Yu X, Wang X, Zhang W, Qian T, Tang G, Guo Y, Zheng C (2008) Antisense suppression of an acid invertase gene (MAI1) in muskmelon alters plant growth and fruit development. J Exp Bot 59(11):2969–2977. https://doi.org/10.1093/jxb/ern158

Zakzeski J, Bruijnincx PCA, Jongerius AL, Weckhuysen BM (2010) The catalytic valorization of lignin for the production of renewable chemicals. Chem Rev 110(6):3552–3599. https://doi.org/10.1021/cr900354u

Zale J, Jung JH, Kim JY, Pathak B, Karan R, Liu H, Chen X, Wu H, Candreva J, Zhai Z, Shanklin J, Altpeter F (2016) Metabolic engineering of sugarcane to accumulate energy-dense triacylglycerols in vegetative biomass. Plant Biotechnol J 14(2):661–669. https://doi.org/10.1111/pbi.12411

Zhao Y, Han J, Tan J, Yang Y, Li S, Gou Y, Luo Y, Li T, Xiao W, Xue Y, Hao Y, Xie X, Liu Y, Zhu Q (2022) Efficient assembly of long DNA fragments and multiple genes with improved nickase-based cloning and Cre/loxP recombination. Plant Biotechnol J 20(10):1983–1995. https://doi.org/10.1111/pbi.13882

Zhong R, Morrison WH, Freshour GD, Hahn MG, Ye ZH (2003) Expression of a mutant form of cellulose synthase AtCesA7 causes dominant negative effect on cellulose biosynthesis. Plant Physiol 132(2):786–795. https://doi.org/10.1104/pp.102.019331

Zhu X, Galili G (2003) Increased lysine synthesis coupled with a knockout of its catabolism synergistically boosts lysine content and also transregulates the metabolism of other amino acids in Arabidopsis seeds. Plant Cell 15(4):845–853. https://doi.org/10.1105/tpc.009647

Zhu YJ, Komor E, Moore PH (1997) Sucrose accumulation in the sugarcane stem 1s regulated by the difference between the activities of soluble acid invertase and sucrose phosphate synthase. Plant Physiol 11:609

Powering Tomorrow: Advancements in Energycane Breeding for Sustainable Bioenergy Solutions

16

P. Govindaraj, M. R. Meena, Ravinder Kumar, Neha Patwa, K. Elayaraja, Amaresh, and V. Sreenivasa

Abstract

Energycane, a hybrid of sugarcane and a wild species, is emerging as a promising bioenergy crop owing to its high biomass yield and adaptability across diverse environments. This chapter synthesizes advancements in energycane breeding and its applications in bioenergy production. Breeding programs exploit energycane's genetic diversity to enhance traits crucial for biofuel production, including biomass yield, sugar content, disease resistance, and stress tolerance. Molecular breeding techniques, such as marker-assisted selection and genomic selection, expedite the identification and introgression of favorable alleles associated with bioenergy traits. Trait selection in energycane breeding prioritizes characteristics optimizing biomass accumulation, enhancing biomass quality, and facilitating efficient biofuel conversion. Furthermore, efforts are directed toward developing energycane cultivars tailored to various agro-climatic regions, ensuring adaptability and productivity across diverse environments. Targeted manipulation of metabolic pathways associated with biomass accumulation and biofuel synthesis could lead to the development of elite energycane varieties optimized for bioenergy production. Looking forward, future research directions encompass harnessing genomic resources for trait manipulation, optimizing cultivation practices for sustainable bioenergy production, and exploring novel biotechnological interventions to enhance biomass yield and quality. A holistic approach integrating breeding and biotechnology will be instrumental in realizing the full potential of energycane as a renewable bioenergy feedstock, contributing to the transition towards a sustainable and carbon-neutral energy economy.

P. Govindaraj · K. Elayaraja · Amaresh · V. Sreenivasa
ICAR-Sugarcane Breeding Institute, Coimbatore, India

M. R. Meena (✉) · R. Kumar · N. Patwa
ICAR-Sugarcane Breeding Institute, Regional Centre, Karnal, India

G. S. Suresha et al. (eds.), *Value Addition and Product Diversification in Sugarcane*, https://doi.org/10.1007/978-981-97-7228-5_16

Keywords

Sugarcane · Energycane · Bioenergy · Biofuel · Biogas · Cogeneration

16.1 Introduction

The *Saccharum* genus of perennial grasses is known for its sugarcane, which is mostly utilized as a raw ingredient in the sugar industry. This perennial grass belongs to the genus *Saccharum* and is mostly used for sugar production. The utilization of sugarcane as an energy source has garnered more attention. Sugarcanes are perennial grasses in the *Saccharum* genus that are mostly utilized for sugar, the production of fossil fuels, and mitigating climate change. The lignocellulosic biomass derived from sugarcane, mostly consisting of bagasse and straw, has the potential to be utilized in multiple energy-related applications, including the production of electricity, second-generation ethanol, and chemical and bio-based product synthesis in the context of a biorefinery. As a result, the desirable characteristics of sugarcane varieties have shifted to reflect the increased focus on biomass utilization. The idea behind "energy cane" first put forth by Alexander in Puerto Rico in 1985 is to breed sugarcane types with higher yields of biomass and fiber. In response to the increased interest in bioenergy, researchers have been examining the physiological source-to-sink mechanisms of sugarcane in recent decades. This phase is essential for boosting the plant's efficiency in absorbing solar energy and storing sugar to maximize its use for diverse human uses. As a C4 plant, sugarcane has one of the greatest rates of dry matter accumulation due to its excellent photosynthetic efficiency (Sage and Zhu 2011; Byrt et al. 2011). To cut greenhouse gas (GHG) emissions, negative emissions technologies (NETs; i.e., technologies that result in the net removal of CO_2/GHGs from the atmosphere) must be widely implemented. Soil carbon sequestration, biochar, and bioenergy carbon capture and storage technology are the three primary forms of net-enabled technology now in use. To offset the current levels of CO_2 emissions brought on by human activity, however, considerable efforts are still required as most of these technologies are still mostly in the demonstration stage. One study indicates that sugarcane plantations can convert more CO_2 into soil carbon than mature or secondary forests since sugarcane is a crop with a high biomass production rate (Haberl et al. 2011).

16.2 Sugarcane: A Promising Energy Crop

Sugarcane occupies a significant portion of the food and energy crops grown in almost 100 tropical and subtropical nations. About 100 tropical and subtropical countries rely heavily on sugarcane as a food and energy crop; these countries' economies cannot function without it. Sugarcane was cultivated on 28 million hectares in more than 100 countries in 2023, accounting for 0.5% of the world's total agricultural area (FAO 2023). Despite making up a significantly greater share of the world's arable land than sugarcane, sugarcane is the most productive crop globally in terms of total tonnes produced (FAO 2023). The output-to-input ratio of first-generation ethanol made from sugarcane plants is 1.6:1 for maize and 8–10:1 for sugarcane plants (Goldemberg 2008).

After the juice is crushed and squeezed, the sugarcane factory uses the stalks to produce steam and electricity these days. If the technology for cellulosic ethanol is taken into consideration, it is predicted that the output of ethanol will increase by 40–50%, from the existing 7500–13,000 L/ha (Fischer and Schrattenholzer 2001). While sugar mills are not yet allowed to produce second-generation ethanol from sugarcane, several ongoing breeding programs are increasing the genetic variety of the crop to enhance its current germplasm in terms of biomass output and sucrose yield in anticipation. When combined with garbage and dry stalk, sugarcane yields 39 tonnes of unit energy per acre annually, which is comparable to other high-yielding bioenergy sources. In comparison to other high-yielding bioenergy crops like *Miscanthus* (29.6 tonnes/ha), the yearly production of sugarcane per hectare, including trash and dry stalk (39 tonnes/ha), is comparable. Traditionally based mostly on petroleum, interest in biomass as a renewable energy source has grown. A variety of biological materials are included in biomass, especially those that come from crops cultivated expressively to provide fuel for energy generation.

16.3 Characteristics of Energy Cane

An ideal energy has the following characteristics:

1. ***High biomass potential***: Energy cane is specifically bred to yield the highest amount of biomass feasible. Biomass yield is a crucial component of bioenergy crops since it directly influences the amount of feedstock available for the production of biofuel. In general, energy cane cultivars yield significantly more biomass per hectare than traditional sugarcane cultivars.
2. ***High fiber content***: The fiber content of energy canes is composed of cellulose, hemicellulose, and lignin. These components are required for the production of cellulosic ethanol. Cane cellulose and hemicellulose can be fermented to produce fermentable sugars, which can subsequently ferment to produce ethanol. Lignin can be processed to produce energy and other bioproducts, but it requires more research to bring it to a commercial level.
3. ***Drought and stress tolerance***: Energy cane is bred to withstand drought and poor soil conditions, among other environmental challenges. Biomass yield is assured even in less-than-ideal farming conditions because of its resilience.
4. ***Persistent crop***: Energycane can be repeatedly harvested from the same planting place over an extended length of time. This results in the ability to keep energycane for several years without the need for repeated planting, which lowers input prices and labor expenses while also lessening soil disturbance.

16.4 Breeding for Bioenergy Canes

These days, *Saccharum officinarum* and *Saccharum spontaneum* are the two principal *Saccharum* species used in the development of commercial sugarcane hybrids. Originating in Papua New Guinea, *S. officinarum* is a plant with thick stalks and a high sucrose content that thrives in a purely tropical climate, and it makes up 70–80% of the genome in modern sugarcane cultivars. While *S. spontaneum*, a grassy wild species, naturally grows in diverse conditions extending from Africa to Southeast Asia and Pacific Island, which is known for its slender stalk and high fiber content, it makes up 10–20% of the genome of modern hybrids and provides resilience to a range of abiotic stressors and diseases, as well as good tillering and ratooning capacities. During early breeding operations in Java and India in the nineteenth century, these features were introgressed into the *Saccharum officinarum* genome. Harvesting high sugar content while introducing disease resistance from wild species was the primary goal. Nonetheless, it was discovered that in the highly selected F_1 group, there was a negative association between fiber characteristics and brix %. An essential component of energycane is high fiber. High fiber content was later found to be a significant characteristic from an energy perspective. Sugarcane fiber contains mainly cellulose, hemicellulose, and lignin. The proportion of these substances in sugarcane bagasse (SCB) is given in Fig. 16.1.

A specific variety of sugarcane cultivated for the production of bioenergy is called energycane. These hybrids, which are a growing trend as fuel for bioenergy, are a cross between commercial cane and its wild ancestors. Utilizing clones of *Erianthus arundinaceus*, which produce high levels of fiber and biomass, along with *S. spontaneum* and other commercial cultivars, is part of the breeding program for energycane (Mohanraj and Nair 2014; Meena et al. 2020). In order to achieve crop diversification and create multipurpose sugarcane variations, it has been essential to accelerate the breeding program for sugarcane cultivars. High biomass productivity should be the focus of selection pressure if ultimate objective is production

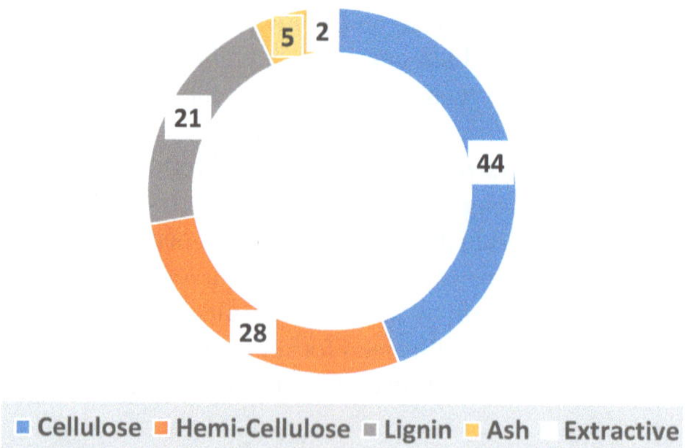

Fig. 16.1 Composition of sugarcane bagasse

of energycane. Several studies have shown that choosing desired clones straight from the segregating F_1 population has the potential to greatly boost biomass. In Wang et al.'s (2008) study, they crossed commercial clones of *S. spontaneum* with *S. officinarum*, and the result was that the progeny of those clones had higher stalk counts than cane and biomass yields of the top 5% of the clones compared to commercial cultivar checks. Tew and Cobil (2008) separated energy canes into Type I and Type II categories. Type I energy canes are less pure, contain less sucrose, and have a higher fiber content than regular sugarcane (Thangavelu 1982). These kinds of canes are utilized for producing sugar and bioenergy. However, Type II energy canes, which are solely intended for the generation of biomass, have an even higher fiber content than Type I canes and just a little amount of sugar. Therefore, Type II offers significant feedstocks for the use of bioenergy without emphasizing the extraction of sugar. *Saccharum spontaneum* adds high fiber content, higher tillering potential, better adaptability to biotic and abiotic stresses, and better ratoonability to Type I energy canes. *Erianthus arundinaceus*, a member of the *Saccharum* family, is frequently utilized in the development of Type II energy canes because of its high biomass production capacity and resilience to the majority of pests and diseases (Meena et al. 2024). As carbon is allocated among storage, respiration, and cell wall production following photosynthesis, it has been noted that an increase in fiber content usually corresponds with a decrease in sucrose levels. The various breeding approaches are used to produce a high biomass and high biofuel energy canes, as depicted in Fig. 16.2.

16.5 Benefits of Growing Energycanes

Energycanes are a practical renewable energy source. Energycane, being a renewable resource, aids in decreasing reliance on fossil fuels and the emission of greenhouse gases. It offers an environmentally friendly substitute for producing electricity. Because of their high drought-tolerant potential (DTP), energycanes can be grown in difficult environments, even on marginal or underutilized soils where other crops cannot be grown. A thorough analysis provides a general understanding of energy canes as a feedstock for bioenergy production (Kane et al. 2022). Carbon emissions can be decreased, and soil health can be enhanced by using energycane for bioenergy. Farmers can diversify their revenue streams and improve their financial security by engaging in the bioenergy market, which is another source of income from energycane.

16.6 Application of Sugarcane as Bioenergy Crop

Energycane is an excellent biomass feedstock with strong biomass-yielding characteristics. It also contains a lot of fiber, which can be converted by biochemical and thermochemical processes into bioenergy. The sugar industry uses bagasse to produce electricity. It is accessible for 6 months during sugar mill crushing seasons. For

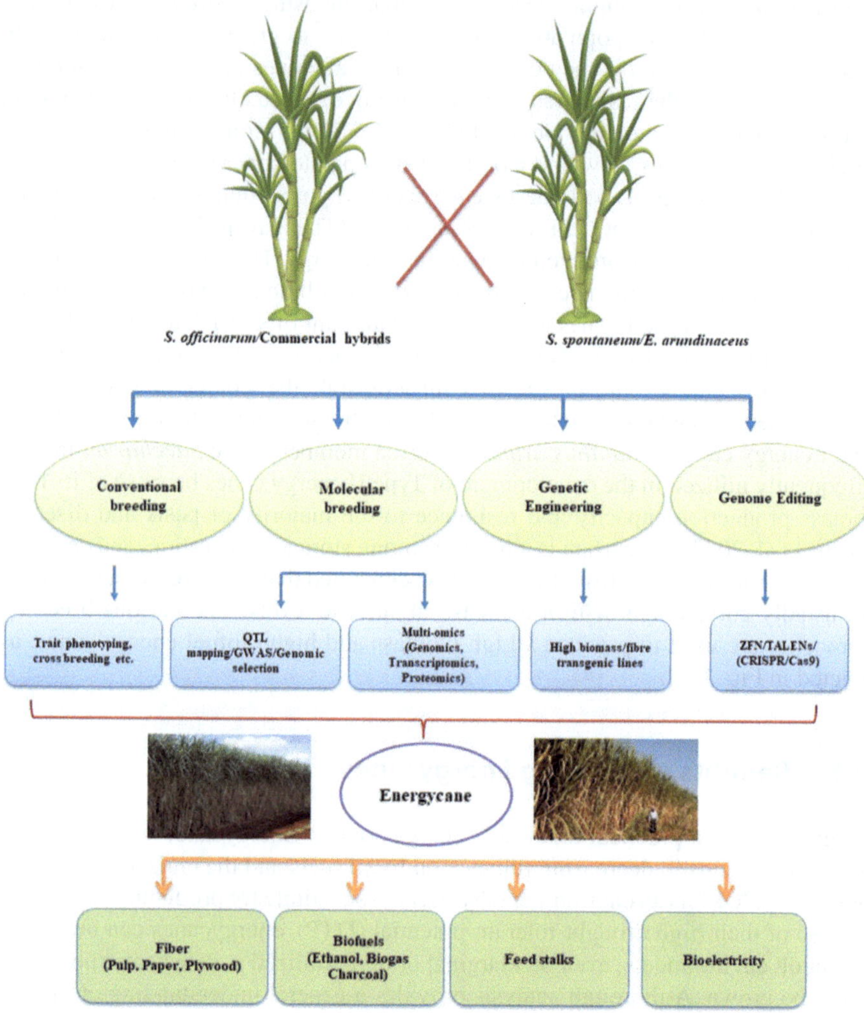

Fig. 16.2 Various breeding approaches for the production of bioenergy canes

the next 6 months, cogeneration units cannot run. Because of this, energy cane might be used as a fallback method of producing electricity in the event that feedstock is unavailable. Energycane is thus a fallback option to keep the cogeneration units operating in the off-season (Thammasittirong et al. 2023). Second-generation biofuels, which are derived from lignocellulosic biomass and have significant challenges in the commercial scale of use, may be crucial to meeting the nation's energy needs. The fibrous material left over after crushing the sugarcane stalk and from the stalk's tops and leaves is called sugarcane bagasse. These are some instances of biomass that is lignocellulosic. Apart from this, wood chips, corn straw, and tree bark can also be used to make lignocellulosic biomass. Despite expectations for the use of second-generation biofuels and the availability of the necessary technology, there are numerous obstacles to their commercial scale application. In a practical scenario, the environmentally friendly and sustainable technique of producing energy is the conversion of biomass from sugarcane to ethanol. This will be a successful approach in line of the utilization of algal biomass in bioenergy production (Chia et al. 2018).

16.7 As Energy Source of Biofuel

Sugarcane emerges as a valuable resource for the creation of biofuels such as bio-ethanol, bio-methane, bio-hydrogen, and bio-butanol, thus fostering sustainable energy generation. Research indicates the efficacy of SCB as a feedstock for bio-methane production, where acid or enzymatic pretreatments are employed to optimize bio-methane yields. This process represents a sustainable avenue for energy production from SCB, aligning with global initiatives toward renewable energy sources and environmental stewardship (Ajala et al. 2021; Ling et al. 2021). Furthermore, bio-ethanol, derived from fermenting reducing sugars extracted from SCB, presents advantages such as renewability, cost-effectiveness, and environmental compatibility compared to conventional fossil fuels. Various applications in energy production and industrial processes of sugarcane including (Hoang et al. 2015; Huang et al. 2020) hydrolyzing the cell wall with enzymes, physical processes, and chemical treatments and sugars generated from lignocellulosic material can also be used to make ethanol. Although the lignocellulose conversion process' efficiency has not yet shown to be profitable, producing second-generation ethanol is a widely desired objective since it might greatly increase the variety of feedstock options. Then, it would be feasible to generate biofuel from noncrop plants or crop biomass fractions that are not intended for human or animal consumption.

- **Generation of electricity:** Bagasse serves as a prevalent fuel source in sugar mills, utilized to produce steam for electricity generation via steam turbines. Surplus electricity produced can either be supplied to the grid or utilized to fulfill on-site energy requirements.
- **Industrial process heat:** Bagasse offers a source of process heat for various industrial operations, encompassing drying, evaporation, distillation, and sterilization across sectors like food processing, textiles, and chemical manufacturing.
- **Production of biofuels:** Biogas or bio-methane derived from bagasse can be employed as a renewable fuel for transportation or blended with natural gas for applications in heating or electricity production.

16.8 Paper Industry

Incorporating sugarcane into the paper industry represents a pioneering strategy that harnesses the fibrous attributes of SCB, the by-product remaining after extracting juice from sugarcane, as a sustainable substitute for conventional wood-derived pulp (Rainey and Covey 2016). Paper products derived from sugarcane present numerous benefits across a variety of uses, including paper made from sugarcane are well suited for printing and writing purposes, encompassing office documents, notebooks, and stationery items. These papers boast excellent printability, smoothness, and opacity, making them optimal for high-quality printing and writing tasks. Sugarcane-based papers find utility in packaging applications, such as corrugated boxes, cartons, and bags. These papers offer robustness, longevity, and resistance to moisture, rendering them suitable for safeguarding and transporting commodities. Sugarcane-based papers can be customized to meet specific needs for specialty applications, including tissue papers, specialized packaging, and industrial papers. These papers can be engineered to fulfill particular performance criteria, such as strength, absorbency, and resistance to chemicals (Sayed et al. 2022; Azmi and Othman 2023).

16.9 Textile Industry

Exploring the use of sugarcane in the textile industry is an emerging area that leverages the fibrous characteristics of SCB and its promise as an eco-friendly substitute for conventional textile materials. Sugarcane fibers are recognized for their softness and comfort, rendering them appropriate for applications in clothing and apparel. The breathability of sugarcane fibers facilitates air circulation and moisture evaporation, thereby improving comfort and functionality in clothing. With their robustness and resilience, sugarcane fibers prove fitting for a range of textile applications, spanning from garments to upholstery, and technical textiles. The biodegradability and compostability of sugarcane fibers present environmental advantages over synthetic fibers sourced from petroleum-based reservoirs (Costa et al. 2013; Sakthivel et al. 2021).

16.10 Application in Industry as Biocoagulants

SCB-based biocoagulants offer versatile and eco-friendly solutions for water treatment across industries. In water treatment plants, they remove suspended solids, organic matter, and contaminants from wastewater for industrial reuse or safe discharge (Iwuozor et al. 2021, 2022). These biocoagulants enhance drinking water quality by eliminating turbidity, pathogens, and impurities, particularly beneficial in communities lacking clean water access (Sillanpää et al. 2018). In textile manufacturing, SCB-based biocoagulants mitigate environmental impacts by treating wastewater from dyeing and finishing processes (Ewuzie et al. 2022). Similarly, they aid in pharmaceutical wastewater treatment, removing suspended particles and organic matter (Alazaiza et al. 2022). In the paper and pulp industry, these biocoagulants clarify effluents before discharge, reducing pollution (Grötzner et al. 2018). They also treat mining wastewater, agricultural runoff, aquaculture water, landfill leachate, and food and beverage manufacturing effluents, demonstrating their wide-ranging applicability and environmental benefits. Overall, SCB-based biocoagulants represent a promising solution for various water treatment challenges, offering efficiency and sustainability across diverse industries.

16.11 As a Biocompostable Material

Sugarcane bioenergy enables the production of biocompostable materials, including bioplastics and textiles, offering eco-friendly alternatives to conventional products. Bioplastics derived from sugarcane, such as polylactic acid (PLA) and polyhydroxyalkanoates (PHA), find use in packaging and single-use items, while sugarcane fibers are transformed into textiles for clothing and home accessories. These materials biodegrade naturally, reducing plastic pollution and promoting a circular economy. From agricultural mulch films to utensils and tableware, sugarcane bioenergy contributes to sustainable solutions across diverse industries, addressing environmental concerns in packaging, textiles, and beyond (Narancic and O'Connor 2017; Kusuma et al. 2023).

16.12 Cogeneration

Energycane has a greater potential to be employed as a continuous supply of raw materials for the biofuel industries throughout the year. It is anticipated that energy consumption will increase by more than 200% by 2030. Since India's economy is primarily based in rural areas, one of the most crucial factors influencing its growth and development is the availability of power. An estimated 700 tonnes of bagasse with 50% moisture per day, we are needed to build a 10 MW/h power business. A total of 3700 ha of energycane acreage with a harvestable biomass potential of 150 tonnes/ha are needed for the power plant to provide a year-round supply of feedstock. The amount of electricity generated by these plants could meet the needs of the surrounding agro-based industries for several communities and the energy needs of residential use and agriculture. It might raise rural farmers' standards of living and strengthen their economic empowerment. The sugar mills in the same energy plantation area have cogeneration units with a 20 MW/h capacity, which is enough to power the plant for 6 months during the crushing offseason. The energy canes that have been designed up to this point can therefore be tested in several locations and environments. Sugar factories and the power industry would use this for large-scale commercial cultivation, and power plants would use it as feedstock.

But it is important to appropriately handle the difficulties that existed before the energycanes entered into the Indian agricultural sector. The main issue will be preventing the unintentional or deliberate mixing of energycane with sugarcane, which could cause issues with the sugar mills' ability to recuperate. It must therefore be properly grown in isolation and kept out of fields where sugarcane is grown. Energycane faced difficulties when it was first introduced into the Indian farming system and then commercially cultivated in farmer's fields. The main restriction is the management of energycane and sugarcane mixing, which can purposefully or unintentionally result in inadequate sugar recovery in sugar mills. That is, energycane should be grown in areas other than sugarcane fields and in wastelands and marginal areas of barren land (Govindaraj 2017). SBIEC 14006, a derivation of *E. arundinaceus*, is the current commercially cultivable energycane clone from ICAR-Sugarcane Breeding Institute, Coimbatore, India, and it can be easily distinguished from commercial sugarcanes (Govindaraj 2014; SBI 2013). It lacks the morphologically obvious characteristic known as "dewlap," which is the point where the leaf and leaf sheath join and absent in commercial sugarcane cultivars (Govindaraj 2020). The agronomy of the energycane cultivation is not yet worked out as huge biomass production per unit area may deplete the nutritional status of the soils. Thus, it would be wise to research energycane agronomy, production costs, energycane economics, and energycane price fixing in the near future. It is expected that harvestable biomass will have a moisture level of over 50%; hence, techniques for lowering the moisture content before boiler feeding need to be developed, since they could affect the boiler efficiency in cogeneration units used in the sugar business. Despite its drawbacks, this biomass source boosts sugar industry profits and aids in the production of environmentally friendly, sustainable green power through biomass-based industries. Three industry partners have received licenses to use SBIEC 14006 for commercial purposes, and Telangana, Rajasthan, and Maharashtra are currently conducting evaluations of the technology. It may be possible to determine whether energycane is suitable for continued expansion in various agroclimatic and agricultural settings by testing the crop in various environments and input circumstances. Evaluation of this energy cane in various environments and input conditions may help determine its appropriateness for various agro-climatic conditions and cultivation for future commercial usage in waste, marginal, and uncultivable areas to meet energy requirements in the future.

16.13 Conclusion and the Way Forward

Energy canes had shown to be remarkably resilient permitting it to flourish in a variety of climates and surroundings. Its resilient nature and rapid growth allow it to tolerate a variety of stresses, including drought, pathogens, and pests, guaranteeing dependable and long-term biomass production. The energy cane selection criteria differ considerably from those used for commercial types of sugarcane. When selecting energy canes, the emphasis should focus more on characteristics that affect biomass yield, fiber content, and potential energy than features connected with sugar production. In conclusion, sugarcane bioenergy presents versatile applications across industries, from biofuel production to bio-compostable materials, contributing to sustainable development and environmental conservation. Prospects include advancements in technology for enhanced efficiency, expanded utilization in various sectors, and continued innovation toward a greener future.

References

Ajala EO, Ighalo JO, Ajala MA, Adeniyi AG, Ayanshola AM (2021) Sugarcane bagasse: a biomass sufficiently applied for improving global energy, environment and economic sustainability. Bioresour Bioprocess 8:1–25

Alazaiza MY, Albahnasawi A, Ali GA, Bashir MJ, Nassani DE, Al Maskari T, Amr SS, Abujazar MS (2022) Application of natural coagulants for pharmaceutical removal from water and wastewater: a review. Water 14(2):140

Azmi AA, Othman SA (2023) Future of sugarcane bagasse paper: a review. AIP Conf Proc 2544(1):040034

Byrt CS, Grof CP, Furbank RT (2011) C4 plants as biofuel feedstocks: optimising biomass production and feedstock quality from a lignocellulosic perspective free access. J Integr Plant Biol 53(2):120–135

Chia SR, Ong HC, Chew KW, Show PL, Phang SM, Ling TC, Nagarajan D, Lee DJ, Chang JS (2018) Sustainable approaches for algae utilisation in bioenergy production. Renew Energy 129:838–852

Costa SM, Mazzola PG, Silva JC, Pahl R, Pessoa A Jr, Costa SA (2013) Use of sugar cane straw as a source of cellulose for textile fiber production. Ind Crop Prod 42:189–194

Ewuzie U, Saliu OD, Dulta K, Ogunniyi S, Bajeh AO, Iwuozor KO, Ighalo JO (2022) A review on treatment technologies for printing and dyeing wastewater (PDW). J Water Process Eng 50:103273

FAO (2023) World food and agriculture—statistical yearbook. FAO, Rome. https://doi.org/10.4060/cc8166en

Fischer G, Schrattenholzer L (2001) Global bioenergy potentials through 2050. Biomass Bioenergy 20(3):151–159

Goldemberg J (2008) The Brazilian biofuels industry. Biotechnol Biofuels 1:1–7. https://doi.org/10.1186/1754-6834-1-6

Govindaraj P (2014) Paradigm shift in breeding for sugarcane to energycane—an exclusive biofuel. In: National seminar on "challenges and innovative approaches in crop improvement", organized at Agricultural College and Research Institute, Madurai, Tamil Nadu Agricultural University between 16 and 17, December 2014

Govindaraj P (2017) Energy canes as a feed stock for biofuel industries. Indian Farm 67(2):61–63

Govindaraj P (2020) SBIEC 14006—a high biomass energycane for power, alcohol and paper industries. J Sugarcane Res 10:100–106

Grötzner M, Melchiors E, Schroeder LH, dos Santos AR, Moscon KG, de Andrade MA, Martinelli SH, Xavier CR (2018) Pulp and paper mill effluent treated by combining coagulation-flocculation-sedimentation and Fenton processes. Water Air Soil Pollut 229:1–7

Haberl H, Erb KH, Krausmann F, Bondeau A, Lauk C, Müller C, Plutzar C, Steinberger JK (2011) Global bioenergy potentials from agricultural land in 2050: sensitivity to climate change, diets and yields. Biomass Bioenergy 35(12):4753–4769

Hoang NV, Furtado A, Botha FC, Simmons BA, Henry RJ (2015) Potential for genetic improvement of sugarcane as a source of biomass for biofuels. Front Bioeng Biotechnol 3:182

Huang J, Khan MT, Perecin D, Coelho ST, Zhang M (2020) Sugarcane for bioethanol production: potential of bagasse in Chinese perspective. Renew Sust Energ Rev 133:110296

Iwuozor KO, Ighalo JO, Emenike EC, Ogunfowora LA, Igwegbe CA (2021) Adsorption of methyl orange: a review on adsorbent performance. Curr Res Green Sustain Chem 4:100179

Iwuozor KO, Ogunfowora LA, Oyekunle IP (2022) Review on sugarcane-mediated nanoparticle synthesis: a green approach. Sugar Tech 24(4):1186–1197

Kane AO, Pellergini VOA, Espirito MC, Ngom BD, García JM, Acevedo A, Erazzú LE, Polikarpov I (2022) Evaluating the potential of culms from sugarcane and energy cane varieties grown in Argentina for second-generation ethanol production. Waste Biomass Valori 13:329–343. https://doi.org/10.1007/s12649-021-01528-5

Kusuma HS, Permatasari D, Umar WK, Sharma SK (2023) Sugarcane bagasse as an environmentally friendly composite material to face the sustainable development era. Biomass Convers Biorefinery 13:1–14

Ling R, Wu W, Yuan Y, Wei W, Jin Y (2021) Investigation of choline chloride-formic acid pretreatment and Tween 80 to enhance sugarcane bagasse enzymatic hydrolysis. Bioresour Technol 326:124748

Meena MR, Kumar R, Ramaiyan K, Chhabra ML, Raja AK, Krishnasamy M, Kulshreshtha N, Pandey SK, Ram B (2020) Biomass potential of novel interspecific and intergeneric hybrids of Saccharum grown in sub-tropical climates. Sci Rep 10(1):21560. https://doi.org/10.1038/s41598-020-78329-8

Meena MR, Govindaraj P, Kumar RA, Elayaraja K, Appunu C, Kumar R, Chhabra ML, Kulshreshtha N, Hemaprabha G (2024) Biomass and energy potential of *Erianthus arundinaceus* and *Saccharum spontaneum*-derived novel sugarcane hybrids in rainfed environments. BMC Plant Biol 24(1):198. https://doi.org/10.1186/s12870-024-04885-0

Mohanraj K, Nair VV (2014) Biomass potential of novel interspecific hybrids involving improved clones of Saccharum. Ind Crop Prod 53:128–132

Narancic T, O'Connor KE (2017) Microbial biotechnology addressing the plastic waste disaster. Microb Biotechnol 10(5):1232

Rainey TJ, Covey G (2016) Pulp and paper production from sugarcane bagasse. In: Sugarcane-based biofuels and bioproducts. Wiley, New York, pp 259–280

Sage RF, Zhu X (2011) Exploiting the engine of C4 photosynthesis. J Exp Bot 62(9):2989–3000

Sakthivel JC, Brindha S, Gowthamraj G, Sabna J (2021) Studies on the properties of sugarcane bagasse fibre for sustainable textile applications. In: Proceedings of the first international conference on combinatorial and optimization, ICCAP 2021, December 7–8 2021, Chennai, India

Sayed EAM, Hamad KH, Eid Y, Alaa A, Hassan A, Mohsen A, Safwat A, Atta K, Benbella M, Abdel Tawab S (2022) Production of paper using chemical pulping process of sugarcane bagasse. Int J Ind Sustain Dev 3(1):57–65

SBI (2013) Sugarcane Breeding Institute, germplasm collections. https://www.sugarcane.res.in/index.php/resrch/genetic-resources

Sillanpää M, Ncibi MC, Matilainen A, Vepsäläinen M (2018) Removal of natural organic matter in drinking water treatment by coagulation: a comprehensive review. Chemosphere 190:54–71

Tew TL, Cobil RM (2008) Genetic improvement of sugarcane (*Saccharum* spp.) as an energy crop. In: Vermerris W (ed) Genetic improvement of bioenergy crops. Springer, Berlin, pp 249–272

Thammasittirong SNR, Chatwachirawong P, Khemdee K, Thammasittirong A (2023) Evaluating the potential of newly developed energy cane clones for first- and second-generation ethanol production. Fermentation 9(3):267

Thangavelu SCRK (1982) Comparison of Rapi pol extractor and cutex cane shredder methods for direct determination of fiber in Saccharum clones. In: Proceedings of annual convention of Sugar Technologists Association of India, vol 46, pp 15–21

Wang LP, Jackson A, Lu X (2008) Evaluation of sugarcane x *Saccharum spontaneum* progeny for biomass composition and yield components. Crop Sci 48(3):951–961

Sugarcane Fibre as Dietary Supplement, Health Benefits and Applications in Functional Foods

17

Giriyapura Shivalingamurthy Suresha, Hari Kuppusamy, Murali Palanichamy, Lavanya Rajamani, Kalaiselvi Kalaimani, Gopalareddy Krishnappa, Rangappa Thimmaiah Maruthi, Huskur Kumaraswamy Mahadeva Swamy, Amaresh, Vinayaka, and Elayaraja Kandasamy

Abstract

Sugarcane is one of the most important sources of fibre worldwide. Sugarcane bagasse, which is a major by-product of sugar industry, is an excellent raw material for dietary fibre. Bagasse flour can be introduced along with wheat and other routinely used grain flours, to prepare a variety of food products. An array of bakery products enriched with cane fibre is being tested for human consumption. In this context, the present manuscript made a sincere attempt to cover recent developments in the area of dietary fibre with special reference to sugarcane bagasse. Here, we discussed the basics of dietary fibre and classification of dietary fibres and health benefits. More importantly, we have covered the latest developments in sugarcane as potential source of dietary fibre, its physicochemical characteristics, methods for extraction of dietary fibre from sugarcane bagasse and applications of sugarcane-based dietary fibre in food and pharmaceutical industries. Therefore, this chapter would help the readers to enrich their knowledge on recent technological developments in the area of dietary fibre research. This would encourage to take up new avenues of business and entrepreneurship and ultimately the beneficiaries' rural youth, women, small-scale entrepreneurs, food products manufacturers and food industries.

G. S. Suresha (✉) · H. Kuppusamy
Crop Production Division, ICAR-Sugarcane Breeding Institute, Coimbatore, Tamil Nadu, India

R. T. Maruthi · Amaresh · E. Kandasamy · H. K. M. Swamy
Crop Improvement Division, ICAR-Sugarcane Breeding Institute, Coimbatore, Tamil Nadu, India

M. Palanichamy · L. Rajamani · K. Kalaimani · G. Krishnappa
ICAR-Sugarcane Breeding Institute, Coimbatore, Tamil Nadu, India

G. S. Suresha et al. (eds.), *Value Addition and Product Diversification in Sugarcane*, https://doi.org/10.1007/978-981-97-7228-5_17

Keywords

Dietary fibre · Foods · Pre-treatment · Health benefits · Bagasse

17.1 Introduction

Fibre is the most essential part of human health. Our body needs around 30 g daily intake of fibre in our diet (Elleuch et al. 2011). Dietary fibre is also called as roughage which is a part of our food sourced mainly from plant and not digested by human digestive system mostly in the upper intestinal tract. Dietary fibre is chemically grouped under naturally synthesised carbohydrates having distinctive intramolecular linkages and physicochemical properties. Digestive enzymes present in the body have access to these compounds and regulate the digestion process. Dietary fibre is naturally present in the form of non-starch polysaccharides, oligosaccharides, sugar alcohols, slow digesting starch and glyco-conjugates which are bonded chemically to non-carbohydrate compounds (Schultz et al. 2011). Dietary fibre has been defined in different ways. According to European Union's Codex Alimentarius, dietary fibre is defined as polymers of carbohydrate molecules made up of ten or more monomeric units, which are resistant to hydrolysis by endogenous enzymes of human digestive system. These groups of compounds are categorised into viz. edible carbohydrate polymers present in our foods naturally; carbohydrate polymers, which have been isolated physically, chemically or enzymatically from the ingredients of food which is having health benefits; and chemically synthesised carbohydrate polymers having evidence of health benefits and accepted by the competent authorities (Codex Alimentarius Commission 2008). According to The Commission of the European Communities, fibre is a polymer of carbohydrate which is connected to lignin and other compounds like phenols, waxes, saponins, phytates, cutin and phytosterols. These are considered as fibre only for extraction and analytical purpose not for adding fibre into food (EC 2008). In early century, fibres of plant origin which cannot be digested by human body are called as dietary fibre. This fibre mainly constitutes lignin and other polysaccharides. But later resistant starch and other oligosaccharides are also included under this category. However, the most accepted ones are lignin and most of the polysaccharides like cellulose which are not digested by the digestive enzymes in our body (Anderson et al. 2009; Jha and Mishra 2021).

17.2 Classification of Dietary Fibres and Health Benefits

Based on the source of food, dietary fibre is divided into soluble and insoluble fibres. Fibre derived from the legumes, whole grains and cereals, vegetables, fruits, nuts and seeds contains both soluble and insoluble fibres. These fibres present in different proportions based on the viscosity and fermentability (Institute of Medicine 2001). Major non-starch dietary fibres present in the plant and other living sources are cellulose, resistant starch, resistant dextrin, inulin, lignins, chitins, pectins, beta-glucans and oligosaccharides (Consensus Study Report 2005).

Benefits of dietary fibre for human health depend on the type of fibre one consumes, and the larger part of this effect is on the gastrointestinal system (Gallaher 2007). Bulking fibres can be soluble or insoluble in nature. These include cellulose, hemicellulose and psyllium. Insoluble fibre mainly helps our body to absorb and hold water and regularity. The fibres like betaglucan and psyllium is more viscous, which helps to thicken the faecal mass (Institute of Medicine 2001). Fibres that are metabolised in our body and produce short-chain fatty acids are called as fermentable fibres. These fibres are resistant starch, xanthan gum and inulin which mainly help in the metabolism in large intestine by providing food to the microorganisms, which ultimately benefits gastrointestinal system (Bedford and Gong 2018; Cummings 2001; Ostrowski et al. 2022). These fermentable fibres are also called as soluble fibres or prebiotic fibres due to the property of solubility in water and easily fermented in the intestinal tract and help in the digestion of food by gut bacteria. These soluble fibres mainly include beta-glucans and guar gum. Beta-glucans are mainly obtained from the oats, barley and mushrooms. Another bulking fibre is psyllium, which is a soluble, viscous, nonfermented fibre that helps to hold water when the food passages through the digestive tract, enabling easy soft movement. Bulking fibres helps in digestion process by inducing slow release of gastric from the stomach and keeps the body with a feeling of prolonged fullness. The major sources for soluble non-viscous fibres are from inulin of chicory root, dextrin from the wheat and resistant starches present in legumes and bananas (Keenan et al. 2015). Oats and barley are rich sources of beta-glucans, and adding this into our daily diet helps to reduce bad fat or LDL cholesterol in the blood (Jovanovski et al. 2018; Ho et al. 2017; Ghavami et al. 2023).

The role of insoluble fibres on the human health has been extensively studied. Types of dietary fibres which are not dissolving in water and resistant for digestion in our upper gastrointestinal tract are grouped as insoluble fibres. Plant biomass that is made of lignin, cellulose and hemicellulose and wheat bran are the best examples for insoluble fibre, cellulose and lignin. This fibre can be ground into a fine powder and used as the source of fibre. These fibres have many health benefits. Fibre processed from the known source can be ground into a powder of a different particle size and can be used as ingredient in our food. There are studies which revealed that insoluble fibre gives bulking effect to our body by secreting mucus in the intestine gut, but some part of insoluble fibre can also induce constipation depending on the particle size. Finer the particle size can induce more constipation. Fermentable insoluble fibre, like resistant starches, helps in digestion of food in colon (Lockyer and Nugent 2017).

Edible plant parts are known to contain contain both soluble and insoluble fibres. Based on the fermentability, fibres are divided into highly fermentable or slowly fermentable fibres. These fibres are majorly carbohydrates which are made up of simple sugars bonded in combination to form a complex structure. These complex polysaccharides or oligosaccharides are the source of soluble fermentative sugars. Fibres that are not digested and absorbed in the intestines where these compounds are not digested by gastric juice and digestive enzymes present in the intestine and stomach are called as insoluble fibre. This ability of fermentation occurs in large intestine where colon bacteria acted on the partially digested food to yield short-chain fatty acids like butyric, acetic, propionic and valeric acids which are known to have beneficial effects on the health (Wong et al. 2006). As per the study by Tungland and Meyer (2002), fibres, namely cellulose, methyl cellulose, hemicellulose, lignin and waxes, are grouped as low fermentable fibres whereas compounds like resistant starch, beta glucans, pectins, gums, inulins and other oligosaccharides are grouped as highly fermentable fibres.

Plants accumulate significant amount of fibre both in soluble and insoluble form. Legumes such as a peas, soya beans, lupins and other beans; cereals such as oats, rye, chia and barley; fruits such as figs, avocados, plums, prunes, berries, ripe bananas and the skin of apples, quinces and pears; vegetables such as broccoli, carrots and Jerusalem artichokes; root tubers and root vegetables like sweet potatoes and onions; psyllium seed husks (a mucilage soluble fibre) and flax seeds and nuts with almonds reported to have high amounts of soluble fibres. Whole grain foods; wheat and corn bran; legumes such as beans and peas; nuts and seeds; potato skins; lignans; vegetables such as green beans, cauliflower, zucchini, celery and nopal; some fruits including avocado, unripe bananas and the skin of some fruits, including kiwifruit, grapes and tomatoes are the best-known source for insoluble fibres (Stacewicz-Sapuntzakis et al. 2001; Alvarado et al. 2001).

Health benefits of soluble fibres have been studied by various researchers around the world. It is reported that soluble fibre helps to combat inflammatory bowel diseases (IBDs) mainly diarrhoea, constipation and abdominal related disease symptoms (Friedman 1989). Soluble fibres are being used as prebiotic supplements in which inulin and oligosaccharides are the main constituents, and these products have been tested against IBD (Ewaschuk and Dieleman 2006), Crohn's disease (Guarner 2005), ulcerative colitis (Seidner et al. 2005; Rodrıguez-Cabezas et al. 2003) and *Clostridium difficile* (Ward and Young 1997). These fibres function as anti-inflammatory medicines in the bowel through the formation of short-chain fatty acids (Saemann et al. 2002; Cavaglieri et al. 2003). Unlike soluble fibres, insoluble fibres have a different role on human health. Resistance starch is the most common insoluble fibre source from corn. There are studies reported that resistance starch helps to fight against diabetes (Robertson et al. 2012; Maki et al. 2012; Johnston et al. 2010), promotes regularity (Phillips et al. 1995), diarrhoea (Ramakrishna et al. 2002) and ulcerative colitis (Raghupathy et al. 2006).

17.3 Sugarcane Fibre and Its Characteristics

Sugarcane is one of the most important sources of fibre worldwide. Sugarcane bagasse, an abundant agricultural lignocellulosic by-product, is a fibrous residue of cane stalks left over after extraction of the juice from the sugarcane. Bagasse, obtained as a by-product of sugarcane processing, is composed of fibre, pith, insoluble solids and water. Sugarcane bagasse is the most potential by-product of sugar industries. Chemically, bagasse is made up of lignocellulosic complex where cellulose and hemicellulose are bonded with the lignin polymer (Yu et al. 2015; Batalha et al. 2015). Bagasse has been extensively analysed for its composition which comprises cellulose of 45–55%, hemicelluloses of 20–25%, lignin of 18–24% and ash content of 1–4% (Paroha et al. 2020; Shivalingamurthy and Kuppusamy 2021). Cellulose is a glucose polymer, hemicellulose is a polymer of xylose, arabinose, galactose, glucose and mannose and lignin is a complex phenolic ester made up of coniferyl sinapyl and paracoumaryl alcohols. The morphological structure of bagasse fibre is different and not much stronger than wood fibre. The major advantage of bagasse is easy availability and can be directly obtained and concentrated in the sugar factory as a process of by-product, thus simplifying its handling and transport operations. Thousands of tonnes of waste by-products from the sugar industry are generated annually in the form of bagasse. When compared to other plant sources, bagasse offers cheap and high-valued excellent source of dietary fibre in the form of both soluble and insoluble fibres. Bagasse flour can be introduced along with wheat flour and can be used to prepare a variety of food products. Bread, Rusk and Cookies are considered as a vehicle for combating malnutrition, especially in the developing world. It is to be fortified with fibre of acceptable cooking quality. Considering the projected rise in diabetic population in India to 80 million by 2030 and in an attempt to diversify the use of sugarcane bagasse flour for fibre enrichment, it is highly essential to explore the potential of bagasse as a source of dietary fibre. Besides, bakery products getting wide popularity among the young Indians and in the metros as a convenient food, transformation to new health and wellness foods is essential to add value to the bakery products. In this direction, studies are needed to evaluate the feasibility and benefits of bagasse along with composite flour containing wheat and other important ingredients.

17.4 Bagasse Dietary Fibre Extraction Methods

The extraction of dietary fibre from any agro-industrial wastes requires pre-treatment process to remove lignin from the fibre component. This is mainly due to the non-digestibility of lignin in our body. This process of removal of lignin from the biomass fibre has been well established by employing various physical, chemical and biological methods (Al Arni 2018; Maniet et al. 2017; Zhang et al. 2019). Chemical pre-treatment methods involve the use of strong acids like sulphuric acid and hydrochloric acid or alkali like NaOH, whereas physical treatment method like steam explosion is most commonly used (Mancera et al. 2010; Motaung and Anandjiwala 2015; Justiz-Smith et al. 2008; Al Arni 2018; Wahba et al. 2015; Zhang and Lynd 2004). In general, acid treatment process requires high temperature and pressure with less incubation time. This would affect the equipment, and also this process requires residual effluent treatment. Acids that are left in the fibre as residue are not ideal for food applications. Therefore, alkaline treatment is more advantageous due to low temperature and pressure operating conditions (Suksombat 2004). However, this process requires longer incubation time. So, choosing the right method to extract fibre from sugarcane bagasse is very important to use it as dietary supplement (Mugica-Álvarez et al. 2016; Sahu 2018; Aguilar-Rivera et al. 2012). Alkaline treatments using microwave method are reported as efficient process as it requires less reaction time (Ya-Wei et al. 2015; Wang et al. 2013). Microwave-assisted pretreatment would separate the cellulose, hemicellulose and lignin in the fibre by rupturing the cell wall due to friction of molecules at high speeds and generation of heat due these diploe forces (Amini et al. 2018; Kim et al. 2015; Aguilar-Reynosa et al. 2017; Kappe et al. 2012).

Paroha et al. (2020) reported that bagasse can be a potential source of dietary fibre which can be easily fortified in food products. Among the pre-treatment methods, steam explosion and alkali treatment methods are better than acid pre-treatment method to get good fibre which.is having better solubility, swelling property, water holding capacity and oil absorption capacity. Another study on microwave pretreatment combined with alkaline hydrolysis of bagasse and sugarcane tops reported that processed sugarcane fibre can be easily used as an ingredient in human foods like bread, chapatti and noodles (Gil-Lopez et al. 2019). Form this study, it is revealed that alkaline treatment in combination with microwave processing can extract lignin better way than simple incubation with the alkali. This method was proven more effective in deconstructing the lignocellulose complex structure. Using this method, they achieved the highest lignin removal which was highly suitable for incorporation into wheat flour thereby used in chapatti, bread and noodles (Gil-Lopez et al. 2019). Afrazeh et al. (2021) reported that bagasse has nearly 50% of dietary fibre which is significantly higher than date seed (40%). They also evaluated the physicochemical properties, antioxidant activity and prebiotic activity. From these studies, it is revealed that bagasse dietary fibre had higher minerals content, oil binding capacity, water holding capacity and swelling capacity. Bagasse dietary fibre showed significantly higher antioxidant activity and prebiotic activity which provided much scope for using it as a source of dietary fibre in food and medicinal applications (Afrazeh et al. 2021). Another study on the soluble fibre of sugarcane bagasse revealed that alkaline extraction of dieter fibre and subsequent macromolecular characterisation showed soluble arabinoxylans which can be used as a novel source of dietary fibre. The composition of this type of dietary fibre is mainly sugars like xylose arabinose along with glucose, galactose and uronic acids (Dimopoulou and Kontogiorgos 2020). Another study reports alkaline hydrogen peroxide treatment of bagasse removed lignin of 62.23% with recovery of highly pure insoluble dietary fibre. Further alkaline hydrogen peroxide combined with higher temperature and pressure followed by enzymatic treatment improved the physicochemical characteristics and functional properties like water holding capacity, oil holding capacity, bile salt adsorption capacity and nitrite ion adsorption capacity. These results further validated the potentiality of bagasse as a source of dietary fibre (Luo et al. 2023). In general, bagasse dietary fibre extracted from various treatment methods has a crude fibre content of approximately around 30%, crude fat of 2.0%, carbohydrate content of around 55.0%, highest dietary fibre content of nearly 80.0 g/100 g and insoluble dietary fibre content of 75.0 g/100 g along with significantly higher oil holding capacity and organic molecule absorption capacity. These data clearly indicated that bagasse is a cheap and most potential source of dietary fibre which can be positively used in food industries (Leang and Saw 2011).

17.5 Health Benefits of Bagasse Dietary Fibre and Its Applications

Due to the advancement in various pre-treatment methods, it is well possible to convert sugarcane fibre into edible fibre. Despite its application in paper, biopolymer and composites industries, sugarcane fibre has the potential to use in food and medicinal applications (Andrade and Colodette 2014; El-Fattah et al. 2015) (Fig. 17.1). Sangnark and Noomhorm (2004) reported that using 10% bagasse fibre in bread products is not sufficient as per the customers' feedback which indicates the demand for dietary fibre sources. Phenolic compounds present in the sugarcane bagasse exhibited significant anti-microbial activity against food-borne pathogens of *Staphylococcus aureus*, *Listeria monocytogenes* and *Escherichia coli* (Zhao et al. 2015). More recently, the study on physicochemical and sensory properties of chicken sausages with the incorporation of sugarcane fibre revealed significant increase in the total phenolic content and free radical scavenging activity. These results further confirmed the role of sugarcane fibre in improving the quality of food products and also add more value due to its health benefits (Fang et al. 2019). Sugarcane fibre has been studied for its physiological role through incorporation in different foods. This was supported by the report which revealed the successful development of bread incorporated with sugarcane bagasse dietary fibre (Sangnark

Fig. 17.1 Sugarcane dietary fibre and its health benefits

and Noomhorm 2003, 2004). World health bodies like Food and Agriculture Organization (FAO) and the World Health Organization (WHO) have stressed upon for improving health through good diet and physical activity. More importantly, diet containing less carbohydrates or low glycaemic index would help to control diabetes (Frost et al. 1998). This depends on the degree of starch digestion which results in spiking of blood sugar levels along with different health conditions like decrease in insulin sensitivity and increased cholesterol accumulation and also inducing cancer (Frost et al. 1998; Bjorck et al. 2000). In this direction, incorporation of fibre in the foods containing high starch breads and pasta would provide balanced nutrition and maintains low-glycaemic diet. Lau et al. (2022) evaluated the noodles that were incorporated with sugarcane bagasse fibre for its suitability to use it as high dietary fibre food. They observed that 5% bagasse dietary fibre incorporated noodles were the most optimum ratio and had good colour, texture and fibre content with low glycaemic index. Therefore, sugarcane fibre offers tremendous scope to add in the variety of food products due its high dietary fibre content and also low in glycaemic index (GI) (Gil-Lopez et al. 2019). Using bagasse in foods could give natural source for low GI diet. In recent time, there is a significant increase in the diabetic patients due to more consumption of junk foods, improper dietary habits and sedentary life. As per the International Diabetes Federation, around 463 million adults are suffering from diabetes worldwide, and it is likely to surge up to 700 million by 2045 (International Diabetes Federation 2019). Therefore, fibre-rich foods play an important role in controlling many health issues in near future. It was reported the addition of sugarcane dietary fibre in meat products increased the faucal bulk and helps digestion through fermentation in colon (Zhuang et al. 2017; Sangnark and Noomhorm 2003, 2004). Sugarcane dietary fibre was proved as safer herbal source of medicine and effective for curing patients with different health problems when it is administered orally as an adjuvant therapy. This was showed when bagasse fibre was used as an adjuvant therapy to treat chronic obstructive pulmonary disease (COPD) (Miao et al. 2016). Wang et al. (2007) studied the effect of sugarcane fibre for using in the human food. They revealed that sugarcane fibre contains some amount of soluble fibre which can benefit for intestinal regularity. Based on the animal experiments, it is revealed that sugarcane fibre can help in controlling type 2 diabetes. Compounds like lignoceric and cerotic acids are present in sugarcane fibre (Nnabugwu and Uchenna 2019). Zidan et al. (2021) reported that the xylooligosaccharides isolated from sugarcane bagasse exhibited prebiotic activity which was further validated by incorporating it in noodles. Licona-Aguilar et al. (2023) reported that bagasse dietary fibre was incorporated in wheat flour and prepared cookies and biscuits and studied their feasibility to use in bakery products. The results from this study revealed that cookies and biscuits prepared with the incorporation of bagasse dietary fibre showed good physicochemical characteristics like texture colour, proximal and sensory features along with greater acceptability by the consumers. Another study by Vijerathna et al. (2019) developed the cookies fortified with bagasse dietary fibre without the addition of sugars. Results from this study revealed that 5% bagasse dietary fibre incorporation showed maximum acceptability for manufacturing cookies and other bakery products. Ojha and Verma

(2023) reported that sugarcane bagasse is a rich source for insoluble dietary fibre. In this study, biscuits added with 15% bagasse dietary fibre showed the highest sensory acceptance.

17.6 Conclusion

Sugarcane bagasse is a rich source of fibre. Due to its chemical composition and high energy value, bagasse has been widely used as energy raw material in many industries. In the current scenario, there is a more demand for natural fibre-rich foods; therefore, bagasse offers tremendous potential to explore as a source of dietary fibre. In this context, this chapter has been discussed about sugarcane dietary fibre and its importance and various extraction methods and its applications in food industries and pharmaceutical sector. This manuscript would encourage the readers to gain more insights into the latest research developments in the area of dietary fibre, especially bagasse-based value-added food products and its health benefits.

References

Afrazeh M, Tadayoni M, Abbasi H, Sheikhi A (2021) Extraction of dietary fibers from bagasse and date seed, and evaluation of their technological properties and antioxidant and prebiotic activity. J Food Meas Charact 15(2):1949–1959. https://doi.org/10.1007/s11694-020-00774-w

Aguilar-Reynosa A, Romaní A, Rodríguez-Jasso RM, Aguilar CN, Garrote G, Ruiz HA (2017) Microwave heating processing as alternative of pretreatment in second-generation biorefinery: an overview. Energy Convers Manag 136:50–65

Aguilar-Rivera N, Rodríguez L DA, Enríquez RV, Castillo MA, Herrera SA (2012) The Mexican sugarcane industry: overview, constraints, current status and long-term trends. Sugar Tech 14:207–222

Al Arni S (2018) Extraction and isolation methods for lignin separation from sugarcane bagasse: a review. Ind Crop Prod 115:330–339

Alvarado A, Pacheco-Delahaye E, Hevia P (2001) Value of a tomato byproduct as a source of dietary fiber in rats. Plant Foods Hum Nutr 56:335–348

Amini A, Ohno K, Maeda T, Kunitomo K (2018) Effect of particle size and apparent density on the initial stages of temperature increase during the microwave heating of Fe3O4. Powder Technol 338:101–109

Anderson JW, Baird P, Davis RH Jr, Ferreri S, Knudtson M, Koraym A, Waters V, Williams CL (2009) Health benefits of dietary fiber. Nutr Rev 67(4):188–205

Andrade MF, Colodette JL (2014) Dissolving pulp production from sugar cane bagasse. Ind Crop Prod 52:58–64

Batalha LA, Han Q, Jameel H, Chang HM, Colodette JL, Gomes FJ (2015) Production of fermentable sugars from sugarcane bagasse by enzymatic hydrolysis after autohydrolysis and mechanical refining. Bioresour Technol 180:97–105

Bedford A, Gong J (2018) Implications of butyrate and its derivatives for gut health and animal production. Anim Nutr 4(2):151–159

Bjorck I, Liljeberg H, Ostman E (2000) Low glycaemic-index foods. Br J Nutr S1:S149–S155

Cavaglieri CR, Nishiyama A, Fernandes LC, Curi R, Miles EA, Calder PC (2003) Differential effects of short-chain fatty acids on proliferation and production of pro- and anti-inflammatory cytokines by cultured lymphocytes. Life Sci 73(13):1683–1690

Codex Alimentarius Commission (2008) Report of the 30th session of the Codex Committee on nutrition and foods for special dietary uses: para 27–54, p 46

Cummings JH (2001) The effect of dietary fiber on fecal weight and composition, 3rd edn. CRC Press, Boca Raton, p 184. ISBN 978-0-8493-2387-4

Dimopoulou M, Kontogiorgos V (2020) Soluble dietary fibres from sugarcane bagasse. Int J Food Sci Technol 55(5):1943–1949. https://doi.org/10.1111/ijfs.14445

EC–European Commission (2008) Commission Directive 2008/100/EC of 28 October 2008 amending Council Directive 90/496/EEC on nutrition labelling for foodstuffs as regards recommended daily allowances, energy conversion factors and definitions. Off J Eur Union L 285(9)

El-Fattah AA, El Demerdash AG, Alim Sadik WA, Bedir A (2015) The effect of sugarcane bagasse fiber on the properties of recycled high density polyethylene. J Compos Mater 49(26):3251–3262

Elleuch M, Bedigian D, Roiseux O, Besbes S, Blecker C, Attia H (2011) Dietary fibre and fibre-rich by-products of food processing: characterisation, technological functionality and commercial applications: a review. Food Chem 124(2):411–421

Ewaschuk JB, Dieleman LA (2006) Probiotics and prebiotics in chronic inflammatory bowel diseases. World J Gastroenterol: WJG 12(37):5941

Fang Z, Lin P, Ha M, Warner RD (2019) Effects of incorporation of sugarcane fibre on the physico-chemical and sensory properties of chicken sausage. Int J Food Sci Technol 54(4):1036–1044

Friedman G (1989) Nutritional therapy of irritable bowel syndrome. Gastroenterol Clin N Am 18(3):513–524

Frost G, Leeds A, Trew G, Margara R, Dornhorst A (1998) Insulin sensitivity in women at risk of coronary heart disease and the effect of a low glycemic diet. Metabolism 47(10):1245–1251

Gallaher DD (2007) "7 Influence of Dietary" In: Sports Nutrition: Energy Metabolism and Exercise, p. 193.

Ghavami A, Ziaei R, Talebi S, Barghchi H, Nattagh-Eshtivani E, Moradi S, Rahbarinejad P, Mohammadi H, Ghasemi-Tehrani H, Marx W, Askari G (2023) Soluble fiber supplementation and serum lipid profile: a systematic review and dose-response meta-analysis of randomized controlled trials. Adv Nutr 14(3):465–474

Gil-Lopez DL, Lois-Correa JA, Sanchez-Pardo ME, Dominguez-Crespo MA, Torres-Huerta AM, Rodríguez-Salazar AE, Orta-Guzmán VN (2019) Production of dietary fibers from sugarcane bagasse and sugarcane tops using microwave-assisted alkaline treatments. Ind Crop Prod 135:159–169

Guarner F (2005) Inulin and oligofructose: impact on intestinal diseases and disorders. Br J Nutr 93(S1):S61–S65

Ho HVT, Jovanovski E, Zurbau A, Blanco MS, Sievenpiper JL, Au-Yeung F, Jenkins AL, Duvnjak L, Leiter L, Vuksan V (2017) A systematic review and meta-analysis of randomized controlled trials of the effect of konjac glucomannan, a viscous soluble fiber, on LDL cholesterol and the new lipid targets non-HDL cholesterol and apolipoprotein B. Am J Clin Nutr 105(5):1239–1247. https://doi.org/10.3945/ajcn.116.142158. ISSN 1938-3207

Institute of Medicine (2001) Dietary reference intakes, proposed definition of dietary fiber. Institute of Medicine Press, Washington, DC, p 25. ISBN 978-0-309-07564-0

International Diabetes Federation (2019) IDF diabetes atlas, 9th edn. https://diabetesatlas.org/. Accessed 20 Aug 2020

Jha R, Mishra P (2021) Dietary fiber in poultry nutrition and their effects on nutrient utilization, performance, gut health, and on the environment: a review. J Anim Sci Biotechnol 12:1–6

Johnston KL, Thomas EL, Bell JD, Frost GS, Robertson MD (2010) Resistant starch improves insulin sensitivity in metabolic syndrome. Diabet Med 27(4):391–397. https://doi.org/10.1111/j.1464-5491.2010.02923.x. S2CID 27570039. PMID 20536509

Jovanovski E, Yashpal S, Komishon A, Zurbau A, Mejia SB, Ho HV, Li D, Sievenpiper J, Duvnjak L, Vuksan V (2018) Effect of psyllium (Plantago ovata) fiber on LDL cholesterol and alternative lipid targets, non-HDL cholesterol and apolipoprotein B: a systematic review and meta-analysis of randomized controlled trials. Am J Clin Nutr 108(5):922–932

Justiz-Smith NG, Virgo GJ, Buchanan VE (2008) Potential of Jamaican banana, coconut coir and bagasse fibres as composite materials. Mater Charact 59(9):1273–1278

Kappe CO, Stadler A, Dallinger D (2012) Microwaves in organic and medicinal chemistry. John Wiley & Sons, New York

Keenan MJ, Zhou J, Hegsted M, Pelkman C, Durham HA, Coulon DB, Martin RJ (2015) Role of resistant starch in improving gut health, adiposity, and insulin resistance. Adv Nutr 6(2):198–205

Kim D, Ward DR, Simmons CB, Gamble JK, Blume-Kohout R, Nielsen E, Savage DE, Lagally MG, Friesen M, Coppersmith SN, Eriksson MA (2015) Microwave-driven coherent operation of a semiconductor quantum dot charge qubit. Nat Nanotechnol 10(3):243–247

Lau KQ, Sabran MR, Shafie SR, Chan SW (2022) Incorporation of sugarcane bagasse in the development of high dietary fibre noodles. Int J Food Sci Technol 57(7):4586–4593

Leang YH, Saw HY (2011) Proximate and functional properties of sugarcane bagasse. Agro Food Ind Hi Tech 22(2):5–8

Licona-Aguilar ÁI, Lois-Correa JA, Torres-Huerta AM, Domínguez-Crespo MA, Urdapilleta-Inchaurregui V, Rodríguez-Salazar AE, Brachetti-Sibaja SB (2023) Production of dietary cookies based on wheat-sugarcane bagasse: determination of textural, proximal, sensory, physical and microbial parameters. LWT 184:115061. https://doi.org/10.1016/j.lwt.2023.115061

Lockyer S, Nugent AP (2017) Health effects of resistant starch. Nutr Bull 42(1):10–41

Luo M, Wang C, Wang C, Xie C, Hang F, Li K, Shi C (2023) Effect of alkaline hydrogen peroxide assisted with two modification methods on the physicochemical, structural and functional properties of bagasse insoluble dietary fiber. Front Nutr 9:1110706

Maki KC, Pelkman CL, Finocchiaro ET, Kelley KM, Lawless AL, Schild AL, Rains TM (2012) Resistant starch from high-amylose maize increases insulin sensitivity in overweight and obese men. J Nutr 142(4):717–723

Mancera A, Fierro V, Pizzi A, Dumarçay S, Gérardin P, Velásquez J, Quintana G, Celzard A (2010) Physicochemical characterisation of sugar cane bagasse lignin oxidized by hydrogen peroxide. Polym Degrad Stab 95(4):470–476

Maniet G, Schmetz Q, Jacquet N, Temmerman M, Gofflot S, Richel A (2017) Effect of steam explosion treatment on chemical composition and characteristic of organosolv fescue lignin. Ind Crop Prod 99:79–85

Miao L, Fengjie Z, Lei N, Yan S, Ruohan W, Tianyu Z, Jinchao Z, Xianggen Z, Yuhang L (2016) Sugarcane bagasse dietary fiber as an adjuvant therapy for stable chronic obstructive pulmonary disease: a four-center, randomized, double-blind, placebo-controlled study. J Tradit Chin Med 36(4):418–426

Motaung TE, Anandjiwala RD (2015) Effect of alkali and acid treatment on thermal degradation kinetics of sugar cane bagasse. Ind Crop Prod 74:472–477

Mugica-Álvarez V, Ramos-Guízar S, Santiago-De La Rosa N, Torres-Rodríguez M, Noreña-Franco L (2016) Black carbon and particulate organic toxics emitted by sugarcane burning in Veracruz, México. Int J Environ Sci Dev 7(4):290–294

Nnabugwu AE, Uchenna AP (2019) Nutrient and antioxidant properties of oils from bagasses, agricultural residues, medicinal plants, and fodders. J Am Coll Nutr 38(2):132–140

Ojha K, Verma A (2023) Physico-chemical parameters and sensory characteristics of biscuits incorporated with sugarcane bagasse powder. Pharm Innov J 12(5):1491–1493

Ostrowski MP, La Rosa SL, Kunath BJ, Robertson A, Pereira G, Hagen LH, Varghese NJ, Qiu L, Yao T, Flint G, Li J (2022) Mechanistic insights into consumption of the food additive xanthan gum by the human gut microbiota. Nat Microbiol 7(4):556–569

Paroha S, Srivastava VP, Chaturvedi N (2020) Sugarcane bagasse as dietary fibre. Indian J Pure Appl Biosci 8(6):590–597. https://doi.org/10.18782/2582-2845.8442

Phillips J, Muir JG, Birkett A, Lu ZX, Jones GP, O'Dea K, Young GP (1995) Effect of resistant starch on fecal bulk and fermentation-dependent events in humans. Am J Clin Nutr 62(1):121–130

Raghupathy P, Ramakrishna BS, Oommen SP, Ahmed MS, Priyaa G, Dziura J, Young GP, Binder HJ (2006) Amylase-resistant starch as adjunct to oral rehydration therapy in children with diarrhea. J Pediatr Gastroenterol Nutr 42(4):362–368

Ramakrishna BS, Venkataraman S, Srinivasan P, Dash P, Young GP, Binder HJ (2002) Amylase-resistant starch plus oral rehydration solution for cholera. N Engl J Med 342(5):308–313

Robertson MD, Wright JW, Loizon E, Debard C, Vidal H, Shojaee-Moradie F, Russell-Jones D, Umpleby AM (2012) Insulin-sensitizing effects on muscle and adipose tissue after dietary fiber intake in men and women with metabolic syndrome. J Clin Endocrinol Metabol 97(9):3326–3332

Rodrıguez-Cabezas ME, Galvez J, Camuesco D, Lorente MD, Concha A, Martinez-Augustin O, Redondo L, Zarzuelo A (2003) Intestinal anti-inflammatory activity of dietary fiber (Plantago ovata seeds) in HLA-B27 transgenic rats. Clin Nutr 22(5):463–471

Saemann MD, Böhmig GA, Zlabinger GJ (2002) Short-chain fatty acids: bacterial mediators of a balanced host-microbial relationship in the human gut. Wien Klin Wochenschr 114(8–9):289–300

Sahu O (2018) Assessment of sugarcane industry: suitability for production, consumption, and utilization. Ann Agrar Sci 16(4):389–395

Sangnark A, Noomhorm A (2003) Effect of particle sizes on functional properties of dietary fibre prepared from sugarcane bagasse. Food Chem 80(2):221–229

Sangnark A, Noomhorm A (2004) Effect of dietary fiber from sugarcane bagasse and sucrose ester on dough and bread properties. LWT-Food Sci Technol 37(7):697–704. https://doi.org/10.1016/j.lwt.2004.02.015

Schultz C, Meier M, Schmid HP (2011) Nutrition, dietary supplements and adenocarcinoma of the prostate. Maturitas 70(4):339–342

Seidner DL, Lashner BA, Brzezinski A, Banks PL, Goldblum J, Fiocchi C, Katz J, Lichtenstein GR, Anton PA, Kam LY, Garleb KA (2005) An oral supplement enriched with fish oil, soluble fiber, and antioxidants for corticosteroid sparing in ulcerative colitis: a randomized, controlled trial. Clin Gastroenterol Hepatol 3(4):358–369

Shivalingamurthy SG, Kuppusamy H (2021) Comparative evaluation of xylan extraction methods and product characterization from sugarcane bagasse. J Sugarcane Res 11(2):158–171. https://doi.org/10.37580/JSR.2021.2.11.158-171

Stacewicz-Sapuntzakis M, Bowen PE, Hussain EA, Damayanti-Wood BI, Farnsworth NR (2001) Chemical composition and potential health effects of prunes: a functional food? Crit Rev Food Sci Nutr 41(4):251–286

Suksombat W (2004) Comparison of different alkali treatment of bagasse and rice straw. Asian Australas J Anim Sci 17(10):1430–1433

Tungland BC, Meyer D (2002) Nondigestible oligo- and polysaccharides (dietary fiber): their physiology and role in human health and food. Compr Rev Food Sci Food Saf 1(3):90–109

Vijerathna MP, Wijesekara I, Perera R, Maralanda SM, Jayasinghe M, Wickramasinghe I (2019) Physico-chemical characterization of cookies supplemented with sugarcane bagasse fibres. Vidyodaya J Sci 22(1):29

Wahba SM, Darwish AS, Shehata IH, Abd Elhalem SS (2015) Sugarcane bagasse lignin, and silica gel and magneto-silica as drug vehicles for development of innocuous methotrexate drug against rheumatoid arthritis disease in albino rats. Mater Sci Eng C 48:599–610

Wang ZQ, Zuberi AR, Zhang XH, Macgowan J, Qin J, Ye X, Son L, Wu Q, Lian K, Cefalu WT (2007) Effects of dietary fibers on weight gain, carbohydrate metabolism, and gastric ghrelin gene expression in mice fed a high-fat diet. Metabolism 56(12):1635–1642

Wang X, Gao N, Zhou Q, Dong H, Yu H, Feng Y (2013) Acidic and alkaline pretreatments of activated carbon and their effects on the performance of air-cathodes in microbial fuel cells. Bioresour Technol 144:632–636

Ward PB, Young GP (1997) Dynamics of Clostridium difficile infection: control using diet. In: Mechanisms in the pathogenesis of enteric diseases. Springer, Berlin, pp 63–75

Wong JM, De Souza R, Kendall CW, Emam A, Jenkins DJ (2006) Colonic health: fermentation and short chain fatty acids. J Clin Gastroenterol 40(3):235–243

Ya-Wei W, Cheng-Min G, Xiao-Tang N, Mei-Xue C, Yuan-Song W (2015) Multivariate analysis of sludge disintegration by microwave–hydrogen peroxide pretreatment process. J Hazard Mater 283:856–864

Yu H, You Y, Lei F, Liu Z, Zhang W, Jiang J (2015) Comparative study of alkaline hydrogen peroxide and organosolv pre-treatments of sugarcane bagasse to improve the overall sugar yield. Bioresour Technol 187:161–166

Zhang YH, Lynd LR (2004) Toward an aggregated understanding of enzymatic hydrolysis of cellulose: noncomplexed cellulase systems. Biotechnol Bioeng 88(7):797–824

Zhang N, Tao P, Lu Y, Nie S (2019) Effect of lignin on the thermal stability of cellulose nanofibrils produced from bagasse pulp. Cellulose 26:7823–7835

Zhao Y, Chen M, Zhao Z, Yu S (2015) The antibiotic activity and mechanisms of sugarcane (Saccharum officinarum L.) bagasse extract against food-borne pathogens. Food Chem 185:112–118

Zhuang X, Zhang W, Liu R, Liu Y, Xing L, Han M, Kang ZL, Xu XL, Zhou GH (2017) Improved gel functionality of myofibrillar proteins incorporation with sugarcane dietary fiber. Food Res Int 100:586–594

Zidan D, Sabran MR, Ramli NS, Shafie SR, Fikry M (2021) Prebiotic properties of xylooligosaccharide extracted from sugarcane wastes (pith and rind): a comparative study. Int J Food Sci Technol 56(5):2175–2181

[illegible]

18 Technological Advancements in Enzyme Production for 2G Ethanol

Rajeev K. Sukumaran, Meena Sankar, V. P. Adarsh, Reshma M. Mathew, AthiraRaj Sreeja-Raju, Athulya, P. S. Neetha, Biji Raphy, and Valan Rebinro Gnanaraj

Abstract

Bioethanol is projected as one of the major important renewable fuels with a pivotal role in reducing greenhouse gas emissions in the short term, as countries worldwide are adopting ethanol blending in gasoline due to the low-carbon footprint of 2G ethanol (2GE). While 1G ethanol technologies are mature and running, 2G has the unique advantage of better carbon capture and release figures and does not have the food versus fuel conflict. With 1GE poised to be phased out, there is huge pressure to have mature and cost-efficient 2GE technologies. The major bottleneck and cost contributor to 2GE production is the enzyme catalysts used in lignocellulose hydrolysis. While there are a few efficient biomass hydrolyzing enzyme cocktails in the market, the costs involved in using them for 2GE production do not seem to justify the returns, and onsite production of the enzyme is proposed for cost efficiency. Technological advancements in the production of enzymes for lignocellulose digestion need to be understood to identify the gaps and possibilities of further improvements in their hydrolytic efficiencies and address the lacunae of current production technologies and the enzymes themselves. The enzymes hydrolyzing biomass and their modes of action are introduced first, and the current status of cellulolytic enzyme production is elaborated focusing on the fugal sources of enzyme, production technologies, parameters affecting production, downstream operations, and creation of enzyme blends, specifically highlighting the challenges and the possibilities of improvement.

R. K. Sukumaran (✉) · M. Sankar · V. P. Adarsh · R. M. Mathew · A. Sreeja-Raju · Athulya · P. S. Neetha · B. Raphy · V. R. Gnanaraj
Microbial Processes and Technology Division, CSIR-National Institute for Interdisciplinary Science and Technology, Thiruvananthapuram, Kerala, India

Academy of Scientific and Innovative Research (AcSIR), Ghaziabad, India
e-mail: rajeevs@niist.res.in

G. S. Suresha et al. (eds.), *Value Addition and Product Diversification in Sugarcane*, https://doi.org/10.1007/978-981-97-7228-5_18

Keywords

Bioethanol · Cellulase · Glucosidase · LPMO · Bioeconomy · Fermentation

18.1 Introduction

The steady rise of global temperatures over the past several decades and the recent extreme weather change events have been a cause of concern, prompting global nations to take urgent actions to limit the increases in greenhouse gas (GHG) emissions. Fossil fuel burning is one of the major contributors to GHG emissions. Recently, at the 28th Conference of the Parties (COP28) to the United Nations Framework Convention on Climate Change (UNFCCC) in 2023, more than 130 countries signed a pledge to increase global renewable energy capacity by 3 times and energy efficiency by 2 times by 2030 (IEA 2023). This included a call on governments to expedite the transition from fossil fuels to renewables in their next round of climate commitments. The transportation sector is a major contributor to GHG emissions through the direct combustion of fossil fuels, and fuels like ethanol have a low carbon intensity (CI) in comparison to gasoline (Scully et al. 2021). Also, it has been indicated that diverting the biomass feedstock to the production of compounds with a low emission potential in their entire life cycle can further reduce GHG emissions (Winter et al. 2022). While some reports have projected second-generation (2G) ethanol to have a slightly increased CO_2 emission compared to the first generation, this is mentioned to be due to the lack of a high protein by-product, which can negate the carbon emission and, of course, the transport cost of feedstock (Buchspies and Kaltschmitt 2018). However, in countries like India, where diverting the 1G feedstock having food/feed value to ethanol is not a feasible option and those with large resources of agro-residues or other lignocellulosic feedstock, ethanol from lignocellulosic biomass is a more practical solution to renewable biofuels. Also, several reports have proposed a better (lower) CO_2 emission status for 2G ethanol (Sharma et al. 2020). Apparently, 2G ethanol is a better option to avoid the food versus fuel conflict and is preferred in the long term over 1G ethanol. The European Commission had conclusively decided to limit the use of 1G biofuels and support a wider use of 2G ethanol. This could eventually help in the withdrawal of 1G ethanol from the market. The biofuel demand is set to expand worldwide by 38 billion litres by 2028, nearly a 30% increase from the last 5 years. In fact, the bioethanol demand alone is set to grow from 108 billion litres per annum in 2023 to 145.1 billion litres in 2028 (IEA 2023), and most of this demand must be met through second generation feedstock, i.e., the lignocellulosic biomass.

While the 2G ethanol (2GE) or cellulosic ethanol has been proven to be difficult to produce at scale in comparison to 1GE, the climate change implications seem to be much better as the carbon sequestration potential is much better with 2G feedstock and the fuel itself leaves less carbon on combustion (Bettenhausen 2023). There have been several attempts in the past decade to build commercial cellulosic ethanol plants, from the likes of Poet-DSM, Dupont, Abengoa, etc., most of which did not take off due to claimed administrative and policy reasons. At the same time, several other companies like Clariant, Praj, and Raizen have gone ahead with the commissioning of new 2GE plants. Apart from the major problem of economic feedstock transportation faced by all the companies, there are technical challenges leading to increased production costs. The operation cost for lignocellulosic ethanol forms 27–30% of the estimated 0.6–0.9 €/L of ethanol (Susmozas et al. 2020). It is therefore recognized that reduction in cost of 2GE production is critical for the development and implementation of 2G ethanol (Brown et al. 2020). Understanding the operation cost of 2G ethanol should require an understanding of the production pipeline of lignocellulosic biomass (LCB) to ethanol. LCB typically contains polymeric sugars, cellulose, and hemicellulose as the major components and a significant portion of the remaining as lignin, a polymer of aromatic compounds. Both the polymers—cellulose and hemicellulose—can be fragmented into their monomeric sugar units, which can then be used in ethanol production by fermentation. Cellulose and hemicellulose hydrolysis can be performed using chemical agents (e.g., acids) or enzymes. The enzymes are considered a more efficient and environmentally benign option and require only ambient conditions for the reactions. In comparison, chemical-mediated hydrolysis requires higher temperatures, and there are issues including inhibitor and acidic waste stream generation (Visser et al. 2015).

The cost of biomass hydrolyzing enzymes has been one of the major, if not the major, contributors to the overall cost of cellulosic ethanol (Klein-Marcuschamer et al. 2012; Liu et al. 2016). It was estimated that the enzyme cost could be 22–28% of the operation cost of cellulosic ethanol, with onsite- and offsite-produced enzymes, respectively (Johnson 2016), but it can go as high as 48% of the process cost (Liu et al. 2016). While high biomass conversion efficiencies have been achieved by some of the enzymes currently in the market for specific feedstock, it seems that a generic solution is difficult because the enzymatic conversion depends on feedstock properties, type of preconditioning and pretreatment employed, etc., in addition, the enzyme properties itself. Often, an integrated approach is essential, and onsite production of the enzyme is recommended by all the studies that have investigated the operation cost of 2G Ethanol. The diversity of biomass hydrolyzing enzymes and the microbes producing them are very high, though we are exploiting only a few of them for enzyme production at a commercial scale. The choice of microorganisms, production processes, including both upstream and downstream processes, and formulations, including the blending of various heterologous activities either in pure or in crude form, all determine the suitability and efficiency of the enzyme cocktail preparation in biomass hydrolysis. The current chapter describes the concept of lignocellulolytic enzyme cocktails and the strategies and advances in their production for 2G ethanol bio-refineries.

18.2 Biomass Hydrolyzing Enzymes

Understanding the basics of biomass hydrolyzing enzymes is essential to appreciate the technologies for their production and formulations. This requires an understanding of the structure of the plant cell wall. Plant cells are enclosed by primary and secondary cell walls, the latter deposited internally to the former after the cell expansion is stopped (De Souza et al. 2023). While the primary cell wall composition is of an array of different compounds, including cellulose, hemicellulose, structural proteins, pectin, and phenolic compounds (in grasses), the secondary contains fibres of cellulose and hemicellulose in a matrix of lignin, resulting in a complex and rigid structure which is recalcitrant to attack from microbes and enzymatic hydrolysis (Meents et al. 2018; Zhao et al. 2012). Lignocellulose is almost 75% carbohydrates, and the recalcitrance is related to the spatial interlinks between lignin, hemicellulose, pectin, etc., creating physical barriers protecting cellulose from degradation (Zhao et al. 2012). Cellulose, a linear polysaccharide of β (1 → 4), linked glucose units, with a degree of polymerization (DP) of up to 10,000. The chains of cellulose are usually organized as microfibrils, cross-linked by hydrogen bonds. This results in their amorphous and crystalline structure (Pu et al. 2013). Hemicellulose, in comparison, is a diverse group of polysaccharides with short chain, substituted and branched chains of sugars like xylose, arabinose, mannose, galactose, and rhamnose. The degree of polymerization is much lower at 40–200 (Gírio et al. 2010). Lignin is composed of cross-linked aromatic units, mainly *p*-coumaryl alcohol, coniferyl alcohol, and sinapyl alcohols. Typically, plant cell wall structure consists of microfibrils of cellulose embedded in a lignin matrix interspersed with fibres of hemicellulose and thus forms the organized rigid structure of the cell wall, which is quite difficult to break (Fig. 18.1).

Woody biomass contains 40–50% of cellulose, 8–25% xylan, and 18–35% lignin, whereas the grass biomass typically contains 25–50% cellulose, 20–50% xylan, and 10–30% of lignin (Zhao et al. 2012).

Nature's inbuilt machinery for breaking open the plant cell wall structure comprises enzymes that can hydrolyze all these polymers, and these bio-catalysts have a prominent role in the global carbon cycle (Sukumaran et al. 2017). Microorganisms, mostly fungi and bacteria, are involved in organic carbon recycling, effected through the production of enzymes that break down cellulose, hemicellulose, and lignin. These organisms may produce either one or many of these enzymes (cellulases, hemicellulases, and lignin-degrading enzymes), collaboratively participating in the mineralization of lignocellulose in nature. The same activities (cell-free) are harnessed for deriving sugars from biomass under controlled conditions to be fermented to ethanol or to other value-added products. While a plethora of microorganisms have the capability to degrade biomass, efficient synthesis and secretion of enzymes is limited to a few genera of fungi, including *Trichoderma*, *Penicillium*, and *Aspergillus*, species of which are commonly employed for producing lignocellulose degrading enzymes at a commercial scale.

Quite often, in nature, the microorganisms producing lignocellulolytic enzymes exercise strict regulatory controls on the production and secretion of these enzymes to avoid competition and to exercise energy economy. Cellulases need to be expressed by the organisms only when they encounter complex polymers like cellulose as their only source of carbon and energy. A repression mechanism known as Carbon Catabolite Repression (CCR) represses the expression of cellulases in the presence of simple sugars like glucose that can be metabolized easily by the

Fig. 18.1 Graphical representation of lignocellulosic biomass structure. (*Reproduced from* Langan et al. (2014), *with permission from the Royal Society of Chemistry*). Clusters of long fibres (green) represent cellulose, individual fibres (light green) hemicellulose, and the brown structures lignin

organism. A complex network of cellular signaling involving transporters, receptors, and transcription factors is tightly regulated to facilitate the expression of genes and protein secretion to degrade a cellulosic substrate (Christopher et al. 2023).

While this economy practised by the microorganism is crucial for maintaining the carbon cycle and beneficial to the host organism, this is not the ideal situation for commercial production of the enzyme for bio-refinery applications, as we do not want the enzyme synthesis to stop after the hydrolysis produces a threshold concentration of glucose. Understanding the enzymes involved and their mode of action are important in addressing such issues.

18.3 Lignocellulolytic Enzymes: Classes and Modes of Action

The types and numbers of enzymes and proteins that aid the deconstruction of lignocellulosic biomass are as wide and diverse as the biomass structure and complex. These may be generally grouped into cellulases, hemicellulases, and lignin-degrading enzymes. However, proteins like swollenins, expansins, etc. are also involved in biomass hydrolysis but do not fall under these general categories. Most of the enzymes discussed in relation to the commercial biomass hydrolyzing cocktails are extracellular/secreted fungal enzymes, though cellulose degrading cell wall bound enzyme complexes called cellulosomes are common in cellulolytic bacteria

(Fontes and Gilbert 2010). The subsequent discussions are on the free/non-complexed enzymes from filamentous fungi, as they are considered superior for biomass hydrolysis in the bio-refinery context, as the operations can be cell-free allowing higher glucose concentrations to be achieved and to operate under non-sterile conditions.

18.3.1 Cellulases

Cellulases are enzymes that hydrolyze the β (1 → 4) glycosidic linkages of cellulose (insoluble in nature) at the solid/liquid interface, generating oligosaccharides that are soluble (Igarashi et al. 2012). These enzymes belong to a broad family of glycoside hydrolases (GHs) acting on the β(1 → 4) glycosidic bonds through an acid hydrolysis mechanism (Thoresen et al. 2021). Three major types of enzymes are recognized under cellulases, which bring about the complete digestion of native cellulose—i.e., endoglucanases (EG, EC 3.2.1.4), which cleaves from within the chain, acting randomly on the cellulose polymer, creating nicks and exposes reducing and non-reducing ends (Sheehan and Himmel 1999); exoglucanase/cellobiohydrolase (CBH) attacks the reducing (CBHI, EC 3.2.1.176) or non-reducing (CBHII, EC 3.2.1.91) ends of the exposed cellulose chain to generate cellobiose units (Cantarel et al. 2009); and beta-glucosidases (BGL, EC 3.2.1.21) hydrolyzes the terminal non-reducing β-D-glucosyl residues/cellobiose by releasing β-D-glucose (Leah et al. 1995).

Beta-glucosidases require a special mention in the context of biomass hydrolyzing enzyme cocktails due to multiple reasons. First of all, these are the enzymes that catalyse the last step in cellulose hydrolysis and are important for regulating the cellulose hydrolysis cascade as they act as a rate-limiting factor by removing the cellobiose, which can inhibit both endo and exoglucanases through a feedback mechanism (Singhania et al. 2017). Also, in most cases, BGLs are feedback inhibited by their own product glucose, thereby allowing glucose concentrations to control the entire cellulolytic cascade (Costa et al. 2019). Many of the BGLs are also inhibited by their substrates as they can also catalyse transglycosylation reactions, which compete with the hydrolytic activity (Bohlin et al. 2013). It is important to note that most of the filamentous fungi employed currently for the commercial production of cellulases have limited BGL activity, and hence, the commercial biomass hydrolyzing enzyme cocktails are often supplemented with BGL sourced from other high BGL producers (Teugjas and Väljamäe 2013; Fernandes et al. 2022). Glucose-tolerant BGLs assume importance in this context, as their use in biomass hydrolyzing enzyme cocktails allows higher glucose concentrations to be achieved without inhibition issues. The relevance of glucose-tolerant BGLs in biomass hydrolysis has been reviewed by Salgado et al. (2018).

Apart from the three major types of cellulolytic enzymes mentioned above, there are multiple other enzymes/proteins aiding cellulose hydrolysis. Saloheimo et al. (2002) described swollenin, a protein like the plant expansins, which was believed to act by weakening the hydrogen bonds in cellulose in a non-hydrolytic manner

(Jäger et al. 2011; Gourlay et al. 2012). More recently, *Trichoderma reesei* swollenin was demonstrated to have hydrolytic activity releasing cellobiose from substrates containing β-1 → 4 glucosidic bonds (Andberg et al. 2015). Studies on such activities have revealed that several other proteins, like expansins, bacterial and fungal expansin-like proteins and loosenins, are active in hydrolyzing cellulose (Arantes and Saddler 2010; Gourlay et al. 2013).

Another important class of enzymes that participate in cellulose hydrolysis was discovered more recently, which followed an oxidative mechanism for breaking of the β-1,4 linkages in cellulose and was termed as Lytic Polysaccharide Mono Oxygenases (LPMOs) (Vaaje-Kolstad et al. 2010; Quinlan et al. 2011). These enzymes have a characteristic ability to attack crystalline regions of the cellulose, unlike the endoglucanases (EGs), which hydrolyzes the linkages at the amorphous region of cellulose. Their action is synergistic with the glycosyl hydrolases, aiding the initiation of cellulose breakdown by opening the structure similar to EGs, allowing subsequent action by the exoglucanases. LPMOs require an electron donor and molecular oxygen along with a copper ion in their active sites (Payne et al. 2015). Cellulose dehydrogenase (CDH) is another important enzyme in cellulose hydrolysis and has been known since long to be involved in cellulose depolymerization (Csarman et al. 2020). CDH is the only known secreted hemo-flavoenzyme (Sützl et al. 2018) and is proposed to have multiple functions, including oxidation of cellobiose preventing inhibition of the endo and exoglucanases, production of cellobionic acid that aids the action of peroxidases (manganese dependent) by chelation of manganese ions, reduction of iron into ferrous form providing substrates for Fenton reaction to generate hydroxyl radicals for non-enzymatic oxidative degradation of cellulose, etc. (Csarman et al. 2020). The heme B containing cytochrome (CYT) domain of CDH is now demonstrated to shuttle electrons between the FAD-containing dehydrogenase (DH) domain of the enzyme and the copper ion in the active site and therefore acts as an electron donor of LPMOs (Kracher et al. 2016).

Understanding cellulose hydrolysis also requires an understanding of the cellulose structure. Cellulose structure has a crystalline cellulose microfibrils array with the polymeric glucose chains interlinked by hydrogen bonds. The disorderly arranged glucan chains are called the amorphous regions. The need for high enzyme concentrations at the surface for efficient hydrolysis is addressed by the unique feature of having carbohydrate-binding modules (CBM) in most of the exo- and endoglucanases. They have a three-domain structure comprising a catalytic domain, which performs the hydrolysis; the CBM, which helps in anchoring enzyme on cellulose; and a linker region, flexible in nature, connecting these two, allowing free movement of the catalytic domain (Payne et al. 2015). This structure allows the enzymes for consecutive reactions without detaching from the cellulose, making them processive. Thus, the hydrolytic mechanism involves recognition of a free chain end by the CBHs followed by threading of the cellulose fibre into a tunnel in the catalytic domain, forming a catalytically active complex (CAC) structure. Hydrolysis proceeds, releasing the product, all the while, the catalytic domain moves along the fibre, and finally, the enzyme is dissociated, followed by reinitiating the cycle at a new site (Payne et al. 2015). In the case of endoglucanases, the chain threading and product release are lacking, but the processive cycle proceeds at the catalytic site. EGs are now believed to act not just in generating reducing and non-reducing ends that help in the attachment of CBH but also by helping the moving CBH to dissociate when they encounter amorphous regions of cellulose chain during their processive action and get stalled (Jalak et al. 2012).

The classical concept of cellulose hydrolysis is redefined to account for the newly discovered roles of these enzymes. Endoglucanases and LPMOs are the initiators of cleaving glycosidic linkages in the cellulose chain in an endo-acting fashion in amorphous and crystalline regions, respectively, leaving the reducing and non-reducing ends for CBHI and II, respectively. The exoglucanases act processively to liberate cellobiose, which is then fragmented by beta-glucosidases (Sukumaran et al. 2017). The simplified mechanism of action is given in Fig. 18.2.

18.3.2 Hemicellulases

Hemicellulose refers to all non-cellulose polysaccharides in plant cell walls. It exhibits considerable variation between and within plants (Horn et al. 2012). Commonly recognized hemicelluloses include xylans, abundant in hardwood and grasses, mannans that are abundant in gymnosperms which are the major softwoods, xyloglucans, glucomannans ferulate esters, β-(1,3 and 1,4) glucans, etc. (Scheller

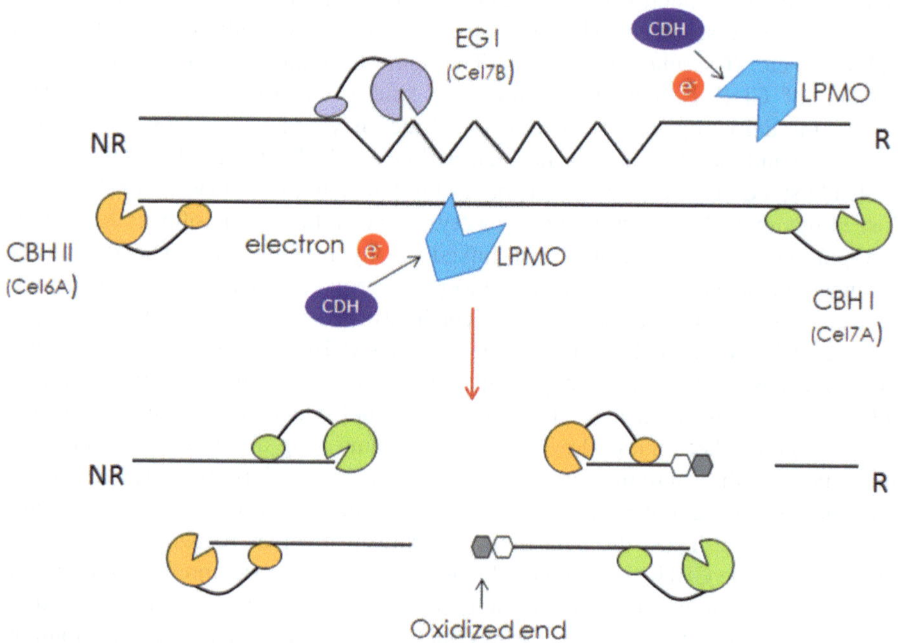

Fig. 18.2 Simplified mechanism of the enzymatic hydrolysis of cellulose. (*Reproduced from* Sukumaran et al. *(2017), with permission from Springer Nature*)

and Ulvskov 2010; Horn et al. 2012). Hemicelluloses are branched chain heteropolymers with different degrees of branching. Hardwood xylans have a β-(1 → 4) linked xylopyranose backbone with acetyl esterifications and, to a lesser extent, α-(1 → 2) linked glucuronic acid or substituted 4-*O*-Methyl-glucuronic acid (Dodd and Cann 2009). In comparison, grass xylans have a more complex structure having more arabinose substitutions and esterification, including acetyl esterification and esterification with non-flavonoid phenols like *p*-coumaric acid and ferulic acid (Biely et al. 2011). A detailed description of hemicellulose structure may be obtained from Scheller and Ulvskov (2010).

The diverse structure of hemicellulose requires the combined action of many enzymes acting synergistically or sequentially for its hydrolysis. These include GHs, which are depolymerizing enzymes like xylanases, xylosidases, mannases, mannosidases, and debranching enzymes, including carbohydrate esterases (CEs) and several accessory enzymes. Most biomass-degrading filamentous fungi secrete a cocktail of hemicellulases that act synergistically on the hemicellulose, facilitating the conversion of hemicellulose to their monomers. Different hemicellulolytic enzymes with their modes of action and reaction products are described in Juturu and Wu (2013) and listed in Table 18.1.

18.3.3 Lignin Degrading Enzymes

Lignin is an aromatic complex heteropolymer primarily comprised of the monomeric units *p*-coumaryl alcohol, coniferyl alcohol, and sinapyl alcohols incorporated as the structural units *p*-hydroxyphenyl (H), guaiacyl (G), and syringyl (S), respectively (Horn et al. 2012). Lignin is cross-linked with the carbohydrate polymers through ester or ether linkages in the lignocellulosic biomass (Takahashi and Koshijima 1988). This makes it important to remove lignin for access of the enzymes to the sugar polymers and to prevent futile binding of enzymes on lignin. The efficient hydrolysis of plant biomass is often achieved by the pretreatment of biomass (Placido and Capareda 2015). Lignin degradation is complex and brought about mostly by oxidoreductases like laccase, lignin peroxidase, and manganese peroxidase, often requiring cofactors, which makes it difficult for large-scale use in biorefineries to aid biomass hydrolysis (Horn et al. 2012). Lignin-degrading enzymes can be considered as a biological agent for lignin removal, facilitating the action of carbohydrate hydrolyzing enzymes (Gutiérrez et al. 2012; Wang et al. 2013), but for large-scale applications, it may be advisable to use ligninase producing organisms than their isolated enzymes. These enzymes may have applications in degradation of inhibitors like furan derivatives, phenolic compounds, and weak acids produced during thermo-chemical pretreatments (Carro et al. 2015). Apparently, lignin degrading enzymes are not a prime target as of now, for development of biomass degrading enzymes to be used for lignocellulosic ethanol or other products from lignocellulose feedstock.

Table 18.1 Hemicellulose degrading enzymes[a]

Enzyme	Action sites	Hydrolysis products	CAZyme families
I. Glycosyl hydrolases			
Endo-1,4-β-xylanase (EC 3.2.1.8)	β-1,4 xylan	Xylo-oligomers	GH (5, 7, 8, 10, 11, 30, 43, 51, 98, 141)
β-Xylosidase (EC 3.2.1.37)	Exo β-1,4 xylo-oligomers	Xylose	GH (1, 3, 30, 39, 43, 51–52, 54, 116, 120)
Endo-1,4-mannanase (EC 3.2.1.78)	β-1,4 mannan	Mannans oligomers	GH (5, 26, 113, 134)
β-Mannosidase (EC 3.2.1.25)	β-1,4 mannans oligomers	Mannose	GH (1–2, 5, 26)
α-L-Arabinofuranosidase (EC 3.2.1.55)	O-2 and O-3 positions of α-L-arabinofuranosides	L-arabinose	GH (2–3, 51, 54, 43, 62)
α-L-Arabinanase (EC 3.2.1.99)	α-1,5-arabinan	L-arabinose	GH43
α-D-galactosidases (EC 3.2.1.22)	Alpha-galactosyl group containing oligosaccharides	Galactose and sucrose	GH (4, 27, 31, 36, 57, 97, 110)
α-D-Glucuronidase (EC 3.2.1.139)	α-1,2-glycoside between 4-*O*-methyl-glucuronic acid and xylo-oligomers	Glucuronic acid	GH (4, 67, 115)
II. Carbohydrate esterases			
Acetyl xylan esterase (EC 3.1.1.72)	2- or 3-position of -*O*-acetyl xylan	Acetic acid	CE (1–7, 12, 16)
Feruloyl xylan esterase (EC 3.1.1.73)	Ferulic acid substitutions	Ferulic acid	CE1
Glucuronyl esterase (EC 3.1.1.117)	Ester bonds between 4-*O*-methyl-D-glucuronic acid of glucuronoxylan and lignin aromatic alcohols	Glucuronic acid and aromatic alcohols	CE15

[a] Reproduced from Houfani et al. (2020), with permission from Elsevier

18.4 Production of Enzymes for 2G Ethanol Bio-refineries

Biomass hydrolyzing enzymes—cellulases and hemicellulases—are probably the most critical and expensive component going into the production of 2G ethanol or any other product which has biomass derived sugars as the starting material. While multiple studies have estimated the cost of cellulases to be 20–50% of the operation cost in lignocellulosic ethanol production (Klein-Marcuschamer et al. 2012; Liu et al. 2016), the intricacies and cost distribution of individual operations in enzyme production have never been in the public domain, making it difficult to gauge the real challenges and also opportunities in improving the enzyme production. Over the past few decades, focused R&D in captive labs of the enzyme industry has remarkably improved the hydrolytic efficiency of commercial cellulase preparations, and efficiencies nearing the theoretical maximum are no longer a challenge. However, the same enzyme preparation giving near 100% conversion of cellulose with one type of pretreated substrate may perform less efficiently with a different substrate or a different pretreatment of the same substrate, further consolidating the need for tailor-made enzymes.

Filamentous fungi have naturally been the choice as the enzyme source for commercial production, as fungi are recognized as one of the most efficient plant biomass deconstructing organisms in nature, contributing to the carbon cycle. Moreover, they are known to thrive under aerobic conditions, tolerate a wide range of temperatures, and, more importantly, can secrete copious amounts of the enzyme into their medium, making them ideal to produce cellulases at an industrial scale (Paul et al. 2021; Yang et al. 2023). At the same time, it may be noted that a single organism may not elaborate the entire array of cellulolytic and hemicellulolytic enzymes in sufficient quantities and proportions. Also, the type and quantity of different activities secreted by a fungus can vary depending upon the production strategy and culture growth conditions (Sankar et al. 2023). This necessitates the formulation of blends, mixing heterologous component enzymes from other fungi or recombinant enzymes produced in a compatible host system for mass production (Moya et al. 2024; Zhang et al. 2022; Agarwal et al. 2021), creating efficient cocktails.

Strategies to produce (ligno) cellulolytic enzymes still follow the time-tested solid-state and submerged fermentation routes, each with its advantages and disadvantages and technical challenges, which are continuously being addressed and improved. Then there are the challenges associated with the strong regulatory controls on cellulase expression, encountered in the fungal producers of the enzyme cocktail, and which requires careful selection of the inducers, media, and operation conditions. Pure cellulose forms like Avicell, microcrystalline cellulose, cotton, natural and treated lignocellulosic material, and soluble low molecular weight inducers have all been tried as inducers to produce cellulases (Liu et al. 2021; Verma and Kumar 2022; Li et al. 2022). The choice of medium, conditions of fermentation, reactor designs, fungal morphology under cultivation, and mass and oxygen transfer limitations are other important parameters that affect the yield of enzymes in both solid-state fermentation and submerged fermentations for producing cellulases.

Downstream processing of enzymes is yet another factor adding to the cost of biomass hydrolyzing enzyme cocktails. However, most of the cellulolytic enzymes are used as such in the crude form with minimal downstream processing, which is typically filtration/removal of mycelia and concentration, since a purification step would remove several of the accessory activities that are equally significant as the major enzyme activities by cellulase and hemicellulase (Sukumaran et al. 2017). Crude cellulase preparations have shown a better performance for the hydrolysis of pretreated biomass since there are multiple proteins that aid biomass hydrolysis which may be present in quantities that are often undetectable or they may even be unknown, considering that we are still discovering new activities/proteins which may be involved directly or indirectly in lignocellulose breakdown. While it is now well recognized that onsite production of the enzymes may be advantageous with respect to better cost economics (Johnson 2016), bio-refineries may still rely on third-party produced enzymes because of their superior performance, the lack of expertise on enzyme technology/access to enzyme production technology, or simply out of convenience. In such cases, the enzyme producer would have to do the stabilization and formulation of the enzyme cocktail, which involves the addition of multiple additives, including stabilizers, surfactants, and even antibacterial agents, for better shelf life and to prevent spoilage during transportation and/or storage. Enzyme production for bioethanol/bio-refineries would have to address all these factors while still maintaining a cost which would allow economical production of 2G ethanol. The following discussion tries to elaborate on the production strategies and other concerns encountered in the manufacturing of biomass hydrolyzing enzymes.

18.5 Organisms for Production of Lignocellulolytic Enzymes and Parameters Affecting Production

18.5.1 Microorganisms for Cellulase Production

Production of enzymes for degrading plant cell walls is a trait exhibited by several microorganisms, including bacteria, archaea, actinomycetes, filamentous fungi, and even microalgae (Korsa et al. 2023). The types and levels of the components of the lignocellulolytic enzyme complexes produced by these organisms are as diverse as the organism, and the biomass types. Also, the choice of organism for industrial production of cellulase is often made on the hydrolytic ability of their enzyme cocktails on chosen feedstock, and whether the organism can elaborate the minimal set of enzymes essential for complete hydrolysis. Efficient hydrolysis of biomass requires a myriad of different activities, including endo- and exoglucanases, LPMOs, beta-glucosidases, xylanases, xylosidases, and several accessory enzymes and proteins and filamentous fungi are the best-known producers of all these activities in a single organism and in secreted form. Ascomycete fungi like *Trichoderma, Aspergillus,* and *Penicillium* have been the major focus as the industrial sources of these enzymes (Hansen et al. 2015). Not surprisingly, the major enzyme manufacturers across the globe, including the enzyme majors Novozymes, Genencor, and DSM, have used ascomycete fungi, especially strains of *Trichoderma reesei* and *Aspergillus niger*, *Humicola insolens*, *Chrysosporium lucknowense,* and their engineered derivatives to produce lignocellulolytic enzymes at commercial scale. Among these, the mutant strain *Trichoderma reesei* RUT C30 enjoys a pivotal status in enzyme production for cellulosic ethanol, since close to 80% of the currently active capacity for cellulosic ethanol production is assumed to utilize enzymes from this fungal source (Bischof et al. 2016). The fungus is known to produce and secrete very high titres of cellulase, reaching about 30 g/L (Durand et al. 1988). High cellulase activities (measured as filter paper units) of 10–15 FPUs/g substrate under solid-state fermentation (SSF) and activity of about 13 FPUs/mL for submerged fermentation have been reported from various fungi (Hansen et al. 2015). Comparison of cellulase activities reported from fungi and available in the public domain can be obtained from several studies (Sukumaran et al. 2005; Mathew et al. 2008; Chandel et al. 2012; Hansen et al. 2015, b, c; Cunha et al. 2016; Singhania et al. 2021). More recently, different species of *Penicillium* have emerged as mighty contestants for cellulase production due to their enhanced yields of the enzyme and their unique enzyme compositions and advantages with respect to cultivation and hydrolytic performance in comparison to *T. reesei* (Sreeja-Raju et al. 2020; Singhania et al. 2021; Christopher et al. 2023; Sankar et al. 2023). With several decades of research going to the optimization of cellulase production, arguably, the current industrial strains are at their peak performance levels, and further improvements are expected only through genetic modifications which target multiple aspects like deregulation of the controls like carbon catabolite repression, the introduction of efficient promoters for key enzymes, overexpression of native and heterologous enzymes, enhancing the secretion, etc. Strategies for addressing these have been reviewed by Sukumaran et al. (2021) and Yang et al. (2023).

18.5.2 Parameters Affecting Production of Cellulases

Commercial production of cellulases is accomplished by solid or submerged state fermentation, and regardless of the strategy employed, certain parameters influence the microbial growth and the enzyme production. Cellulases are inducible/adaptive enzyme that are produced more in the presence of positive stimuli such as cellulose/lignocellulose (Iram et al. 2021; Yang et al. 2023). Most of the cellulase producing fungi are also subjected to strong carbon catabolite repression limiting the range of substrates that can be used for fermentation. Thus, glucose, which is an ideal carbon source for growth of the fungus, can repress cellulase expression and is not preferred in the production medium for cellulase due to the strong repression of cellulase expression (Singhania et al. 2021). However, it may be used for building the fungal biomass essential for maintaining high levels of production, either in the growth phase or at controlled doses to maintain the cell biomass. Pure cellulose added to the media can serve as both a carbon source and an inducer. Studies conducted by Dashtban et al. (2011) with different mutant strains of *Trichoderma reesei* evaluating soluble carbon sources, insoluble pure cellulose and lignocellulosic biomass indicated that the cellulase production was the highest with pure celluloses, and they served the dual purpose as inducers and energy sources. Defined soluble carbon sources, including cellobiose, lactose, xylose, carboxymethyl cellulose, etc., have also been tried giving variable results. While lactose and sophorose can stimulate high levels of cellulase production at lower concentrations by virtue of their being recognized inducers of the cellulolytic system in *Trichoderma reesei*, their use on a larger scale is limited by cost concerns. While pure celluloses fare better in this aspect, even they are not cost effective when operated at bio-refinery scales. The most economical and environmentally friendly carbon sources to be used are probably the lignocellulosic waste materials which can act as both carbon and nitrogen sources (Hansen et al. 2015). Agro-residues including wheat bran, rice straw, corn stover, sugar bagasse, etc. are already being used for cellulase production (Bansal et al. 2012; Visser et al. 2013; Rana et al. 2014; Dey et al. 2023). Pretreatment of the substrate is often necessary for their use in cellulase production (Brijwani and Vadlani 2011; Hansen et al. 2015). Industrial waste materials or by-products offer a relatively pure and homogenous substrate for production of cellulases even though they can be complex and ensure minimal variability in comparison to the agro-residues. Typical examples include the kraft pulp residues from paper industry (Li et al. 2019). Being inducible enzymes, the cellulolytic machinery requires complex cellulosic substrates to have all the relevant enzymes expressed, as the use of defined sugars/carbohydrates often results in the production of only few components of the cellulolytic system or with disproportionate increase or decrease in some of the enzymes, rendering the digestion of target biomass inefficient. While the use of pure celluloses and lignocellulosic substrates is cost efficient and can probably induce the production of a more balanced cellulolytic cocktail, they are also not free from problems. This includes the necessity for pretreatment to remove lignin, operational limitations due to the insoluble nature of the substrate—including problems in mixing, mass transfer, and oxygen transfer and the challenges in downstream processing—especially the choking of lines and filters in solid liquid separation and concentration steps (Gabelle et al. 2012). Also, since the substrate is insoluble, the organism must first break it apart using its constitutive enzymes for obtaining the carbon for its metabolism and cellulase production resulting in process lag and increased fermentation time. The use of additional soluble inducers/carbon sources like lactose, cellobiose, sophorose, galactose, and sorbose is proposed to circumvent this issue (Singhania et al. 2021). While this can probably increase the cellulase titres, the cost of most of these soluble inducers is prohibitive for large-scale operations. The possible work around is the use of a crude inducer preparation generated from concentrated glucose solutions through transglycosylation by beta-glucosidases (England et al. 2015). The range of different feedstock used for cellulase production has been documented in Iram et al. (2021).

Apart from the carbon source, the other important components of the cellulose production media include nitrogen source(s) and mineral nutrients. Most of the lignocellulosic substrates used in cellulase production have low nitrogen content, and it is often necessary to supplement a nitrogen source in the medium. Both organic nitrogen sources like sulphite pulp, corn steep liquor, peptones, yeast extract, soya flour, etc., and inorganic nitrogen sources like ammonium sulphate, urea, etc. are used as nitrogen supplements in the medium (Iram et al. 2021). Some of these organic nitrogen supplements are also capable of providing the organism with other essential nutrients including vitamins, amino acids, co-factors, and even minerals. The nutrients and minerals supplemented in the medium can influence the growth and enzyme production both positively or negatively, and it is often necessary to optimize their levels for the optimal yield of product (Nazir et al. 2024). There are several reports focusing on media engineering and optimization of process parameters for improving cellulase production by filamentous fungi, mostly using statistical methods of the Design of Experiments (DOEs) (Mekala et al. 2008; Acharya et al. 2008; Idris et al. 2017; Ellilä et al. 2017). For submerged fermentation, proper mixing and optimal dissolved oxygen content have to be ensured, and it is important to optimize these parameters as well. Agitation of the fungal cultures, especially while using the insoluble substrates, presents multiple challenges including the media viscosity and shear stress imparted to the fungal mycelia (Iram et al. 2023). The choice of impeller becomes important to address these, and often classical impeller designs may not be sufficient. There have been studies on improved impeller designs which can reduce the shear stress on fungal mycelia (Yang et al. 2012), which may be adopted for cellulase production using filamentous fungi.

18.6 Fermentation Systems for Fungal Cellulase Production

Commercially, the classical fermentation mode applied for cellulase production is the submerged fermentation (SmF), while some manufacturers follow the SSF route. Each strategy has its merits and demerits and comparison is mostly irrelevant, at least in the case of cellulase production by filamentous fungi, as the choice is based on the growth patterns of the organism employed, substrate features and limitations, mass transfer and oxygen transfer requirements, existing infrastructure, down streaming requirements etc., among the different other factors. SSF processes are similar to the natural mode of fungal growth and the macro-morphology of the cultivated fungus varies significantly from SmF where it can assume shapes like loose mycelial clumps to dense pellets (Papagianni 2004). In comparison, the typical morphology of SSF grown fungi are aerial and substrate penetrating mycelial hyphae. Although SSF is known to be a cheaper method and produces a more concentrated enzyme (e.g., Tengerdy 1996; Castilho et al. 2000; Viniegra-González et al. 2003; Krishna 2005), there are several reports which reported otherwise (Nakeeran et al. 2012). It is also of interest to observe that the types and quantities of enzymes secreted by the same fungus can vary significantly when grown in solid or submerged state (Hansen et al. 2015; Sankar et al. 2023). Sankar et al. (2023) reported that the enzyme cocktail secreted by *Penicillium janthinellum* cultivated in SSF mode produce a cocktail with better hydrolytic efficiency compared to that obtained from SmF. Nevertheless, SmF has the advantages of better process control and lower contamination possibility and easiness in downstream processing.

18.6.1 Solid-State Fermentation

In solid-state fermentation, instead of liquid medium, solid substrates are used with just enough moisture required for the growth and metabolism of microorganisms. This solid surface acts as a support for the microbial growth. Agricultural residues, grains, and agro-industrial by-products are used as solid substrate that contains a mixture of carbohydrates, proteins, and other nutrients that serve as the energy and carbon sources for microbial growth (Singhania et al. 2006). The major advantages of SSF include higher enzyme titres, concentrated enzymes, lesser catabolite repression, reduced contamination risk etc. In SSF, the process typically uses commonly used substrate such as wheat bran, which is moistened or impregnated with a medium containing essential nutrients, particularly a nitrogen source. Since the substrate usually acts also as the carbon source, the addition of a nitrogen source is crucial to support the growth and metabolism of microbes. The moistening or impregnation of the substrate with the nutrient medium serves to provide the necessary moisture and nutrients for microbial growth while creating an environment conducive to fermentation (Fig. 18.3). This allows the microorganisms to colonize the substrate, metabolize the available nutrients, and produce the desired bioproducts. The addition of inducers in SSF stimulates the production of cellulases by the microbial culture, leading to enhanced enzyme yields. These inducers can either be simple disaccharides like lactose/cellobiose, which are small molecules or can be larger polysaccharides like pure cellulose. The addition of an inducer serves to upregulate the expression of genes coding for cellulase enzymes, thereby increasing their production by the microorganisms. Typically, trays made of steel/aluminium are used, wherein solid substrate along with media and inoculum is added and fermentation is carried out in rooms having temperature and humidity control (Idris et al. 2017). This fermented substrate in either its wet or dry form can be used directly as enzyme, which is an added advantage (Zhuang et al. 2007; Singhania et al. 2015). In this case, the elimination of downstream operations can significantly reduce cost, but certain disadvantages also co-occur like contamination and lower productivity of biomass hydrolysis step due to the carryover of spore/mycelia from fermentation stage and increase in solid content in the hydrolysis reactor which leads to inefficient mixing. Thus, the choice of fermented material as such or enzyme extracted with no fungal spores/mycelia depends mainly on the conditions of enzymatic saccharification. As SSF mimics the natural growth conditions of filamentous fungi, which grows by deriving nutrients from degrading plant materials, it is expected that the enzyme cocktail produced by solid-state fermentation will be more efficient in digestion of lignocellulosic substrates. The cellulase cocktail

Pilot scale tray fermentation system with steel racks to hold SSF trays

Cellulase production trays with mature fungal culture

Fig. 18.3 SSF Pilot Plant (Koji Room). (Courtesy—CSIR-NIIST, Trivandrum, India)

produced by SSF has shown to contain enzymes like beta-glucosidase, LPMO, xylanases in higher titres which increases the initial rate of reaction, and improved sugar yields (Sankar et al. 2023).

18.6.2 Submerged Fermentation

Industrial production of enzymes like cellulases uses the most preferred technique of submerged fermentation as the process parameters and their controls can be easily monitored online (Hansen et al. 2015). Most commonly, the submerged fermentation is done with dissolvable substrates along with essential salts mixed in a submerged liquid medium (Namnuch et al. 2021). One of the major advantages of submerged fermentation is better heat and mass transfer, which is probably since the cells grow in very conducive environment in regulated temperature conditions with better mixing. To improve cellulase production, agricultural residues like straws, bagasse, etc. are used, which act as inducers owing to their cellulose and hemicellulose content (Gupte et al. 2023). But such inducers require some form of pretreatment to remove lignin, which might reduce the accessibility of cellulose and thus result in lower enzyme titres. Better mixing and aeration can be achieved in stirred tank reactors for submerged fermentation of cellulases.

Cellulase production with high productivities and titres can be achieved by limiting carbon flux in fed batch or continuous mode (Han et al. 2017). Submerged fermentation can be carried out in either batch, fed batch, or continuous mode. Several inducers have been tried till date, out of which lactose seems to be a cheaper soluble inducer used in commercial production of cellulases using *Trichoderma*. Fed batch is the most employed mode for commercial production of cellulase with inducers being dosed carefully at appropriate intervals. Insoluble substrate/inducers like lignocellulosic biomass, paper pulp, etc. pose a serious challenge of mixing in the production reactors. In contrast, soluble substrates offer several advantages such as better mixing, aeration, and proper control over the fermentation process. In case of *Trichoderma reesei,* sophorose has been identified as the best inducer, but it is costly and difficult to manufacture (Sternberg and Mandels 1979). Lately, a process patented by Danisco Inc. had produced substantial quantities of sophorose using concentrated glucose (England et al. 2015). The cellulase preparation from *T. reesei* was used to generate sophorose by transglycosylation activity from a mixture of sugars.

Although SSF and SmF have their own benefits and drawbacks, the method applied for commercial cellulase production must consider the efficacy of enzyme production, cost analysis, and its end application.

18.7 Downstream Processing of Cellulases

Downstream processing of biomass hydrolyzing enzyme cocktails elaborated by fungi is quite different for SmF and SSF produced enzymes. Often the cost of recovery of the enzyme is reported to be up to 70% of the enzyme production cost (Bhatt and Shilpa 2019). Typically, the downstream operations for SmF produced enzyme are limited to a solid–liquid (S/L) separation stage post-fermentation followed by a microfiltration to remove suspended fine particles and ultrafiltration for concentrating the enzyme. Around 74% recovery was attained using membrane filtration (Mores et al. 2001). This is much different from classical protein purification operations where multiple operations include S/L separation, solvent extraction/precipitation, salting out, dialysis, ultrafiltration, and chromatographic techniques to obtain pure enzymes (Fig. 18.4). Biomass hydrolyzing enzymes are required in their crude form as elaborated by the source organism to avoid removal of any activity that aids cellulolytic. In several instances, the concentration of enzyme is achieved through a solvent precipitation step involving acetone as solvent followed by re-dissolving of the enzyme precipitate in an appropriate buffer/water. Enzymes produced by SSF can have more steps to their downstream operations, especially if the final formulation is in a liquid form. The fermented substrate from SSF is usually extracted with an appropriate buffer or water to recover the enzyme (Marín et al. 2019).

Fig. 18.4 Pilot Scale Ultrafiltration unit. (Courtesy—CSIR-NIIST, Trivandrum, India)

Ultra-sonication of the substrate/water slurry can improve the recovery of enzyme by more than twofold (Szabo et al. 2015). Once extracted into the liquid, the downstream operations are like the SmF which involves concentration of the enzyme by solvent precipitation or ultrafiltration. In several instances, SSF produced enzymes can also be used directly in hydrolysis of biomass without extraction (Zhuang et al. 2007; Singhania et al. 2015).

18.8 Enzyme Formulations

Cellulolytic enzyme cocktails elaborated by the fungi are often not complete with respect to all the major activities and accessory enzymes. Many of these enzymes and accessory proteins would be suboptimal with respect to their concentrations and relative proportions, and it is highly essential to have an idea about the optimal enzyme dosage required for efficient biomass hydrolysis. For example, the cellulase cocktail elaborated by *T. reesei* RUT C30, though rich in cellobiohydrolases, has sub-optimal concentrations of beta-glucosidase, and generally beta-glucosidases from other fungal sources such as *Aspergillus niger* are blended in large quantities by the enzyme manufacturers (Knauf and Moniruzzaman 2004; Rana et al. 2014). For efficient biomass saccharification, the optimal combination and concentration of individual enzyme components is unknown, so the supplementation requirements are derived empirically through trial-and-error methods by performing experiments. This necessitates the evaluation of multiple combinations of component enzymes experimentally, so as to achieve the optimal one (Sukumaran et al. 2017).

Enzyme cocktails for biomass hydrolysis are also supplemented with several other additives that serve different functions like stabilization, preventing contamination, proteolytic digestion of enzymes, surface activity, etc., and the additives are often proprietary information of the manufacturer (Sukumaran et al. 2017). Surfactants are a major additive in such formulations, which performs multiple roles in enhancing hydrolytic performance such as prevention of non-specific enzyme binding to lignin, changes in biomass chemical structure, viscosity and surface tension reduction and subsequently reduction in enzyme contact with air–liquid interface, etc. (Eckard et al. 2013). Other major additive is the protease inhibitors which prevent the dame to enzyme proteins by the native proteases elaborated by the source organism. An example of such additions is presented in Lenoir (2011).

18.9 Conclusion

Biomass hydrolyzing enzymes are the critical components in 2G ethanol production and considered to be the major bottleneck in sugar platform-based bio-refineries. As enzyme technology for production of these enzymes has improvised a lot over the years and cocktails with improved efficiency are available in the market from global enzyme majors like Novozymes and Dupont Genencor, there are still limitations with respect to their suitability across multiple substrates and above all the cost of enzymes. Cellulolytic enzyme production is best performed with filamentous fungi, and the complexities associated with the need to have the myriads of different activities present in optimal concentrations in the final preparation require research and development to have a better understanding about enzymatic saccharification, for formulating the optimal enzyme cocktails. There are multiple opportunities for cost reduction involving the use of cheap lignocellulosic substrates as carbon sources and inducers for enzyme production, improving the efficiency of enzyme production by the organism through media and process engineering, genetic modifications, and fine-tuning of the fermentation strategies to name a few. Hydrolytic efficiencies may also be improved by creating enzyme cocktails by blending of heterologous enzymes from different organisms or recombinant enzymes to the base cellulase. The chapter has tried to address these issues and to present to the reader an overview of the current developments in the domain providing the feed of thought for improving the most important component in 2G ethanol production—the biocatalyst.

Acknowledgements The authors acknowledge funding by 34/1/TD-AgriNutriBiotech/NCP-FBR 2010-RPPBDD-TMDSeMOI (MLP 0053) for research on cellulolytic enzymes.

References

Acharya PB, Acharya DK, Modi HA (2008) Optimization for cellulase production by Aspergillus niger using saw dust as substrate. Afr J Biotechnol 7:4147–4152

Agarwal D, Tsang A, Chadha BS (2021) Economizing the lignocellulosic hydrolysis process using heterologously expressed auxiliary enzymes feruloyl esterase D (CE1) and β-xylosidase (GH43) derived from thermophilic fungi *Scytalidium thermophilum*. Bioresour Technol 339:125603. https://doi.org/10.1016/j.biortech.2021.125603

Andberg M, Penttilä M, Saloheimo M (2015) Swollenin from *Trichoderma reesei* exhibits hydrolytic activity against cellulosic substrates with features of both endoglucanases and cellobiohydrolase. Bioresour Technol 181:105–113

Arantes V, Saddler JN (2010) Access to cellulose limits the efficiency of enzymatic hydrolysis: the role of amorphogenesis. Biotechnol Biofuels 3:4

Bansal N, Tewari R, Soni R, Soni R (2012) Production of cellulases from *Aspergillus niger* NS-2 in solid state fermentation on agricultural and kitchen waste residues. Waste Manag 32:1341–1346

Bettenhausen C (2023) Ethanol rising. C&EN Global Enterprise 101(6):20–26. https://doi.org/10.1021/cen-10106-cover

Bhatt SM, Shilpa (2019) Downstream processing technology for cellulase production. In: New and future developments in microbial biotechnology and bioengineering. Elsevier, Amsterdam, pp 195–208. https://doi.org/10.1016/b978-0-444-64223-3.00012-6

Biely P, Mastihubova M, Tenkanen M, Eyzaguirre J, Li XL, Vrsanska M (2011) Action of xylan deacetylating enzymes on monoacetyl derivatives of 4-nitrophenyl glycosides of beta-D-xylopyranose and alpha-L-arabinofuranose. J Biotechnol 151:137–142. https://doi.org/10.1016/j.jbiotec.2010.10.074

Bischof RH, Ramoni J, Seiboth B (2016) Cellulases and beyond: the first 70 years of the enzyme producer *Trichoderma reesei*. Microb Cell Factories 15:106

Bohlin C, Praestgaard E, Baumann MJ, Borch K, Praestgaard J, Monrad RN, Westh P (2013) A comparative study of hydrolysis and transglycosylation activities of fungal β-glucosidases. Appl Microbiol Biotechnol 97:159–169. https://doi.org/10.1007/s00253-012-3875-9

Brijwani K, Vadlani PV (2011) Cellulolytic enzymes production via solid-state fermentation: effect of pretreatment methods on physicochemical characteristics of substrate. Enzyme Res 2011:860134

Brown A, Waldheim L, Landälv I, Saddler J, Ebadian M, McMillan JD, Bonomi A, Klein B (2020) Advanced biofuels—potential for cost reduction. IEA Bioenergy, Paris. https://www.ieabioenergy.com/wp-content/uploads/2020/02/T41_CostReductionBiofuels-11_02_19-final.pdf

Buchspies B and Kaltschmitt M (2018) A consequential assessment of changes in greenhouse gas emissions due to the introduction of wheat straw ethanol in the context of European legislation. Appl. Energy 211:368–381

Cantarel BL, Coutinho PM, Rancurel C, Bernard T, Lombard V, Henrissat B (2009) The Carbohydrate-Active EnZymes database (CAZy): an expert resource for glycogenomics. Nucleic Acid Res 37(Suppl 1):D233–D238

Carro J, Ferreira P, Rodríguez L, Prieto A, Serrano A, Balcells B, Ardá A, Jiménez-Barbero J, Gutiérrez A, Ullrich R, Hofrichter M, Martínez AT (2015) 5-Hydroxymethylfurfural conversion by fungal aryl-alcohol oxidase and unspecific peroxygenase. FEBS J 282(16):3218–3229

Castilho LR, Polato CM, Baruque EA, Sant Anna GI Jr, Freire DM (2000) Economic analysis of lipase production by *Penicillium restrictum* in solid-state and submerged fermentations. Biochem Eng J 4:239–247

Chandel AJ, Chandrasekhar G, Silva MB, da Silva SS (2012) The realm of cellulases in biorefinery development. Crit Rev Biotechnol 32(3):187–202

Christopher M, Sreeja-Raju A, Kooloth-Valappil P, Gokhale DV, Sukumaran RK (2023) Cellulase hyper-producing fungus *Penicillium janthinellum* NCIM 1366 elaborates a wider array of proteins involved in transport and secretion, potentially enabling a diverse substrate range. Bioenergy Res 16:61–73. https://doi.org/10.1007/s12155-022-10407-3

Costa LSC, Mariano DCB, Rocha REO, Kraml J, Silveira CH, Liedl KR, de Melo-Minardi RC, Lima LHF (2019) Molecular dynamics gives new insights into the glucose tolerance and inhibition mechanisms on β-glucosidases. Molecules 24(18):3215. https://doi.org/10.3390/molecules24183215

Csarman F, Wohlschlager L, Ludwig R (2020) Cellobiose dehydrogenase. Enzymes 47:457–489. https://doi.org/10.1016/bs.enz.2020.06.002

Cunha FM, Vasconcellos VM, Florencio C, Badino AC, Farinas CS (2016) On-site production of enzymatic cocktails using a non-conventional fermentation method with agro-industrial residues as renewable feedstocks. Waste Biomass Valori 8:517. https://doi.org/10.1007/s12649-016-9609-y

Dashtban M, Buchkowski R, Qin W (2011) Effect of different carbon sources on cellulase production by *Hypocrea jecorina* (*Trichoderma reesei*) strains. Int J Biochem Mol Biol 2(3):274–286

De Souza WR, Mitchell RAC, Cesarino I (2023) Editorial: the plant cell wall: advances and current perspectives. Front Plant Sci 14:1235749. https://doi.org/10.3389/fpls.2023.1235749

Dey P, Chakrabortty S, Haldar D, Rangarajan V, Ashok S (2023) On-site enriched production of cellulase enzyme using rice straw waste and its hydrolytic performance evaluation through systematic dynamic modeling. Environ Sci Pollut Res 30:36710–36727. https://doi.org/10.1007/s11356-022-24797-6

Dodd D, Cann IK (2009) Enzymatic deconstruction of xylan for biofuel production. Glob Change Biol Bioenergy 1(1):2–17

Durand H, Clanet M, Tiraby G (1988) Genetic improvement of *Trichoderma reesei* for large scale cellulase production. Enzym Microb Technol 10:341–346

Eckard AD, Muthukumarappan K and Gibbons W (2013) A review of the role of amphiphiles in biomass to ethanol conversion. Appl. Sci. 3(2):396–419.

Ellilä S, Fonseca L, Uchima C, Cota J, Goldman GH, Saloheimo M, Sacon V, Siika-aho M (2017) Development of a low-cost cellulase production process using Trichoderma reesei for Brazilian biorefineries. Biotechnol Biofuels 10:30. https://doi.org/10.1186/s13068-017-0717-0

England GR, Kelley A, Mitchinson C (2015) Induction of gene expression using a high concentration sugar mixture. US Patent 9034628 B2 issued 19/05/2015

Fernandes CG, Sawant SC, Mule TA, Khadye VS, Lali AM, Odaneth AA (2022) Enhancing cellulases through synergistic β-glucosidases for intensifying cellulose hydrolysis. Process Biochem 120:202–212. https://doi.org/10.1016/j.procbio.2022.06.011

Fontes CM, Gilbert HJ (2010) Cellulosomes: highly efficient nanomachines designed to deconstruct plant cell wall complex carbohydrates. Annu Rev Biochem 79:655–681

Gabelle JC, Jourdier E, Licht RB, Chaabane FB, Henaut I, Morchain J, Augier F (2012) Impact of rheology on the mass transfer coefficient during the growth phase of Trichoderma reesei in stirred bioreactors. Chem Eng Sci 75:408–417

Gírio FM, Fonseca C, Carvalheiro F, Duarte LC, Marques S, Bogel-Łukasik R (2010) Hemicelluloses for fuel ethanol: a review. Bioresour Technol 101:4775–4800. https://doi.org/10.1016/j.biortech.2010.01.088

Gourlay K, Arantes V, Saddler J (2012) Use of substructure-specific carbohydrate binding modules to track changes in cellulose accessibility and surface morphology during the amorphogenesis step of enzymatic hydrolysis. Biotechnol Biofuels 5:51

Gourlay K, Hu J, Arantes V, Andberg M, Saloheimo M, Penttilä M, Saddler J (2013) Swollenin aids in the amorphogenesis step during the enzymatic hydrolysis of pretreated biomass. Bioresour Technol 142:498–503

Gupte A, Prajapati D, Bhatt A, Pandya S, Raghunathan M, Gupte S (2023) Agro-industrial residues: an eco-friendly and inexpensive substrate for fungi in the development of white biotechnology. In: Fungi and fungal products in human welfare and biotechnology. Springer Nature Singapore, Singapore, pp 571–603

Gutiérrez A, Rencoret J, Cadena EM, Rico A, Barth D, del Río JC, Martínez ÁT (2012) Demonstration of laccase-based removal of lignin from wood and non-wood plant feedstocks. Bioresour Technol 119:114–122

Han X et al (2017) Improving cellulase productivity of *Penicillium oxalicum* RE-10 by repeated fed-batch fermentation strategy. Bioresour Technol 227:155–163

Hansen GH, Lubeck M, Frisvad JC, Lubeck PS, Andersen B (2015) Production of cellulolytic enzymes from ascomycetes: comparison of solid state and submerged fermentation. Process Biochem 50(9):1327–1341. https://doi.org/10.1016/j.procbio.2015.05.017

Horn SJ, Vaaje-Kolstad G, Westereng B, Eijsink VGH (2012) Novel enzymes for the degradation of cellulose. Biotechnol Biofuels 5:45. https://doi.org/10.1186/1754-6834-5-45

Houfani AA, Anders N, Spiess AC, Baldrian P, Benallaaua S (2020) Insights from enzymatic degradation of cellulose and hemicellulose to fermentable sugars—a review. Biomass Bioenergy 134:105481. https://doi.org/10.1016/j.biombioe.2020.105481

Idris ASO, Pandey A, Rao SS, Sukumaran RK (2017) Cellulase production through solid-state tray fermentation, and its use for bioethanol from sorghum stover. Bioresour Technol 242:265–271

IEA (2023) Renewables 2023—analysis and forecasts to 2028. https://www.iea.org/reports/renewables-2023

Igarashi K, Uchihashi T, Koivula A, Wada M, Kimura S, Penttilä M, Ando T, Samejima M (2012) Visualization of cellobiohydrolase I from Trichoderma reesei moving on crystalline cellulose using high-speed atomic force microscopy. In: Gilbert HJ (ed) Methods in enzymology, vol 510. Academic Press, New York, pp 169–182. https://doi.org/10.1016/B978-0-12-415931-0.00009-4

Iram A, Cekmecelioglu D, Demirci A (2021) Ideal feedstock and fermentation process improvements for the production of lignocellulolytic enzymes. Processes 9:38. https://doi.org/10.3390/pr9010038

Iram A, Cekmecelioglu D, Demirci A (2023) The effects of dilution, aeration, and agitation on fungal cellulase and xylanase production in the fermentation media based on distillers dried grains with solubles using stirred tank bioreactors. Ind Biotechnol 19:229–236

Jäger G, Girfoglio M, Dollo F, Rinaldi R, Bongard H, Commandeur U, Fischer R, Spiess AC, Büchs J (2011) How recombinant swollenin from *Kluyveromyces lactis* affects cellulosic substrates and accelerates their hydrolysis. Biotechnol Biofuels 4:33

Jalak J, Kurašin M, Teugjas H, Väljamäe P (2012) Endo-exo synergism in cellulose hydrolysis revisited. J Biol Chem 287(34):28802–28815

Johnson E (2016) Integrated enzyme production lowers the cost of cellulosic ethanol. Biofuels Bioprod Biorefin 10(2):164–174. https://doi.org/10.1002/bbb.1634

Juturu V, Wu JC (2013) Insight into microbial hemicellulases other than xylanases: a review. J Chem Technol Biotechnol 88(3):353–363

Klein-Marcuschamer D, Oleskowicz-Popiel P, Simmons BA, Blanch HW (2012) The challenge of enzyme cost in the production of lignocellulosic biofuels. Biotechnol Bioeng 109(4):20121083. https://doi.org/10.1002/bit.24370

Knauf M, Moniruzzaman M (2004) Lignocellulosic biomass processing: a perspective. Int Sugar J 106:147–150

Korsa G, Konwarh R, Masi C, Ayele A, Haile S (2023) Microbial cellulase production and its potential application for textile industries. Ann Microbiol 73:13. https://doi.org/10.1186/s13213-023-01715-w

Kracher D, Scheiblbrandner S, Felice AKG, Breslmayr E, Preims M, Luwicka C, Haltrich D, Eijsink VGH, Ludwig R (2016) Extracellular electron transfer systems fuel cellulose oxidative degradation. Science 352:1098–1101. https://doi.org/10.1126/science.aaf3165

Krishna C (2005) Solid-state fermentation systems—an overview. Crit Rev Biotechnol 25:1–30

Langan P, Petridis L, O'Neill HM, Pingali SV, Foston M, Nishiyama Y, Schulz R, Lindner B, Leif Hanson B, Harton S, Heller WT, Urban V, Evans BR, Gnanakaran S, Ragauskas AJ, Smith JC, Davison BH (2014) Common processes drive the thermochemical pretreatment of lignocellulosic biomass. Green Chem 16:63–68. https://doi.org/10.1039/C3GC41962B

Leah R, Kigel J, Svendsen I, Mundy J (1995) Biochemical and molecular characterization of a barley seed beta-glucosidase. J Biol Chem 270:15789–15797

Lenoir PM (2011) Enzyme stabilization. US Patent 7928052B2, To Dow Global Technologies

Li J, Shi Sm Kang L, Wang W, Guan W (2019) Cellulase production from kraft hardwood pulp by Trichoderma reesei rut C-30. Biofuels Bioprod Biorefin 13(5):1160–1168. https://doi.org/10.1002/bbb.2008

Li Y, Zhang P, Zhu D, Yao B, Hasunuma T, Kondo A, Zhao X (2022) Efficient preparation of soluble inducer for cellulase production and saccharification of corn stover using in-house generated crude enzymes. Biochem Eng J 178:108296. https://doi.org/10.1016/j.bej.2021.108296

Liu G, Zhang J, Bao J (2016) Cost evaluation of cellulase enzyme for industrial-scale cellulosic ethanol production based on rigorous Aspen Plus modeling. Bioprocess Biosyst Eng 39(1):133–140. https://doi.org/10.1007/s00449-015-1497-1

Liu P, Li A, Wang Y, Cai Q, Yu H, Li Y, Peng H, Li Q, Wang Y, Wei X, Zhang R, Tu Y, Xia T, Peng L (2021) Distinct *Miscanthus* lignocellulose improves fungus secreting cellulases and xylanases for consistently enhanced biomass saccharification of diverse bioenergy crops. Renew Energy 174:799–809. https://doi.org/10.1016/j.renene.2021.04.107

Marín M, Artola A, Sánchez A (2019) Optimization of down-stream for cellulases produced under solid-state fermentation of coffee husk. Waste Biomass Valori 10:2761–2772. https://doi.org/10.1007/s12649-018-0327-5

Mathew GM, Sukumaran RK, Singhania RR, Pandey A (2008) Progress in research on fungal cellulases for lignocellulose degradation. J Sci Ind Res 67:898–907

Meents MJ, Watanabe Y, Samuels AL (2018) The cell biology of secondary cell wall biosynthesis. Ann Bot 121:1107–1125. https://doi.org/10.1093/aob/mcy005

Mekala NK, Singhania RR, Sukumaran RK, Pandey A (2008) Cellulase production under solid-state fermentation by Trichoderma reesei RUT C30: statistical optimization of process parameters. Appl Biochem Biotechnol 151(2–3):122–131

Mores WD, Knutsen JS, Davis RH (2001) Cellulase recovery via membrane filtration. In: Davison BH, McMillan J, Finkelstein M (eds) Twenty-second symposium on biotechnology for fuels and chemicals. ABAB symposium. Humana Press, Totowa. https://doi.org/10.1007/978-1-4612-0217-2_25

Moya EB, Syhler B, Dragone G, Mussatto SI (2024) Tailoring a cellulolytic enzyme cocktail for efficient hydrolysis of mildly pretreated lignocellulosic biomass. Enzym Microb Technol 175:110403. https://doi.org/10.1016/j.enzmictec.2024.110403

Nakeeran E, Gowthaman MK, Umesh-Kumar S, Subramanian R (2012) Techno-economic analysis of processes for *Aspergillus carbonarius* polygalacturonase production. J Biosci Bioeng 113:634–640

Namnuch N, Thammasittirong A, Thammasittirong SNR (2021) Lignocellulose hydrolytic enzymes production by *Aspergillus flavus* KUB2 using submerged fermentation of sugarcane bagasse waste. Mycology 12(2):119–127

Nazir M, Iram A, Cekmecelioglu D, Demirci A (2024) Approaches for producing fungal cellulases through submerged fermentation. Front Biosci 16(1):5. https://doi.org/10.31083/j.fbe1601005

Papagianni M (2004) Fungal morphology and metabolite production in submerged mycelial processes. Biotechnol Adv 22:189–259

Paul M, Mohapatra S, Mohapatra KDP, Thatoi H (2021) Microbial cellulases—an update towards its surface chemistry, genetic engineering and recovery for its biotechnological potential. Bioresour Technol 340:125710. https://doi.org/10.1016/j.biortech.2021.125710

Payne CM, Knott BC, Mayes HB, Hansson H, Himmel ME, Sandgren M, Ståhlberg J, Beckham GT (2015) Fungal cellulases. Chem Rev 115(3):1308–1448

Placido J, Capareda S (2015) Ligninolytic enzymes: a biotechnological alterative for bioethanol production. Bioresour Bioprocess 2:23

Pu Y, Hun F, Huang F, Davison BH, Ragauskas AJ (2013) Assessing the molecular structure basis for biomass recalcitrance during dilute acid and hydrothermal pretreatments. Biotechnol Biofuels 6:15. https://doi.org/10.1186/1754-6834-6-15

Quinlan RJ, Sweeney MD, Lo Leggio L, Otten H, Poulsen JCN, Johansen KS, Krogh K, Jorgensen CI, Tovborg M, Anthonsen A, Tryfona T, Walter CP, Dupree P, Xu F, Davies GJ, Walton PH (2011) Insights into the oxidative degradation of cellulose by a copper metalloenzyme that exploits biomass components. Proc Natl Acad Sci U S A 108:15079–15084. https://doi.org/10.1073/pnas.1105776108

Rana V, Eckard AD, Teller P, Ahring BK (2014) On-site enzymes produced from Trichoderma reesei RUT-C30 and *Aspergillus saccharolyticus* for hydrolysis of wet exploded corn stover and loblolly pine. Bioresour Technol 154:282–289

Salgado JCS, Meleiro LP, Carli S, Ward RJ (2018) Glucose tolerant and glucose stimulated β-glucosidases—a review. Bioresour Technol 267:704–713. https://doi.org/10.1016/j.biortech.2018.07.137

Saloheimo M, Paloheimo M, Hakola S, Pere J, Swanson B, Nyyssönen E, Bhatia A, Ward M, Penttilä M (2002) Swollenin, a *Trichoderma reesei* protein with sequence similarity to the plant expansins, exhibits disruption activity on cellulosic materials. Eur J Biochem 269(17):4202–4211

Sankar M, Mathew RM, Puthiyamadam A, Sreej-Raju A, Christopher M, Gokhale DV, Sukumaran RK (2023) Comparison of the solid-state and submerged fermentation derived secretomes of hyper-cellulolytic *Penicillium janthinellum* NCIM 1366 reveals the changes responsible for differences in hydrolytic performance. Bioresour Technol 371:128602. https://doi.org/10.1016/j.biortech.2023.128602

Scheller HV, Ulvskov P (2010) Hemicelluloses. Annu Rev Plant Biol 61:263–289

Scully MJ, Norris GA, Falconi TMA, MacIntosh DL (2021) Carbon intensity of corn ethanol in the United States: state of the science. Environ Res Lett 16(4):043001. https://doi.org/10.1088/1748-9326/abde08

Sharma B, Larroche C, Dussap CG (2020) Comprehensive assessment of 2G bioethanol production. Bioresour Technol 313:123630. https://doi.org/10.1016/j.biortech.2020.123630

Sheehan J, Himmel M (1999) Enzymes, energy and the environment: a strategic perspective on the US Department of Energy's research and development activities for bioethanol. Biotechnol Prog 15:817–827

Singhania RR, Sukumaran RK, Pillai A, Prema P, Szakacs G, Pandey A (2006) Solid-state fermentation of lignocellulosic substrates for cellulase production by Trichoderma reesei NRRL 11460. Indian J Biotechnol 5(3):332–336

Singhania RR, Saini R, Adsul M, Saini JK, Mathur M, Tuli DK (2015) An integrative process for bio-ethanol production employing SSF produced cellulase without extraction. Biochem Eng J 102:45–48

Singhania RR, Patel A, Pandey A, Ganansounou E (2017) Genetic modification: a tool for enhancing beta-glucosidase production for biofuel application. Bioresour Technol 245:1352–1361. https://doi.org/10.1016/j.biortech.2017.05.126

Singhania RR, Ruiz HA, Awasthi MK, Dong CD, Chen CW, Patel AK (2021) Challenges in cellulase bioprocess for biofuel applications. Renew Sust Energ Rev 151:111622. https://doi.org/10.1016/j.rser.2021.111622

Sreeja-Raju A, Christopher M, Kooloth-Valappil P, Kuni-Parambil R, Gokhale DV, Sankar M, Abraham A, Pandey A, Sukumaran RK (2020) *Penicillium janthinellum* NCIM1366 shows improved biomass hydrolysis and a larger number of CAZymes with higher induction levels over Trichoderma reesei RUT-C30. Biotechnol Biofuels 13:196. https://doi.org/10.1186/s13068-020-01830-9

Sternberg DAVID, Mandels G (1979) Induction of cellulolytic enzymes in Trichoderma reesei by sophorose. J Bacteriol 139(3):761–769

Sukumaran RK, Singhania RR, Pandey A (2005) Microbial cellulases—production, applications and challenges. J Sci Ind Res 64:832–844

Sukumaran RK, Abraham A, Mathew AK (2017) Enzymes for bioenergy. In: Sugathan S, Pradeep N, Abdulhameed S (eds) Bioresources and bioprocess in biotechnology. Springer, Singapore. https://doi.org/10.1007/978-981-10-4284-3_1

Sukumaran RK, Christopher M, Kooloth Valappil P, Sreeja-Raju AR, Mathew RM, Sankar M, Puthiyamadam A, Adarsh VP, Aswathi A, Rebinro V, Abraham A, Pandey A (2021) Addressing challenges in production of cellulases for biomass hydrolysis: targeted interventions into the genetics of cellulase producing fungi. Bioresour Technol 329:124746. https://doi.org/10.1016/j.biortech.2021.124746

Susmozas A, Martín-Sampedro R, Ibarra D, Eugenio ME, Iglesias R, Manzanares P, Moreno AD (2020) Process strategies for the transition of 1G to advanced bioethanol production. Processes 8(10):1310. https://doi.org/10.3390/pr8101310

Sützl L, Laurent CVFP, Abrera AT, Schutz G, Ludwig R, Haltrich D (2018) Multiplicity of enzymatic functions in the CAZy AA3 family. Appl Microbiol Biotechnol 102:2477–2492. https://doi.org/10.1007/s00253-018-8784-0

Szabo OE, Csiszar E, Koczka B, Kiss K (2015) Ultrasonically assisted single stage and multiple extraction of enzymes produced by *Aspergillus oryzae* on a lignocellulosic substrate with solid-state fermentation. Biomass Bioenergy 75:161–169. https://doi.org/10.1016/j.biombioe.2015.02.028

Takahashi N, Koshijima T (1988) Ester linkages between lignin and glucuronoxylan in a lignin-carbohydrate complex from beech (*Fagus crenata*) wood. Wood Sci Technol 22:231–241. https://doi.org/10.1007/BF00386018

Tengerdy RP (1996) Cellulase production by solid substrate fermentation. J Sci Ind Res 55:313–316

Teugjas H, Väljamäe P (2013) Selecting β-glucosidases to support cellulases in cellulose saccharification. Biotechnol Biofuels 6:105. https://doi.org/10.1186/1754-6834-6-105

Thoresen M, Malgas S, Mafa MS, Pletschke BI (2021) Revisiting the phenomenon of cellulase action: not all endo- and exo-cellulase interactions are synergistic. Catalysts 11(2):170. https://doi.org/10.3390/catal11020170

Vaaje-Kolstad G, Westereng B, Horn SJ, Liu Z, Zhai H, Sorlie M, Eijsink VGH (2010) An oxidative enzyme boosting the enzymatic conversion of recalcitrant polysaccharides. Science 330(6001):219–222

Verma N, Kumar V (2022) Influence of sugars, sugar alcohols and their combinations on environmentally significant cellulase production under liquid state fermentation. Nat Environ Pollut Technol 21:127–139. https://doi.org/10.46488/NEPT.2022.v21i01.014

Viniegra-González G, Favela-Torres F, Aguilar CN, Rómero-Gomez SDJ, Diaz-Godinez G, Augur C (2003) Advantages of fungal enzyme production in solid state over liquid fermentation systems. Biochem Eng J 13:57–167

Visser EM, Falkoski DL, de Almeida MN, Maitan-Alfenas GP and Guimarães VM (2013) Production and application of an enzyme blend from Chrysoporthe cubensis and Penicillium pinophilum with potential for hydrolysis of sugarcane bagasse. Bioresour technol 144:587–594

Visser EM, Leal TF, de Almeida MN and Guimarães VM (2015) Increased enzymatic hydrolysis of sugarcane bagasse from enzyme recycling. Biotechnol Biofuels 8:1–9

Wang F, Xie H, Chen W, Wang E, Du F, Song A (2013) Biological pretreatment of corn stover with ligninolytic enzyme for high efficient enzymatic hydrolysis. Bioresour Technol 144:572–578

Winter B, Meys R, Sternberg A, Bardow A (2022) Sugar-to-what? An environmental merit order curve for biobased chemicals and plastics. ACS Sustain Chem Eng 10(48):15648–15659. https://doi.org/10.1021/acssuschemeng.2c03275

Yang Y, Xia J, Li J, Chu J, Li L, Wang Y, Zhuang Y, Zhang S (2012) A novel impeller configuration to improve fungal physiology performance and energy conservation for cephalosporin C production. J Biotechnol 161:250–256. https://doi.org/10.1016/j.jbiotec.2012.07.007

Yang J, Yue HR, Pan LY, Feng LY, Zhao S, Swannarangsee S, Champreda V, Liu CG, Zhao XQ (2023) Fungal strain improvement for efficient cellulase production and lignocellulosic biorefinery: current status and future prospects. Bioresour Technol 385:129449. https://doi.org/10.1016/j.biortech.2023.129449

Zhang Y, Wang R, Yang J, Liu L, Yang Y, Wang E, Yuan H (2022) Enzymatic formulation strategies unlock highly-efficient saccharification of distinct pretreated corncobs. Ind Crop Prod 187(Part A):115320. https://doi.org/10.1016/j.indcrop.2022.115320

Zhao X, Zhang L, Liu D (2012) Biomass recalcitrance. Part I: the chemical compositions and physical structures affecting the enzymatic hydrolysis of lignocellulose. Biofuels Bioprod Biorefin 6(4):465–482. https://doi.org/10.1002/bbb.1331

Zhuang J, Marchant M, Nokes S, Strobel H (2007) Economic analysis of cellulase production methods for bio-ethanol. Appl Eng Agric 23:679–687

Correction to: Transgenic Sugarcane–Oilcane: An Alternative Feedstock for the Production of Drop-in Fuel and Value-Added Bioproducts

Shraddha Maitra and Vijay Singh

Correction to:
Chapter 9 in: G. S. Suresha et al. (eds.),
***Value Addition and Product Diversification in Sugarcane*,**
https://doi.org/10.1007/978-981-97-7228-5_9

The original version of this chapter was inadvertently published without an acknowledgement.

The following acknowledgement statement has been included in the chapter:

This work was funded by the DOE Center for Advanced Bioenergy and Bioproducts Innovation (U.S. Department of Energy, Office of Science, Biological and Environmental Research Program under Award Number DE-SC0018420). Any opinions, findings, and conclusions or recommendations expressed in this publication are those of the author(s) and do not necessarily reflect the views of the U.S. Department of Energy or the U.S. Department of Agriculture.

The updated version of this chapter can be found at
https://doi.org/10.1007/978-981-97-7228-5_9

G. S. Suresha et al. (eds.), *Value Addition and Product Diversification in Sugarcane*, https://doi.org/10.1007/978-981-97-7228-5_19

MIX
Papier aus verantwortungsvollen Quellen
Paper from responsible sources
FSC® C105338

If you have any concerns about our products,
you can contact us on
ProductSafety@springernature.com

In case Publisher is established outside the EU,
the EU authorized representative is:
Springer Nature Customer Service Center GmbH
Europaplatz 3, 69115 Heidelberg, Germany

Printed by Libri Plureos GmbH
in Hamburg, Germany